Sustainable Environmental Waste Management Strategies

Series Editors

Alok Prasad Das, Department of Life Sciences
Rama Devi Women's University
Bhubaneswar, Odisha, India

Eric van Hullebusch, Équipe ACE
Institut de Physique du Globe de Paris
Paris Cedex 05, France

Sustainable waste management is an absolute imperative in a world that is already dealing with the effects of climate change. This book series seeks to publish an extensive collection of scientific books targeting researchers, scientists, professionals, academics, students, and anyone else involved in the wide, diverse field of sustainable waste management. The series accepts monographs, edited volumes, textbooks, and conference proceedings. Topics covered include but are not limited to environmental pollution and remediation, sources of environmental waste, environmental and human health, sustainable waste management, circular economy, nature-inspired environmental solutions, pollution prevention, and energy-efficient technologies for waste management. The series puts a particular emphasis on developing countries/ the majority world but accepts proposals globally as long as the topic is in scope.

Alok Prasad Das · Kosala Sirisena · Dipalee Bhanja
Editors

Sustainable Municipal Waste Management with a Zero Waste Approach

Springer

Editors
Alok Prasad Das
Department of Life Sciences
Rama Devi Women's University
Bhubaneswar, Odisha, India

Dipalee Bhanja
All India Institute of Medical Sciences
Bhubaneswar, Odisha, India

Kosala Sirisena
Department of Environmental Technology
Faculty of Technology
University of Colombo
Colombo, Sri Lanka

ISSN 3005-1622 ISSN 3005-1630 (electronic)
Sustainable Environmental Waste Management Strategies
ISBN 978-3-032-05949-9 ISBN 978-3-032-05950-5 (eBook)
https://doi.org/10.1007/978-3-032-05950-5

This Springer imprint is published by the registered company Springer Nature Switzerland AG
The registered company address is: Gewerbestrasse 11, 6330 Cham, Switzerland

Contents

An Introduction to Municipal Solid Waste Pollution and Its Sustainable Management

Shireen Khan, Sailaja Priyadarsini, Amel Gacem, Ipsita Dipamitra Behera, Maha Awjan Alreshidi, Krishna Kumar Yadav, and Alok Prasad Das

1 Introduction

Waste, or undesirable or undesired by-products, is an unavoidable consequence of most actions (Das et al., 2024; Dey et al., 2023; Mishra et al., 2019b). On an annual basis, the world produces 7–9 billion tons of waste (Wilson & Velis, 2015); if existing consumption and waste management practices continue, this amount is predicted to increase by 70% by 2050 (Kaza et al., 2018). Municipal rubbish, also known as municipal solid waste, includes a broad range of products made by homes, businesses, institutions, and non-industrial activities. Since it contains residuals, recyclables, poisonous compounds, and biodegradable components, managing this trash is essential and difficult. The issue of municipal waste pollution has gotten exponentially

S. Khan · S. Priyadarsini · A. P. Das (✉)
Department of Life Sciences, Rama Devi Women's University, Bhubaneswar, Odisha, India
e-mail: alok1053@gmail.com

A. Gacem
Department of Physics, Faculty of Sciences, University 20 Août 1955, Skikda, Algeria

I. D. Behera
Department of Chemical Engineering, Indira Gandhi Institute of Technology, Sarang, Dhenkanal, Odisha, India

M. A. Alreshidi
Department of Chemistry, College of Science, University of Ha'il, Ha'il, Saudi Arabia
e-mail: m.alrashedi@uoh.edu.sa

K. K. Yadav
Department of Environmental Science, Parul Institute of Applied Sciences, Parul University, Vadodara, Gujarat, India

Environmental and Atmospheric Sciences Research Group, Scientific Research Center, Al-Ayen University, Nasiriyah, Thi-Qar, Iraq

A. P. Das et al. (eds.), *Sustainable Municipal Waste Management with a Zero Waste Approach*, Sustainable Environmental Waste Management Strategies,
https://doi.org/10.1007/978-3-032-05950-5_1

(Herbert, 2007). The most common types of solid waste in developed areas with low recycling activity are yard waste, food trash, market garbage, product packaging materials, plastic containers and other solid wastes from institutional, commercial, residential, and industrial sources (Chen et al., 2021). According to EPA estimation, 254 million tons of MSW were generated in the US in 2013 (Pennsylvania et al., 2013) and around 214 million tons of MSW were generated in China in 2017 (Zhou et al., 2014). Such MSW's material classification and composition are provided in Figs. 1 and 2. Household or municipal waste is usually generated from a range of sources where different human activities occur. Several studies indicate that between 55 and 80% of MSW generated in developing countries comes from households, with the remaining 10–30% coming from markets or commercial sectors. The latter comes from various sources, including industries, institutions, streets, and many more (Nabegu, 2010). The composition of solid waste from these sources is usually somewhat varied. Because of this, their chemical and physical characteristics differ depending on their origin. Food waste, yard garbage, plastics, paper, wood, metals, leather, rubbers, batteries, inert materials, paint containers, textiles, demolition and building debris, and many other items are among the numerous objects that would be difficult to classify. Figures 2 and 3 represent the overall composition and classification of MSW produced by United State and China.

Building a facility in a municipality to convert solid waste to energy and deciding how best to handle and manage waste depend heavily on the amount and composition of MSW. Designing an effective waste-to-energy system, maximizing energy recovery, and reducing environmental effects are all made easier with an understanding of the kinds and quantities of waste produced, including organics, plastics, paper, glass, and metals. Ultimately, by aiding in the development of effective waste

Fig. 2 Composition and classification of MSW produced by the United States in 2013

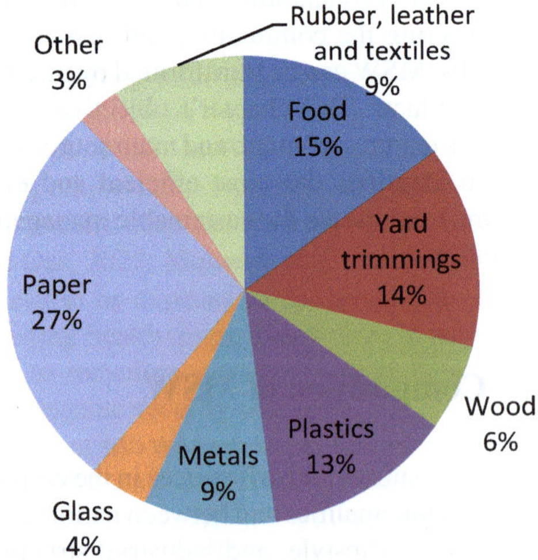

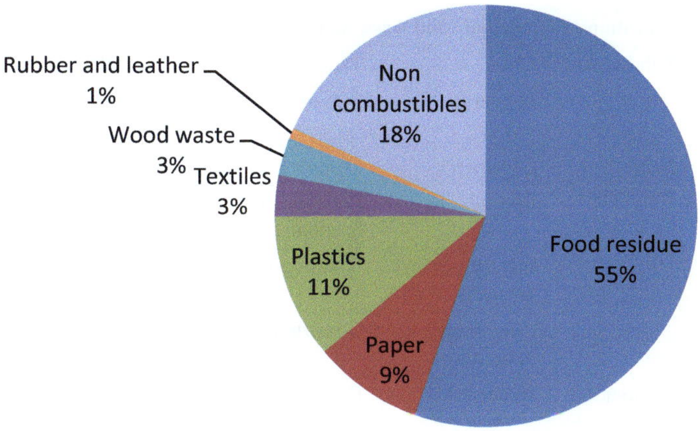

Fig. 3 Composition and classification of MSW produced by China in 2017

sorting and processing methods and supporting the selection of suitable conversion technologies, such as gasification, anaerobic digestion, or incineration, this information promotes a sustainable waste management system (Buenrostro et al., 2001).

3 Source and Their Classification

MSW is generated from a wide array of sources, including households, commercial establishments, institutions, and industries. Developing focused waste management plans, strengthening recycling and resource recovery, and improving waste segregation techniques helps in understanding MSW sources (Adedibu, 1985). The source classification of MSW considers three categories: urban, rural, and industrial; each one is represented as a discrete unit. These differences result in the hierarchical derivation of seven classes (Fig. 3). The urban division comprises, regardless of population size, the sources of SW within human settlements; the urban division is divided into two subdivisions: residential and non-residential (commercial, construction/demolition, institutional/services, and special); the industrial division is composed of a single industrial class; all facilities, regardless of size, are included in the industrial division; and the rural division is composed of all sources originating from agricultural and animal husbandry activities (Buenrostro et al., 2001). Table 1 represents the different sources of municipal e-waste.

According to the source of generation, the distinct solid waste produced by each class is separated into the following categories:

Table 1 Classification of municipal solid waste sources

Division	Subdivision	Class	Type of waste	Characteristics of waste
Industrial		Industries	Industrial waste	Non-dangerous; potentially dangerous; dangerous
Urban	Non residential	Commerce	Commercial waste	Non dangerous
		Special	Pollutants from hospital, etc.	Potentially dangerous; biological-infectious
		Institutions and services	Institutional waste	Non dangerous
		Construction/ demolition	Construction waste	Non dangerous
	Residential	Household	Domestic waste	Non-dangerous; potentially dangerous
Rural		Agriculture and animal husbandry	Agriculture and animal husbandry waste	Non-dangerous; Potentially dangerous

Households

Composition of MSW in Indian households and found that food waste (including organic waste) comprised up to 60% of household waste, followed by paper, plastics, and other recyclable materials (Sharma et al., 2020). Packaging materials like paper and plastics contribute significantly to the organic garbage that makes up households in developing countries (Hoornweg & Bhada-Tata, 2012).

Commercial and Institutional Sources

Commercial buildings including shops, eateries, hotels, offices, and schools are another significant source of MSW. These entities contribute a range of waste products including packaging, food waste, office papers and plastics. Specialized waste streams like medical waste, hazardous trash, and e-waste are produced by institutional sources, such as hospitals and educational institutions. However, compared to home and commercial trash, these waste streams make up a lesser percentage of the total volume (Guerra & Galvez, 2021).

Industrial and Construction Waste

Even while industrial trash is sometimes left out of municipal waste management systems, it still has an influence on the total amount of trash produced, particularly in cities. Materials including wood, concrete, plastics and metals are discarded during construction and demolition, which also contributes significantly to MSW (Martínez-Alier et al., 2016).

Agricultural Waste

Although it is not as commonly mentioned as a source of MSW, agricultural activities in rural or peri-urban regions generate agricultural waste like crop residues, manure and food waste, which frequently find their way into municipal garbage systems. The primary producer of agriculture wastes is farm households. Composting and biogas production may be made possible by incorporating farm waste into urban waste management procedures (Mugisha et al., 2019).

Electronic Waste (E-Waste)

E-waste includes discarded electronic devices like phones, computers, and televisions, which is an increasingly significant source of MSW. A significant portion of the 44.8 million metric tons of e-waste produced worldwide in 2016 ended up in municipal garbage streams. E-waste contributes significantly to MSW in industrialized nations as users replace equipment more frequently (Baldé et al., 2017).

Special Waste

Waste that must be managed using particular methods, either due to its state, relative hazard, or the requirement of environmental legislation. Airports, terrestrial transportation hubs, medical facilities, industrial and car repair shops, pharmaceutical stores for humans and animals, and scientific research are some of the industries that produce this garbage (Buenrostro et al., 2001).

4 Characteristics of MSW

Physical Characteristics

When determining how to handle, store, and dispose of MSW, its physical characteristics are crucial. Compaction, density, moisture content, and particle size are some of the physical characteristics that can be used to describe MSW. The densities of MSW vary from 200–400 kg/m^3 (Tchobanoglous et al., 1993) to 400–800 kg/m^3 (Chen et al., 2018), according to research. The moisture content of MSW is another important physical characteristic; it has been shown to range from 20–50% (Tchobanoglous et al., 1993) to 40–70% (Chen et al., 2018). Additionally, studies have demonstrated that MSW has a significant particle size distribution, with sizes ranging from 10–30 mm (Chen et al., 2018) to 50–100 mm (Tchobanoglous et al., 1993). When designing garbage disposal and storage systems, it is also necessary to comprehend density, energy, and compaction ratio are examples of compaction attributes (Kumar et al., 2017). Understanding these physical characteristics is essential for designing effective waste-to-energy conversion systems and efficiently managing MSW.

Chemical Characteristics

Effective waste management necessitates an understanding of the chemical behavior and characterization of MSW. If solid waste is to be used as fuel or for any other purpose, we need to be aware of its chemical characteristics. Understanding MSW's composition, behavior, and environmental impacts requires an understanding of its chemical characteristics. According to studies, MSW frequently contains organic and inorganic compounds, including carbon, hydrogen, oxygen, nitrogen, sulfur, and heavy metals (Tchobanoglous et al., 1993). The chemical composition of MSW can vary significantly depending on factors including trash formation rates, waste management strategies, and population density (Kumar et al., 2017). For instance, MSW from high-income nations tends to contain more glass, paper, and plastics, while MSW from low-income nations tends to contain more organic waste. Another important chemical characteristic that influences MSW's potential for energy recovery is its heating value, which varies between 8 and 12 MJ/kg (Yu et al., 2013). Designing efficient waste management techniques, such as landfilling, composting, and waste-to-energy conversion, requires an understanding of the chemical properties of MSW.

5 Effects of MSW on Human Health and Environment

Solid waste will unavoidably accumulate as a result of human activities, and how this garbage is managed has a direct effect on the environment and general public health. The output of MSW has increased to previously unheard-of levels due to on-going urbanization and population growth, creating serious obstacles to public health and sustainable development (Choudhary et al., 2024). The amount of waste that people worldwide are throwing away is growing, and as plastics and electronic consumer goods become more accessible, the waste's composition is getting more complicated. These two tendencies pose a challenge to the cities' responsibility to protect their citizens from their waste. Uncontrolled trash disposal in urban dumping sites has proven to be an annoyance for the residents in several developing cities. Despite environmental rules, open garbage disposal poses a notable concern to environment and human health in urban settings (Frosch, 1996). Studies recommended that because of the dumpsite's close vicinity to the communities of residents, both nearby and distant people may have contracted diseases like cholera, diarrhea, and malaria (Narayana, 2009). Furthermore, it is believed that the air pollutants from rubbish incinerators cause cancer (Ziraba et al., 2016). Poor waste management endangers both human health and the environment. Workers in this industry are largely affected by direct health dangers, thus they should be kept as far away from waste as feasible. Managing garbage from hospitals and clinics presents additional challenges. The population has the highest indirect health hazards due to the expansion of the disease vector, particularly flies and rats (Adnan & others, 2020). When industrial hazardous waste that is not regularly managed mixes with municipal garbage, it causes a risk to human health. The connection between MSWs, heavy metal liquid industrial effluents discharged to storm water systems or open MSW dumpsites, and the wastes discharged as a result, maintains a vicious cycle that includes these wastes, highlighting the on-going risk of heavy metal concentration in the food chain (El-Fadel et al., 1997). Figure 4 represents the various anthropogenic sources of MSW and its environmental and health impact.

One common way that the environment gets contaminated is when trash breaks down into its component chemicals. Few impoverished countries now have landfill

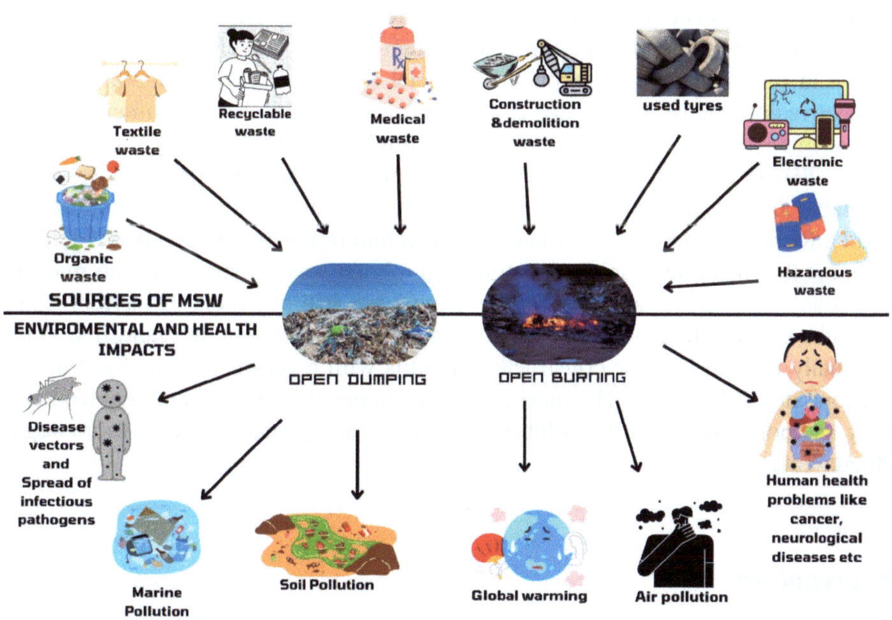

Fig. 4 Sources of MSW and its environmental and health impacts

Tchobanoglous, G., et al. (1993). *Integrated solid waste management: Engineering principles and management issues*. McGraw-Hill.

[illegible]

[illegible]

[illegible]

[illegible]

[illegible]

[illegible]

[illegible]

[illegible]

[illegible]

[illegible] (2017). Solid waste management in Saudi Arabia [illegible] municipal solid waste management [illegible]. Accessed 14 Apr 2018.

[illegible]

[illegible]

[illegible]

[illegible]

Algae in Municipal Waste Treatment

M. A. P. C. Piyathilaka and A. R. Pilapitiya

1 Introduction

Introduction to Municipal Wastewater Treatment

Municipal wastewater treatment is a vital process that ensures the safe disposal or reuse of water contaminated with domestic and industrial waste. It is a multi-stage procedure designed to remove physical, chemical, and biological pollutants from wastewater, protecting public health, and ecosystems (Tchobanoglous et al., 2022). The increasing urban population and industrialization have amplified the demand for effective treatment technologies, making wastewater treatment a cornerstone of modern urban infrastructure (Metcalf & Eddy, 2013; Mishra & Das, 2021, 2023; Mishra et al., 2024).

There are several key processes to be completed in any municipal wastewater treatment plant namely: Preliminary Treatment, Primary Treatment, Secondary Treatment, Tertiary, and Advanced Treatment (Fig. 1). Preliminary treatment focuses on removing large solids and debris to protect downstream processes. It involves mechanical operations such as screening and grit removal (Spellman, 2018). During primary treatment, sedimentation tanks are used to remove settleable solids and floating materials. This step typically reduces about 50–60% of suspended solids (Tchobanoglous et al., 2022). Secondary treatment involves biological processes to degrade organic matter using microorganisms. Common methods include activated sludge processes, trickling filters, and lagoons (Metcalf & Eddy, 2013). Tertiary treatment removes residual pollutants like nutrients, pathogens, and heavy metals. Techniques in tertiary treatment include filtration, disinfection (chlorination, UV),

M. A. P. C. Piyathilaka (✉) · A. R. Pilapitiya
Department of Environmental Technology, Faculty of Technology, University of Colombo, Technocity, Homagama, Sri Lanka
e-mail: poorna@et.cmb.ac.lk

A. P. Das et al. (eds.), *Sustainable Municipal Waste Management with a Zero Waste Approach*, Sustainable Environmental Waste Management Strategies,
https://doi.org/10.1007/978-3-032-05950-5_2

and advanced oxidation processes. These methods are essential for achieving water quality standards for reuse or discharge (Mara, 2004). Wastewater also contains electronic waste, discarded mobile batteries, and releases harmful heavy metals like lithium to contaminate the aquatic ecosystem (Priyadarsini & Das, 2024, 2025).

Modern wastewater treatment plants face challenges such as energy consumption, climate change impacts, and emerging contaminants like pharmaceuticals and microplastics (Naidoo & Olaniran, 2014). Innovative approaches, including decentralized systems, energy recovery, and resource recycling, are being explored to make treatment processes more sustainable (Verstraete et al., 2009). Municipal wastewater treatment is crucial for sustainable water management, safeguarding aquatic ecosystems, and meeting regulatory standards. Its role in resource recovery, such as water reuse and nutrient extraction, underscores its importance in circular economies (Smith et al., 2018).

Role of Algae in Wastewater Treatment Systems

Algae plays a pivotal role in wastewater treatment systems, offering a sustainable and efficient approach to removing contaminants while facilitating resource recovery. Their unique physiological characteristics make them well-suited for treating various types of wastewater, including municipal, industrial, and agricultural effluents. The prime role of algae in wastewater treatment is contaminant removal and they perform several mechanisms to utilize contaminants in wastewater (El-Sheekh et al., 2022).

Nutrient Uptake: Algae assimilate inorganic nutrients such as nitrogen and phosphorus for their growth, effectively reducing nutrient concentrations in wastewater. This process helps prevent eutrophication in receiving water bodies. Studies have demonstrated that certain microalgae species can remove up to 98% of chemical oxygen demand (COD), 94% of inorganic nitrogen, and 100% of phosphate from industrial wastewater (Arashiro et al., 2020b).

Photosynthetic Oxygenation: Through photosynthesis, algae produce oxygen, which supports aerobic bacteria in degrading organic pollutants. This symbiotic relationship enhances the overall efficiency of the treatment process (El-Sheekh et al., 2022).

Heavy Metal Sequestration: Certain algae species can adsorb heavy metals from wastewater, thereby reducing toxicity and facilitating the recovery of these metals (García et al., 2018; Mishra et al., 2022b; Parial & Dey, 2023).

Various system designs incorporate algae for wastewater treatment facilities. High-Rate Algal Ponds (HRAPs) are shallow, mixed ponds that provide light exposure and nutrient uptake, leading to high algal productivity and efficient contaminant removal. Research indicates that HRAPs can achieve significant reductions in nutrient concentrations while producing valuable biomass (Parial & Dey, 2023). Photobioreactors (PBRs) are enclosed systems that provide controlled environments

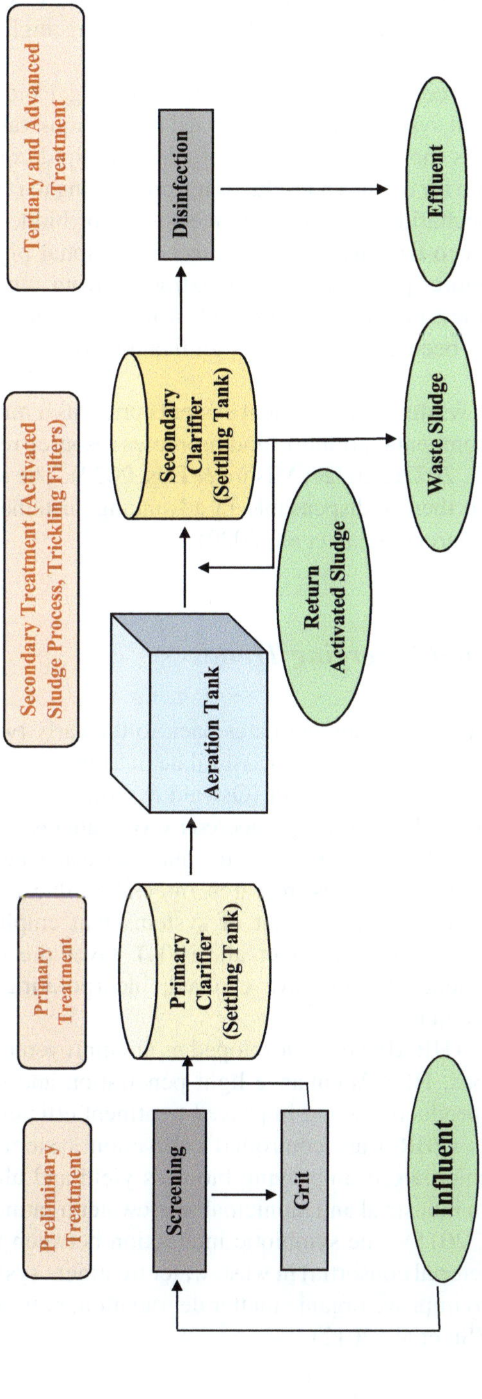

Fig. 1 Key processes in municipal waste water treatment plant. Adapted From Daphtary (2024)

for algal cultivation. PBRs are effective in treating specific waste streams and recovering high-value products. Full-scale tubular PBRs have been successfully employed to remove nutrients from agricultural runoff, demonstrating high efficiency and reliability (García et al., 2018).

In the aspect of resource recovery and sustainability, algal biomass harvested from wastewater treatment systems can be processed into various value-added products. The best example is the production of biofuels. The lipid content of certain algae makes them suitable for biodiesel production, contributing to renewable energy resources. Algal polysaccharides can serve as feedstock for biodegradable plastic production, offering an eco-friendly alternative to conventional plastics (Arashiro et al., 2020b). Furthermore, phycobiliproteins extracted from algae are valuable as natural pigments in the food and cosmetic industries. Cultivating microalgae in industrial wastewater has been shown to yield significant amounts of these pigments (Arashiro et al., 2020a).

Integrating algae into wastewater treatment systems presents a multifaceted solution that addresses environmental pollution and promotes resource recovery (Kumar et al., 2023; Mishra et al., 2022a, 2022c; Mishra & Das, 2022). The adaptability and efficiency of algae make them indispensable in advancing sustainable wastewater management practices (García-Galán et al., 2020).

Historical Context and Emerging Trends

The use of algae in wastewater treatment dates back to the early twentieth century when researchers recognized their ability to assimilate nutrients and produce oxygen for microbial degradation of organic matter (Oswald & Gotaas, 1957). The concept of High-Rate Algal Ponds (HRAPs) was pioneered by William Oswald in the 1950s, integrating microalgae with bacterial consortia to enhance wastewater treatment efficiency (Oswald, 1988). By the 1970s, research intensified due to concerns over eutrophication, leading to the development of systems that emphasized nutrient removal and biomass recovery (Craggs et al., 2004). Over the decades, algae-based wastewater treatment systems have evolved, incorporating many modern biotechnological advancements.

High-rate algal Ponds (HRAPs) have developed as an improvement over conventional wastewater lagoons, HRAPs enhance light penetration and nutrient uptake, leading to higher algal productivity and improved treatment efficiency (Park et al., 2011). Photobioreactors (PBRs) are controlled cultivation systems that optimize growth conditions for microalgae, improving biomass yield and allowing for efficient nutrient recovery in industrial and municipal wastewater treatment applications (Morales-Amaral et al., 2015). The symbiotic interaction between microalgae and bacteria (Microalgal-bacterial consortia) in wastewater treatment systems has gained attention for its ability to improve organic matter degradation, nutrient removal, and biomass harvestability (Su et al., 2012).

With a close look into these systems, several prominent emerging trends in algae-based wastewater treatments can be identified. Recent studies highlight the potential for resource recovery from algal biomass, including biofuels, fertilizers, and high-value bioproducts such as pigments and bioplastics (Chisti, 2007). The integration of wastewater treatment with algal biorefinery concepts supports sustainability goals (Pittman et al., 2011). Combining algae-based treatment with anaerobic digestion, constructed wetlands, and membrane bioreactors has shown promise in improving efficiency and energy recovery while reducing operational costs (Ruiz-Martinez et al., 2012). Advances in genetic engineering have enabled the modification of microalgae to enhance nutrient uptake, lipid accumulation, and biomass productivity, making wastewater-based algae cultivation more economically viable (Guschina & Harwood, 2006). AI-driven monitoring and automation in algal cultivation systems help optimize light exposure, CO_2 supply, and harvesting strategies, enhancing the scalability of algae-based wastewater treatment (Chew et al., 2017).

Algae-based wastewater treatment has evolved from early experimental studies to sophisticated biotechnological applications. With growing concerns over water scarcity, pollution, and energy sustainability, the integration of algae-based technologies with resource recovery and smart system management is shaping the future of wastewater treatment (Guschina & Harwood, 2006; Pittman et al., 2011).

2 Algae as a Sustainable Solution for Waste Treatment

Algal Species in Wastewater Treatment

Microalgae play a crucial role in wastewater treatment due to their ability to assimilate nutrients, degrade organic pollutants, and generate valuable biomass as given in Table 1. Various algal species have been explored for wastewater treatment based on their growth rate, tolerance to pollutants, and efficiency in nutrient uptake (Pittman et al., 2011). These species belong primarily to the genera *Chlorella, Scenedesmus, Spirulina, Nannochloropsis*, and *Dunaliella*, among others.

Chlorella spp.: *Chlorella* is one of the most widely used microalgae in wastewater treatment due to its high nutrient removal efficiency and adaptability to various wastewater conditions. It is known to efficiently uptake nitrogen, phosphorus, and heavy metals while producing high-value biomass for biofuels and bioproducts (Abdel-Raouf et al., 2012). Studies show that *Chlorella vulgaris* can remove up to 80% of total nitrogen and 90% of phosphorus from municipal wastewater (Park et al., 2011).

Scenedesmus spp.: *Scenedesmus* species, including *Scenedesmus obliquus* and *Scenedesmus dimorphus*, are highly efficient in nutrient removal and biomass production. They have been reported to achieve over 90% nitrogen and phosphorus removal in wastewater treatment ponds (Ruiz-Martinez et al., 2012). These species are also

Table 1 Key characteristics of algal species to be used in municipal wastewater treatment

Algal species	Key characteristics in wastewater treatment
Chlorella spp.	• High nutrient removal efficiency (N, P, heavy metals) • Adaptable to various wastewater conditions • Produces biomass for biofuels and bioproducts • Can remove up to 80% of total nitrogen and 90% of phosphorus
Scenedesmus spp.	• High nitrogen and phosphorus removal (over 90%) • Resistant to environmental fluctuations • Thrives in different wastewater types • Efficient in biomass production
Spirulina (Arthrospira) spp.	• Rapid growth and high protein content • Tolerates high pH and salinity • Useful in industrial and agricultural wastewater with high nitrogen concentrations
Nannochloropsis spp.	• Effective in removing ammonium, nitrate, and phosphate • Thrives in saline and brackish environments • Accumulates lipids for biofuel production
Dunaliella spp.	• Halotolerant • Used in food and beverage industry wastewater • Potential applications in pigment and biofuel production

Sources Park et al. (2011), Pittman et al. (2011), Ruiz-Martinez et al. (2012)

known for their resistance to environmental fluctuations and ability to thrive in different wastewater types.

Spirulina (*Arthrospira*) spp.: *Spirulina* species, particularly *Arthrospira platensis*, are used in wastewater treatment due to their rapid growth, high protein content, and ability to tolerate high pH and salinity (Olguín, 2012). They are particularly useful in treating industrial and agricultural wastewater with elevated nitrogen concentrations.

Nannochloropsis spp.: *Nannochloropsis* species are primarily used for nutrient removal and lipid accumulation for biofuel production. They are effective in removing ammonium, nitrate, and phosphate from wastewater and can thrive in saline and brackish environments (Ma et al., 2016).

Dunaliella spp.: *Dunaliella* species, particularly *Dunaliella salina*, are halotolerant microalgae that can grow in high-salinity wastewater. They are often used in wastewater from food and beverage industries and have potential applications in pigment and biofuel production (Borowitzka & Borowitzka, 1990).

Different algal species exhibit unique capabilities in wastewater treatment, including nutrient uptake, heavy metal sequestration, and biomass generation. Selecting the appropriate algal species depends on wastewater composition, environmental conditions, and desired end-products. Advances in algal biotechnology and genetic engineering continue to enhance the efficiency and applicability of algae-based wastewater treatment systems (Chew et al., 2017).

Algal Metabolism and Nutrient Uptake

Microalgae metabolic processes facilitate the uptake and removal of pollutants such as nitrogen, phosphorus, and heavy metals. Algal metabolism is primarily driven by photosynthesis, which not only supports biomass growth but also enhances oxygenation in wastewater treatment systems, benefiting bacterial communities involved in organic matter degradation (Pittman et al., 2011). Microalgae utilize sunlight and CO_2 to produce organic compounds through photosynthesis. This process generates oxygen, which supports the aerobic degradation of organic matter by heterotrophic bacteria, improving the efficiency of biological wastewater treatment systems (Park et al., 2011). The interplay between algae and bacteria creates a self-sustaining ecosystem that enhances pollutant removal.

Microalgae assimilate nitrogen primarily in the form of ammonium (NH_4^+), nitrate (NO_3^-), and nitrite (NO_2^-). Ammonium is the preferred nitrogen source due to its lower energy requirement for assimilation (Lau et al., 1995). Algae convert nitrogen into cellular proteins, amino acids, and nucleotides, reducing the nitrogen load in wastewater and preventing eutrophication (Ruiz-Martinez et al., 2012). Phosphorus removal in wastewater treatment occurs through algal uptake in the form of orthophosphate (PO_4^{3-}). Some microalgae, such as *Scenedesmus* and *Chlorella* species, can store excess phosphorus as intracellular polyphosphates, which can later be harvested and utilized as biofertilizers (Park et al., 2011).

Algae have demonstrated high metal-binding capacity, facilitating the bioremediation of heavy metals such as cadmium, lead, and arsenic from industrial wastewater (Mehta & Gaur, 2005). The mechanism involves biosorption through cell surface interactions and intracellular bioaccumulation, making microalgae a viable option for treating metal-contaminated wastewater. Algae-based wastewater treatment systems also contribute to carbon sequestration by utilizing CO_2 from the atmosphere or industrial flue gases. Carbon availability influences algal productivity and nutrient uptake rates, making CO_2 enrichment a strategy for optimizing wastewater treatment efficiency (Chisti, 2007).

Advantages of Algal Systems Over Conventional Methods

Algae-based wastewater treatment technologies offer several advantages over conventional treatment methods, including reduced energy consumption, enhanced carbon sequestration, and valuable biomass production for resource recovery as discussed in Table 2. These advantages make algal systems a sustainable and eco-friendly alternative to conventional wastewater treatment approaches such as activated sludge and chemical precipitation (Pittman et al., 2011).

Table 2 Pros and cons of algal systems vs conventional methods

Aspect	Algal systems	Conventional methods
Energy consumption	Lower energy demand since algae utilize sunlight for growth and nutrient removal, reducing aeration costs. Example: High-rate algal ponds (HRAPs) consume ~ 50% less energy than activated sludge systems	High energy demand due to mechanical aeration and pumping in systems like activated sludge. Example: Activated Sludge Process (ASP) requires 0.3–1.5 kWh/m^3 for aeration
Carbon sequestration potential	High CO_2 sequestration as algae fix CO_2 during photosynthesis, mitigating greenhouse gas emissions. Example: Algal ponds can sequester 1.8 kg CO_2 per kg of biomass	Limited carbon sequestration; some systems produce CO_2 instead. Example: ASP releases CO_2 due to microbial respiration and fossil fuel use
Biomass Production for Resource Recovery	High-value biomass can be harvested for biofuels, fertilizers, or bioplastics. Example: Chlorella and Spirulina produce lipid-rich biomass suitable for biofuel production	Limited or no biomass recovery; sludge is typically treated as waste or used in low-value applications like landfilling or anaerobic digestion. Example: Waste sludge from ASP is often incinerated or landfilled
Nutrient removal efficiency	Efficient removal of nitrogen and phosphorus through assimilation into algal biomass. Example: Microalgae-based systems remove 80–90% of nitrogen and phosphorus	Effective but energy-intensive nutrient removal using chemical precipitation and nitrification–denitrification processes. Example: Chemical phosphorus removal via alum dosing
Operational complexity and land requirement	Requires large land areas and controlled environmental conditions for optimal performance. Example: HRAPs need extensive pond space	More compact systems are suitable for urban settings. Example: Membrane Bioreactors (MBRs) require less space but consume high energy

Sources Chisti (2007), Craggs et al. (2014a, 2014b), Pittman et al. (2011)

Reduced Energy Consumption

Traditional wastewater treatment methods, such as activated sludge systems, require significant energy inputs for aeration, sludge processing, and chemical addition. In contrast, algal-based treatment systems rely on photosynthesis to naturally oxygenate wastewater, reducing the need for external aeration (Craggs et al., 2014a, 2014b). High-rate algal ponds (HRAPs) have demonstrated significant energy savings compared to conventional secondary treatment processes, making them more energy-efficient in municipal and industrial wastewater treatment applications (Park et al.,

2011). Additionally, algal systems produce less sludge compared to conventional methods, reducing sludge handling and disposal costs (Olguín, 2012).

Carbon Sequestration Potential

One of the key environmental benefits of algae-based wastewater treatment is its ability to capture and utilize atmospheric and industrial CO_2 emissions. Microalgae use CO_2 as a carbon source for photosynthesis, contributing to greenhouse gas mitigation (Chisti, 2007). Wastewater treatment facilities that integrate algal systems with carbon capture technologies can significantly reduce their carbon footprint while enhancing biomass productivity (Singh & Olsen, 2011). Studies have shown that microalgae such as *Chlorella* and *Scenedesmus* can effectively assimilate CO_2 from industrial flue gases while simultaneously removing nutrients from wastewater (Morales-Amaral et al., 2015).

Biomass Production for Resource Recovery

Unlike conventional wastewater treatment methods, which generate waste sludge requiring disposal, algal systems produce biomass that can be valorized for various applications, including biofuel production, animal feed, bioplastics, and fertilizers (Pittman et al., 2011). The harvested algal biomass is rich in lipids, proteins, and carbohydrates, making it a valuable feedstock for bioenergy generation (Rawat et al., 2011). Additionally, algal biomass can be used in agriculture as a biofertilizer, contributing to circular economy principles by recycling nutrients back into food production systems (Ma et al., 2016).

These benefits make algal systems a promising solution for sustainable wastewater treatment and resource recovery, aligning with global efforts to promote environmentally friendly and economically viable wastewater management strategies. Continued research and technological advancements will further optimize the efficiency and scalability of these systems for broader implementation.

3 Mechanisms of Algal Wastewater Treatment

Algae-based wastewater treatment is an eco-friendly and cost-effective alternative to conventional treatment methods, utilizing the natural metabolic processes of microalgae to remove pollutants. Algal systems facilitate wastewater treatment through several key mechanisms, including nutrient removal (nitrogen and phosphorus), organic matter degradation, heavy metal and toxin absorption, and pathogen reduction. These processes contribute to improved water quality while enabling resource recovery and environmental sustainability (Pittman et al., 2011).

Nutrient Removal (Nitrogen and Phosphorus)

One of the primary mechanisms in algal wastewater treatment is the assimilation of nitrogen (N) and phosphorus (P), which are major contributors to water pollution and eutrophication. Microalgae absorb nitrogen primarily in the form of ammonium (NH_4^+), nitrate (NO_3^-), and nitrite (NO_2^-), incorporating these compounds into proteins, nucleic acids, and other cellular components (Lau et al., 1995). Some algal–bacterial consortia also promote nitrogen removal through nitrification and denitrification processes, enhancing treatment efficiency (Ruiz-Martinez et al., 2012). Algae take up phosphorus in the form of orthophosphate (PO_4^{3-}) and store it as intracellular polyphosphates. Certain microalgal species, such as *Chlorella* and *Scenedesmus*, are particularly effective at phosphorus accumulation, which can later be recovered for use as biofertilizers (Hawrot-Paw et al., 2020).

Organic Matter Degradation

Algal systems contribute to the breakdown of organic pollutants by promoting oxygenation through photosynthesis. The oxygen produced by algae supports aerobic bacteria in degrading organic matter, reducing biochemical oxygen demand (BOD) and chemical oxygen demand (COD; Park et al., 2011). In high-rate algal ponds (HRAPs), synergistic interactions between algae and bacteria accelerate organic matter decomposition, improving wastewater treatment efficiency (Olguín, 2012).

Heavy Metal and Toxin Absorption

Microalgae are known for their ability to remove heavy metals and toxins from wastewater through biosorption and bioaccumulation. Algae possess functional groups such as carboxyl, hydroxyl, and amino groups that bind to heavy metal ions, facilitating their sequestration (Mehta & Gaur, 2005). Studies have shown that microalgae can effectively remove metals like cadmium (Cd), lead (Pb), arsenic (As), and mercury (Hg) from industrial wastewater (Mata et al., 2010). Additionally, certain algal strains can metabolize and detoxify organic pollutants, including pesticides and pharmaceuticals, further enhancing wastewater treatment outcomes.

Pathogen Reduction

Algal wastewater treatment systems also contribute to pathogen reduction through mechanisms such as high oxygenation, pH increase, and antimicrobial compounds produced by algae. The elevated pH levels in algal ponds, often reaching above 9 due to CO_2 uptake, create an unfavorable environment for bacteria such as *Escherichia coli* and *Salmonella* (Craggs et al., 2014a, 2014b). Furthermore, some microalgae produce biocidal compounds that inhibit pathogenic bacteria, viruses, and protozoa, enhancing the disinfection process (Mehta & Gaur, 2005).

4 Algal Wastewater Treatment Technologies

High-Rate Algal Ponds (HRAPs)

High-Rate Algal Ponds (HRAPs) are an advanced form of pond-based wastewater treatment systems designed to optimize algal growth and nutrient removal (Fig. 2). Unlike conventional wastewater lagoons, HRAPs incorporate controlled hydrodynamic and environmental conditions to enhance algal productivity and wastewater treatment efficiency (Craggs et al., 2014a, 2014b). These systems are gaining popularity due to their dual functionality in pollutant removal and biomass generation, which can be utilized for bioenergy production or other valuable bioproducts (Park et al., 2011).

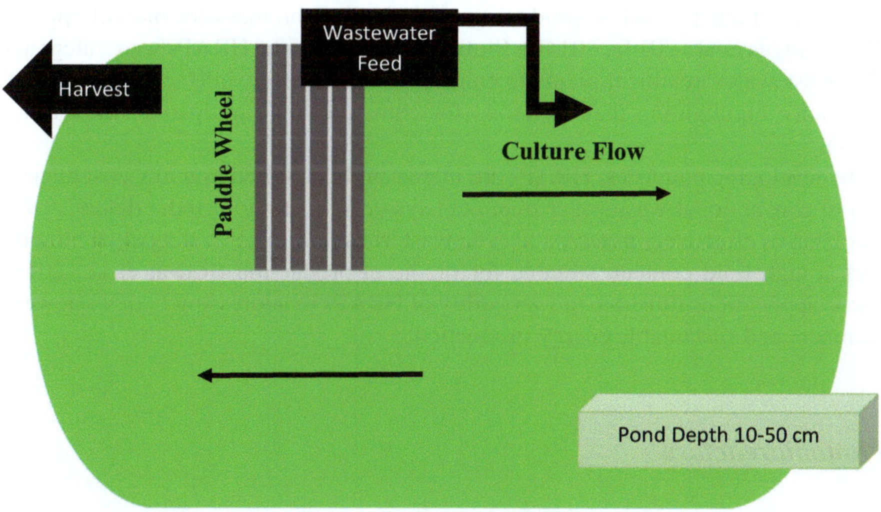

Fig. 2 Schematic diagram of a high-rate algal pond. Adapted from Ferro (2019)

Design and Operational Principles

HRAPs are engineered open pond systems that consist of shallow raceway channels, typically with a depth ranging from 20 to 50 cm, to facilitate optimal light penetration for algal photosynthesis (Craggs et al., 2014a, 2014b). These ponds are equipped with paddle wheels that continuously circulate the wastewater, ensuring even distribution of nutrients, CO_2, and oxygen while preventing sedimentation (Park et al., 2011). The hydraulic retention time in HRAPs generally ranges from 4 to 10 days, depending on the pollutant load and treatment objectives (Cai et al., 2013).

The core mechanism of HRAPs is based on the symbiotic relationship between microalgae and bacteria. Algae assimilate nitrogen and phosphorus from wastewater while releasing oxygen through photosynthesis, which supports aerobic bacterial degradation of organic matter. This interaction enhances pollutant removal efficiency while simultaneously reducing greenhouse gas emissions compared to conventional treatment methods (Pittman et al., 2011).

Case Studies and Applications

HRAPs have been successfully implemented in various regions for municipal and industrial wastewater treatment. In New Zealand, a pilot-scale HRAP system demonstrated high nutrient removal efficiencies, achieving up to 90% nitrogen and 80% phosphorus reduction while producing valuable algal biomass for biofuel applications (Craggs et al., 2014a, 2014b). Similarly, in California, HRAPs were integrated with wastewater treatment plants to enhance water quality while sequestering CO_2 emissions, highlighting their potential for climate change mitigation (Park et al., 2011).

In developing countries, HRAPs are increasingly explored as a low-cost alternative to conventional wastewater treatment systems. A study in India demonstrated that HRAPs effectively treated domestic wastewater while generating algal biomass with a high lipid content, suitable for biodiesel production (Rawat et al., 2011). These applications illustrate the versatility of HRAPs in addressing both wastewater treatment and sustainable energy production.

Photobioreactors

Photobioreactors (PBRs) are closed, controlled systems used for cultivating algae under optimal conditions to enhance wastewater treatment efficiency. Unlike open pond systems, PBRs provide higher biomass productivity, better contamination control, and efficient light utilization, making them suitable for large-scale algal wastewater treatment (Razzaka et al., 2024). These reactors allow precise regulation of environmental parameters such as temperature, pH, CO_2 concentration, and nutrient levels, leading to improved pollutant removal and biomass generation (Gupta et al., 2024).

Types and Configurations

PBRs come in various designs to optimize algal growth and wastewater treatment, including:

Tubular PBRs: These consist of transparent tubes through which wastewater circulates. They provide high surface area for light absorption and efficient CO_2 utilization but require external cooling to prevent overheating (Posten, 2009).

Flat-Panel PBRs: These have large, flat surfaces that enhance light penetration and gas exchange, making them ideal for high-density algal cultivation (Zittelli et al., 2013).

Airlift and Bubble Column PBRs: These rely on aeration to circulate wastewater and mix algae, providing a cost-effective and energy-efficient method for nutrient removal (Razzaka et al., 2024).

Hybrid PBRs: Combining features of different designs, hybrid PBRs aim to optimize algal productivity while maintaining low operational costs and high pollutant removal efficiency (Gupta et al., 2024).

Efficiency and Scale-Up Potential

PBRs demonstrate high nutrient removal efficiencies, with studies reporting over 90% nitrogen and phosphorus reduction in municipal wastewater treatment applications (Cai et al., 2013). The controlled conditions within PBRs allow for consistent algal biomass production, which can be harvested for biofuels, animal feed, and other valuable products (Razzaka et al., 2024). Despite their advantages, scaling up PBRs for industrial wastewater treatment remains challenging due to high capital and operational costs. However, advancements in automation, material science, and biorefinery integration are driving improvements in cost-effectiveness and commercial feasibility (Gupta et al., 2024). Large-scale demonstration projects in Europe and North America are showcasing the potential of PBRs to complement existing wastewater treatment technologies while producing valuable algal biomass (Posten, 2009).

Hybrid Systems: Integration with Conventional Treatment

Hybrid wastewater treatment systems integrate algae-based methods with conventional processes to enhance efficiency and sustainability. These systems leverage the

Biomass Harvesting Techniques

Efficient biomass harvesting is a critical component in the operation of algae-based wastewater treatment plants, as it directly influences both the quality of treated effluent and the economic viability of the process. The selection of an appropriate harvesting technique depends on factors such as algal species, cell density, and desired end-products.

Sedimentation, Flotation, and Filtration

Among the various harvesting methods, sedimentation, flotation, and filtration are commonly employed due to their effectiveness and adaptability to different operational conditions. Sedimentation relies on gravity to separate algal biomass from water (Sharma, 2023). In this process, algal cells settle at the bottom of a containment unit over time, allowing for the collection of concentrated biomass. However, the natural settling velocity of microalgae is often low due to their small size and negative surface charge, which can make sedimentation a slow and sometimes inefficient method (Kucmanova & Gerulová, 2019). To enhance efficiency, chemical flocculants are often added to aggregate the cells into larger flocs, increasing their settling rate. While this approach improves sedimentation efficiency, the cost and potential environmental impact of chemical additives must be considered (Arashiro et al., 2020b).

Flotation techniques separate algal biomass by introducing fine air bubbles into the wastewater, which attach to algal cells and lift them to the surface, forming a concentrated algal-rich froth that can be easily removed. Dissolved Air Flotation (DAF) is a widely used method in which air is dissolved under pressure in the water and then released at atmospheric pressure in a flotation tank, producing the necessary bubbles for separation (Ruiz et al., 2024). DAF is effective in removing suspended solids and oils, making it suitable for harvesting algal biomass. However, the energy requirements for air dissolution and pressurization can be significant, impacting the overall cost-effectiveness of the process (Sharma, 2023).

Filtration involves passing the algal suspension through a porous medium to separate biomass from the liquid phase. Techniques such as microfiltration and ultrafiltration utilize membranes with specific pore sizes to retain algal cells while allowing water to pass through (Kucmanova et al., 2019). While effective, membrane fouling caused by the accumulation of algal cells and extracellular substances on the membrane surface can reduce filtration efficiency and increase maintenance requirements. Regular cleaning and membrane replacement are necessary to sustain performance, which can add to operational costs (Arashiro et al., 2020a).

Flocculation Methods

Flocculation is a critical step in algae-based wastewater treatment, facilitating the aggregation and subsequent removal of microalgal cells from treated water. Efficient harvesting of microalgae not only enhances water quality but also enables the recovery of biomass for biofuel production and other applications. Various flocculation methods have been developed, each with distinct mechanisms and efficiencies (Yang et al., 2024).

Chemical Flocculation—Chemical flocculation involves the addition of chemical agents to neutralize the negative surface charge of microalgal cells, promoting aggregation. Commonly used chemical flocculants include inorganic compounds such as aluminum sulfate ($Al_2(SO_4)_3$), ferric chloride ($FeCl_3$), and polyaluminum chloride (PAC), as well as organic polymers like polyacrylamide (PAM). These agents have demonstrated effectiveness in enhancing flocculation efficiency. However, concerns regarding their environmental impact, potential toxicity, and the generation of chemical residues have prompted the exploration of alternative methods (Luo et al., 2024).

Bioflocculation—Bioflocculation leverages biological agents to induce flocculation, offering an eco-friendly alternative to chemical methods. One approach involves the use of microbial flocculants (MBFs), which are biopolymers produced by microorganisms such as bacteria, fungi, and algae (Muradov et al., 2015). These MBFs, comprising polysaccharides, proteins, and glycoproteins, facilitate the aggregation of microalgal cells through mechanisms like charge neutralization and bridging. Studies have highlighted the potential of MBFs in achieving high flocculation efficiencies while being biodegradable and environmentally benign (Yang et al., 2024). Another bioflocculation strategy involves fungal-assisted algal flocculation. In this method, filamentous fungi co-cultivated with microalgae form bioflocs through natural aggregation. This symbiotic interaction not only enhances biomass recovery but also improves the overall efficiency of wastewater treatment. Research has demonstrated that fungal-assisted flocculation can significantly reduce harvesting costs and energy consumption, addressing major challenges in the commercialization of microalgal biotechnology (Muradov et al., 2015).

Plant-Derived Flocculants—Plant-derived flocculants have emerged as sustainable and cost-effective alternatives in wastewater treatment. Extracts from various plant sources, including seeds, leaves, and roots, contain natural polymers capable of inducing flocculation. These bioflocculants are biodegradable, non-toxic, and readily available, making them attractive for large-scale applications. Recent studies have explored the efficacy of plant-based flocculants in removing pollutants from wastewater, highlighting their potential in green chemistry approaches for environmental remediation (Das et al., 2021).

The selection of an appropriate flocculation method in algae-based wastewater treatment depends on various factors, including cost, environmental impact, and efficiency. While chemical flocculants are effective, their potential adverse effects have led to increased interest in bioflocculation techniques and plant-derived flocculants. These alternative methods offer sustainable solutions, aligning with the goals of environmental conservation and resource recovery (Yang et al., 2024).

Algal Growth Optimization Strategies

Optimizing algal growth in wastewater treatment systems is essential for enhancing pollutant removal efficiency and maximizing biomass yield for resource recovery. Several strategies have been developed to achieve these objectives, focusing on cultivation methods, operational parameters, and system configurations (Arashiro et al., 2020a).

Cultivation Methods are important in achieving optimum algae growth in wastewater treatment plants. There both open pond systems and high-rate algal pond systems are commonly employed. Open pond systems are shallow, open-air ponds where algae are cultivated using natural sunlight and atmospheric carbon dioxide. Open ponds are cost-effective and commonly used in large-scale operations. High-Rate Algal Ponds (HRAPs) are designed to enhance algal productivity by improving water circulation and nutrient distribution. They offer efficient wastewater treatment and facilitate resource recovery, such as biogas or biofertilizer production (Arias et al., 2018a).

Operational Parameters are another important aspect to be considered in optimizing algae growth in wastewater treatment plants. Adjusting Hydraulic Retention Time (HRT) and Solids Retention Time (SRT) influence nutrient loading and microbial community composition. Studies have shown that shorter HRTs can enhance nutrient removal rates and biomass production. Light Intensity and Quality are two critical factors for photosynthetic organisms. Optimizing light conditions, including intensity and photoperiod, can significantly impact algal growth and nutrient uptake (Arias et al., 2018b).

System Configurations are also matter in this aspect. Photo-Sequencing Batch Reactors (PSBRs) operate in cycles of filling, reacting, settling, and decanting, allowing for controlled growth conditions. This configuration has been effective in selecting desired algal species and improving nutrient removal efficiencies. The co-cultivation of microalgae and bacteria in microalgal-bacterial flocs (MaB-Flocs) leads to the formation of flocs that settle easily, simplifying biomass harvesting. This symbiotic relationship enhances nutrient removal and supports high-density cultures (Arashiro et al., 2020a).

6 Applications and Case Studies

Municipal Wastewater Treatment Facilities Using Algae

Algae-based wastewater treatment systems have been implemented in various municipal settings worldwide, demonstrating their potential for effective nutrient removal and resource recovery. These case studies underscore the versatility and effectiveness of algae-based systems in municipal wastewater treatment, highlighting their potential for nutrient removal, resource recovery, and environmental sustainability.

1. **Integrated Algae Pond Systems (IAPS) in South Africa**

In South Africa, the implementation of Integrated Algae Pond Systems (IAPS) has been explored for municipal wastewater treatment. These systems utilize a series of ponds where algae and bacteria work symbiotically to treat wastewater. Research indicates that IAPS can effectively reduce organic matter and nutrient concentrations, offering a sustainable alternative to conventional treatment methods (Cowan et al., 2019).

2. **High-Rate Algal Ponds (HRAPs) in Spain**

A study conducted in Spain assessed the performance of pilot-scale High-Rate Algal Ponds (HRAPs) treating urban wastewater. The research focused on the effect of primary treatment on HRAP efficiency, biomass productivity, and biogas production potential. Findings revealed that HRAPs, both with and without primary treatment, achieved high ammonium removal rates (91–93%) and significant chemical oxygen demand (COD) reduction (62–65%). Additionally, co-digestion of microalgal biomass with primary sludge enhanced methane yields, indicating the viability of HRAPs for wastewater treatment and bioenergy recovery (Arashiro et al., 2020b).

3. **Microalgae-Bacteria Consortium in Ethiopia**

In Ethiopia, researchers investigated the use of native microalgae-bacteria consortia for treating primary-treated municipal wastewater. The study demonstrated that these consortia could effectively remove nutrients and organic pollutants, presenting a cost-effective and environmentally friendly alternative to traditional wastewater treatment technologies (Fito & Alemu, 2019; Luo et al., 2024).

4. **Municipal Lagoon Treatment in Oswego, Kansas**

The City of Oswego, Kansas, operates a lagoon-based municipal wastewater treatment system that reclaims treated effluent for irrigating a local golf course. This approach not only addresses wastewater disposal challenges but also provides a sustainable water source for irrigation. The system includes measures to manage algae growth, which is typical in lagoon treatments, ensuring compliance with environmental standards (USEPA, 2024).

Success Stories from Around the World

Algae-based wastewater treatment systems have been successfully implemented worldwide, showcasing their potential for effective nutrient removal and resource recovery. A few notable global success stories are as follows.

1. **Gross-Wen Technologies (GWT) in the United States**

Gross-Wen Technologies, based in Iowa, has developed the Revolving Algal Biofilm (RAB) system, an innovative approach to wastewater treatment. This system utilizes vertically oriented conveyor belts on which algae grow, absorbing nutrients such as nitrogen and phosphorus from the wastewater. The harvested algae can then be processed into sustainable fertilizers. A demonstration of this technology validated its effectiveness in retrofitting existing aerated lagoons, achieving significant nutrient removal while potentially saving over $1 million in capital expenses. Following these promising results, the city pursued a commercial-scale RAB facility, with plant upgrades expected to treat the full flow of wastewater and meet stringent effluent requirements (Algae Biomass Organization, 2020).

2. **ALL-GAS Project in Spain**

The ALL-GAS project in Spain has demonstrated the feasibility of large-scale algae cultivation for wastewater treatment and biofuel production. By utilizing wastewater as a nutrient source, the project has achieved an algae harvest of approximately 100 tons per hectare annually, yielding up to 15,000 kg of methane per year through anaerobic digestion. This initiative highlights the potential of integrating algae-based systems into existing wastewater infrastructure to enhance sustainability and energy recovery (IEA Bioenergy, 2019).

3. **Microalgae Biorefinery 2.0 in the European Union**

The Microalgae Biorefinery (MAB) 2.0 project, supported by Climate-KIC in the European Union, focuses on developing systems that cultivate algae in the side streams of wastewater treatment plants. This integration significantly reduces the number of treatment cycles required, leading to savings in time, energy, and operational costs. Additionally, the produced algae biomass can be utilized for various applications, including biofuels, fertilizers, and bioplastics, contributing to a circular economy (https://www.climate-kic.org/success-stories/microalgae-biorefinery/).

4. **Alicante Port Biofilter Project in Spain**

The Port of Alicante in Spain initiated a pilot project employing a biofilter based on Mediterranean algae to purify over 30,000 m^3 of port waters. This was developed by Mediterranean Algae Technologies, the 'Bioremed Algae' system aims to eliminate excess nutrients and heavy metals, thereby improving water quality. This project exemplifies the application of algae-based technologies in enhancing marine environmental health within port infrastructures. The actual project is scheduled to commence operations in March 2025. (https://cadenaser.com/comunitat-valenciana/2025/02/19/un-programa-piloto-purificara-30000-metros-cubicos-de-agua-del-puerto-de-alicante-radio-alicante/).

The above examples underscore the versatility and effectiveness of algae-based systems in municipal and industrial wastewater treatment, highlighting their potential for nutrient removal, resource recovery, and environmental sustainability.

Lessons Learned and Best Practices

Selecting appropriate cultivation methods, such as open ponds, raceway ponds, or photobioreactors, is crucial for maximizing nutrient removal and biomass production. Design considerations should account for local climate conditions, land availability, and wastewater characteristics to enhance treatment efficiency. Understanding the mechanisms of nutrient uptake by algae is essential for effective treatment. Manipulating environmental factors like light intensity, temperature, and nutrient concentrations can enhance the assimilation of nitrogen and phosphorus, leading to improved water quality (Arashiro et al., 2020b).

Retrofitting existing wastewater treatment facilities with algae-based systems can be cost-effective and improve overall performance. For instance, incorporating high-rate algal ponds (HRAPs) can enhance nutrient removal and facilitate resource recovery, such as biogas or biofertilizer production. Efficient harvesting techniques, including sedimentation, flotation, or filtration, are vital for recovering algal biomass (Ugwuanyi et al., 2024). The harvested biomass can be valorized into biofuels, fertilizers, or other value-added products, contributing to the economic viability of the treatment system. Conducting life cycle assessments (LCA) and economic analyses helps in evaluating the sustainability and feasibility of algae-based treatments. Studies have shown that HRAP systems coupled with biogas production can be more environmentally friendly, while those combined with biofertilizer production may offer better economic prospects (Nguyen et al., 2022).

Operators of this type of facilities should be prepared to manage issues such as algal blooms, system fouling, and seasonal variations in algal growth. Implementing regular monitoring and adaptive management strategies can mitigate these challenges and maintain system efficiency. By incorporating these best practices, algae-based wastewater treatment plants can enhance their performance, sustainability, and economic viability (Arashiro et al., 2020b).

7 Challenges and Limitations

Algae-based wastewater treatment plants (WWTPs) offer a sustainable method for nutrient removal and biomass production. However, several challenges and limitations affect their efficiency and broader application (Lavrinovičs & Juhna, 2017).

Efficiently separating microalgal biomass from treated water is a significant hurdle due to the small size and similar density of algal cells compared to water. This process can be energy-intensive and costly, impacting the overall feasibility of the technology (Zhou et al., 2018). Adequate light is essential for photosynthesis in microalgae. In dense cultures, self-shading can occur, reducing light availability to cells deeper in the medium and limiting growth rates. Optimizing reactor design and operational parameters is necessary to address this issue (Vaz et al., 2023). Wastewater composition can vary significantly, affecting nutrient availability for microalgal growth. Fluctuations in nutrient concentrations can lead to inconsistent treatment performance and biomass yields. Developing robust algal strains and adaptive management strategies is crucial to handle such variability (Lavrinovičs & Juhna, 2017).

The operational costs associated with algae-based systems, including energy for mixing, CO_2 supplementation, and biomass harvesting, can be substantial. Economic viability often depends on the valorization of algal biomass into high-value products, which requires additional processing and market development. Addressing these challenges through technological innovations and process optimizations is essential for the successful implementation of algae-based wastewater treatment systems (Gonçalves, 2020).

Climate and Geographic Constraints

Microalgae growth is highly dependent on temperature, with optimal ranges typically between 20 and 30 °C. In regions with cooler climates, especially during winter months, lower temperatures can significantly reduce algal metabolic rates, leading to decreased treatment efficiency (Zhang et al., 2023b). Conversely, excessively high temperatures can cause thermal stress, adversely affecting algal viability. Implementing temperature control measures or selecting thermotolerant algal strains may mitigate these challenges. Adequate light is essential for photosynthesis in microalgae (Lavrinovičs & Juhna, 2017). Geographic locations with limited sunlight, due to high latitudes or frequent cloud cover, can experience reduced algal productivity. Seasonal variations also impact light availability, with shorter daylight hours in winter months further constraining algal growth. Innovative reactor designs that maximize light exposure or the use of artificial lighting can help address these limitations (Arashiro et al., 2020b).

The establishment of algae-based WWTPs requires sufficient land area and access to water resources. In densely populated or arid regions, securing the necessary space and water for large-scale algal cultivation can be challenging. Integrating algal systems with existing wastewater infrastructure or utilizing non-arable lands may offer viable solutions (Zhang et al., 2023a). Seasonal changes can lead to fluctuations in wastewater composition and volume, affecting nutrient availability for algal growth. For instance, increased storm water runoff during rainy seasons can dilute nutrient concentrations, while dry periods may concentrate pollutants. Adaptive management strategies and flexible system designs are essential to accommodate these variations (Zhang et al., 2023b).

Addressing these climate and geographic constraints is crucial for the successful implementation of algae-based wastewater treatment systems. Tailoring system designs to local environmental conditions, selecting appropriate algal species, and incorporating technological innovations can enhance resilience and efficiency (Algae Biomass Organization, 2020; Oberholster & Botha, 2018).

Scalability and Cost-Effectiveness

The integration of algae-based treatment technologies into full-scale WWTPs has gained momentum globally (Arashiro et al., 2020a). Various process designs are available, differing in species selection, reactor configuration, and biomass growth modes, allowing for adaptability to diverse operational scales and treatment requirements. Algae-based systems are recognized for their sustainability and energy efficiency in treating wastewater (Clippinger & Davis, 2021). Utilizing natural processes, these systems can reduce operational costs associated with conventional treatments. For instance, algal turf scrubbers have demonstrated phosphorus removal for $24 per kilogram, compared to $77 per kilogram using engineered wetland processes (Li et al., 2019).

Coupling algae cultivation with wastewater treatment offers economic benefits. A techno-economic assessment indicated that integrating algae farming with wastewater treatment can enhance algal biomass production, contributing to cost offsets and improved scalability (Molinuevo-Salces et al., 2019). Life Cycle Assessments (LCAs) have shown that high-rate algal ponds (HRAPs) are environmentally friendly and economically feasible, especially when combined with biofertilizer production. These systems can offer a viable alternative to traditional activated sludge processes, particularly in small communities (Lavrinovičs & Juhna, 2017).

Fouling and Algal Bloom Management

Fouling, particularly biofouling, occurs when microorganisms accumulate on surfaces, leading to the formation of biofilms. In membrane-based wastewater treatment processes, biofouling can cause a decline in membrane flux, increased feed pressure, and higher energy consumption. This necessitates frequent membrane cleaning or replacement, thereby escalating operational costs. Implementing effective fouling control strategies, such as optimizing operational parameters and employing advanced cleaning techniques, is essential to maintain system performance (Cairone et al., 2024).

While controlled algal growth is beneficial within treatment systems, uncontrolled algal blooms can pose significant problems. Excessive algal proliferation can lead to elevated effluent total suspended solids (TSS), increased biological oxygen demand (BOD), and pH imbalances, complicating compliance with discharge regulations (Molinuevo-Salces et al., 2019). Moreover, algal blooms can deplete dissolved oxygen levels, adversely affecting aerobic treatment processes and leading to odor issues. To mitigate these effects, it is crucial to monitor and regulate nutrient inputs, particularly phosphorus and nitrogen, as their overabundance can trigger harmful algal blooms (Cairone et al., 2024).

Effective management of nutrient levels is vital to prevent undesirable algal blooms. Wastewater treatment plants can implement strategies to limit phosphorus concentrations, as even levels as low as 0.1 mg/L can promote algal growth. By optimizing nutrient removal processes and employing advanced treatment technologies, facilities can reduce the risk of algal overgrowth and maintain ecological balance within the treatment system (Cairone et al., 2024). Continuous monitoring of algal populations and water quality parameters enables early detection of fouling and bloom events. Implementing adaptive management practices, such as adjusting hydraulic retention times, modifying light exposure, and controlling nutrient dosing, can help maintain optimal algal growth and prevent system disruptions. Additionally, integrating real-time data analytics can enhance decision-making and operational efficiency in algae-based WWTPs (Clippinger & Davis, 2021).

Integration with Existing Wastewater Infrastructure

Integrating algae-based wastewater treatment systems with existing wastewater infrastructure offers a sustainable approach to enhancing treatment efficiency and resource recovery. Combining algae-based processes with conventional methods, such as activated sludge or membrane bioreactors, can improve nutrient removal and reduce operational costs (Algae Biomass Organization, 2020). In these hybrid systems, algae assimilate nutrients like nitrogen and phosphorus, while producing oxygen through photosynthesis, which supports aerobic bacterial degradation of organic pollutants. This symbiotic relationship enhances overall treatment performance (Lavrinovičs & Juhna, 2017).

Existing wastewater treatment plants can be retrofitted with algal cultivation units, such as open ponds or photobioreactors, to treat effluents further. This approach allows for the utilization of existing infrastructure, minimizing capital expenditures (Arashiro et al., 2020a). For instance, integrating high-rate algal ponds can enhance nutrient removal and enable the recovery of valuable biomass for biofuel or fertilizer production. Employing mixed cultures of algae and bacteria can enhance treatment efficiency. These consortia can adapt to varying wastewater compositions and environmental conditions, leading to more robust and resilient treatment processes. The synergistic interactions between algae and bacteria facilitate the breakdown of complex organic compounds and improve nutrient uptake. Algae-based systems offer opportunities for energy and resource recovery (Zhou et al., 2018). The biomass produced can be processed into biofuels, bioplastics, or fertilizers, contributing to a circular economy. Additionally, the oxygen generated by algal photosynthesis can reduce the energy requirements for aeration in conventional treatment processes (Gonçalves, 2020).

8 Resource Recovery and Circular Economy

Algae-based wastewater treatment technologies offer a multifaceted approach to resource recovery, aligning with circular economy principles. These systems not only purify wastewater but also facilitate the extraction of valuable resources, thereby transforming waste streams into sustainable inputs for various industries. Through nutrient recycling, energy production, and the generation of valuable biomass, these systems transform wastewater treatment from a disposal process into a resource-generating endeavor.

Algal Biomass Utilization

Microalgae efficiently assimilate nutrients such as nitrogen and phosphorus from wastewater, leading to significant nutrient removal and the production of algal biomass. This biomass serves as a versatile resource, with applications ranging from biofuel production to use as fertilizers and animal feed, thereby contributing to a circular bioeconomy (Nguyen et al., 2022).

Biofuel and Biogas Production

The harvested algal biomass can be processed through anaerobic digestion to produce biofuel or biogas, renewable energy sources. Studies have demonstrated that co-digesting microalgae with primary sludge enhances methane yields, offering a viable method for bioenergy recovery in wastewater treatment facilities (Arashiro et al., 2019).

Fertilizers and Animal Feed

Algae-based wastewater treatment plants offer a sustainable approach to wastewater management by not only purifying water but also generating valuable byproducts such as fertilizers and animal feed. The integration of microalgae cultivation in these systems facilitates the recovery of nutrients and the production of biomass rich in proteins, lipids, and other essential compounds. Microalgae assimilate nutrients

Contribution to Carbon Capture and Climate Mitigation

Algae-based wastewater treatment plants (WWTPs) offer a multifaceted approach to environmental management by integrating wastewater purification with carbon capture, thereby contributing to climate change mitigation (Kumar & Ghangrekar, 2024). These systems utilize microalgae's natural photosynthetic capabilities to assimilate carbon dioxide (CO_2) and nutrients from wastewater, resulting in reduced greenhouse gas emissions and the production of valuable biomass (Zhang et al., 2023a). Microalgae, such as *Chlorella vulgaris*, are proficient in sequestering CO_2 through photosynthesis, converting it into organic compounds and oxygen (Lavars, 2019). When cultivated in wastewater, these algae not only treat the effluent by removing contaminants but also absorb CO_2 from the atmosphere or flue gases introduced into the system (Demirbas, 2011). This dual function enhances the sustainability of wastewater treatment processes. Studies have demonstrated that microalgae can achieve carbon capture efficiencies of up to 90%, making them effective agents for CO_2 mitigation (Singh et al., 2011).

The integration of algae-based systems in WWTPs contributes to climate mitigation in several ways (Demirbas, 2011):

Reduction of Greenhouse Gas Emissions: Traditional wastewater treatment processes can be significant sources of greenhouse gases (GHGs). Algae-based treatments can lower these emissions by capturing CO_2 and reducing the energy consumption associated with conventional methods (Kumar & Ghangrekar, 2024; Wang et al., 2013).

Biomass Utilization: The harvested algal biomass can be processed into biofuels, fertilizers, or other bioproducts, providing renewable alternatives to fossil fuel-based products and further reducing the carbon footprint (Zhang et al., 2023a).

However, the implementation of algae-based carbon capture in WWTPs faces challenges such as Economic Viability (The costs associated with large-scale cultivation and processing of algal biomass can be high (Wang et al., 2013). Innovations in technology and economies of scale are needed to make these systems more cost-effective) and System Optimization (Factors such as light availability, temperature, and CO_2 concentration must be optimized to maximize algal growth and carbon sequestration efficiency; Singh et al., 2011).

9 Future Prospects and Innovations

Genetic engineering has significantly advanced the application of algae in solid waste management, enhancing their efficiency in bioremediation and resource recovery.

Advances in Genetic Engineering for Algae

Microalgae, such as *Scenedesmus* species, have been genetically modified to improve their ability to treat wastewater by removing contaminants like ammonia and phosphorus. These engineered strains not only purify wastewater but also produce biomass that can be utilized in various applications, contributing to economic sustainability (Calatrava et al., 2024).

In the context of solid waste management, algae have been employed to treat landfill leachate, which is often rich in organic pollutants and heavy metals. Studies have demonstrated that certain microalgae can effectively reduce the concentration of these contaminants, thereby mitigating the environmental impact of landfill sites (Satpati, 2021). Furthermore, advancements in genetic engineering have enabled the development of algae strains with enhanced capabilities for bioremediation (Demirbas, 2011). For instance, genetically modified algae have been designed to express specific enzymes that degrade hazardous compounds more efficiently, offering a sustainable solution for the treatment of industrial waste. These innovations in algal biotechnology not only improve waste treatment processes but also facilitate the recovery of valuable resources from waste streams, aligning with the principles of a circular economy (Jafari et al., 2013).

Automation and AI in Algal Waste Treatment

The integration of automation and artificial intelligence (AI) into algal waste treatment processes has significantly enhanced efficiency and effectiveness in wastewater management. Algal-based systems are recognized for their potential in nutrient uptake and resource recovery. The symbiotic relationship between algae and bacteria in these systems is complex, and AI-driven automation plays a crucial role in optimizing these interactions. By analyzing vast datasets, AI can optimize process parameters and predict outcomes in real time, leading to improved treatment efficiency and reduced operational costs (Sakkaravarthy et al., 2024). AI-driven process automation enables continuous monitoring of water quality parameters, allowing for immediate adjustments to treatment processes and ensuring compliance with environmental regulations. Machine learning algorithms can predict equipment failures before they occur, minimizing downtime and maintenance costs. Additionally, AI systems analyze historical data to identify trends and anomalies, facilitating informed decision-making regarding operational adjustments (Restack.io, 2024).

The application of AI in wastewater treatment is not without challenges. The successful implementation of AI systems heavily relies on high-quality data; inaccurate or incomplete data can lead to incorrect predictions and suboptimal decisions. Despite these challenges, the integration of AI and automation in algal waste treatment represents a significant advancement, offering sustainable solutions for efficient wastewater management (Nagpal et al., 2024).

Anand, U., Dey, S., Parial, D., Federici, S., Ducoli, S., Bolan, N. S., Dey, A., & Bontempi, E. (2023). Algae and bacteria consortia for wastewater decontamination and transformation into biodiesel, bioethanol, biohydrogen, biofertilizers and animal feed: A review. *Environmental Chemistry Letters, 21*, 1585–1609. https://doi.org/10.1007/s10311-023-01562-w

Arashiro, L. T., Boto-Ordonez, M., Van Hulle, S. W., Ferrer, I., Garfi, M., & Rousseau, D. P. (2020a). *Natural pigments from microalgae grown in industrial wastewater*. arXiv preprint arXiv:2003.06192

Arashiro, L. T., Ferrer, I., Rousseau, D. P. L., Van Hulle, S. W. H., & Garfí, M. (2019). The effect of primary treatment of wastewater in high rate algal pond systems: Biomass and bioenergy recovery. *Bioresource Technology, 280*, 27–36.

Arashiro, L. T., Josa, I., Ferrer, I., Van Hulle, S. W. H., Rousseau, D. P. L., & Garfí, M. (2022). Life cycle assessment of microalgae systems for wastewater treatment and bioproducts recovery: Natural pigments, biofertilizer and biogas. *Science of the Total Environment, 15*(847), Article 157615. https://doi.org/10.1016/j.scitotenv.2022.157615

Arashiro, L. T., Montero, N., Ferrer, I., Acien, F. G., Gomez, C., & Garfi, M. (2020b). *Life cycle assessment of high rate algal ponds for wastewater treatment and resource recovery*. arXiv preprint arXiv:2003.06194. https://arxiv.org/abs/2003.06194

Arias, D. M., Rueda, E., García-Galán, M. J., Uggetti, E., & García, J. (2018a). *Selection of cyanobacteria over green algae in a photo-sequencing batch bioreactor fed with wastewater*. arXiv preprint arXiv:1812.09559. https://arxiv.org/abs/1812.09559

Arias, D. M., Uggetti, E., García-Galán, M. J., & García, J. (2018b). *Nutrients and biomass dynamics in photo-sequencing batch reactors treating wastewater with high nutrients loadings*. arXiv preprint arXiv:1806.03882

Blánquez, P., Hom-Díaz, A., Vicent, T., Guieysse, B. (2020). Microalgae-based processes for the removal of pharmaceuticals in wastewater. In: S. Rodriguez-Mozaz, P. Blánquez Cano & M. Sarrà Adroguer (Eds.), *Removal and degradation of pharmaceutically active compounds in wastewater treatment. The handbook of environmental chemistry*, 108. Springer, Cham. https://doi.org/10.1007/698_2020_682

Borowitzka, M. A., & Borowitzka, L. J. (1990). *Micro-algal biotechnology*. Cambridge University Press.

Cai, T., Park, S. Y., & Li, Y. (2013). Nutrient recovery from wastewater streams by microalgae: Status and prospects. *Renewable and Sustainable Energy Reviews, 19*, 360–369.

Cairone, S., Hasan, S. W., Choo, K. H., Li, C. W., Zorpas, A. A., Ksibi, M., Zarra, T., Belgiorno, V., & Naddeo, V. (2024). Enhancing membrane fouling control in wastewater treatment processes through artificial intelligence modeling: Research progress and future perspectives. *Euro-Mediterranean Journal for Environmental Integration, 9*, 1887–1905. https://doi.org/10.1007/s41207-024-00659-0

Calatrava, V., Ballester, G. D., & Dubini, A. (2024). Microalgae for bioremediation: advances, challenges, and public perception on genetic engineering. *BMC Plant Biology, 24*, 1261. https://bmcplantbiol.biomedcentral.com/articles/10.1186/s12870-024-05995-5

Chew, K. W., Yap, J. Y., Show, P. L., Suan, N. H., Juan, J. C., Ling, T. C., & Lee, D. J. (2017). Microalgae biorefinery: High-value products perspectives. *Bioresource Technology, 229*, 53–62.

Chisti, Y. (2007). Biodiesel from microalgae. *Biotechnology Advances, 25*(3), 294–306.

Clippinger, J., & Davis, R. (2021). Techno-economic assessment for opportunities to integrate algae farming with wastewater treatment. National Renewable Energy Laboratory. https://www.nrel.gov/docs/fy21osti/75237.pdf

Cowan, A. K., Mambo, P. M., Westensee, D. K., & Render, D. S. (2019). *Evaluation of integrated algae pond systems for municipal wastewater treatment*. Institute for Environmental Biotechnology, Rhodes University (EBRU), Rhodes University, Grahamstown.

Craggs, R., Park, J., Heubeck, S., & Sutherland, D. (2014). High rate algal pond systems for low-energy wastewater treatment, nutrient recovery and energy production. *New Zealand Journal of Botany, 52*(1), 60–73. https://doi.org/10.1080/0028825X.2013.861855

Craggs, R., Sutherland, D., & Campbell, H. (2014). Hectare-scale demonstration of high rate algal ponds for enhanced wastewater treatment and biofuel production. *Journal of Applied Phycology, 24*, 329–337. https://doi.org/10.1007/s10811-012-9810-8

Craggs, R. J., Zwart, A., Nagels, J. W., & Davies-Colley, R. J. (2004). Modelling sunlight disinfection in a high rate pond. *Ecological Engineering, 22*(2), 1, 113–122. https://doi.org/10.1016/j.ecoleng.2004.03.001

Daphtary, N. (2024). *Wastewaster treatment process demystified*. Available at: https://chemtech-us.com/the-complete-guide-to-co2-check-valves-2/. Accessed February 27, 2025.

Das, N., Ojha, N., & Mandal, S. K. (2021). Wastewater treatment using plant-derived bioflocculants: Green chemistry approach for safe environment. *Water Science and Technology, 83*(8), 1797–1812. https://doi.org/10.2166/wst.2021.100

Demirbas, M. F. (2011). Biofuels from algae for sustainable development. *Applied Energy, 88*(10), 3473–3480. https://doi.org/10.1016/j.apenergy.2011.01.059

El-Sheekh, M., El-Dalatony, M. M., Thakur, N., Zheng, Y., & Salama, E. (2022). Role of microalgae and cyanobacteria in wastewater treatment: Genetic engineering and omics approaches. *International Journal of Environmental Science and Technology, 19*, 2173–2194. https://doi.org/10.1007/s13762-021-03270-w

El-Sherif, D. M., Mahmoud, A., & Saber, A. N. (2024). The application of the algal approach in the technological, socioeconomic, and wastewater treatment domains. In T. A. Kurniawan & A. Anouzla (Eds.), *Algae as a natural solution for challenges in water-food-energy Nexus. Environmental science and engineering*. Springer, Singapore. https://doi.org/10.1007/978-981-97-2371-3_11

Environmental XPRT. (2023). *Decentralised wastewater treatment: The future is now*. Retrieved from https://www.environmental-expert.com/articles/decentralised-wastewater-treatment-the-future-is-now-1149256

Ferro, L. (2019). Wastewater treatment and biomass generation by Nordic microalgae: growth in subarctic climate and microbial interactions (Doctoral dissertation, Umeå University).

Fito, J., & Alemu, K. (2019). Microalgae–bacteria consortium treatment technology for municipal wastewater management. *Nanotechnology for Environmental Engineering, 4*, 4. https://doi.org/10.1007/s41204-018-0050-2

García, J., Ortiz, A., Álvarez, E., Belohlav, V., García-Galán, M. J., Díez-Montero, R., Álvarez, J. A., & Uggetti, E. (2018). *Nutrient removal from agricultural run-off in demonstrative full-scale tubular photobioreactors for microalgae growth*. arXiv preprint arXiv:1807.05189. https://arxiv.org/abs/1807.05189

García-Galán, M. J., Arashiro, L., Santos, L. H. M. L. M., Insa, S., Rodríguez-Mozaz, S., Barceló, D., Ferrer, I., & Garfi, M. (2020). Fate of priority pharmaceuticals and their main metabolites and transformation products in microalgae-based wastewater treatment systems. *Quantitative Biology: Quantitative Methods*. https://doi.org/10.48550/arXiv.2003.06882

Gimondo, J. A., Currey, C. J., Jarboe, D. H., Gross, M., & Graves, W. R. (2019). Wastewater-grown algae pellets and paste as fertilizers for containerized crops. *HortScience, 54*(3), 528–536. https://doi.org/10.21273/HORTSCI13474-18

Gonçalves, A. L. (2020). Microalgae-based wastewater treatment: methodologies, challenges, and developments. *Journal of Environmental Management, 246*, 51–67. https://atlasofscience.org/microalgae-based-wastewater-treatment-for-nutrients-recovery/

Gupta, R., Mishra, N., Singh, G., Mishra, S., & Lodhiyal, N. (2024). Microalgae cultivation and value-based products from wastewater: Insights and applications. *Blue Biotechnology, 1*, 20. https://doi.org/10.1186/s44315-024-00019-1

Guschina, I. A., & Harwood, J. L. (2006). Lipids and lipid metabolism in eukaryotic algae. *Progress in Lipid Research, 45*(2), 160–186.

Hawrot-Paw, M., Koniuszy, A., Gałczyńska, M., Zając, G., & Szyszlak-Bargłowicz, J. (2020). Production of microalgal biomass using aquaculture wastewater as growth medium. *Water, 12*(1), 106. https://doi.org/10.3390/w12010106

https://cadenaser.com/comunitat-valenciana/2025/02/19/un-programa-piloto-purificara-30000-metros-cubicos-de-agua-del-puerto-de-alicante-radio-alicante/. Accessed February 12, 2025.

https://www.climate-kic.org/success-stories/microalgae-biorefinery/. Accessed February 10, 2025.

https://www.epa.gov/sites/default/files/2015-06/documents/mou-intro-paper-081712-pdf-adobe-acrobat-pro.pdf. Accessed February 10, 2025.

IEA Bioenergy. (2019). *Success stories of advanced biofuels for transport: ALL-GAS: Algae biofuels*. Retrieved from https://www.ieabioenergy.com/wp-content/uploads/2020/11/16-AFF_IEABio_SuccessStories_Allgas_AlgaeBiofuels.pdf

Jafari, M., Danesh, Y. R., Goltapeh, E. M., & Varma, A. (2013). Bioremediation and genetically modified organisms. *Fungi as Bioremediators, Soil Biology*. https://doi.org/10.1007/978-3-642-33811-3_19

Kucmanova, A., & Gerulová, K. (2019). Microalgae harvesting: A review. *Research Papers Faculty of Materials Science and Technology Slovak University of Technology, 27*, 129–143. https://doi.org/10.2478/rput-2019-0014

Kumar, M., Pandit, S., Patel, V., Khanna, N., Nag, M., Lahiri, D., Ray, R. R., Das, A., & Das, D. (2023). Maximization of energy recovery from starch processing wastewater by thermophilic dark fermentation coupled with microbial fuel cell technology. *Geomicrobiology Journal*. https://doi.org/10.1080/01490451.2023.2209555

Kumar, S., & Ghangrekar, M. M. (2024). Algae-based bioenergy production as a carbon mitigation technology. In: T. A. Kurniawan & A. Anouzla (Eds.), *Algae as a natural solution for challenges in water-food-energy nexus. Environmental science and engineering*. Springer, Singapore. https://doi.org/10.1007/978-981-97-2371-3_29

Lau, P. S., Tam, N. F. Y., & Wong, Y. S. (1995). Effect of algal density on nutrient removal from primary settled wastewater. *Environmental Pollution, 89*(1), 59–66.

Lavars, N. (2019). Algae-fueled bioreactor soaks up CO_2 400x more effectively than trees. *New Atlas*. https://newatlas.com/environment/algae-biological-carbon-sequestration/. Accessed on January 04, 2025.

Lavrinovičs, A., & Juhna, T. (2017). Review on challenges and limitations for algae-based wastewater treatment. *Scientific Journal of Riga Technical University, 17*, 45. https://www.researchgate.net/publication/324492336_Review_on_Challenges_and_Limitations_for_Algae-Based_Wastewater_Treatment

Li, Y., Chen, Y. F., Chen, P., Min, M., Zhou, W., Martinez, B., Zhu, J., & Ruan, R. (2011). Characterization of a microalga *Chlorella* sp. well adapted to highly concentrated municipal wastewater for nutrient removal and biodiesel production. *Bioresource Technology, 102*(8), 5138–44. https://doi.org/10.1016/j.biortech.2011.01.091

Li, K., Liu, Q., Fang, F., Luo, R., Lu, Q., Zhou, W., Huo, S., Cheng, P., Liu, J., Addy, M., Chen, P., Chen, D., & Ruan, R. (2019). Microalgae-based wastewater treatment for nutrients recovery: A review. *Bioresource Technology, 291*. https://doi.org/10.1016/j.biortech.2019.121934

Li, W., Peng, S., Jiao, H., Zhang, J., Huang, Y., & Sun, L. (2021). Performance assessment of microalgae-based wastewater treatment (MBWT) system in response to operation modes, hydraulic retention time (HRT) and cyclical light. *Desalination and Water Treatment, 229*, 79–92. https://doi.org/10.5004/dwt.2021.27368

Luo, B., He, H., Yan, Y., Wang, Y., Yang, X., Liu, Y., Xu, J., & Huang, W. (2024). Flocculants for the high-concentration activated sludge method and the effectiveness of urban wastewater treatment. *Water, 16*, 2281. https://doi.org/10.3390/w16162281

Ma, X. N., Chen, T. P., Yang, B., Liu, J., & Chen, F. (2016). Lipid production from *Nannochloropsis*. *Marine Drugs, 25*, 14(4), 61. https://doi.org/10.3390/md14040061

Mara, D. D. (2004). *Domestic wastewater treatment in developing countries*. Earthscan.

Mata, T. M., Martins, A. A., & Caetano, N. S. (2010). Microalgae for biodiesel production and other applications: A review. *Renewable and Sustainable Energy Reviews, 14*(1), 217–232.

Mehta, S. K., & Gaur, J. P. (2005). Use of algae for removing heavy metal ions from wastewater: Progress and prospects. *Critical Reviews in Biotechnology, 25*(3), 113–152.

Metcalf and Eddy. (2013). *Wastewater engineering: Treatment and reuse* (5th ed.). McGraw-Hill.

Mishra, S., & Das, A. P. (2021). Current treatment technologies for removal of microplastic and microfiber pollutants from wastewater. *Wastewater Treatment: Cutting-Edge Molecular Tools, Techniques and Applied Aspects, 237–251*. https://doi.org/10.1016/B978-0-12-821881-5.00011-8

Mishra, S., & Das, A. P. (2022). Treatment of the wastewater polluted with synthetic microfiber released from washing machine. In B. B. Das, H. Hettiarachchi, P. K. Sahu, & S. Nanda (Eds.), *Lecture notes in civil engineering* (Vol. 207, pp. 109–117). Springer Science and Business Media Deutschland GmbH. https://doi.org/10.1007/978-981-16-7509-6_9/COVER

Mishra, S., & Das, A. P. (2023). Microbial remediation of synthetic microfiber contaminated wastewater. In M. P. Shah (Ed.), *Microbial technologies in industrial wastewater treatment* (pp. 337–350). Springer, Singapore. https://doi.org/10.1007/978-981-99-2435-6_17

Mishra, S., Dash, D., & Das, A. P. (2022). Detection, characterization and possible biofragmentation of synthetic microfibers released from domestic laundering wastewater as an emerging source of marine pollution. *Marine Pollution Bulletin, 185*, Article 114254. https://doi.org/10.1016/J.MARPOLBUL.2022.114254

Mishra, S., Dash, D., & Das, A. P. (2024). Isolation and molecular characterization of polyester-degrading bacterial strains from municipal wastewater and their application in microplastic degradation in contaminated wastewater. *Geomicrobiology Journal*. https://doi.org/10.1080/01490451.2024.2376101

Mishra, S., Dash, D., Al-Tawaha, A. R. M. S., & Das, A. P. (2022). A review on heavy metal ion adsorption on synthetic microfiber surface in aquatic environments. *Applied Biochemistry and Biotechnology, 194*(10), 4639–4654. https://doi.org/10.1007/S12010-022-04029-W/METRICS

Mishra, S., Singh, R. P., Rout, P. K., & Das, A. P. (2022). Membrane bioreactor (MBR) as an advanced wastewater treatment technology for removal of synthetic microplastics. *Development in Wastewater Treatment Research and Processes, 45–60*. https://doi.org/10.1016/B978-0-323-85583-9.00022-3

Molinuevo-Salces, B., Riaño, B., Hernández, D., & García-González, M. C. (2019). In M. Alam & Z. Wang (Eds.), *Microalgae biotechnology for development of biofuel and wastewater treatment*. Springer, Singapore. https://doi.org/10.1007/978-981-13-2264-8_20

Morales-Amaral, M. M., Gómez-Serrano, C., Acién, F. G., Fernández-Sevilla, J. M., & Molina-Grima, E. (2015). Outdoor production of *Scenedesmus* sp. in thin-layer and raceway reactors using wastewater. *Algal Research, 12*, 99–108.

Muradov, N., Taha, M., Miranda, A. F., Wrede, D., Kadali, K., Gujar, A., Stevenson, T., Ball, A. S., & Mouradov, A. (2015). Fungal-assisted algal flocculation: Application in wastewater treatment and biofuel production. *Biotechnology for Biofuels, 8*, 24. https://doi.org/10.1186/s13068-015-0210-6

Nagpal, M., Siddique, M. A., Sharma, K., Sharma, N., & Mittal, A. (2024). Optimizing wastewater treatment through artificial intelligence: Recent advances and future prospects. *Water Science and Technology, 90*(3), 731–757. https://doi.org/10.2166/wst.2024.259

Naidoo, S., & Olaniran, A. O. (2014). Treated wastewater effluent as a source of microbial pollution of surface water resources. *International Journal of Environmental Research and Public Health, 23*, 11(1), 249–270. https://doi.org/10.3390/ijerph110100249

Nguyen, L. N., Aditya, L., Vu, H. P., Johir, A. H., Bennar, L., Ralph, P., Hoang, N. B., Zdarta, J., & Nghiem, L. D. (2022). Nutrient removal by algae-based wastewater treatment. *Current Pollution Reports, 8*, 369–383. https://doi.org/10.1007/s40726-022-00230-x

Oberholster, P. J., & Botha, A. M. (2018). The environmental feasibility of low cost algae-based sewage treatment in developing countries: A case study from South Africa. *CSIR Research Space*. https://researchspace.csir.co.za/bitstream/handle/10204/10459/Oberholster1_2018.pdf

Olguín, E. J. (2012). Dual purpose microalgae–bacteria-based systems that treat wastewater and produce biodiesel and chemical products within a biorefinery. *Biotechnology Advances, 30*(5), 1031–1046.

Oswald, W. J. (1988). Micro-algae and waste-water treatment. In M. A. Borowitzka & L. J. Borowitzka (Eds.), *Micro-algal biotechnology* (pp. 305–328). Cambridge University Press.

Oswald, W. J., & Gotaas, H. B. (1957). Photosynthesis in sewage treatment. *Transactions of the American Society of Civil Engineers, 122*(1), 73–105.

Öztürk, S., & Satpati, G. G. (2024). Extremophilic algae-based wastewater treatment, nutrient recovery, and animal feed production. In Mp P. Shah & S. Dey (Eds.), *Trends in biotechnology of polyextremophiles*. Springer, Cham. https://doi.org/10.1007/978-3-031-55032-4_11

Parial, D., & Dey, S. (2023). Role of microalgae in wastewater treatment and their role in nutrient recovery. In M. P. Shah (Eds.), *Sustainable industrial wastewater treatment and pollution control*. Springer, Singapore. https://doi.org/10.1007/978-981-99-2560-5_3

Park, J. B. K., Craggs, R. J., & Shilton, A. N. (2011). Wastewater treatment high rate algal ponds for biofuel production. *Bioresource Technology, 102*(1), 35–42.

Pittman, J. K., Dean, A. P., & Osundeko, O. (2011). The potential of sustainable algal biofuel production using wastewater resources. *Bioresource Technology, 102*(1), 17–25.

Posten, C. (2009). Design principles of photobioreactors for cultivation of microalgae. *Engineering in Life Sciences, 9*(3), 165–177. https://doi.org/10.1002/elsc.200900003

Priyadarsini, S., & Das, A. P. (2024). Analytical and structural characterization of waste lithium-ion batteries for their effective recycling strategy. *Environmental Chemistry and Ecotoxicology*. https://doi.org/10.1016/j.enceco.2024.12.004

Priyadarsini, S., & Das, A. P. (2025). Lithium bioleaching: A review on microbial-assisted sustainable technology for lithium bio-circularity. *Journal of Water Process Engineering, 69*, Article 106744. https://doi.org/10.1016/j.jwpe.2024.106744

Rawat, I., Ranjith Kumar, R., Mutanda, T., & Bux, F. (2011). Dual role of microalgae: Phycoremediation of domestic wastewater and biomass production for sustainable biofuels production. *Applied Energy, 88*(10), 3411–3424.

Razzaka, S. A., Bahara, K., Islama, K. M. O., Haniffaa, A. K., Faruquea, M. O., Hossainc, S. M. Z., & Hossain, M. M. (2024). Microalgae cultivation in photobioreactors: Sustainable solutions for a greener future. *Green Chemical Engineering, 5*, 418–439. https://doi.org/10.1016/j.gce.2023.10.004

Restack.io. AI solutions for wastewater management. Restackio. https://www.restack.io/p/ai-in-waste-management-optimization-answer-wastewater-cat-ai. Accessed February 12, 2025.

Ruiz, A. O., García, J., Sampaio, I. C. F., & Montero, R. D. (2024). Harvesting microalgae: Overcoming the bottlenecks in microalgae-based wastewater treatment through industrial-scale gravity sedimentation and thickening. *Microalgal bioengineering* (pp. 83–102). https://doi.org/10.1007/978-3-031-61253-4_4

Ruiz-Martinez, A., Martin Garcia, N., Romero, I., Seco, A., & Ferrer, J. (2012). Microalgae cultivation in wastewater: Nutrient removal from anaerobic membrane bioreactor effluent. *Bioresource Technology, 126*, 247–253.

Sakkaravarthy, S., Jano, N. A., & Vijayakumar, A. (2024). Overcoming challenges in traditional wastewater treatment through AI-driven innovation. In *The AI cleanse: Transforming wastewater treatment through artificial intelligence* (pp. 53–81). Springer. https://link.springer.com/chapter/10.1007/978-3-031-67237-8_3

Satpati, G. G. (2021). Solid waste management by algae: Current applications and future perspectives. *Pollution Research, 40*(1), 259–264.

Sharma, P. (2023). *Cultivation of algae in wastewater treatment*. [PDF]. Retrieved from https://www.theseus.fi/bitstream/handle/10024/790388/Sharma_Pratik.pdf

Singh, A., Nigam, P. S., & Murphy, J. D. (2011). Renewable fuels from algae: An answer to debatable land based fuels. *Bioresource Technology, 102*(1). 10–16 pp. https://doi.org/10.1016/j.biortech.2010.06.032

Singh, J., & Olsen, S. I. (2011). A critical review of biochemical conversion, sustainability and life cycle assessment of algal biofuels. *Applied Energy, 88*(10), 3548–3555.

Smith, A. L., Stadler, L. B., Cao, L., Love, N. G., Raskin, L., & Skerlos, S. J. (2018). Navigating wastewater energy recovery strategies: A life cycle comparison of anaerobic membrane bioreactor and conventional treatment systems. *Environmental Science & Technology, 52*(1), 268–277.

Spellman, F. R. (2018). *Handbook of water and wastewater treatment plant operations* (3rd ed., pp. 97–110). CRC Press.

Su, Y., Mennerich, A., & Urban, B. (2012). Synergistic cooperation between wastewater-born algae and activated sludge for wastewater treatment: Influence of algae and sludge inoculation ratios. *Bioresource Technology, 105*, 67–73.

Tchobanoglous, G., Kenny, J., Leverenz, H., & Oliveri, A. (2022). Establishing log reduction values for wastewater treatment processes from ambient influent and effluent pathogen monitoring data. *Frontiers in Environmental Science, 10*, Article 940014. https://doi.org/10.3389/fenvs.2022.940014

Tian, X., Lin, X., Xie, Q., Liu, J., & Luo, L. (2024). Effects of temperature and light on microalgal growth and nutrient removal in turtle aquaculture wastewater. *Biology, 13*(11), 901. https://doi.org/10.3390/biology13110901

Ugwuanyi, E. D., Nwokediegwu, Z. Q. S., Dada, M. A., Majemite, M. T., & Obaigbena, A. (2024). The role of algae-based wastewater treatment systems: A comprehensive review. *World Journal of Advanced Research and Reviews, 21*(2), 937–949.

USEPA. (2024). Water reuse case study: Oswego, Kansas. https://www.epa.gov/waterreuse/water-reuse-case-study-oswego-kansas?utm_source=chatgpt.com. Accessed February 10, 2025.

Van Den Hende, S., Vervaeren, H., & Boon, N. (2012). Flue gas compounds and microalgae: (bio)chemical interactions leading to biotechnological opportunities. *Biotechnology Advances, 30*(6), 1405–1424. https://doi.org/10.1016/j.biotechadv.2012.02.015

Vaz, B. S., Moreira, J. B., Morais, M. G., & Costa, J. A. V. (2023). Challenges and potential solutions of microalgae-based wastewater treatment systems: A review. *Frontiers in Bioengineering and Biotechnology, 11*, 1210228. https://www.frontiersin.org/articles/10.3389/fbioe.2023.1210228/full

Verstraete, W., Van de Caveye, P., & Diamantis, V. (2009). Maximum use of resources present in domestic "used water." *Bioresource Technology, 100*(23), 5537–5545.

Wang, K. G., Brown, R. C., Homsy, S., Martinez, L., & Sidhu, S. S. (2013). Fast pyrolysis of microalgae remnants in a fluidized bed reactor for bio-oil and biochar production. *Bioresource Technology, 127*, 494–499. https://doi.org/10.1016/j.biortech.2012.09.095

Wang, Y., Liu, J., Kang, D., Wu, C., & Wu, Y. (2017). Removal of pharmaceuticals and personal care products from wastewater using algae-based technologies: A review. *Reviews in Environmental Science and Biotechnology, 16*, 717–735. https://doi.org/10.1007/s11157-017-9446-x

Wu, S., Wallace, S., Brix, H., Kuschk, P., Kirui, W. K., Masi, F., & Dong, R. (2015). Treatment of industrial effluents in constructed wetlands: Challenges, operational strategies and overall performance. *Environmental Pollution, 192*, 245–262.

Yang, Y., Jiang, C., Wang, X., Fan, L., Xie, Y., Wang, D., Yang, T., Peng, J., Zhang, X., & Zhuang, X. (2024). Unraveling the potential of microbial flocculants: Preparation, performance, and applications in wastewater treatment. *Water, 16*(14), 1995. https://doi.org/10.3390/w16141995

Zhang, X., Lei, Z., Ge, S., Ji, B., & Zhang, B. (2023a). Editorial: Algae in advanced wastewater treatment. *Frontiers in Microbiology*. https://algaebiomass.org/blog/11590/algae-in-advanced-wastewater-treatment/

Zhang, X., Lei, Z., Ge, S., Ji, B., & Zhang, B. (2023). Editorial: Algae and microalgae-bacteria based technology for sustainable wastewater treatment. *Frontiers in Microbiology, 14*, 1263955. https://doi.org/10.3389/fmicb.2023.1263955

Zhou, W., Li, Y., Min, M., Hu, B., Zhang, H., Ma, X., Li, L., Cheng, Y., Liu, Y., Chen, P., & Ruan, R. (2018). Microalgae and wastewater treatment. In *Biofuels from algae* (pp. 143–156). Elsevier. https://link.springer.com/chapter/10.1007/978-981-13-2264-8_20

Zittelli, G. C., Biondi, N., Rodolfi, L., & Tredici, M. R. (2013). Photobioreactors for mass production of microalgae. *Handbook of microalgal culture: Applied phycology and biotechnology*, Second Edition, Chapter 13. https://doi.org/10.1002/9781118567166.ch13

Brooks, 2015). If solid waste is removed from public life immediately, as it was before, no one will care where it goes. To ensure social peace, local government governors spend a lot of effort and high costs on these services to avoid problems.

Sustainability can only be implemented if waste generates profitable income (Bala et al., 2024; Mohanty et al., 2024; Singh Jadaun et al., 2022). Through proper separation, processing waste can largely solve the waste problem and even make profits (Soares et al., 2013). Biomedical waste is any type of solid or liquid waste containing infectious or potentially infectious material from medical, laboratory, or research origin, arising from activities such as diagnosis, prevention, and treatment of disease (Khatun et al., 2023; Sadasivuni et al., 2019; WHO, 2018). Biomedical waste has a high potential to cause infection or injury to medical staff, patients, and bystanders. It consists of human body material, animal waste, microbiological and biotechnological waste, sharps waste, discarded drugs and cytotoxic drugs, contaminated waste, liquid waste, incineration ash, and chemical waste. Common sources of biomedical waste include hospitals, clinics, medical and veterinary colleges, blood banks, mortuaries, autopsy centers, biotechnology facilities, laboratories, home health care, and funeral homes. Hazardous chemicals and radioactive waste, although not infectious, still need to be disposed of properly. According to the World Health Organization, 10% of hospital waste is infectious waste and 5% is non-infectious but hazardous waste (Prüss-Üstün et al., 2013).

The World Health Organization classifies medical waste into eight categories including general waste, pathological waste, radioactive waste, chemical waste, infectious waste, sharps waste, pharmaceutical waste, and pressurized waste (Chartier et al., 2014). In India, the Biomedical Waste (Management and Handling) Rules, 1998, and other amendments regulate the handling of biomedical waste. They consist of six lists, which include biomedical waste categories, color coding and container types, labels for biomedical waste containers or bags (which should be non-washable and clearly visible), labels for transport of biomedical waste containers or bags, and standards (Patil & Pokhrel, 2005). For treatment and disposal, see the list of waste treatment systems such as incinerators, autoclaves, and microwave systems. Operating standards such as combustion efficiency and emission standards are laid down. This review article focuses on basic issues such as definition, categories, issues, handling procedures, and disposal methods in biomedical waste management.

The definition of waste can be very subjective. What is waste to one person can be a valuable resource to another (Williams, 2005). This attitude can be advantageous in the waste management process. Materials that are thrown in the trash are not considered waste. Some of them can be considered important and used as raw materials for other groups.

Technically, "solid waste" means trash, refuse, sludge, and other discarded solid materials, including solid waste from industrial, commercial, agricultural, and municipal activities, but does not include solid or dissolved materials in domestic wastewater or other significant pollutants in water (EPA, 2012). Resource water pollutants such as sludge, sand, and gravel, dissolved or suspended solids in industrial wastewater, fly ash, cement kiln dust waste, dissolved or other common materials in irrigation canals, drilling fluids, fuel from compacted waste, and any toxic or radioactive waste are not included. Regardless of the definition, solid waste should be considered a major problem now in our society to be handled by licensed and trained personnel. Waste management (WM) means handling all kinds of waste technically and systematically. Waste should be removed from people's daily lives as soon as possible. This phase requires collaboration between people and businesses. Waste is considered one of the biggest problems in our society and needs to be disposed of by authorized and trained people. The importance of waste disposal only becomes apparent when technical or social problems occur, such as system disruptions or strikes by garbage collection workers (Kumar et al., 2023; Mishra & Das, 2022; Mishra et al., 2024).

On the contrary, through an integrated waste management system (IWM), it is possible to make money from waste or at least reduce high disposal costs (Alagöz & Kocasoy, 2008). This chapter discusses the sustainability possibilities of integrated waste management. Which is more important: costs or public health? This question can be discussed from different perspectives. However, the authors focus on public health from the perspective of waste disposal. Of course, improper waste disposal always poses a risk to public health. On the other hand, industrial waste is always a potential threat to public health and is more dangerous than municipal waste. It is true that for a better life, we need to reduce environmental problems, diseases, and solid waste. To reduce solid waste, we need to address issues such as source reduction, separate collection, reuse, and recycling. These preventive measures can be taken by both individuals and governments.

2 Background

Waste disposal reflects people's lifestyles and social activities. Waste is generated in daily life and social activities. Therefore, differences between regions can be attributed to industrial structure, logistics, people's awareness of waste, etc. (Zimmermann, 2017). In general, the amount of waste in urban areas is greater than that in rural areas. Communities with higher economic levels generate less organic matter, ash, ceramics, etc. Unless we start reducing waste sources, the amount of urban solid waste generated will continue to increase. On the other hand, "waste" can also be defined as any kind of material that is no longer needed by its owner. This means that there may be waste that is useful to others. This leads to the idea of "reuse" before the process of "recycling" which we will explain in more detail below. There is a trend called "3R" (reduce, reuse, recycle). Reducing the amount of waste should be the first step. Then the process of reuse and recycling begins. In recent times, this trend has already changed to the "4Rs" with the addition of "recovery" as energy recovery (Windfeld & Brooks, 2015).

can spread infectious diseases including HIV, hepatitis, tetanus, and septicemia (Kümmerer, 2009). Biomedical waste includes hazardous items like sharps, pathological waste, genotoxic waste, and pharmaceutical waste. Several illnesses, including avian influenza, AIDS, meningitis, gastrointestinal, respiratory, skin, and eye infections, can also be brought on by biomedical waste (Patwary et al., 2011).

Inappropriate biomedical waste management poses serious threats to the environment and public health. The soil contamination that occurs when untreated waste is disposed of in unlined landfills introduces pathogens, heavy metals, and pharmaceutical compounds into the terrestrial environment. Studies have documented soil microbial community disruptions and reduced soil fertility near improperly managed disposal sites (Coutinho et al., 2006). Aquatic ecosystems face vulnerability to biomedical waste contamination, with research detecting concentrations of antibiotics, hormones, and disinfectants in water bodies receiving inadequately treated medical effluents (Verlicchi et al., 2012). Atmospheric pollution represents another significant environmental impact, particularly from the incineration of medical waste without adequate emission controls. The climate change implications have recently gained attention within the biomedical waste management discourse, with the healthcare sector contributing approximately 4.4% of global greenhouse gas emissions (Karliner et al., 2019).

6 Classification and Categories of Biomedical Waste

Biomedical waste classification systems provide structured frameworks essential for proper segregation, handling, and disposal. These systems vary globally but share common principles aimed at minimizing risks to human health and the environment.

WHO Classification System

The World Health Organization has created an extensive classification system that has served as the basis for many national frameworks. This system divides biomedical waste into eight categories according to their potential hazards and respective handling (WHO, 2018).

General waste makes up around 85% of total healthcare waste and consists of materials, such as office paper, packing, and food waste that is not critical to being treated specifically. Pathological waste deals with human tissues, organs, body parts, and body fluids, and there can be a risk of infection from it, as well as a moral component in handling human remains. In contrast, radioactive waste, which is produced in much lower volumes—almost exclusively as a result of nuclear medicine procedures—requires specific protocols and lengthy storage times to permit radioactive decay prior to disposal (Chartier et al., 2014).

Chemical waste—leftover solids, liquids, and gases from diagnostic, experimental, and cleaning processes. This includes disinfectants, solvents, and reagents which can be toxic, corrosive, flammable, or genotoxic. Infectious waste has pathogens, which could cause disease if in sufficient quantities; it comprises cultures from laboratories, waste from surgery, and items that have come in contact with infected patients (WHO, 2018).

Sharps waste—needles, scalpels, broken glass, and any objects that may cause cuts or puncture wounds—are a major mode of disease transmission through injury, and this requires special consideration. Pharmaceutical waste (PW) is involved in classes of expired, unused, and contaminated pharmaceutical products which require specific treatment for disposal to prevent the deterioration of the environment. Explosions can occur from pressurized containers such as gas cylinders or aerosol cans when improperly punctured or incinerated (Prüss-Üstün et al., 2013).

Country-Specific Classifications with Focus on India

The policies concerning the management of biomedical waste in India have seen evolution over the years, the most notable being the formulation of the biomedical waste (management and handling) rules in 1998. Medical waste management rules and its amendments: (2016, 2018) signifies a major step toward the waste management in the country (Ministry of Environment, Forest & Climate Change, 2016).

According to the Indian classification system, biomedical waste is classified into four categories, which have their respective color codes and methods for disposal. Biological waste which needs treatment like incineration or deep burial forms yellow category waste; it includes human anatomical waste, animal anatomical waste, soiled waste, expired/discarded medicines, and chemical waste. The red category includes recyclable contaminated waste, tubing, bottles and intravenous sets, and laboratory waste which needs autoclaving or microwaving and then recycling (Patil & Pokhrel, 2005).This is the white (translucent) category, reserved for sharps only—needles, syringes, broken glass, all of which must be disinfected and destroyed before final disposal. Blue category waste, such as glass bottles and metallic body parts, consists of broken or discarded glassware that needs disinfection and cleaning before recycling (Kumar et al, 2017).

Such a color-coded system allows for appropriate segregation at source—a vital first step in the waste chain. India's system adds more emphasis on recycling potential while also staying focused on infection control compared to the WHO classification. It is developed to be applicable for implementation in diverse healthcare settings from advanced urban hospitals to rural primary healthcare facilities (CPCB, 2018). Challenges with these classifications' implementation are insufficient with infrastructure, not complying to segregation rules and clinician limitations. Recently, studies showed an improvement in compliance at least to a certain extent, especially in tertiary care hospitals and urban centers (Mandal & Dutta, 2019).

These classification systems, if understood and applied correctly, provide the basis of biomedical waste management. It is imperative for healthcare facilities to implement the components of these frameworks into their programs, with training and monitoring as the accompaniment. Correct waste segregation at source lowers recycling and disposal risks and optimizes the disposal and treatment process (Sharma et al., 2020).

7 Implications of Inappropriate Biomedical Waste Management

Human Health Risks

Improper management of biomedical waste poses significant threat to human health by different exposure routes. Healthcare workers are in the frontline of damage waiting to happen, reports that needle-stick injuries occur among up to 3.5 million healthcare workers every year across the globe, resulting in approximately 66,000 new hepatitis B infections, 16,000 new hepatitis C infections, and 1000 new HIV infections (WHO, 2019). Proper sharps management in healthcare settings is critical to prevent the staggering statistics above.

In addition to occupational exposure, patients whose healthcare is provided in facilities with poor waste management practices are at increased risk of healthcare-associated infections (HAIs). Improper treatment and disposal of infectious waste may serve as a reservoir for the spread of antimicrobial-resistant organisms in healthcare settings and impede patient's recovery and increase the death rate (Allegranzi et al., 2017). This relationship between waste management and patient outcomes highlights that adherence to waste protocols is a clinical endeavor, not just a regulatory concern.

Workers who handle waste, including sanitation workers and waste transporters and disposers, are uniquely at risk, as they may be directly exposed to inadequately treated or not completely segregated waste. Data from developing nations reveal that waste handlers frequently operate without adequate personal protective equipment, not given sufficient training on the dangers they may encounter (Kumar et al., 2015). As a result, this group has an increased prevalence and incidence of infectious disease, dermatological conditions, and respiratory disease when compared to control populations.

Biomedical waste poses a health risk for the general public when it leaves controlled management systems, especially in the water resources. Various pharmaceuticals, pathogens, and chemical pollutants from poorly treated medical waste have been found in surface waters, groundwater, and even drinking water sources in multiple regions (Kümmerer, 2009). Such contaminants have been associated with endocrine disruption, development of antimicrobial resistance, and increased cancer risks to surrounding communities.

This is especially true for children, a particularly vulnerable population, who may reside near healthcare facilities or disposal sites where scavenging takes place informally. Their immature immune systems, mouthing behaviors, and relative body weight-to-exposure further put them at increased risk for adverse effects from biomedical waste contaminants (Nzeadibe & Anyadike, 2012).

Environmental Impact

The environmental implications of inadequate biomedical waste disposal vary through many environmental compartment, forming multi-compartmental and persistent troubles (Akter, 2000; Batterman, 2004). Untreated waste disposed in landfills, which are not lined, or on uncontrolled dumps is responsible for the introduction of pathogens, heavy metals, and pharmaceutical compounds into the terrestrial environment (Altin et al., 2003). Studies have shown disturbance of soil microbial-communities and decreased soil fertility and bio-magnification of pollutants in soil-resident organisms surrounding sub-optimally disposed waste (Patwary et al., 2011).

Biomed and biomedical waste, including solid, liquid, and gaseous wastes, especially liquid wastes of varying hardness pose threats to aquatic ecosystems due to the significant bio-radar of most of these waste products, and research has shown that antibiotics, hormones, and disinfectants were measurable concentrations in the waters receiving inadequately treated medical effluents. Such compounds are associated with reproductive abnormalities in fish, algal bloom formation, and disruption of aquatic food webs (Verlicchi et al., 2012). Some pharmaceutical compounds are resistant to the degradation process of activated sludge (the most used technology to treat wastewater); hence, they may contaminate the drinking-water sources.

Atmospheric pollution is another potential source of environmental impact, especially from the uncontrolled burning of medical waste. Chlorinated materials often found in healthcare waste produce a group of highly toxic compounds upon combustion known as dioxins (Coutinho et al., 2006) and poorly managed incinerators also emit particulate matter, heavy metals, and acid gases. They cause respiratory diseases, participate in acid rain creation, and lead to the deposition of persistent organic pollutants.

Impacts of climate change have recently been introduced into the discussion about biomedical waste management. Waste management practices from the healthcare sector—including transport, treatment processes, and landfill decomposition—account for certified ~ 4.4% of the greenhouse gas emissions for the planet (Karliner et al., 2019). And the creation of single-use medical items that get disposed adds embodied carbon whose temperature effects compound the climate footprint of healthcare activities.

Biomedical waste contaminating natural habitats leads to loss of biodiversity through various pathways. Endocrine-disrupting compounds originating from pharmaceutical waste have been shown to alter reproduction in wildlife, while pathogens from infectious waste can spillover to naïve animal populations. These direct effects are compounded by habitat degradation adjacent to disposal sites that creates ecological dead zones where few species can survive (Kümmerer, 2009).

The environmental justice aspect of biomedical waste effects should not be neglected. Disposal facilities and the environmental burdens associated with them tend to be located in low-income communities and other populations that are vulnerable and lack the political power needed to resist such developments. This unequal sharing of environmental burdens poses a profound ethical challenge for waste management planning/policy design.

The complexities of the situation arise from the interdependency of different aspects of environmental and human health impacts. Contaminants traveling across ecological systems cycle back to humans via exposure pathways (e.g., food ingestion, drinking water, air inhalation). This contamination cycle demonstrates in concrete terms that the protection of the environment is a critical part of the protection of public health.

To significantly reduce these negative impacts, you can utilize specific management techniques such as proper segregation of these wastes or advanced treatment technologies and environmentally sound disposal methods. Generalized management systems do not only mitigate extant populations but also halts the build-up of bioaccumulatable environmental contaminants that may have adverse effects on future generations.

8 Regulatory Frameworks and Compliance

International Agreements and Conventions

This has drawn global attention toward medical waste management, and several global frameworks have been formed to uphold standards for the safe disposal of medical waste. May 6, 1998: The Basel Convention on the Control of Transboundary Movements of Hazardous Wastes and Their Disposal is one of the earliest and most comprehensive international agreements that deal with biomedical waste. First adopted in 1989 and taking effect from 1992, this convention codifies the movement of harmful waste across borders—in particular from wealthy to poorer nations—and lays down procedures for a more environmentally friendly management of such waste (Krueger, 1999). Within this framework, biomedical waste is classified as hazardous, requiring special notification procedures, and consent mechanisms to import for its cross-border transportation.

The Stockholm Convention on Persistent Organic Pollutants (POPs) is also relevant to medical waste, though it is not biomedical waste-specific, as it covers the unintentional release of dioxins and furans a byproduct of medical waste incineration. This included the application of best available techniques and best environmental practices as necessary to minimize dioxins emissions, which has direct impacts on the design and operation of biomedical waste treatment facilities (Weber et al., 2008). The convention guidelines are steering the development of more advanced incineration processes, as well as modifying treatment approaches to achieve lower emission profiles.

The World Health Organization has created authoritative technical guidelines, which despite their nonbinding character, exert enormous influence on the design and implementation of national laws and institutional (policies and practices). Guidelines on the Safe Management of Wastes from Health-Care Activities (the first edition was issued in 1999 and updated in 2014) were developed, providing comprehensive recommendations along the entire waste management chain, from generation through to final disposal (Chartier et al., 2014). These guidelines set the minimum standards for healthcare facilities globally and act as a guidepost for countries developing their own regulatory frameworks.

Far from being isolated to formal treaties, international cooperation (also) encompasses measures to build capacity around the world and to transfer technology. Various organizations such as the United Nations Development Programme (UNDP), the Global Environment Facility (GEF), and the World Bank have financed projects aimed at reinforcing biomedical waste management systems in developing nations, emphasizing infrastructure development, training of personnel, and policy formulation strategy (Emmanuel, 2007). These collaborative initiatives highlight the transnational aspects of health and environmental hazard risks posed by inappropriate waste management.

Across countries with disparate resources, capabilities, and healthcare systems, however, there remain major obstacles to standardization despite these international agreements. In contexts where there has been little implementation, the gaps remain particularly pronounced in terms of competing with other public health priorities or resource limitations (Harhay et al., 2009).

National Regulations with Focus on India

Biomedical waste management (BMWM) in India has evolved significantly over the last three decades, in response to the growing awareness of risks associated with biomedical waste and the need for organized management systems. The first comprehensive legislation addressing health care was the Biomedical Waste (Management and Handling) Rules of 1998, which placed basic segregation, treatment, and disposal requirements on medical establishments in the country (Patil & Pokhrel, 2005).

As the rules formulated in 2016 with amendments in 2018 under the title medical waste Management was a remarkable step in regulating the management of medical waste in India. These revised regulations follow contemporary waste management practices, implement stricter criteria to ensure compliance, and tackle voids observed in the previous policy. Dogs have just become popular to track wastes with the help of bar code systems, thereby indicating the areas that are making the responsible use of bio wastes higher (Ministry of Environment, Forest & Climate Change, 2016).

labeling with text and universal symbols, secure closures to prevent spillage, adequate sizes to fit their location, and materials compatible with what they are intended to hold. Containers themselves must be puncture-proof and give no access of hands to the contents deposited inside (Caniato et al., 2015), especially for sharps waste.

Training of staff is another key component of successful segregation programs. Training should include the reasons behind segregation requirements, the many categories of waste and their identifying features, health and environmental impacts of improper segregation, and practical demonstrations of good practices. Regular refresher training can enhance completion of segregation protocols in the long-term (Blenkharn, 2006).

Collection frequency needs to be balanced with operational capability, as waste should not pile up beyond storage space or sit long enough to decompose and pose odor, pest or infection hazards. For infectious and general waste, best practice recommends collection at least once daily, and more frequent collection in areas of high-volume generation, e.g., operating theaters and emergency departments (Olaniyi et al., 2018).

Storage and Transportation Standards

However, provisional storage units in medical centers should be properly designed to ensure segregation integrity, to prevent unintentional access, and to control environmental danger. Some international standards advise that fully dedicated storage areas with impermeable floors and proper drainage systems, pest control, good ventilation, restricted access, and protection against weather elements should be provided. Pathological waste must be stored under controlled temperature conditions for more than 24 h to avoid decomposition and the associated risks (Singh et al, 2014).

Facilities that generate large volumes of waste often rely on refrigerated storage as a necessary measure to avoid decomposition of pathological and infectious waste, especially in warm climates and when there is a lag in the availability of treatment facilities. Nevertheless, this strategy demands a dependable supply of electrical energy and a suitable backup system to avoid instrument malfunctions that could pose serious risks to public health (Xin, 2015).

Biomedical waste should be transported in dedicated trolleys or carts, clearly labeled for biomedical waste use, and regularly disinfected to avoid cross-contamination. Waste handlers at a minimum should wear gloves, aprons, closed footwear, and respiratory protection if necessary. Here, pathways of movement should eschew areas of patient care, food preparation, and areas subject to public access where practicable (Sapkota et al., 2014).

Transporting patients to treatment facilities creates new challenges involving specialized vehicles and operating protocols. Best practices should involve compartmentalized vehicles that have a separate compartment for each waste type, secure mechanisms to prevent shifting during transportation, effective spill containment materials and tools, proper hazard signage, and vehicle disinfection protocols. Driver training covers not only operational procedures but also emergency preparedness in the event of an accident or spill (Zamparas et al., 2019).

Documentation and tracking systems have undergone a significant evolution, with contemporary methodologies leveraging digital technologies such as barcode scanning, RFID (radio-frequency identification) tags, and blockchain authentication. Such systems establish an accountability culture along the waste management chain, making it possible to verify that waste arrives at its correct destination and receives adequate treatment. Waste registration will provide clear information on the amount of waste by source, its generation period, its collection, transportation details, and treatment (Singh et al., 2020).

Chain-of-custody documentation is useful for more than just meeting regulations; it can also deliver data needed to conduct waste audits, identify trends within the system, and make key decisions around optimization. These records enable the healthcare facilities to detect abnormal generation trends, assess the success of waste minimization projects, and assign costs correctly to the departments (Alagöz & Kocasoy, 2008).

Treatment Technologies and Methods

Treatment technologies strive to make biomedical waste safe by destroying infectious agents, neutralizing chemical hazards, and destroying recognition factors for pathological waste. The choice of technology must take into account the characteristics of the waste, the volume of the waste, the local available infrastructure, regulations, environmental impact, and economic feasibility (Diaz et al., 2008).

The most widely deployed non-incineration treatment technologies globally have involved steam-based systems, such as autoclaving and hydroclave systems. These systems use high-pressure steam to achieve thermal disinfection of waste, generating microbial inactivation with a minimal amount of air emissions. Mechanical processes such as shredding and mixing have been crying for the advancement of modern-day autoclave systems to improve the penetration of the steam and reduce the volume of waste. They are particularly appropriate for infectious waste and laboratory cultures but are not effective for some pharmaceutical and chemical wastes (Batterman, 2004).

The chemical disinfection systems inactivate pathogens by using chlorine compounds, peracetic acid, alkaline hydrolysis, or other disinfectants. While these methods have benefits in low-infrastructure contexts, issues of chemical residues and effluent disposal are raised. Recent innovations within this category include enclosed treatment systems equipped with neutralization capabilities, designed to mitigate these environmental concerns (Gielar & Helios-Rybicka, 2013).

Microbial treatment systems often use temperature generation with moisture for disinfection and generally include shredder functions for efficiency. These systems provide relatively low environmental footprint but must be well-engineered to achieve total pathogen inactivation throughout the waste mass (Ravikant et al., 2002).

The incineration of some waste streams, such as cytotoxic pharmaceuticals, anatomical waste, and some laboratory chemicals, is still needed in spite of environmental concerns. Thermal treatment systems for health care refuse with advanced air pollution control, real-time temperature feedback, and reaction engineering to optimize for low emissions and capture of hazardous materials as well as pathogens. Pyrolysis and gasification are advanced thermal technologies that may have a lower environmental impact than incineration (Lee & Huffman, 1996).

Promising technologies in this category are plasma-based systems, where waste is reduced to its constituent elemental parts using extreme temperatures with minimal emissions. Other innovative methods that are currently being developed or are in limited deployment include electron beam technology, steam reforming, and biological treatment systems (Datta et al., 2018).

Radiation treatment efficacy verification is an important aspect of best practice, which can include biological indicators (for example, Bacillus stearothermophilus spores), chemical indicators, parametric monitoring, and the microbial testing of treated waste. Routine verification reduces the risk of equipment malfunction, operator errors, and other failures that can allow inadequately treated waste to flow to final disposal (Hossain et al., 2012).

Segregation + storage + transportation + treatment best practices = system (KPIs); a system generates the following key performance indicators (KPIs) to protect healthcare workers from clients, clients from waste handlers, waste handlers from the public and the environment. The tools are only as effective as the layers of institutional commitment, resources, training programs, monitoring, and continuous improvement processes used to implement them. By following these established best practices, healthcare facilities can not only meet their obligations for safe waste management, they can also lower the costs and environmental impact associated with dealing with the waste.

10 Advanced Treatment Technologies

Conventional Methods

Incineration has long been the primary treatment option for biomedical waste, especially for pathological waste, pharmaceutical residues, and highly infectious material. Currently, modern healthcare waste incinerators are two-stage combustion systems with primary chambers operating at 800–900 °C and secondary at 1000–1200 °C, which additionally ensures complete decomposition of organic compounds and pathogenic microorganisms. "These conditions, coupled with enough residence time (usually 2 s in the secondary chamber) effectively destroy infectious agents and reduce the waste volume by as much as 95%" (Chartier et al., 2014).

Despite its strategic value, incineration is under increasing scrutiny for potential environmental effect, particularly creation of atmospheric pollutants, dioxins, furans, heavy metals, and listing (or maybe P) sort atoms. Modern emission control technologies—electrostatic precipitators, fabric filters, wet scrubbers, and activated carbon injection systems—can reduce these emissions dramatically, but they demand sophisticated engineering, regular maintenance, and huge capital investment. These setups restrict successful execution in resource-challenged contexts (Inyang, 2009).

The method that is another conventional technology most commonly used is autoclaving, where saturated steam is used under pressure (often under 121 °C at 15 psi), which leads to microbial inactivation by thermal processes. Commonly used gravity-displacement autoclaves depend on the displacement of air directly by steam; however, more modern machines, such as vacuum autoclaves, directly evacuate air from the chamber prior to steam penetration: as a result, penetration is more efficient. The autoclaving process for biomedical waste requires an exposure time of 30–60 min depending on the type of waste characteristics and equipment specification (Hossain et al., 2012).

The successful sterilization of waste in an autoclave depends critically on steam penetrating the entire mass of waste, which requires careful attention to waste preparation, container shape and size, and loading configurations. Modern autoclave systems include shredding systems, rotating drums, or mixing systems to increase the contact of steam with all waste components. These improvements are especially vital for dense material, large volumes, and waste containing a significant amount of organic matter that may prevent steam from penetrating (Diaz et al., 2008).

Microwave treatment systems are a relatively new conventional technology that occurs through thermal effects and moisture for disinfection. Microwaves are usually 2450 MHz systems and work by irradiating water in the waste which consequently heats it up. Unlike typical microwaves commonly found in households, microwave systems used for waste treatment often integrate pre-shredder components, moisture presence, and conveyor systems that facilitate homogeneous treatment within the entire waste mass (Oliveira et al., 2013) with the potential benefits of fast heating rates, lower energy consumption, and lower emissions, although limitations related to capacity and waste species are anticipated (Oliveira et al., 2013).

Emerging Technologies

Plasma pyrolysis is one of the most attractive advanced techniques for the treatment of biomedical waste, in which an electric arc is used to generate a plasma (ionized state of gas) with very high temperatures of 5000–15,000 °C, allowing for 100% molecular dissociation of organic components in a waste stream into their respective elemental constituents, and vitrification of inorganic components into stable slag. The resulting synthesis gas (mostly hydrogen and carbon monoxide) can then be used for energy recovery, which may keep the system self-sufficient energetically. Plasma systems destroy all pathogens, reduce the end product volume significantly, can prevent dioxins formation, and have the option of energy recovery but with the high capital costs and manufacturing complexity (Nema & Ganeshprasad, 2002).

Hydrothermal treatment technologies such as alkaline hydrolysis and supercritical water oxidation make use of water under conditions of high temperature and pressure for the decomposition of organic materials. Alkaline hydrolysis uses a combination of heated alkali solutions (most often KOH or NaOH) and heat at a temperature of 110–150 °C to solubilize tissues, denature proteins, and hydrolyze nucleic acids. This technology is ideal for decomposition of anatomical waste, animal carcasses, and microbiological cultures resulting in a sterile aqueous solution that can be safely discharged to sewage system after neutralization of pH (Thacker, 2004).

Supercritical water oxidation functions at more extreme conditions—temperatures higher than 374 °C and pressures higher than 22.1 MPa—where water is a strong oxidizing medium. In this way, under these conditions, organic compounds are rapidly oxidized to carbon dioxide and water and complete destruction of pathogens and organic pollutants is ensured. This technology has great potential for pharmaceutical and cytotoxic wastes that are difficult to treat by conventional approaches (Xu et al., 2012).

Enzymatic and biological treatment systems, and also composting for waste fractions such as food, represent green alternatives to treat specific components of biomedical waste. Enzymatic systems involve the use of specific enzymes or microbial consortia to facilitate organic component degradation and pathogen inactivation under controlled conditions. These approaches offer low energy consumption and minimal environmental impact but require careful waste segregation and face limitations regarding treatment time and applicable waste types (Saini et al., 2004).

Comparative Analysis of Treatment Methods

When evaluating different methods of treatment technologies, a number of factors besides pathogen inactivation efficacy need to be taken into consideration. The capital cost varies widely depending on technology, with simpler autoclave systems requiring the least capital investment and plasma and supercritical water systems representing the greatest upfront capital requirement. The patterns of operational costs are different, where incineration usually entails the highest continuing costs related to fuel (burning), emissions control, and maintenance needs (Soares et al., 2013).

Environmental impact assessment suggests substantial differences across technologies, with conventional incineration typically presenting the most significant environmental issues even with modern emission control. Autoclaving, microwave treatment, and enzymatic processes have low air emissions but can impact wastewater streams if not controlled. Plasma systems and supercritical water oxidation are examples of advanced thermal technologies with potentially better environmental performance if properly designed and operated (Windfeld & Brooks, 2015).

Another important aspect of comparison is the relative treatment effectiveness of the different waste fractions. Pathological waste can warrant the need for destruction, but must also be considered under the plane of recognition factor elimination requiring specific technologies (such as incineration, alkaline hydrolysis, or plasma systems) to cover both pathogen destruction and reducing recognition factor. Waste disposal for pharmaceutical and cytotoxic waste typically requires high-temperature destruction methods to eliminate all active constituents of chemical compounds in the waste stream, with incineration, plasma treatment, or supercritical water oxidation preferable. Similar to other healthcare waste, sharps waste not only needs disinfection to be processed but it also requires the reduction of its volume along with its deformation, which means mechanical processing is necessary in addition to microbial inactivation (Lee et al., 2004).

Technical appropriateness of the various contextual implementation types should include infrastructural requirements, operational and maintenance complexity, and local competency. While more resource-constrained settings may be better served by robust, low-technology options that require minimal maintenance and technical know-how, urban centers with more resources and technical capacity can introduce higher-technology systems with better environmental performance. Decentralized treatment models embedded in less intensive technologies may be more appropriately suited to rural healthcare networks while goals for economies of scale can be achieved with centralized facilities and more high-level technologies in urban areas (Harhay et al., 2009).

Energy and material recovery is an emerging consideration in technology selection, highlighting the integrated approach for treatment technologies. Examples are advanced thermal systems such as plasma pyrolysis, gasification and controlled incineration that can recover energy from waste, allowing to potentially decrease operational costs. Material recovery opportunities are feasible mostly for the non-hazardous fraction of healthcare waste, highlighting the need for effective segregation systems upstream of the facilities treating waste (Zimmermann, 2017).

Medical waste management is evolving, and selection of technologies increasingly necessitates a systems approach associated to the overall waste chain from generation to final disposal. Since biomedical waste consists of various components, there is no one size fits all solution, and the best method is a combination of different technologies tackling different waste fractions. This contributes to the effectiveness of the treatment while also lowering environmental impacts and others operational costs, especially when backed by solid segregation and waste reduction programs.

11 Challenges in Biomedical Waste Management

Infrastructure and Resource Limitations

Biomedical waste disposal involves segregation, storage, transport, and treatment facilities. Many regions (especially developing) suffer massive deficits of these essential building blocks. Resource constrained settings often have insufficient spaces for waste storage, through studies have shown that makeshift storage places have been set up where safety requirements for access limitation, ventilation, and protection from elements are not satisfied (Hossain et al., 2012).

Another main gap comes with transportation infrastructure, as many healthcare facilities do not have dedicated vehicles for waste transport. In many countries, biomedical waste is delivered in non-specific vehicles without dedicated containment systems, which has high cross-contamination potential. The challenges are quite unique for rural healthcare facilities due to scattered locations, road conditions and lack of transport networks which make it harder to establish centralized treatment systems (Caniato et al., 2015).

Limitations in treatment technology vary from one region to another, but they often come in the form of limits to capacity, outdated equipment, or technology ill-suited for local environmental conditions. It was found that in a survey of treatment facilities across South Asia, over 68% of incinerators operated below the minimum temperature threshold, as well as 92% of facilities lacking emission control systems (Diaz et al., 2008). Even relatively straightforward technologies such as autoclaves often run without critical monitoring instruments, validation protocols, or occupying regular maintenance programs, jeopardizing effective treatment (Singh et al., 2014).

Most of the infrastructure deficiencies have financial implications since healthcare waste management usually has little budgeted allocation relative to patient care services. Cost recovery mechanisms are still poorly developed in many areas, and waste producers are often reluctant or unable to pay a fee that covers costs of adequate treatment systems. This economic reality results in a continuous gap of funds, inducing an infrastructure deficit throughout the entire waste management chain (Caniato et al., 2015).

Adding to the infrastructure issues, human resource constraints exist, as numerous facilities are staffed by individuals without specialized training in waste management. Single full-time dedicated waste management positions are still rare in many healthcare systems, and the responsibility for managing waste is often added onto staff primarily responsible for other functions. Specialists in waste management are in short supply in the regulatory agencies, treatment facilities, and academic institutions that are responsible for effective oversight, operation, and innovation of waste facilities (Kumar et al., 2017).

Training and Awareness Gaps

One of the main issues in this regard is the lack of knowledge among health professionals which has been documented in several studies indicating insufficient knowledge about the types of waste, segregation, and associated risks. In a multi-country survey, it was reported that only 43% of healthcare workers correctly recognized all the categories of biomedical waste, and only 36% of healthcare workers showed complete awareness of the color-coding systems of segregation (Hayleeyesus & Cherinete, 2016). This gap in knowledge directly relates to segregation practice, with observational studies documenting mis-segregation rates as high as 40% or more in some facilities (Mathur et al., 2011).

Limited training data accounts for these knowledge gaps, in fact many facilities only implement basic orientation training once, with limited if any refresher courses and/or performance evaluations. Training of this sort typically focuses more on theory than on practice, with insufficient treatment of the behavioral elements of compliance with waste management legislation. Further, trainings often focus on clinical staff but do not address support staff, administrative staff, and patients who also create or handle waste (Ozder et al., 2013).

Inadequate awareness about the wider ramifications of biomedical waste on the environment and public health is especially acute. Healthcare workers sometimes think of waste management as mainly an occupational safety issue and a regulatory burden, not seeing ties to water quality, air pollution, climate change, and community health. Such a limited focus can reduce the drive to comply beyond what is tangibly present in the work environment and workplace (Tabash et al., 2016).

As waste management issues as peripheral to more core healthcare functions, institutional leadership is often less engaged. This mindset translates into weak policy formulation, ineffective resource allocation, and little to no traceability for waste management systems. A lack of visible commitment from leadership sends a message to staff that waste management is a low organizational priority, impeding compliance (Sapkota et al., 2014).

Public awareness of proper BM waste management is also poor, and most settlements have limited knowledge of the related dangers, appropriate handling methods, or even the importance of their involvement in the management system. A gap in knowledge is particularly problematic for home healthcare waste, expired medications and community healthcare programs where public participation influences waste management (Nzeadibe & Anyadike, 2012).

Compliance Issues

According to regulatory frameworks in many jurisdictions remain underdeveloped, with many facing significant challenges resulting from fragmented authorities, overlapping jurisdictions, outdated standards, and implementation gaps. Some countries, for example, have separate regulations for each waste category or healthcare setting, without clear reference to which part of the regulations apply. On the other hand, they set a laborious standard on the technologies/practices that are beyond the local ability and result in a large number of non-compliance instead of incremental improvements (Thakur & Ramesh, 2015).

The irony is that enforcement mechanisms often are insufficiently resourced when regulatory agencies struggle with limited staff, finite resources, and competing priorities. It is not uncommon for inspection frequencies to be drastically lower than they need to be, with some facilities reporting years between inspections by regulatory bodies. At times, penalty structures can generate perverse incentives with fines or sanctions being set either too low to encourage compliance or too high that they are never enacted in practice (Abah & Ohimain, 2011).

Documentation and monitoring systems remain underdeveloped in many facilities, limiting accountability and improvement opportunities. Record-keeping often focuses on meeting minimum regulatory requirements rather than generating useful data for system optimization. Waste audits, when conducted, frequently emphasize volume metrics over qualitative assessments of segregation effectiveness or treatment adequacy. These documentation limitations complicate efforts to identify specific compliance challenges or evaluate improvement initiatives (Akter, 2000).

And that is because compliance levels are heavily influenced by cultural and behavioral factors, including staff attitudes, perceived risk and established routines tend to be hardest to change. The medical professionals often backslide into "optimistic bias" about waste-related risk, believing them to be less severe than other workplace hazards. The time constraints inherent to clinical settings further exacerbate the challenges of waste management, with many waste-related efforts perceived as breathing room that detracts from more immediate responsibilities toward patient care (Alagöz & Kocasoy, 2008).

Special Challenges in Developing Countries

The challenges of coexistence: Traditional and modern in developing countries biomedical waste is generated outside established management systems by traditional practitioners, community health workers, and informal healthcare providers. Thc distribution of these providers and their limited connection to formalized healthcare systems highlight the challenges in rolling out integrative waste management programs (Harhay et al., 2009).

Significant disparities in waste management capacity exist between urban and rural areas as a result of resource inequities. In this regard, large cities can afford advanced treatment technologies with multispecialty management systems, whereas rural health facilities work with minimal infrastructure. This disparity creates a two-tier system in which waste management standards differ widely based on geographic location and not necessarily their risk profiles (Abah & Ohimain, 2011).

Informal waste handling sectors such as waste pickers and recyclers interact with healthcare waste streams in much of the developing world. They often are not aware of the related dangers and possible protective methods, and they have significant occupational exposures while retrieving potentially valuable items. The economic realities constitutive of this informal sector reveal tensions between the need to make a livelihood and the need to protect public health that complicate policy development (Nzeadibe & Anyadike, 2012).

Added complexities arise in many developing areas from climate considerations, where elevated temperatures may speed up breakdown of waste, potentially prolonging the survival of pathogenic microbes or the activity of the vectors that carry them. Seasonal variations such as monsoon or the incidence of extreme weather conditions can potentially affect waste transportation systems and also cause damage in storage facilities. Such climate elements demand tailored approaches to storage facility design, collection schedules, and treatment technology selection (Diaz et al., 2008).

In developing countries, this simultaneous management of public health challenges diverts both human and financial resources, and attention was diverted. Healthcare systems facing more urgent immediate health threats from outbreaks of infectious disease, maternal mortality, malnutrition, and related health issues may relegated the critical importance of waste management despite its complex and multilayered relationships with these broader health challenges. This prioritization realty calls for embedding waste management in integrated public health frameworks rather than isolated technical responses (Hossain et al., 2012).

Reliance on foreign donors or development agencies for waste management infrastructure leads to sustainability challenges. Most of these projects rely on external sources of funding that focus on dealing with infrastructure constraints at an early stage but fall short in creating adequate capacity locally for operational and maintenance functions. Technology transfer often happens without proper incorporation into local context, leading to a system that is not sustainable beyond the lifetime of the project. These historical patterns highlight the importance of locally appropriate solutions developed with substantive community involvement (Emmanuel, 2007).

12 Sustainable Approaches and Future Directions

Circular Economy Principles in Biomedical Waste

The circular economy model provides a compelling framework for rethinking biomedical waste management beyond conventional linear systems. Although some waste fractions need to be destroyed for safety reasons, large parts of healthcare waste streams could potentially enter circular pathways through careful segregation and appropriate processing (Windfeld & Brooks, 2015). Circular principles must be applied with the nuances of healthcare waste streams in mind, including the balancing of safety requirements with resource recovery opportunities.

Material substitution is a primary circular strategy, replacing harmful or hard-to-manage materials with safer, more recyclable options. We have witnessed great strides in healthcare institutions through mercury free programs, non-PVC medical devices, and compostable food service items. These substitutions help lower hazardous waste volumes and open routes for recovery of residual materials (Karliner et al., 2019).

Reusable medical devices are a promising circular pathway, and modern sterilization technologies now make it possible to safely reprocess increasingly complex instruments. Rich quality control systems with electronic tracking and testing will ensure reprocessed items perform and are as safe as the originals. Economic analysis summarizes cost advantages of reusable systems, despite requiring higher initial investments, especially when taking environmental externalities into consideration (Voudrias, 2018).

Waste-to-resource projects have increasing opportunities for non-hazardous elements of the healthcare waste, such as new technologies enabling the transformation of organic fractions into commodity products. The production of bioplastics from organic waste is commercially viable in multiple geographical areas, and food waste reclamation systems process healthcare waste into soil amendments. Such methods redirect waste disposal from being a cost center to a business opportunity (Zamparas et al., 2019).

Where safeguards are adopted, recovery of material from non-infectious waste streams is economical, environmental, and recovery of an environmental burden. Sterilized plastics and glass from healthcare settings can be picked up by conventional recycling streams, while some metals have a very high recovery value (Priyadarsini & Das, 2024, 2025). The potential for resource recovery is improved by advanced sorting technologies, such as optical systems and even artificial intelligence applications, automating the segregation process that used to be done manually (Soares et al., 2013).

Technological Innovations

Advanced treatment technologies continue evolving toward higher efficiency, reduced environmental impact, and enhanced resource recovery potential. Hybrid systems combining multiple treatment mechanisms demonstrate particular promise, addressing diverse waste fractions through integrated processes. Steam-chemical systems incorporating mechanical shredding achieve high disinfection levels while preparing materials for potential recycling. Microwave-pyrolysis combinations enhance energy recovery while ensuring pathogen destruction. These integrated approaches often achieve superior performance compared to single-technology solutions (Zimmermann, 2017).

Digital tracking systems transform waste management visibility through the application of barcoding, RFID technology, and blockchain verification. These systems create accountability throughout the waste chain while generating valuable data for system optimization. Real-time monitoring allows early identification of non-compliance or operational issues, enabling prompt intervention. The data collected through these systems supports evidence-based policy development, resource allocation, and continuous improvement initiatives (Singh et al., 2020).

Decentralized treatment technologies offer alternatives to centralized models, potentially reducing transportation requirements and associated costs. Portable or modular systems scaled for individual facilities or small healthcare clusters enable on-site treatment with minimal infrastructure requirements. These approaches prove particularly valuable in rural areas, temporary healthcare operations, or regions with transportation limitations. Advanced engineering has reduced the operational complexity of these systems, making them viable in settings with limited technical capacity (Hossain et al., 2012).

Artificial intelligence applications demonstrate growing potential for waste management optimization through predictive analytics, computer vision, and process automation. AI-powered waste audit systems can identify segregation errors through visual recognition, providing immediate feedback and generating training recommendations. Predictive algorithms optimize collection schedules and treatment operations based on historical patterns and real-time data. These technologies enhance system efficiency while reducing human exposure to potentially hazardous materials (Vaccari et al., 2017).

Sensor technologies enable continuous monitoring of critical parameters throughout the waste management chain, from storage conditions to treatment effectiveness. Temperature and humidity sensors in storage areas provide early warning of decomposition risks, while emission monitoring systems ensure treatment technologies operate within environmental parameters. Microbial detection systems verify disinfection effectiveness through rapid testing methodologies. These monitoring capabilities enhance both safety assurance and regulatory compliance (Xin, 2015).

Policy Recommendations

Regulatory harmonization represents a critical policy priority, aligning standards across jurisdictions while accommodating local implementation realities. International frameworks provide valuable foundations, but effective policies must reflect regional capacities, healthcare systems, and environmental conditions. Harmonized approaches should emphasize performance outcomes rather than prescriptive technologies, allowing flexibility in implementation while maintaining consistent safety standards. Collaborative policy development involving healthcare providers, waste managers, and regulatory authorities enhances both appropriateness and compliance (Emmanuel, 2007).

Economic instruments offer powerful tools for steering waste management behaviors while supporting sustainable infrastructure development. Extended producer responsibility programs for medical products create incentives for design improvements while generating funding for end-of-life management. Differential pricing for waste treatment based on segregation quality rewards compliance while penalizing poor practices. Green procurement policies favor products with reduced waste impacts, creating market incentives for manufacturer innovation (Inyang, 2009).

Education and training programs require strategic enhancement, moving beyond basic awareness toward comprehensive competency development. Academic integration of waste management into healthcare professional curricula establishes foundational knowledge before practice patterns become established. Continuing education requirements ensure knowledge remains current as regulations and technologies evolve. Practical skills assessment and demonstration opportunities translate theoretical knowledge into effective practice (Ozder et al., 2013).

Community engagement strategies recognize the essential role of public participation in comprehensive waste management systems. Educational initiatives targeting patients and their families address home healthcare waste challenges while building broader environmental awareness. Community advisory mechanisms provide valuable perspectives for facility waste management planning, particularly regarding off-site impacts. These participatory approaches enhance system legitimacy while leveraging community resources and knowledge (Nzeadibe & Anyadike, 2012).

Research priorities include developing technologies appropriate for resource-constrained settings, quantifying environmental impacts of different management approaches, and evaluating intervention effectiveness. Implementation science methodologies offer particular value for understanding how theoretical best practices translate into real-world settings with various constraints. Cross-disciplinary research involving healthcare providers, environmental scientists, engineers, and social scientists provides the comprehensive perspective necessary for developing truly sustainable solutions (Harhay et al., 2009).

13 Conclusion

Comprehensive biomedical waste management is a critical intersection of public health protection, environmental stewardship, and health system delivery. The research explored the complex problems and evolving strategies in biomedical waste management while some of the challenges continue. International bodies and national governments are known to establish classifications systems highlighting how waste should be segregated and handled. However, the implementation is highly heterogeneous between different healthcare institutions and regions. The rules have been updated as recently as 2016 for biomedical waste reflecting the changing needs of the country and advances in technologies, etc. which are recently updated compared to the original guidelines of 1998. Improper biomedical waste management has far-reaching implications beyond healthcare facilities, resulting in multiple pathways of exposure for waste handlers, community members, and ecosystems. All these consequences underscore the intrinsic relationship between healthcare practices and public environmental health, which demonstrates that waste management practices are more than a regulatory burden; they represent a critical element of the ethical winery of the healthcare sector to "first, do no harm."

However, the potential of these advanced treatment technologies will only be fully realized if they are implemented via suitable regulatory frameworks and institutional systems. The continuing gaps in infrastructure, training, and awareness mentioned across multiple healthcare settings highlight the fact that no amount of technological innovation can be seen as a panacea to the complex realities of biomedical waste management. The move toward applying circular economic principles to healthcare waste management is a promising paradigm shift, redefining the way waste is perceived from being primarily a disposal problem to a potential resource. This transition necessitates optimal accommodation between health imperatives and resource recovery opportunities, together with an appropriate integration of technological innovations and policy frameworks.

The sustainable management of biomedical waste in future should take integrated approaches based on vigorous regulatory systems, suitable technologies, adequate training programs, and community participation. Developing these integrated systems will require collaboration across disciplines, sectors, and stakeholder groups and an integration of traditionally sold activity, including healthcare delivery, environmental management, and public health. Just as healthcare systems are constantly evolving in response to the changing spectrum of diseases, technological advancements and societal demands, biomedical waste management also needs an evolving approach through continuous innovation, adoption of evidence-based practice and systems thinking. By finding adaptive pathways like this, the healthcare sector will be able to realize its mission to heal while doing as little damage as possible to the environment and protecting the health of current and future generations.

References

Abah, S. O., & Ohimain, E. I. (2011). Healthcare waste management in Nigeria: A case study. *Journal of Public Health and Epidemiology, 3*(3), 99–110.

Akter, N. (2000). *Medical waste management: A review.* Asian Institute of Technology.

Alagöz, A. Z., & Kocasoy, G. (2008). Determination of the best appropriate management methods for the health-care wastes in Istanbul. *Waste Management, 28*(7), 1227–1235.

Allegranzi, B., Bagheri Nejad, S., Combescure, C., Graafmans, W., Attar, H., Donaldson, L., & Pittet, D. (2017). Burden of endemic health-care-associated infection in developing countries: Systematic review and meta-analysis. *The Lancet, 377*(9761), 228–241.

Altin, S., Altin, A., Elevli, B., & Cerit, O. (2003). Determination of hospital waste composition and disposal methods: A case study. *Polish Journal of Environmental Studies, 12*(2), 251–255.

Bala, K., Anshul, Verma, S., Dogra, A., Das, A. P., Kumar, R., & Nandini. (2024). Sustainable management for urban plastic waste generation. In *Sustainable management of urban plastic waste through circular economic approaches* (pp. 211–236). https://doi.org/10.1201/9781003465447-13

Batterman, S. (2004). *Findings on an assessment of small-scale incinerators for health-care waste.* Water, Sanitation and Health Protection of the Human Environment, World Health Organization.

Blenkharn, J. I. (2006). Standards of clinical waste management in UK hospitals. *Journal of Hospital Infection, 62*(3), 300–303.

Caniato, M., Tudor, T., & Vaccari, M. (2015). International governance of healthcare waste management: Present and future innovations. *Waste Management & Research, 33*(7), 573–584.

Chartier, Y., Emmanuel, J., Pieper, U., Prüss, A., Rushbrook, P., Stringer, R., Townend, W., Wilburn, S., & Zghondi, R. (2014). *Safe management of wastes from health-care activities* (2nd ed.). World Health Organization.

Coutinho, M., Pereira, M., Rodrigues, R., & Borrego, C. (2006). Impact of medical waste incineration in the atmospheric PCDD/F levels of Porto, Portugal. *Science of the Total Environment, 362*(1–3), 157–165.

CPCB. (2018). *Guidelines for management of healthcare waste as per Biomedical Waste Management Rules, 2016*. Central Pollution Control Board, Ministry of Environment, Forest and Climate Change, Government of India.

Das, A., & Ghosh, S. (2022). Role of microorganisms in extenuation of mining and industrial wastes. *Geomicrobiology Journal, 39*, 173–175. https://doi.org/10.1080/01490451.2022.2038953

Das, A., Ray, S. K., Mohanty, M., Mohanty, J., Dey, S., & Das, A. P. (2024). Ecotoxicity of microplastic wastes and their sustainable management: A review. *Environmental Chemistry and Ecotoxicology, 6*, 144–152. https://doi.org/10.1016/J.ENCECO.2024.05.003

Datta, P., Mohi, G. K., & Chander, J. (2018). Biomedical waste management in India: Critical appraisal. *Journal of Laboratory Physicians, 10*(1), 6–14.

Dey, S., Das, A., Mallick, K., Sahu, A., & Das, A. P. (2023). Environmental petroleum waste: Pollution. *Toxicity, Sustainable Remediation*. https://doi.org/10.1007/978-3-031-48220-5_7

Diaz, L. F., Savage, G. M., & Eggerth, L. L. (2008). Alternatives for the treatment and disposal of healthcare wastes in developing countries. *Waste Management, 28*(7), 1138–1145.

Emmanuel, J. (2007). Best environmental practices and alternative technologies for medical waste management. In *Proceedings of the 8th international waste management and landfill symposium*.

EPA. (2012). *Municipal solid waste generation, recycling, and disposal in the United States: Facts and figures*. United States Environmental Protection Agency.

Gielar, A., & Helios-Rybicka, E. (2013). Distribution of pollutants in the floor sediments of healthcare institutions. *Environmental Monitoring and Assessment, 185*(11), 9621–9628.

Harhay, M. O., Halpern, S. D., Harhay, J. S., & Olliaro, P. L. (2009). Health care waste management: A neglected and growing public health problem worldwide. *Tropical Medicine & International Health, 14*(11), 1414–1417.

Hayleeyesus, S. F., & Cherinete, W. (2016). Healthcare waste generation and management in public healthcare facilities in Adama, Ethiopia. *Journal of Health & Pollution, 6*(10), 64–73.

Hossain, M. S., Santhanam, A., Norulaini, N. A. N., & Omar, A. K. M. (2012). Clinical solid waste management practices and its impact on human health and environment–A review. *Waste Management, 31*(4), 754–766.

Inyang, H. I. (2009). Emerging regulations in healthcare waste management. *Journal of Material Cycles and Waste Management, 11*(3), 223–229.

Karliner, J., Slotterback, S., Boyd, R., Ashby, B., & Steele, K. (2019). Health care's climate footprint: How the health sector contributes to the global climate crisis and opportunities for action. Health Care Without Harm.

Khatun, R., Singh, S., Dubey, N. K., & Prasad Das, A. (2023). A review on marine-based phytochemicals and their application in biomedical research. In *Recent frontiers of phytochemicals: Applications in food, pharmacy, cosmetics, and biotechnology* (pp. 383–395). Elsevier. https://doi.org/10.1016/B978-0-443-19143-5.00013-X

Krueger, J. (1999). *International trade and the Basel Convention*. Earthscan Publications.

Kumar, M., Pandit, S., Patel, V., Khanna, N., Nag, M., Lahiri, D., Ray, R. R., Das, A., & Das, D. (2023). Maximization of energy recovery from starch processing wastewater by thermophilic dark fermentation coupled with microbial fuel cell technology. *Geomicrobiology Journal*. https://doi.org/10.1080/01490451.2023.2209555

Kumar, R., Shaikh, B. T., Somrongthong, R., & Chapman, R. S. (2015). Practices and challenges of infectious waste management: A qualitative descriptive study from tertiary care hospitals in Pakistan. *Pakistan Journal of Medical Sciences, 31*(4), 795–798.

Kumar, R., Somrongthong, R., & Shaikh, B. T. (2017). Knowledge, attitude and practices of health staff regarding infectious waste handling of tertiary care health facilities at metropolitan city of Pakistan. *Journal of Ayub Medical College Abbottabad, 29*(1), 108–112.

Kümmerer, K. (2009). Antibiotics in the aquatic environment—A review—Part I. *Chemosphere, 75*(4), 417–434.

Lee, B. K., Ellenbecker, M. J., & Moure-Eraso, R. (2004). Alternatives for treatment and disposal cost reduction of regulated medical wastes. *Waste Management, 24*(2), 143–151.

Lee, C. C., & Huffman, G. L. (1996). Medical waste management/incineration. *Journal of Hazardous Materials, 48*(1–3), 1–30.

Mandal, S. K., & Dutta, J. (2019). Integrated bio-medical waste management plan for Patna city. *Interdisciplinary Environmental Review, 10*(5), 112–130.

Mathur, P., Patan, S., & Shobhawat, A. S. (2011). Need of biomedical waste management system in hospitals—An emerging issue—A review. *Current World Environment, 7*(1), 117–124.

Ministry of Environment, Forest and Climate Change. (2016). Bio-Medical Waste Management Rules, 2016. The Gazette of India, Extraordinary, Part II, Section 3, Sub-section (i).

Mishra, S., & Das, A. P. (2022). Treatment of the wastewater polluted with synthetic microfiber released from washing machine. In B. B. Das, H. Hettiarachchi, P. K. Sahu, & S. Nanda (Eds.), *Lecture notes in civil engineering* (Vol. 207, pp. 109–117). Springer Science and Business Media Deutschland GmbH. https://doi.org/10.1007/978-981-16-7509-6_9/COVER

Mishra, S., & Das, A. P. (2023). Microbial remediation of synthetic microfiber contaminated wastewater. In M. P. Shah (Ed.), *Microbial technologies in industrial wastewater treatment* (pp. 337–350). Springer. https://doi.org/10.1007/978-981-99-2435-6_17

Mishra, S., Dash, D., & Das, A. P. (2024). Isolation and molecular characterization of polyester-degrading bacterial strains from municipal wastewater and their application in microplastic degradation in contaminated wastewater. *Geomicrobiology Journal*. https://doi.org/10.1080/01490451.2024.2376101

Mishra, S., Ghosh, S., van Hullebusch, E. D., Singh, S., & Das, A. P. (2023). A critical review on the recovery of base and critical elements from electronic waste-contaminated streams using microbial biotechnology. *Applied Biochemistry and Biotechnology, 2023*, 1–30. https://doi.org/10.1007/S12010-023-04440-X

Mishra, S., Rout, P. K., & Das, A. P. (2019). Solar photovoltaic panels as next generation waste: a review. *Biointerface Research in Applied Chemistry, 9*(6).

Mohanty, J., Dey, S., Bala, K., & Das, A. P. (2024). Recent advances in microplastic pollution for its sustainable management. *Sustainable Management of Urban Plastic Waste through Circular Economic Approaches*. https://doi.org/10.1201/9781003465447-3

Nema, S. K., & Ganeshprasad, K. S. (2002). Plasma pyrolysis of medical waste. *Current Science, 83*(3), 271–278.

Nzeadibe, T. C., & Anyadike, R. N. (2012). Social participation in city governance and urban livelihoods: Constraints to the informal recycling economy in Aba, Nigeria. *City, Culture and Society, 3*(4), 313–325.

Olaniyi, F. C., Ogola, J. S., & Tshitangano, T. G. (2018). A review of medical waste management in South Africa. *Open Environmental Sciences, 10*, 34–45.

Oliveira, E. A., Nogueira, N. G. P., Innocentini, M. D. M., & Pisani, R. (2013). Microwave inactivation of Bacillus atrophaeus spores in healthcare waste. *Waste Management, 33*(3), 646–652.

Ozder, A., Teker, B., Eker, H. H., Altindis, S., Kocaakman, M., & Karabay, O. (2013). Medical waste management training for healthcare managers—A necessity? *Journal of Environmental Health Science & Engineering, 11*(1), 20.

Patil, G. V., & Pokhrel, K. (2005). Biomedical solid waste management in an Indian hospital: A case study. *Waste Management, 25*(6), 592–599.

Patwary, M. A., O'Hare, W. T., & Sarker, M. H. (2011). Assessment of occupational and environmental safety associated with medical waste disposal in developing countries: A qualitative approach. *Safety Science, 49*(8–9), 1200–1207.

Priyadarsini, S., & Das, A. P. (2024). Analytical and structural characterization of waste lithium-ion batteries for their effective recycling strategy. *Environmental Chemistry and Ecotoxicology*. https://doi.org/10.1016/j.enceco.2024.12.004

Priyadarsini, S., & Das, A. P. (2025). Lithium bioleaching: A review on microbial-assisted sustainable technology for lithium bio-circularity. *Journal of Water Process Engineering, 69*, Article 106744. https://doi.org/10.1016/j.jwpe.2024.106744

Prüss-Üstün, A., Emmanuel, J., Rushbrook, P., Zghondi, R., Stringer, R., Pieper, U., Townend, W., Wilburn, S., & Chartier, Y. (2013). *Safe management of wastes from health-care activities* (2nd ed.). World Health Organization.

Ravikant, C., Jaiswal, S. P., Vaidya, K., & Chitnis, D. S. (2002). Effluents from a hospital: A breeding ground for the spread of disease. *Journal of the Academy of Hospital Administration, 14*(1), 33–35.

Sadasivuni, K. K., Rattan, S., Waseem, S., Brahme, S. K., Kondawar, S. B., Ghosh, S., Das, A. P., Chakraborty, P. K., Adhikari, J., Saha, P., & Mazumdar, P. (2019). Silver nanoparticles and its polymer nanocomposites—Synthesis, optimization, biomedical usage, and its various applications. *Lecture Notes in Bioengineering*. https://doi.org/10.1007/978-3-030-04741-2_11

Saini, S., Das, B. K., Kapil, A., Nagarajan, S. S., & Sarma, R. K. (2004). The study of bacterial flora of different types in hospital waste: Evaluation of waste treatment at AIIMS Hospital, New Delhi. *Southeast Asian Journal of Tropical Medicine and Public Health, 35*(4), 986–989.

Sapkota, B., Gupta, G. K., & Mainali, D. (2014). Impact of intervention on healthcare waste management practices in a tertiary care governmental hospital of Nepal. *BMC Public Health, 14*, 1005.

Sharma, A., Sharma, V., Sharma, S., & Singh, P. (2013). Awareness of biomedical waste management among health care personnel in Jaipur, India. *Oral Health and Dental Management, 12*(1), 32–40.

Sharma, V., Sharma, A., & Bansal, M. (2020). A review of the legislative aspect of hospital waste in India. *International Journal of Applied Engineering Research, 15*(1), 1–7.

Singh Jadaun, J., Bansal, S., Sonthalia, A., Rai, A. K., & Singh, S. P. (2022). Biodegradation of plastics for sustainable environment. *Bioresource Technology, 347*, Article 126697. https://doi.org/10.1016/J.BIORTECH.2022.126697

Singh, A., Agarwal, A., Agarwal, V. K., & Saxena, S. (2014). Waste management in healthcare facilities: A review. *Journal of Environmental Science, Toxicology and Food Technology, 8*(5), 14–17.

Singh, Z., Bhalwar, R., Jayaram, J., & Tilak, V. W. (2020). An introduction to essentials of biomedical waste management. *Medical Journal Armed Forces India, 57*(2), 144–147.

Soares, S. R., Finotti, A. R., da Silva, V. P., & Alvarenga, R. A. (2013). Applications of life cycle assessment and cost analysis in health care waste management. *Waste Management, 33*(1), 175–183.

Tabash, M. I., Hussein, R. A., Mahmoud, A. H., El-Borgy, M. D., & Abu-Hamad, B. A. (2016). Impact of an educational program on knowledge and practice of health care staff toward pharmaceutical waste management in Gaza, Palestine. *Journal of the Air & Waste Management Association, 66*(4), 429–438.

Thacker, H. L. (2004). *Alkaline hydrolysis: A comprehensive review*. National Agricultural Biosecurity Center, Kansas State University.

Thakur, V., & Ramesh, A. (2015). Healthcare waste management research: A structured analysis and review. *Waste Management & Research, 33*(10), 855–870.

Vaccari, M., Tudor, T., & Perteghella, A. (2017). Costs associated with the management of waste from healthcare facilities: An analysis at national and site level. *Waste Management & Research, 35*(1), 1–9.

Verlicchi, P., Al Aukidy, M., & Zambello, E. (2012). Occurrence of pharmaceutical compounds in urban wastewater: Removal, mass load and environmental risk after a secondary treatment—A review. *Science of the Total Environment, 429*, 123–155.

Voudrias, E. A. (2018). Healthcare waste management from the point of view of circular economy. *Waste Management, 75*, 1–2.

Weber, R., Gaus, C., Tysklind, M., Johnston, P., Forter, M., Hollert, H., Heinisch, E., Holoubek, I., Lloyd-Smith, M., Masunaga, S., Moccarelli, P., Santillo, D., Seike, N., Symons, R., Torres, J. P. M., Verta, M., Varbelow, G., Vijgen, J., Watson, A., … Zennegg, M. (2008). Dioxin- and POP-contaminated sites---contemporary and future relevance and challenges. *Environmental Science and Pollution Research, 15*(5), 363–393.

WHO. (2018). *Health-care waste.* World Health Organization. https://www.who.int/news-room/fact-sheets/detail/health-care-waste

WHO. (2019). *Health worker occupational health.* World Health Organization. https://www.who.int/occupational_health/topics/hcworkers/en/

Williams, P. T. (2005). *Waste treatment and disposal* (2nd ed.). John Wiley & Sons.

Windfeld, E. S., & Brooks, M. S. L. (2015). Medical waste management—A review. *Journal of Environmental Management, 163*, 98–108.

Xin, Y. (2015). Comparison of hospital medical waste generation rate based on diagnosis-related groups. *Journal of Cleaner Production, 100*, 202–207.

Xu, D., Wang, S., Tang, X., Gong, Y., Guo, Y., Wang, Y., & Zhang, J. (2012). Design of the first pilot scale plant of China for supercritical water oxidation of sewage sludge. *Chemical Engineering Research and Design, 90*(2), 288–297.

Zamparas, M., Kalavrouziotis, I. K., Koukoulakis, P. H., & Aloupi, M. (2019). Setting controls for hospital medical waste management. *Journal of Environmental Science and Health, Part A, 54*(5), 393–402.

Zimmermann, K. (2017). Microwave as an emerging technology for the treatment of biohazardous waste: A mini-review. *Waste Management & Research, 35*(5), 471–479.

Application of Algal Biotechnology in Municipal Wastewater Treatment

M. Purnima Sudha, Sailaja Priyadarsini, Amel Gacem, Kiran Bala, Maha Awjan Alreshidi, Krishna Kumar Yadav, and Alok Prasad Das

1 Introduction

Managing wastewater is chiefly aimed at decreasing the release of nitrogen (N) and phosphorus (P) methylated species into sensitive waters, and carbonaceous substances normally evaluated as BOD. The presence of these fertilizers can alter the trophic status and dissolved oxygen (O_2), which lowers the health of the aquatic flora and fauna. The need to remove nitrogen and phosphorus from wastewater discharge is growing in the context of both developed and developing countries, as is the need to protect or even improve the ecological integrity of water sources. Eutrophication is a process characterized by increased phytoplankton abundance and decreased water transparency, which can lead to the proliferation of toxic and non-toxic algae blooms (Mohsenpour et al., 2021). During nutrient limitation, a process of heterotrophic bacteria consuming biodegradable organic matter (as dead phytoplankton) leads to

M. Purnima Sudha · S. Priyadarsini · A. P. Das (✉)
Department of Life Sciences, Rama Devi Women's University, Bhubaneswar, Odisha, India
e-mail: alok1053@gmail.com

A. Gacem
Department of Physics, Faculty of Sciences, University 20 Août 1955, Skikda, Algeria

K. Bala
Department of Zoology, Deshbandhu College, University of Delhi, Delhi, India

M. A. Alreshidi
Department of Chemistry, College of Science, University of Ha'il, Ha'il, Saudi Arabia
e-mail: m.alrashedi@uoh.edu.sa

K. K. Yadav
Department of Environmental Science, Parul Institute of Applied Sciences, Parul University, Vadodara, Gujarat, India

Environmental and Atmospheric Sciences Research Group, Scientific Research Center, Al-Ayen University, Nasiriyah, Thi-Qar, Iraq

A. P. Das et al. (eds.), *Sustainable Municipal Waste Management with a Zero Waste Approach*, Sustainable Environmental Waste Management Strategies,
https://doi.org/10.1007/978-3-032-05950-5_4

the release of the O_2 in the biomass: The concentration of liquid O_2 is reduced firstly when the produced biomass decomposes. Traditional wastewater treatment plants separate wastewater into primary and secondary treatment (Mishra & Das, 2023; Mishra et al., 2024). Microbes require oxygen to assimilate the material and mineralize it into a stable form against further biological attack (). The O_2 demand of biodegradable carbonaceous matter present in wastewater is estimated at approximately 2 kg/kg of COD. Such saturation levels cannot be maintained in a bulk solution with conventional activated sludge secondary wastewater treatment that requires costly, energy-intensive materials. Nitrification is also an aerobic process, and it costs just as much to provide the necessary O_2 in traditional systems. The world's population growth has worsened the environmental issues. The environment is highly affected by pollution from harmful substances such as heavy metals (Das & Ghosh, 2023a, 2023b; Das et al., 2011; Ghosh & Das, 2018; Priyadarsini & Das, 2024, 2025), microplastics (Mishra et al., 2021; Pradhan et al., 2024; Tripathy et al., 2024), biomedical waste (Khatun et al., 2023; Sadasivuni et al., 2019), radioactive waste, chemical fertilizers, pesticides, hydrocarbons (Behera et al., 2021; Das et al., 2025; Kanungo et al., 2023; Mohanta et al., 2024), and pharmaceutical leftovers. This is a real difficult challenge for the science with a lot of knowledge to be gained and embraced. Pollution removal or remediation has improved naturally and through biological for over several years now (Dey et al., 2023; Mishra et al., 2019; Priyadarshini et al., 2024; Raj & Das, 2023). Wastewater treatment often involves some combination of the following physical or chemical processes: membrane filtration, ion exchange, osmosis, electrochemical treatment, evaporation, or precipitation of sorbents. However, it must be pointed out that most of these processes are not viable commercially or environmentally. Hence, there is a need for new integrated technologies for decreased cost and energy consumption (Abdel-Raouf et al., 2012). This is the reason why we see an increase in biological wastewater treatment as opposed to chemical treatment. Chemical wastewater treatment is an effective method, but the residue contains synthetic chemicals that could be harmful to the environment and living organisms (Kumar et al., 2023; Mishra & Das, 2021, 2022; Mishra et al., 2022b). Thus, the utilization of biological wastewater treatment seems to be the most environmentally friendly option. Microalgae are classified into native microalgae, genetically modified microalgae, and engineered microalgae. One of the most promising of the encouragements is biological remediation (bioremediation) due to its low cost, environmental friendliness, long-term effectiveness, etc. This is an environmental remediation treatment process that utilizes microorganisms to remove or neutralize pollutants from soil or water. In bioremediation, many microorganisms have been used in the removal of pollutants from the environment, which include bacteria, fungi, and microalgae. Bioremediation occurs primarily through two processes: biosorption and bioaccumulation. A dynamic procedure of bioremediation, "bioaccumulation," occurs when living things absorb poisons through their cell walls. A new method for removing heavy metal ions from acidic liquors uses inert biomass or live algae, which is referred to as biosorption. Algal growth, therefore, depends on the process of sorption. Adsorption is a complex process that occurs as a result of various environmental factors, whereby heavy metal ions are taken up by algae cells as they grow (Mishra et al., 2022a). On the contrary, passive bioremediation relies on physicochemical reactions performed by dead or live microbial cells (Das et al., 2015, 2024; Paul et al., 2023; Pradhan et al., 2023). Among the new resources sought for reducing heavy metal concentrations in the environment, new materials have been identified as safe and efficient methods for wastewater treatment and affordable. Biological materials in this context emerge as green, economical substitutes. Microalgae were highlighted as efficient and versatile biosorbent that are rich in metals uptake ability from polluted waters. Microalgae-based bioremediation technology is a promising solution that has gained attention due to its advantages, despite substantial efforts studying bioremediation by a variety of microorganisms in polluted environments. Microalgae bioremediation for contaminant removal could also be a potential way to convert its biomass into other value-added bioproducts (Abinandan & Shanthakumar, 2015). Biological approach is probably the most promising treatment method of industrial wastewater in the long run. It will also promote application of principles of the circular economy. There is considerable interest in applying green algae as a feedstock for producing biofuel as the price of oil is going sky-high and the natural oil reserve is dwindling fast while there is an inescapable fact that fossil fuel consumption causes ecological adverse effects and leads to global warming. The future will be devoid of fossil fuel resources. At a price competitive with current petroleum prices (Aslam et al., 2018), renewably sourced energy supplies that do not or only emit minimal amounts of carbon dioxide (CO_2) can satisfy most of the world's energy requirements. Soybeans, sugarcane, and vegetable oils are the most common crops from which first-generation biofuels are produced. This has led to future food production vs. energy concerns in an effort to focus on second-generation biofuels. In contrast, grass straws, wood, jatropha, switchgrass, and organic waste as non-food biomass sources have more net energy per acre compared to maize and sugarcane. However, these resources are not low-yielding, cost-effective products since they require costly thermochemical and biochemical methods to convert them into biofuel. The photosynthetic thallophytes may just be the ultimate feedstock for third-generation biofuels. Microalgae offer both environmental and economic advantages over other feedstocks, and they are also accessible in sufficient quantities to ensure meeting growing demand with no adverse environmental impact. Biodiesel, bioethanol, and biohydrogen have all been produced using microalgae. Alternative fuels are relevant in many sectors and particularly transport. Their interest is not only from the production point, but the public is generally interested in biofuels derived from microalgae as an alternative solution to these problems. Many benefits have been attached to the use of microalgal biomass for biofuel purposes and other non-toxic bioproducts (Branco-Vieira et al., 2020).

2 Composition of Municipal Wastewater

The volumes and intensities of various causes of pollution in watercourses vary. The technologies and lifestyles of the civilization from which wastewater originates are reflected in its makeup (Abdelfattah et al., 2023). It is made up of a complex blend of inorganic, natural, and synthetic organic components. Proteins, lipids, carbohydrates, amino acids, and volatile acids contribute about 75% of the organic carbon in sewage. Inorganic materials consist of huge quantities of sodium, calcium, potassium, magnesium, chlorine, sulfur, phosphate, bicarbonate, ammonium salts, and heavy metals (Dai & Wang, 2024). Pollutants can originate from numerous sources, such as the release of untreated or treated sewage from communities, leachates from waste disposal sites, runoff from agricultural regions, and emissions from manufacturing or industrial enterprises (Abid et al., 2017). We are looking into the possibility of recycling wastewater and recovering resources due to problems in water scarcity and the increased demand for food and energy (Amaro et al., 2023).

3 Microbial Composition

Wastewater is an ideal habitat for many microorganisms, including bacteria, viruses, and protozoa. Sewage contains dangerous microbes that are discharged in mass amounts by unwell and asymptomatic carriers; however, most of these bacteria are not dangerous and can be utilized in the biological treatment of sewage. Sewage contains parasitic worm eggs, infectious hepatitis viruses, disease-causing cholera, typhoid, tuberculosis pathogenic bacteria, and diarrheal protozoa (Numberger et al., 2019). The reduction in total coliform organisms is often utilized as an indicator for assessing the efficiency of sewage treatment (Chiu et al., 2015).

4 Sewage Treatment Processes

Coliform bacteria, suspended particles, nutrients ($NO^{-}{}_{3}$–N, $NO^{-}{}_{2}$–N, $NH_4{}^{+}$–N, and $PO_4{}^{-3}$–P), biochemical oxygen demand (BOD), and toxicity from wastewater are removed as the main goal of the wastewater treatment system. BOD is the ability of microorganisms to utilize molecular oxygen as an oxidizing agent to break down organic substrates into CO_2 and water. Since excessive BOD can reduce the dissolved oxygen content in receiving waterways, possibly terminating fish and triggering anaerobic conditions, it is essential to remove BOD from wastewater. Physical sedimentation is a good way to remove suspended particles. Dissolved phosphorus and nitrogen must be treated in wastewater treatment plants constructed for the removal of nutrients. When these fertilizers are released into sensitive waterways, they cause eutrophication, which favors the development of unwanted vegetation like aquatic macrophytes and algae. Other concerns regarding nitrogen compounds in wastewaters involve the toxicity of non-ionized ammonia to fish and aquatic life, interference with disinfection processes requiring a free residual chlorine, and methemoglobinemia of drinking water generated by excessive nitrates (above 45 g/m^3) (Kato & Kansha, 2024).

5 Preliminary Treatment

During the preliminary treatment, large solids that are introduced into the plant through the sewers and may cause blockage or damage to equipment are removed. Floating materials include floating rags and wood, feces, and higher grit. Sewage is passed through bars 20–60 mm apart to remove large floating items, and debris collected is periodically swept off the bars (Ugrina & Milojković, 2024). To remove grit, decrease the velocity of flow to the extent that silt and grit will settle and organic material suspends in water. The velocity of flow can generally be between 0.2 and 0.4 m/s (Jayalekshmi et al., n.d.). Primary treatment: Sewage is pumped into sedimentation tanks, which utilize gravity to remove up to 70% of the total settleable solids once heavier objects are released. Forty % of the biochemical oxygen demand (BOD) may be eliminated by solids in a well-constructed sedimentation tank (Du et al., 2022). Different species may have slightly different capacities to eradicate infections during these initial treatments (Ahmad et al., 2017). Secondary treatment: By eliminating organic debris, the secondary treatment procedure aims to lower BOD. Generally speaking, it is due to the activities of heterotrophic bacteria, which require organic material for both energy as well as reproduction. Many biological processes have been known to assist in the aerobic oxidation of BOD. Procedures that fall into this category are divided into two broad categories: suspended growth systems or fixed film systems, based on the microbial population. Biofilms attached to a stationary support in fixed film reactors aerobically take in and degrade organic substrates. Suspended growth systems, including activated sludge processes, incorporate microbes with sewage and suspend by diffused air mixing or mechanical aeration (Prandini et al., 2016). Organisms that may represent more than 90% of pathogenic bacteria in raw sewage can be destroyed by biological oxidation systems; however, viruses are harder to inactivate. Adsorption has been considered the most important method for virus removal. Though the decreased surface areas of biological adsorption points in film reactors lead to differing levels of decrease, adequate addition of solid flocs and sewage can achieve a 90% removal rate in suspended growth reactors (Mata et al., 2014).

6 Tertiary Treatment

All organic ions have to be eliminated by the tertiary treatment procedure. It is chemically or biologically achievable. Relative to chemical methods, biological tertiary treatment is generally superior because chemical methods are very expensive, thus not feasible in many places; also, sometimes they lead to secondary contamination. Moreover, each step of further treatment in a wastewater system increases the cost of the system drastically (Abeysiriwardana-Arachchige et al., 2022). It is also expected that complete tertiary treatment where phosphate, nitrate, and ammonium ions are removed should cost four times the amount paid for first treatment (Yang et al., 2020). However, quaternary treatment, which targets the removal of organic compounds like toxicants and refractory materials, soluble minerals, and heavy metals, would be eight to sixteen times the cost of primary treatment (Mishra et al., 2018). Sophisticated treatments also involve advanced processes such as carbon adsorption, ozonation, reverse osmosis, and chemical precipitation. These removal strategies target specific nutrients, including nitrogen and phosphorus, which are recognized to cause eutrophication in different streams. By removing fine particles, these processes can help achieve discharge levels below regulatory limits, thereby improving overall effluent quality, especially in smaller applications. Commonly used techniques are land application, lagoon storage, and filtration using sand or gravel filters (Zagklis & Bampos, 2022). Total nitrogen and phosphorus levels in certain industrial and agricultural wastewaters are three orders of magnitude higher than those found in natural water systems (Bhandari & Ranade, 2014). This problem has increased the application of conventional main and secondary treatment methods. During primary treatment, they are used to remove readily settled debris, and during secondary treatment, they are used to oxidize the organic material in the wastewater. Consequently, effluent that appears to be clean and pure is dumped into natural water bodies. However, long-term problems such as eutrophication and inorganic nitrogen and phosphorus high levels might result from the discharge of refractory organics in this secondary effluent. Researchers discovered that both aerobic and anaerobic processes were involved when natural purification methods exposed the chemoorganotrophic microbes that break down organic materials. The same kind of natural self-purification is utilized in modern biological wastewater treatment systems, but these are contained in structures designed to speed up the treatment process.

7 Disinfection of Water

Even with primary, secondary, and tertiary treatment, not all waste load is eliminated, leaving many organisms in the wastewater. Regulatory agencies may require removal of harmful organisms from wastewater as a way to address public health concerns and to halt the spread of waterborne diseases. Although most of these bacteria are harmless, it's important to acknowledge that pathogens may still be present.

When wastewater effluents are released into bodies of water that could be used for drinking, swimming, or shell-fishing, it becomes crucial to lower bacterial levels to minimize health risks. Disinfection is a process in wastewater treatment which gets rid of all the microorganisms. Sterilization can be defined as complete removal of microorganisms. (Lazarova et al., 1999) Complete removal of microorganisms is referred to as sterilization. While disinfection eliminates all pathogenic microorganisms, wastewater treatment does not aim for sterilization. On the other hand, disinfection procedures will drastically lower the number of germs in wastewaters to a safe level by eliminating all microorganisms. Generally speaking, disinfection may be done using any technique that eliminates the bacteria. Microorganisms can be eliminated using a variety of physical or chemical techniques under specific circumstances. For instance, physical methods include burning, boiling, and exposure to UV or X-ray radiation. Chemical processes, on the other hand, involve strong acids, alcohols, oxidizing agents, and surfactants. However, pathogen removal from wastewater treatment requires workable methods that can be used with minimal expense to huge amounts of wastewater that have had different degrees of treatment. Chlorine was a major disinfectant in wastewater treatment procedures in the past (Abarnou & Miossec, 1992). Because of its efficacy as a disinfectant and its historical cost, chlorine has been preferred. However, because of its increasing cost, toxicity to fish and other aquatic life even at low concentrations, and tendency to produce dangerous chlorinated hydrocarbons, chlorine is no longer the disinfectant of choice for wastewater treatment. The increasing use of UV light and ozone (ozonation) as disinfectants in wastewater treatment holds great promise for the future. UV light and ozone both are good disinfectants and leave no dangerous residues behind (Chan & Killick, 1995). Ozone may also increase the amount of dissolved oxygen in water. However, the production of ozone is necessary, and it has only recently started to be a cost-effective alternative to chlorination. Recent studies have also focused on the effectiveness and cost of ultraviolet light when applied in large wastewater treatment plants. Although extensive research has been conducted, UV light seems to be a promising and cost-effective disinfection technique that is more economical than chlorination. We will discuss chlorine and ozone as chemical disinfectants, but focus on the specific characteristics and mechanisms of each. Chlorination, however, will be the primary emphasis since chlorine is still widely used as a disinfectant.

8 Aquatic System Treatment

A system featuring a pond covered in a mat of duckweed seems to effectively purify wastewater with the help of bacteria. It will break down with the result that the water environment becomes anaerobic, while the duckweed mat is a good aero-permeable inhibitor of reaeration. Indeed, some laboratory experiments have confirmed that certain duckweed species, such that *Spirodela* and *Lemna* can reduce the concentration of oxygen in the water, (Tom et al., 2021) but this lack of oxygen does not seem to harm the plants. Duckweed is capable of converting essential minerals such as carbon, nitrogen, and phosphorus into protein. It can also remove organic materials by directly ingesting simple organic substances and converting them into different amino acids and carbohydrates. Suspended solids (SS) and carbonaceous biochemical oxygen demand (BOD) can be removed by two major techniques: bacterial metabolism and physical sedimentation, primarily used in removing suspended solids in urban wastewater systems. Aquatic plants contribute very little to wastewater treatment in these settings (Ansari et al., 2017). For most of human existence, agriculture was an important part of biological water treatment by employing potential pollutants such as animal and human waste for fertilizers in the promotion of plant growth. After treatment, municipal sewage is sometimes applied as a source of nutrients to land carrying on natural vegetation or other crops (Rawat et al., 2011). These wastes continue to be critical for agriculture everywhere, especially where access to commercial fertilizers is restricted.

9 Microalgae Treatment

Algal cultures have been used industrially for over 75 years. They have been used to treat wastewater and for the high-volume production of several strains including *Chlorella* and *Dunaliella* (Caporgno et al., 2015). The biologists within these countries should be able to have a profound knowledge of algal biology and ecology of massive algal culture systems, along with engineering methods of large culture systems and methods of harvesting the algae. These cultures are to produce pharmaceuticals and genetically modified goods, among other high-value goods (Brennan & Owende, 2010). These products contain a number of physiologically beneficial compounds, including antiviral, antibacterial, antitumor/anticancer, and antihistamine (Bilanovic et al., 2009). Microalgae's photosynthetic ability and ability to convert solar energy into usable biomasses and assimilate nutrients such as nitrogen and phosphorus make them highly desirable for bio-treatment (Molina Grima et al., 2003). This remarkable theory, first proposed by some scientists in the USA fifty-five years ago (1957), has now been extensively tested in a number of nations. Because microalgal wastewater treatment systems demand a significant amount of land acreage (Paddock, 2019), Hyper-concentrated algal cultures are being applied in planned treatment systems for wastewater. It has been demonstrated to be especially useful in eliminating N and P in a short timeframe—less than an hour, for instance (Mishra et al., 2022c). The agal system is capable of treating a vast array of industrial, agro-industrial, and animal wastes (Abinandan et al., 2019), as well as human sewage. Certain wastes were tried for treatment by microalgal systems, such as pig dung, food processing plant sewage, and other agricultural wastes. Algal systems can also remove poisonous minerals such as lead, cadmium, mercury, scandium, tin, arsenic, and bromine. Algae are utilized to remove wastewater through various methods, such as the removal of nitrogen, phosphate, and/or heavy metals, reducing chemical and biological oxygen demand, and removing coliform bacteria.

10 Coliform Removal

Environmental factors conducive to algal growth may inhibit coliform. The problems of pathogenic species, particularly *Shigella, Salmonella,* viruses, and protozoa, are significant concerns in wastewater. Bacteria typically constitute the main fraction of microorganisms in any biological wastewater treatment process, approaching concentrations of as high as 106 bacteria/ml of wastewater (Balogun & Ogwueleka, 2021). In reality, viruses seem to survive for longer periods of time in the environment than hazardous bacteria, whose life cycles tend to be briefer than those of coliforms. The performance of sewage treatment is often measured in terms of total coliform elimination rate (Bourouache et al., 2019). Most people accept the fact that high-rate sewage stabilization ponds and sewage stabilization ponds are more effective compared to other sewage treatment methods (Aghalari et al., 2020).

11 Reduction of Chemical and Biochemical Oxygen Demand

Water sources are at risk because of the many chemicals and microbes present in wastewater. The three main categories of wastewater pollution are microbial, inorganic, and organic materials. Any compound containing at least one carbon atom is considered an organic compound in wastewater. Carbon dioxide can be generated by the chemical or biological oxidation of these carbon atoms. Chemical oxygen demand is the term for chemical oxidation, while biochemical oxygen demand is the term for biological oxidation. In other words, BOD employs the oxidizing agent molecular oxygen to help the normal decomposition of organic matter by bacteria into carbon dioxide and water. Thus, biochemical oxygen demand is the amount of oxygen required by bacteria to respire while decomposing organic matter in wastewater. The primary objective of wastewater treatment is to eliminate surplus BOD because it can reduce the dissolved oxygen content in receiving water, which can cause anaerobic conditions and fish kills. (Singh et al., 2024) It has been showed that BOD and COD were completely removed at 68.4% and 67.2%, respectively, during the treatment of wastewater.

12 Removal of Nitrogen and Phosphorus

It has been proposed that the use of algae to eliminate pollutants such as nitrogen and phosphate from wastewater, while simultaneously supplying oxygen for aerobic bacteria. Since their initial suggestion, many laboratory and pilot studies have been carried out on this approach, leading to the construction of several sewage treatment plants that utilize different variations of this system (M. B. Ahmed et al., 2017).

Synthetic detergents contribute to the phosphorus in sewage discharge being 50% or more, whereas most of the nitrogen comes from the metabolic conversions of chemicals that are acquired from sources other than the water. Wastewater contains both the dissolved $PO4^{3-}$ (orthophosphate), NO_2^- (nitrite), NO_3^- (nitrate), and NH_4^+ (ammonia). The removal of both elements is referred to as "nutrient stripping," and both of these elements are referred to as nutrients (Li et al., 2023). The biological degradation, aerobic or anaerobic, is the most used method for treating wastewater. However, the inorganic substances remaining in the water, such as phosphate, ammonium, and nitrate ions, can contribute to the lakes becoming eutrophic and create harmful microalgal blooms (Amann et al., 2018). The phosphorus and nitrogen were also found to be significant constituents of eutrophication. It implies that additional treatments must be enforced to avoid the phenomenon of eutrophication in water ecosystems (Priya et al., 2022). Eutrophication, favoring the propagation of unwanted plant species such as algae and water macrophytes, can lead from nutrient accumulation in sensitive waters (Bhagowati & Ahamad, 2019). Nitrate concentration in the water supply that exceeds 45 gm^{-3} can initiate methemoglobinemia in wastewater due to the presence of nitrogenous compounds and damage fish as well as other aquatic organisms; besides, interfere with disinfection (Camargo & Alonso, 2006). Especially in tertiary treatment, microalgal culture is a productive method for extracting nutrients from wastewater (Jin et al., 2023). These microalgae may be grown in enormous quantities using outdoor solar bioreactors and are very efficient at taking inorganic nutrients (Kholssi et al., 2023). General treatment with chemical and physical treatments often proves to be too expensive for general use and may also result in secondary pollution. Generally, biological techniques are superior to such chemical and physical treatments. Being able to efficiently remove heavy metals (Singh et al., 2024) and utilizing inorganic nitrogen and phosphorus for growth, microalgal cultures may be considered suitable for tertiary and quaternary treatments. *Chlorella vulgaris* removed 86% of the inorganic nitrogen and 78% of the inorganic phosphorus (Wirth et al., 2020). In addition, during the treatment of industrial wastewater, nitrogen and phosphorus were removed at levels of 50.2% and 85.7%, respectively, and in home wastewater treated with algae, phosphorus removal was achieved at 97.8% (Kesari et al., 2021). The growing interest in microalgal cultures is largely due to the notable drawbacks of traditional treatment methods: (a) their efficiency can vary based on the specific nutrients being removed; (b) they often incur high operating costs; (c) chemical treatments can lead to secondary pollution; and (d) they tend to waste valuable nutrients like nitrogen and phosphorus (Hammond et al., 2025). The last point is particularly significant, as conventional treatment methods often result in the inefficient use of natural resources (4). The use of plants or microalgae to remove inorganic compounds offers a renewable solution that harnesses solar energy (4). Under optimal conditions, cyanobacteria can outpace the growth of higher plants, suggesting that systems designed for inorganic nutrient removal using cyanobacteria hold considerable potential (Sarup et al., 2021). There are greater signs of N accumulation in the environment due to the increased application of inorganic nitrogenous fertilizers and waste produced by human and animal populations. The main cause for concern with N pollution is the health hazards associated with nitrite, which can either directly cause methemoglobinemia or indirectly act as a precursor to nitrosamines (Tang et al., 2024). Nitrites, being precursors to N-nitroso compounds, mainly nitrosamines, which are gaining much interest due to their potential mutagenic, carcinogenic, and teratogenic properties, assume a special status. Since currently available treatment technologies cannot reduce the levels of nitrates sufficiently in drinking water, research has been focused on developing new approaches to reduce the levels to tolerable values. It is highly acknowledged that photosynthesis drives nitrate absorption and reduction processes in N-sufficient cyanobacteria, which most probably involve the assimilation of ammonium into carbon structures (Al-Ajeel et al., 2022). Additionally, many nutritional and environmental factors, which include light, temperature, pH, and availability of carbon sources, affect nitrate uptake (Abascal et al., 2022). Because too much phosphorus is useless without nitrogen, and vice versa, the metabolism of phosphorus and nitrogen are intimately related. The ability of cyanobacteria to absorb phosphate, which has been studied in different strains, seems to have a hyperbolic profile in relation to the external phosphate concentration (Romanis et al., 2021). The cyanobacteria can be easily recovered by decanting the filtered water after they have taken the nutrients from the effluents (Bertone et al., 2019). The biomass that is harvested could be utilized to extract pigments that have commercial value.

13 Factors Affecting Algal Growth and Nutrient Removal

Algal development and nutrient uptake are also influenced by complex interactions between biotic and physical conditions, including temperature, light intensity, pH, and nutrient availability (Beuckels et al., 2015). Initial density is one of the important biotic factors that influence algal development; higher algal density generally leads to better growth and greater efficiency of nutrient removal (Sutherland et al., 2014). However, if algal density becomes too high, self-shading, the build-up of auto-inhibitors, and lower photosynthetic efficiency can occur (Sutherland et al., 2014).

14 Removal of Heavy Metals

There are photosynthetic microalgae which can effectively detoxify metals found in mine wastewater (Chugh et al., 2022). Cyanobacteria in a simulated pool and meander system can eliminate 99% of dissolved and particulate metals. *Coelastrum proboscideum* can uptake 100% of the lead from a 1.0 ppm solution after 20 h at 23 °C and about 90% after only 1.5 h at 30 °C (Akshata & Sadaf, n.d.). When compared to the ambient, algae in experimental rice paddles accumulated and concentrated Cd^{2+} by around 1000 times (Abdel-Aty et al., 2013). Algae also happen to be very effective collectors of tributyl tin and organochlorides (Dubey et al., 2023). Some of them have even been shown to biodegrade many of these substances.

The insecticide Deltamethrin has been proven by some researchers that it gets biodegraded by the green algae species known as *Dunaliella bioculata* (Baeza-Squiban et al., 1987). It was further found out that algae break up a vast collection of hydrocarbons in a waste full of greasiness (Avila et al., 2021).

15 Algae as Monitor of Water Quality

Algal bioassays have been employed in a number of investigations during the past three decades to assess ecological shifts and their utility as indicators of water quality (S. Ahmed, 2024). Palmer developed a composite grading system in 1959 for those species that are utilized as indicators of water pollution, including *Euglena, Oscillatoria, Chlamydomonas, Scenedesmus, Chlorella, Nitzschia, and Navicula.* However, the sample may be considered to be free from contamination if it harbors organisms like *Lemanea, Stigeoclonium*, and some species of *Micrasterias, Staurastrum, Pinnularia, Meridion*, and *Surirella* (Sanglyne, 2019).

Recently, there has been a notable increase in interest in wastewater integrated algal biorefineries due to their efficiency. Numerous studies have indicated that various types of wastewater, including CO_2 from both air and exhaust gases, microalgae are rich in essential nutrients and thus can be considered as an economic feedstock material for their cultivation, which may come from home, agricultural, and industrial sources (Chandra et al., 2019). The most significant advantage of a wastewater integrated microalgae-based biorefinery is its potential to solve environmental concerns while also producing biofuels and other valuable products such as pigments, microelements, omega fatty acids, antioxidants, and animal feed (Gupta et al., 2024). The growing number of articles on the issue suggests that microalgae's potential as a source of novel bioactive chemicals is becoming more widely recognized. Microalgae include valuable elements including pigments, fatty acids, and other biochemicals, which indicates their significant importance (Song et al., 2024). Microalgae yield a vast variety of biomolecules for different uses. Specific types of molecules can be used for distinct uses. For instance, lipids, proteins, and carbohydrates can be used by the food and nutraceutical industries, whereas the pharmaceutical and cosmetic industries can use pigments and steroids. According to global microalgae market data, pigments and lipids derived from microalgae are among the most sought-after products available. One notable example is *Dunaliella salina*, a green alga that thrives in high-salinity, sunlit open ponds, which is known for producing the pigment β-carotene in significant quantities (Barros et al., 2015). Another important species is the freshwater green algae *Haematococcus pluvialis*, which is known for its production of astaxanthin, a pigment that holds considerable commercial value (Khoo et al., 2019). Although astaxanthin is produced in smaller amounts, it commands a higher market price compared to β-carotene. Because of their high triglyceride content, microalgae have lately received attention as a potential feedstock for third-generation biofuel. However, the majority of energy production still relies on fossil fuels like coal, oil, and natural gas. The public interest in alternative energy solutions is developing due to the increase in energy needs brought about by the population boom and industrialization. Algae are of particular interest as a source of renewable energy because it possesses a high content of lipid and grows fast with the added advantages of growing them without any use of arable land. Moreover, during photosynthesis, algae can effectively absorb CO_2, resulting in the production of polysaccharides and triacylglycerol (TAG) (Yang et al., 2024).

16 Challenges and Future Perspective

A third-generation feedstock, microalgae have potential as raw material for biorenewables that may replace fossil fuels and first-generation biofuels from sugar cane, maize, and soy. Waste can be decreased today combining microalgae with clean technologies such as biorefineries and the generation of biofuel (Wang et al., 2022). While using microalgae for CO_2 sequestration offers environmental advantages, its implementation is limited by the high costs associated with CO_2 capture and transport, along with considerable CO_2 losses during the production of microalgae (Ighalo et al., 2022). High production costs may be associated with this limitation. The standard compression used in the pipeline process is highly energy-intensive and increases transportation cost. It is more economical to provide carbon to microalgae for the production of biodiesel if CO_2 and $NaHCO_3$ are used in combination (Zhang et al., 2022). Some environmental components to further develop bigger scale production contain some gaps that need to be filled. Ammonia (NH_3) is an inorganic fertilizer that helps promote a healthy growth rate and productivity of algal biomass. However, for every kilogram of NH_3 produced, 1.2 kg of CO_2 are emitted, which contributes to environmental degradation and introduces chemical reagents into the ecosystem during the microalgae production process. This challenge needs to be addressed when managing a microalgal production facility powered by renewable energy to ensure environmental sustainability (Rathinavel et al., 2024). A life cycle assessment (LCA) is an important tool for future environmental enhancements as it can be used to investigate energy and environmental implications (Reijnders, 2020). In terms of economy, removing dissolved organic carbon, phosphorus, and nitrogen from various wastewaters through a microalgae system is significantly more viable compared to the traditional methods. This is primarily because microalgae systems can function outdoors under sunlight, which helps reduce operational costs (Anjum et al., 2019). However, the open nature of these systems can lead to an increased risk of contamination, as more microorganisms may infiltrate the colony. This contamination with other microbes can pose a challenge, as the process occurs in non-sterile environments, leaving only a few strains that are robust, fast-growing, and capable of withstanding harsh conditions to thrive in open reactors. To tackle this challenge, one promising strategy is to co-culture microalgae with various microbes, including bacteria, yeast, and fungi, to boost the efficiency of wastewater treatment. This highlights the need for further research into the role of co-culture in both synthetic and real wastewater scenarios (Fabris et al., 2020). Future investigations should delve into the interactions between microalgae and other microorganisms, as well as the genes and transcription factors that govern their mechanisms in wastewater treatment. Employing omics techniques, especially targeted genome editing methods like CRISPR, is strongly encouraged to develop effective solutions that enhance the potential of microalgae for producing valuable bioproducts with significant commercial prospects (Amini et al., 2017). It poses a significant challenge as the nutritional composition of various municipal wastewaters needs to be considered during microalgae treatment. The different types of wastewater need to be mixed in a manner that allows for a balanced concentration and ratio of nutrients in order to meet the requirements of wastewater discharge. Recovering biomass after culture might incur up to 30% of the overall production costs besides the costs of generating algal biomass. The small size of microalgae cells (2–20 μm) and their density, which is similar to water, are major reasons for this challenge. Large-scale agriculture faces several challenges, one of which is the effective recovery of biomass. The intense energy needed for the removal of microalgae from wastewater produced by various industrial treatment methods is another significant issue (Liu et al., 2020). Flocculation and sedimentation need to be carried out in two stages, as per a recent report, to enhance microalgal biomass concentration. The most efficient method among them is bio-flocculation as it minimizes chemical expenses at the cost of no biomass efficiency (Vandamme et al., 2013). Further studies on microalgae's carbon sequestration capability should focus on the factors that affect their growth, including their overall sensitivity and tolerance to temperature, pH, radiation, and nutrient conditions, as small variations in the growing conditions can have an enormous effect on crop yields. Carbon sequestration in microalgae needs to consider some factors, especially those related to their wide sensitivity and tolerance to temperature, pH, irradiation, and nutritional conditions since minimal alterations to the production environment have an impact on the yield of the final product (Rabiu et al., 2024). All these requirements include the development of the necessary technical infrastructure, growth factor augmentation, and a comprehensive understanding of the photosynthetic process, and are crucial for the long-term and financial feasibility of large-scale microalgae carbon sequestration. Figure 1 represents the advantages of algal biomass in wastewater treatment and the formation of value-added products.

17 Conclusion

In conclusion, photosynthetic microorganisms known as microalgae are crucial to the bioremediation of all wastewater types; hence, their capabilities encompass heavy metals removal, as well as biochemical oxygen demand, nitrogen, phosphorus, and carbon reductions. This feature makes them efficient and environmentally friendly in water treatment. This method of using microalgae has several benefits in comparison to other conventional wastewater treatments, such as lower treatment cost, reduced

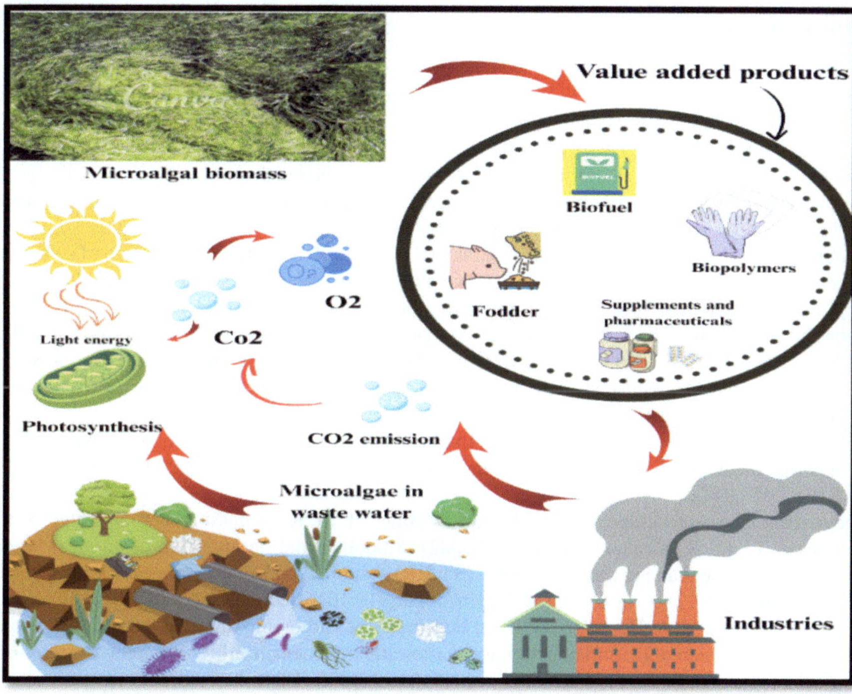

Fig. 1 Represents the advantages of algal biomass in wastewater treatment and the formation of value-added products

energy consumption, and a smaller carbon footprint. It contributes more to environmental sustainability and an effective approach to managing waste. Further, the potential of microalgae for the development of bioenergy is also immense. For the promotion of microalgae bioenergy as a viable and sustainable alternative, it becomes essential to understand the environmental implications of large-scale production. In fact, assessment of such impact is crucial because it will allow us to ascertain whether the positive benefits of using microalgae for bioenergy are not overcome by unintended ecological consequences.

More promisingly, the concept of an integrated microalgal biorefinery is better than the individual units. In fact, not only do these biorefineries treat wastewater but also open the door for handling other environmental problems while creating by-products in the form of biofuels, biodiesel, bioplastics, and numerous other high-value chemicals. This is the idea of a circular economy, in which wastes are minimized, and resources are efficiently used. This trend is likely to continue, opening the door for microalgae to be integrated into environmental sustainability and economic development, which is bound to open up a brighter, greener future.

References

Abarnou, A., & Miossec, L. (1992). Chlorinated waters discharged to the marine environment chemistry and environmental impact. An overview. *Science of The Total Environment, 126*(1–2), 173–197. https://doi.org/10.1016/0048-9697(92)90490-J

Abascal, E., Gómez-Coma, L., Ortiz, I., & Ortiz, A. (2022). Global diagnosis of nitrate pollution in groundwater and review of removal technologies. *Science of the Total Environment, 810*, Article 152233. https://doi.org/10.1016/J.SCITOTENV.2021.152233

Abdel-Aty, A. M., Ammar, N. S., Abdel Ghafar, H. H., & Ali, R. K. (2013). Biosorption of cadmium and lead from aqueous solution by fresh water alga Anabaena sphaerica biomass. *Journal of Advanced Research, 4*(4), 367–374. https://doi.org/10.1016/j.jare.2012.07.004

Abdelfattah, A., Ali, S. S., Ramadan, H., El-Aswar, E. I., Eltawab, R., Ho, S. H., Elsamahy, T., Li, S., El-Sheekh, M. M., Schagerl, M., Kornaros, M., & Sun, J. (2023). Microalgae-based wastewater treatment: Mechanisms, challenges, recent advances, and future prospects. *Environmental Science and Ecotechnology, 13*. https://doi.org/10.1016/j.ese.2022.100205

Abdel-Raouf, N., Al-Homaidan, A. A., & Ibraheem, I. B. M. (2012). Microalgae and wastewater treatment. *Saudi Journal of Biological Sciences, 19*(3), 257–275. https://doi.org/10.1016/J.SJBS.2012.04.005

Abeysiriwardana-Arachchige, I. S. A., Delanka-Pedige, H. M. K., Munasinghe-Arachchige, S. P., Brewer, C. E., & Nirmalakhandan, N. (2022). Modeling and optimization of an algal-based sewage treatment and resource recovery (STaRR) system. *Journal of Environmental Chemical Engineering, 10*(2), Article 107139. https://doi.org/10.1016/J.JECE.2022.107139

Abid, A., Saidane, F., & Hamdi, M. (2017). Feasibility of carbon dioxide sequestration by Spongiochloris sp microalgae during petroleum wastewater treatment in airlift bioreactor. *Bioresource Technology, 234*, 297–302. https://doi.org/10.1016/J.BIORTECH.2017.03.041

Abinandan, S., & Shanthakumar, S. (2015). Challenges and opportunities in application of microalgae (Chlorophyta) for wastewater treatment: A review. *Renewable and Sustainable Energy Reviews, 52*, 123–132. https://doi.org/10.1016/J.RSER.2015.07.086

Abinandan, S., Subashchandrabose, S. R., Pannerselvan, L., Venkateswarlu, K., & Megharaj, M. (2019). Potential of acid-tolerant microalgae, Desmodesmus sp. MAS1 and Heterochlorella sp. MAS3, in heavy metal removal and biodiesel production at acidic pH. *Bioresource Technology, 278*, 9–16. https://doi.org/10.1016/J.BIORTECH.2019.01.053

Aghalari, Z., Dahms, H. U., Sillanpää, M., Sosa-Hernandez, J. E., & Parra-Saldívar, R. (2020). Effectiveness of wastewater treatment systems in removing microbial agents: A systematic review. *Globalization and Health, 16*(1). https://doi.org/10.1186/s12992-020-0546-y

Ahmad, M., Liu, S., Mahmood, N., Mahmood, A., Ali, M., Zheng, M., & Ni, J. (2017). Effects of porous carrier size on biofilm development, microbial distribution and nitrogen removal in microaerobic bioreactors. *Bioresource Technology, 234*, 360–369. https://doi.org/10.1016/J.BIORTECH.2017.03.076

Ahmed, M. B., Zhou, J. L., Ngo, H. H., Guo, W., Thomaidis, N. S., & Xu, J. (2017). Progress in the biological and chemical treatment technologies for emerging contaminant removal from wastewater: A critical review. *Journal of Hazardous Materials, 323*, 274–298. https://doi.org/10.1016/J.JHAZMAT.2016.04.045

Ahmed, S. (2024). *Algae as bioindicators of aquatic pollution.* https://www.researchgate.net/publication/380365699

Akshata, D., & Sadaf, Z. (n.d.). *Bioremediation of heavy metal pollutant lead by microalgae and its impact on their growth and Chlorophyll-a content: A review.* www.ijlsci.in

Al-Ajeel, S., Spasov, E., Sauder, L. A., McKnight, M. M., & Neufeld, J. D. (2022). Ammonia-oxidizing archaea and complete ammonia-oxidizing Nitrospira in water treatment systems. *Water Research X, 15*, Article 100131. https://doi.org/10.1016/J.WROA.2022.100131

Amann, A., Zoboli, O., Krampe, J., Rechberger, H., Zessner, M., & Egle, L. (2018). Environmental impacts of phosphorus recovery from municipal wastewater. *Resources, Conservation and Recycling, 130*, 127–139. https://doi.org/10.1016/J.RESCONREC.2017.11.002

Amaro, H. M., Salgado, E. M., Nunes, O. C., Pires, J. C. M., & Esteves, A. F. (2023). Microalgae systems—Environmental agents for wastewater treatment and further potential biomass valorisation. *Journal of Environmental Management, 337*, Article 117678. https://doi.org/10.1016/J.JENVMAN.2023.117678

Amini, Z., Ilham, Z., Ong, H. C., Mazaheri, H., & Chen, W. H. (2017). State of the art and prospective of lipase-catalyzed transesterification reaction for biodiesel production. *Energy Conversion and Management, 141*, 339–353. https://doi.org/10.1016/J.ENCONMAN.2016.09.049

Anjum, M., Miandad, R., Waqas, M., Gehany, F., & Barakat, M. A. (2019). Remediation of wastewater using various nano-materials. *Arabian Journal of Chemistry, 12*(8), 4897–4919. https://doi.org/10.1016/J.ARABJC.2016.10.004

Ansari, F. A., Singh, P., Guldhe, A., & Bux, F. (2017). Microalgal cultivation using aquaculture wastewater: Integrated biomass generation and nutrient remediation. *Algal Research, 21*, 169–177. https://doi.org/10.1016/J.ALGAL.2016.11.015

Aslam, M., Ahmad, R., Yasin, M., Khan, A. L., Shahid, M. K., Hossain, S., Khan, Z., Jamil, F., Rafiq, S., Bilad, M. R., Kim, J., & Kumar, G. (2018). Anaerobic membrane bioreactors for biohydrogen production: Recent developments, challenges and perspectives. *Bioresource Technology, 269*, 452–464. https://doi.org/10.1016/J.BIORTECH.2018.08.050

Avila, R., Peris, A., Eljarrat, E., Vicent, T., & Blánquez, P. (2021). Biodegradation of hydrophobic pesticides by microalgae: Transformation products and impact on algae biochemical methane potential. *Science of the Total Environment, 754*, Article 142114. https://doi.org/10.1016/J.SCITOTENV.2020.142114

Baeza-Squiban, A., Marano, F., Ronot, X., Adolphe, M., & Puiseux-Dao, S. (1987). Effects of deltamethrin and its commercial formulation DECIS on different cell types in vitro: Cytotoxicity, cellular binding, and intracellular localization. *Pesticide Biochemistry and Physiology, 28*(1), 103–113. https://doi.org/10.1016/0048-3575(87)90118-0

Balogun, S., & Ogwueleka, T. C. (2021). Coliforms removal efficiency of Wupa wastewater treatment plant, Abuja, Nigeria. *Energy Nexus, 4*, Article 100024. https://doi.org/10.1016/J.NEXUS.2021.100024

Barros, A. I., Gonçalves, A. L., Simões, M., & Pires, J. C. M. (2015). Harvesting techniques applied to microalgae: A review. *Renewable and Sustainable Energy Reviews, 41*, 1489–1500. https://doi.org/10.1016/J.RSER.2014.09.037

Behera, I. D., Nayak, M., Biswas, S., Meikap, B. C., & Sen, R. (2021). Enhanced biodegradation of total petroleum hydrocarbons by implementing a novel two-step bioaugmentation strategy using indigenous bacterial consortium. *Journal of Environmental Management, 292*, Article 112746. https://doi.org/10.1016/J.JENVMAN.2021.112746

Bertone, E., Chuang, A., Burford, M. A., & Hamilton, D. P. (2019). In-situ fluorescence monitoring of cyanobacteria: Laboratory-based quantification of species-specific measurement accuracy. *Harmful Algae, 87*, Article 101625. https://doi.org/10.1016/J.HAL.2019.101625

Beuckels, A., Smolders, E., & Muylaert, K. (2015). Nitrogen availability influences phosphorus removal in microalgae-based wastewater treatment. *Water Research, 77*, 98–106. https://doi.org/10.1016/J.WATRES.2015.03.018

Bhagowati, B., & Ahamad, K. U. (2019). A review on lake eutrophication dynamics and recent developments in lake modeling. *Ecohydrology & Hydrobiology, 19*(1), 155–166. https://doi.org/10.1016/J.ECOHYD.2018.03.002

Bhandari, V. M., & Ranade, V. V. (2014). Advanced physico-chemical methods of treatment for industrial wastewaters. *Industrial Wastewater Treatment, Recycling and Reuse*. https://doi.org/10.1016/B978-0-08-099968-5.00002-7

Bilanovic, D., Andargatchew, A., Kroeger, T., & Shelef, G. (2009). Freshwater and marine microalgae sequestering of CO_2 at different C and N concentrations—Response surface methodology analysis. *Energy Conversion and Management, 50*(2), 262–267. https://doi.org/10.1016/J.ENCONMAN.2008.09.024

Bourouache, M., Mimouni, R., Ait Alla, A., Hamadi, F., El Boulani, A., & Bihadassen, B. (2019). Bacteriological and physicochemical quality of treated wastewater of the Mzar treatment plant. *Applied Water Science, 9*(4). https://doi.org/10.1007/s13201-019-0958-0

Branco-Vieira, M., Mata, T. M., Martins, A. A., Freitas, M. A. V., & Caetano, N. S. (2020). Economic analysis of microalgae biodiesel production in a small-scale facility. *Energy Reports, 6*, 325–332. https://doi.org/10.1016/J.EGYR.2020.11.156

Brennan, L., & Owende, P. (2010). Biofuels from microalgae—A review of technologies for production, processing, and extractions of biofuels and co-products. *Renewable and Sustainable Energy Reviews, 14*(2), 557–577. https://doi.org/10.1016/J.RSER.2009.10.009

Camargo, J. A., & Alonso, Á. (2006). Ecological and toxicological effects of inorganic nitrogen pollution in aquatic ecosystems: A global assessment. *Environment International, 32*(6), 831–849. https://doi.org/10.1016/J.ENVINT.2006.05.002

Caporgno, M. P., Taleb, A., Olkiewicz, M., Font, J., Pruvost, J., Legrand, J., & Bengoa, C. (2015). Microalgae cultivation in urban wastewater: Nutrient removal and biomass production for biodiesel and methane. *Algal Research, 10*, 232–239. https://doi.org/10.1016/J.ALGAL.2015.05.011

Chan, Y. Y., & Killick, E. G. (1995). The effect of salinity, light and temperature in a disposal environment on the recovery of E.coli following exposure to ultraviolet radiation. *Water Research, 29*(5), 1373–1377. https://doi.org/10.1016/0043-1354(94)00226-W

Chandra, R., Iqbal, H. M. N., Vishal, G., Lee, H. S., & Nagra, S. (2019). Algal biorefinery: A sustainable approach to valorize algal-based biomass towards multiple product recovery. *Bioresource Technology, 278*, 346–359. https://doi.org/10.1016/J.BIORTECH.2019.01.104

Chiu, S. Y., Kao, C. Y., Chen, T. Y., Chang, Y. B., Kuo, C. M., & Lin, C. S. (2015). Cultivation of microalgal Chlorella for biomass and lipid production using wastewater as nutrient resource. *Bioresource Technology, 184*, 179–189. https://doi.org/10.1016/J.BIORTECH.2014.11.080

Chugh, M., Kumar, L., Shah, M. P., & Bharadvaja, N. (2022). Algal Bioremediation of heavy metals: An insight into removal mechanisms, recovery of by-products, challenges, and future opportunities. *Energy Nexus, 7*, Article 100129. https://doi.org/10.1016/J.NEXUS.2022.100129

Dai, C., & Wang, F. (2024). Potential applications of microalgae–bacteria consortia in wastewater treatment and biorefinery. *Bioresource Technology, 393*, Article 130019. https://doi.org/10.1016/J.BIORTECH.2023.130019

Das, A., Dey, S., Bala, K., Gnanasekaran, R., & Das, A. P. (2024). Microbial bioremediation technology for sustainable treatment and management of synthetic microfiber waste. In *Sustainable microbial technology for synthetic and cellulosic microfiber bioremediation* (pp. 221–238). https://doi.org/10.1007/978-3-031-62600-5_12

Das, A., Dey, S., Gacem, A., Yadav, K. K., Nath, H., Ram, D. K., & Das, A. P. (2025). Revolutionizing solid hydrocarbon sourcing: Plastic's journey from waste to upcycled treasure. *Environmental Science and Engineering, Part, F218*, 331–351. https://doi.org/10.1007/978-3-031-81615-4_14

Das, A., & Ghosh, S. (2022). Role of microorganisms in extenuation of mining and industrial wastes. *Geomicrobiology Journal, 39*, 173–175. https://doi.org/10.1080/01490451.2022.2038953

Das, A. P., & Ghosh, S. (2023a). Microorganisms in manganese biomining processes. In *Manganese mining microorganisms* (pp. 55–61). https://doi.org/10.1016/B978-0-12-822147-1.00013-3

Das, A. P., & Ghosh, S. (2023b). *Manganese mining microorganisms.*

Das, A. P., Ghosh, S., Mohanty, S., & Sukla, L. B. (2015). Advances in manganese pollution and its bioremediation. In L. B. Sukla, N. Pradhan, S. Panda, & B. K. Mishra (Eds.), *Environmental microbe biotechnology* (1st ed., pp. 313–328). Springer. https://doi.org/10.1007/978-3-319-19018-1_16

Das, A. P., Sukla, L. B., Pradhan, N., & Nayak, S. (2011). Manganese biomining: A review. *Bioresource Technology, 102*(16), 7381–7387. https://doi.org/10.1016/J.BIORTECH.2011.05.018

Dey, S., Das, A., Mallick, K., Sahu, A., & Das, A. P. (2023). In *Environmental petroleum waste: Pollution, toxicity, sustainable remediation* (pp. 159–175). https://doi.org/10.1007/978-3-031-48220-5_7

Du, R., Li, C., Liu, Q., Fan, J., & Peng, Y. (2022). A review of enhanced municipal wastewater treatment through energy savings and carbon recovery to reduce discharge and CO_2 footprint. *Bioresource Technology, 364*, Article 128135. https://doi.org/10.1016/J.BIORTECH.2022.128135

Dubey, S., Chen, C. W., Haldar, D., Tambat, V. S., Kumar, P., Tiwari, A., Singhania, R. R., Dong, C. D., & Patel, A. K. (2023). Advancement in algal bioremediation for organic, inorganic, and emerging pollutants. *Environmental Pollution, 317*, Article 120840. https://doi.org/10.1016/J.ENVPOL.2022.120840

Fabris, M., Abbriano, R. M., Pernice, M., Sutherland, D. L., Commault, A. S., Hall, C. C., Labeeuw, L., McCauley, J. I., Kuzhiuparambil, U., Ray, P., Kahlke, T., & Ralph, P. J. (2020). Emerging technologies in algal biotechnology: Toward the establishment of a sustainable, algae-based bioeconomy. *Frontiers in Plant Science, 11*. https://doi.org/10.3389/fpls.2020.00279

Ghosh, S., & Das, A. P. (2018). Metagenomic insights into the microbial diversity in manganese-contaminated mine tailings and their role in biogeochemical cycling of manganese. *Scientific Reports, 8*(1), 1–12. https://doi.org/10.1038/s41598-018-26311-w

Gupta, R., Mishra, N., Singh, G., Mishra, S., & Lodhiyal, N. (2024). Microalgae cultivation and value-based products from wastewater: Insights and applications. *Blue Biotechnology, 1*(1), 20. https://doi.org/10.1186/s44315-024-00019-1

Hammond, C. R., Hernández, M. S. G., & Loge, F. J. (2025). Microalgal-bacterial aggregates for wastewater treatment: Origins, challenges, and future directions. *Water Environment Research, 97*(2). https://doi.org/10.1002/wer.70018

Ighalo, J. O., Dulta, K., Kurniawan, S. B., Omoarukhe, F. O., Ewuzie, U., Eshiemogie, S. O., Ojo, A. U., & Abdullah, S. R. S. (2022). Progress in microalgae application for CO_2 sequestration. *Cleaner Chemical Engineering, 3*, Article 100044. https://doi.org/10.1016/J.CLCE.2022.100044

Jayalekshmi, S. J., Biju, M., Somarajan, J., Ajas, P. M. & Sunny, D. (n.d.). *Wastewater treatment technologies: A review*. www.ijert.org

Jin, L., Sun, X., Ren, H., & Huang, H. (2023). Biological filtration for wastewater treatment in the 21st century: A data-driven analysis of hotspots, challenges and prospects. *Science of the Total Environment, 855*, Article 158951. https://doi.org/10.1016/J.SCITOTENV.2022.158951

Kanungo, J., Sahoo, T., Swain, L. P., & Behera, I. D. (2023). *Toxicity of persistent hydrocarbon pollutants, sources and sustainable remediation process* (pp. 39–65). https://doi.org/10.1007/978-3-031-48220-5_2

Kato, S., & Kansha, Y. (2024). Comprehensive review of industrial wastewater treatment techniques. *Environmental Science and Pollution Research, 31*(39), 51064–51097). https://doi.org/10.1007/s11356-024-34584-0

Kesari, K. K., Soni, R., Jamal, Q. M. S., Tripathi, P., Lal, J. A., Jha, N. K., Siddiqui, M. H., Kumar, P., Tripathi, V., & Ruokolainen, J. (2021). Wastewater treatment and reuse: a review of its applications and health implications. *Water, Air, and Soil Pollution, 232*(5). https://doi.org/10.1007/s11270-021-05154-8

Khatun, R., Singh, S., Dubey, N. K., & Prasad Das, A. (2023). A review on marine-based phytochemicals and their application in biomedical research. In *Recent frontiers of phytochemicals: Applications in food, pharmacy, cosmetics, and biotechnology* (pp. 383–395). Elsevier. https://doi.org/10.1016/B978-0-443-19143-5.00013-X

Kholssi, R., Lougraimzi, H., & Moreno-Garrido, I. (2023). Effects of global environmental change on microalgal photosynthesis, growth and their distribution. *Marine Environmental Research, 184*, Article 105877. https://doi.org/10.1016/J.MARENVRES.2023.105877

Khoo, K. S., Lee, S. Y., Ooi, C. W., Fu, X., Miao, X., Ling, T. C., & Show, P. L. (2019). Recent advances in biorefinery of astaxanthin from Haematococcus pluvialis. *Bioresource Technology, 288*, Article 121606. https://doi.org/10.1016/J.BIORTECH.2019.121606

Kumar, M., Pandit, S., Patel, V., Khanna, N., Nag, M., Lahiri, D., Ray, R. R., Das, A., & Das, D. (2023). Maximization of energy recovery from starch processing wastewater by thermophilic dark fermentation coupled with microbial fuel cell technology. *Geomicrobiology Journal*. https://doi.org/10.1080/01490451.2023.2209555

Lazarova, V., Savoye, P., Janex, M. L., Blatchley, E. R., & Pommepuy, M. (1999). Advanced wastewater disinfection technologies: State of the art and perspectives. *Water Science and Technology, 40*(4–5), 203–213. https://doi.org/10.1016/S0273-1223(99)00502-8

Li, X., Nan, H., Jiang, H., Wang, H., & Wang, C. (2023). Research trends on phosphorus removal from wastewater: A review and bibliometric analysis from 2000 to 2022. *Journal of Water Process Engineering, 55*, Article 104201. https://doi.org/10.1016/J.JWPE.2023.104201

Liu, Y., Gu, J., & Zhang, M. (2020). Energy use and challenges in current wastewater treatment plants. In *A-B processes: Towards energy self-sufficient municipal wastewater treatment* (pp. 1–28). IWA Publishing. https://doi.org/10.2166/9781789060089_0001

Mata, T. M., Mendes, A. M., Caetano, N. S., & Martins, A. A. (2014). Sustainability and economic evaluation of microalgae grown in brewery wastewater. *Bioresource Technology, 168*, 151–158. https://doi.org/10.1016/J.BIORTECH.2014.04.091

Mishra, P., Mandake, M., Shingare, S. P., & Scholar, R. (2018). Tertiary waste water treatment: A review. *International Journal of Creative Research Thoughts, 6*(1). www.ijpub.org

Mishra, S., & Das, A. P. (2021). Current treatment technologies for removal of microplastic and microfiber pollutants from wastewater. In *Wastewater treatment: Cutting-edge molecular tools, techniques and applied aspects* (pp. 237–251). https://doi.org/10.1016/B978-0-12-821881-5.00011-8

Mishra, S., & Das, A. P. (2022). Treatment of the wastewater polluted with synthetic microfiber released from washing machine. In B. B. Das, H. Hettiarachchi, P. K. Sahu, & S. Nanda (Eds.), *Lecture notes in civil engineering* (Vol. 207, pp. 109–117). Springer Science and Business Media Deutschland GmbH. https://doi.org/10.1007/978-981-16-7509-6_9/COVER

Mishra, S., & Das, A. P. (2023). Microbial remediation of synthetic microfiber contaminated wastewater. In M. P. Shah (Ed.), *Microbial technologies in industrial wastewater treatment* (pp. 337–350). Springer. https://doi.org/10.1007/978-981-99-2435-6_17

Mishra, S., Dash, D., Al-Tawaha, A. R. M. S., & Das, A. P. (2022a). A review on heavy metal ion adsorption on synthetic microfiber surface in aquatic environments. *Applied Biochemistry and Biotechnology, 194*(10), 4639–4654. https://doi.org/10.1007/S12010-022-04029-W/METRICS

Mishra, S., Dash, D., & Das, A. P. (2024). Isolation and molecular characterization of polyester-degrading bacterial strains from municipal wastewater and their application in microplastic degradation in contaminated wastewater. *Geomicrobiology Journal*. https://doi.org/10.1080/01490451.2024.2376101

Mishra, S., Rath, Charan, C., & Das, A. P. (2019). Marine microfiber pollution: A review on present status and future challenges. *Marine Pollution Bulletin, 140*, 188–197. https://doi.org/10.1016/J.MARPOLBUL.2019.01.039

Mishra, S., Singh, R. P., Rout, P. K., & Das, A. P. (2022b). Membrane bioreactor (MBR) as an advanced wastewater treatment technology for removal of synthetic microplastics. *Development in Wastewater Treatment Research and Processes*. https://doi.org/10.1016/B978-0-323-85583-9.00022-3

Mishra, S., Singh, V., Cheng, L., Hussain, A., & Ormeci, B. (2022c). Nitrogen removal from wastewater: A comprehensive review of biological nitrogen removal processes, critical operation parameters and bioreactor design. *Journal of Environmental Chemical Engineering, 10*(3), Article 107387. https://doi.org/10.1016/J.JECE.2022.107387

Mishra, S., Swain, S., Sahoo, M., Mishra, S., & Das, A. P. (2021). Microbial colonization and degradation of microplastics in aquatic ecosystem: A review. *Geomicrobiology Journal*. https://doi.org/10.1080/01490451.2021.1983670

Mohanta, S., Pradhan, B., & Behera, I. D. (2024). Impact and remediation of petroleum hydrocarbon pollutants on agricultural land: A review. *Geomicrobiology Journal, 41*(4), 345–359. https://doi.org/10.1080/01490451.2023.2243925

Mohsenpour, S. F., Hennige, S., Willoughby, N., Adeloye, A., & Gutierrez, T. (2021). Integrating micro-algae into wastewater treatment: A review. *Science of the Total Environment, 752*, Article 142168. https://doi.org/10.1016/J.SCITOTENV.2020.142168

Molina Grima, E., Belarbi, E. H., Acién Fernández, F. G., Robles Medina, A., & Chisti, Y. (2003). Recovery of microalgal biomass and metabolites: Process options and economics. *Biotechnology Advances, 20*(7–8), 491–515. https://doi.org/10.1016/S0734-9750(02)00050-2

Nag, M., Lahiri, D., Ghosh, S., Sarkar, T., Pati, S., Das, A. P., Ram, D. K., Bhattacharya, D., & Ray, R. R. (2023). Application of microorganisms in biotransformation and bioremediation of environmental contaminant: A review. *Geomicrobiology Journal*. https://doi.org/10.1080/01490451.2023.2261443

Numberger, D., Ganzert, L., Zoccarato, L., Mühldorfer, K., Sauer, S., Grossart, H. P., & Greenwood, A. D. (2019). Characterization of bacterial communities in wastewater with enhanced taxonomic resolution by full-length 16S rRNA sequencing. *Scientific Reports, 9*(1). https://doi.org/10.1038/s41598-019-46015-z

Paddock, M. (2019). *Microalgae wastewater treatment: A brief history*. https://doi.org/10.20944/preprints201912.0377.v1

Paul, A., Dey, S., Ram, D. K., & Das, A. P. (2023). Hexavalent chromium pollution and its sustainable management through bioremediation. *Geomicrobiology Journal*. https://doi.org/10.1080/01490451.2023.2218377

Pradhan, G., Dey, S., Priyadarsini, S., Kumar, M. S., & Das, A. P. (2024). Characterization and quantification of microplastics pollutants in sediment samples from Daya river of Odisha state in India for their appropriate management. *Environmental Science and Engineering, Part, F2333*, 75–92. https://doi.org/10.1007/978-3-031-51792-1_5

Pradhan, G., Tripathy, B., Ram, D. K., Digal, A. K., & Das, A. P. (2023). Bauxite mining waste pollution and its sustainable management through bioremediation. *Geomicrobiology Journal*. https://doi.org/10.1080/01490451.2023.2235353

Prandini, J. M., da Silva, M. L. B., Mezzari, M. P., Pirolli, M., Michelon, W., & Soares, H. M. (2016). Enhancement of nutrient removal from swine wastewater digestate coupled to biogas purification by microalgae Scenedesmus spp. *Bioresource Technology, 202*, 67–75. https://doi.org/10.1016/J.BIORTECH.2015.11.082

Priya, E., Kumar, S., Verma, C., Sarkar, S., & Maji, P. K. (2022). A comprehensive review on technological advances of adsorption for removing nitrate and phosphate from waste water. *Journal of Water Process Engineering, 49*, Article 103159. https://doi.org/10.1016/J.JWPE.2022.103159

Priyadarshini, S., Jagatee, S., & Das, A. P. (2024). Synthetic fabrics and microfiber pollution–An assessment of their global impact. *Environmental Science and Engineering, Part, F2333*, 137–157. https://doi.org/10.1007/978-3-031-51792-1_8

Priyadarsini, S., & Das, A. P. (2024). Analytical and structural characterization of waste lithium-ion batteries for their effective recycling strategy. *Environmental Chemistry and Ecotoxicology*. https://doi.org/10.1016/j.enceco.2024.12.004

Priyadarsini, S., & Das, A. P. (2025). Lithium bioleaching: A review on microbial-assisted sustainable technology for lithium bio-circularity. *Journal of Water Process Engineering, 69*, Article 106744. https://doi.org/10.1016/j.jwpe.2024.106744

Rabiu, K. O., Khudaida, K., & Das, D. B. (2024). Geological carbon storage technologies. In *Encyclopedia of sustainable technologies* (pp. 378–389). https://doi.org/10.1016/B978-0-323-90386-8.00065-6

Raj, K., & Das, A. P. (2023). Lead pollution: Impact on environment and human health and approach for a sustainable solution. *Environmental Chemistry and Ecotoxicology, 5*, 79–85. https://doi.org/10.1016/J.ENCECO.2023.02.001

Rathinavel, L., Singh, S., Rai, P. K., Chandra, N., Jothinathan, D., Gaffar, I., Pandey, A. K., Choure, K., Waoo, A. A., Joo, J. C., & Pandey, A. (2024). Sustainable microalgal biomass for efficient and scalable green energy solutions: Fueling tomorrow. *Fuels, 5*(4), 868–894. https://doi.org/10.3390/fuels5040049

Rawat, I., Ranjith Kumar, R., Mutanda, T., & Bux, F. (2011). Dual role of microalgae: Phycoremediation of domestic wastewater and biomass production for sustainable biofuels production. *Applied Energy, 88*(10), 3411–3424. https://doi.org/10.1016/J.APENERGY.2010.11.025

Reijnders, L. (2020). Life cycle assessment of microalgae-based processes and products. In *Handbook of microalgae-based processes and products: Fundamentals and advances in energy, food, feed, fertilizer, and bioactive compounds* (pp. 823–840). https://doi.org/10.1016/B978-0-12-818536-0.00030-0

Romanis, C. S., Pearson, L. A., & Neilan, B. A. (2021). Cyanobacterial blooms in wastewater treatment facilities: Significance and emerging monitoring strategies. *Journal of Microbiological Methods, 180*, Article 106123. https://doi.org/10.1016/J.MIMET.2020.106123

Sadasivuni, K. K., Rattan, S., Waseem, S., Brahme, S. K., Kondawar, S. B., Ghosh, S., Das, A. P., Chakraborty, P. K., Adhikari, J., Saha, P., & Mazumdar, P. (2019). Silver nanoparticles and its polymer nanocomposites—Synthesis, optimization, biomedical usage, and its various applications. *Lecture Notes in Bioengineering*. https://doi.org/10.1007/978-3-030-04741-2_11

Sanglyne, M. W. (2019). *Study of organic pollution in three selected water bodies of Meghalaya using Palmer's Algal Pollution Index*. https://www.researchgate.net/publication/348734283

Sarup, R., Behl, K., Joshi, M., & Nigam, S. (2021). Heavy metal removal by cyanobacteria. *New Trends in Removal of Heavy Metals from Industrial Wastewater*. https://doi.org/10.1016/B978-0-12-822965-1.00018-0

Singh, D., Singh, D., Mishra, V., Kushwaha, J., Sengar, M., Sinha, S., Singh, S., & Giri, B. S. (2024). Strategies for biological treatment of waste water: A critical review. *Journal of Cleaner Production, 454*, Article 142266. https://doi.org/10.1016/J.JCLEPRO.2024.142266

Song, X., Liu, B. F., Kong, F., Song, Q., Ren, N. Q., & Ren, H. Y. (2024). New insights into rare earth element-induced microalgae lipid accumulation: Implication for biodiesel production and adsorption mechanism. *Water Research, 251*, Article 121134. https://doi.org/10.1016/J.WATRES.2024.121134

Sutherland, D. L., Turnbull, M. H., Broady, P. A., & Craggs, R. J. (2014). Effects of two different nutrient loads on microalgal production, nutrient removal and photosynthetic efficiency in pilot-scale wastewater high rate algal ponds. *Water Research, 66*, 53–62. https://doi.org/10.1016/J.WATRES.2014.08.010

Tang, M., Du, R., Cao, S., Berry, M., & Peng, Y. (2024). Tracing and utilizing nitrogen loss in wastewater treatment: The trade-off between performance improvement, energy saving, and carbon footprint reduction. *Journal of Environmental Management, 349*, Article 119525. https://doi.org/10.1016/J.JENVMAN.2023.119525

Tom, A. P., Jayakumar, J. S., Biju, M., Somarajan, J., & Ibrahim, M. A. (2021). Aquaculture wastewater treatment technologies and their sustainability: A review. *Energy Nexus, 4*, Article 100022. https://doi.org/10.1016/J.NEXUS.2021.100022

Tripathy, B., Sahoo, P. P., Sundaray, H., & Das, A. P. (2024). In-vitro biodegradation of discarded marine microplastics across the eastern coast of the Bay of Bengal, India using Exiguobacterium sp. *Environmental Chemistry and Ecotoxicology, 6*, 236–247. https://doi.org/10.1016/J.ENCECO.2024.07.003

Ugrina, M., & Milojković, J. (2024). Advances in wastewater treatment, 2024. *Energies, 17*(6). https://doi.org/10.3390/en17061400

Vandamme, D., Foubert, I., & Muylaert, K. (2013). Flocculation as a low-cost method for harvesting microalgae for bulk biomass production. *Trends in Biotechnology, 31*(4), 233–239. https://doi.org/10.1016/J.TIBTECH.2012.12.005

Wang, C., Luo, D., Zhang, X., Huang, R., Cao, Y., Liu, G., Zhang, Y., & Wang, H. (2022). Biochar-based slow-release of fertilizers for sustainable agriculture: A mini review. In *Environmental science and ecotechnology* (Vol. 10). Editorial Board, Research of Environmental Sciences. https://doi.org/10.1016/j.ese.2022.100167

Wirth, R., Pap, B., Böjti, T., Shetty, P., Lakatos, G., Bagi, Z., Kovács, K. L., & Maróti, G. (2020). Chlorella vulgaris and its phycosphere in wastewater: Microalgae-bacteria interactions during nutrient removal. *Frontiers in Bioengineering and Biotechnology*. https://doi.org/10.3389/fbioe.2020.557572

Yang, J., Liu, X., Ying, L., Chen, X., & Li, M. (2020). Correlation analysis of environmental treatment, sewage treatment and water supply efficiency in China. *Science of the Total Environment, 708*, Article 135128. https://doi.org/10.1016/J.SCITOTENV.2019.135128

Yang, Y., Tang, S., & Chen, J. P. (2024). Carbon capture and utilization by algae with high concentration CO_2 or bicarbonate as carbon source. *Science of the Total Environment, 918*, Article 170325. https://doi.org/10.1016/J.SCITOTENV.2024.170325

Zagklis, D. P., & Bampos, G. (2022). Tertiary wastewater treatment technologies: A review of technical, economic, and life cycle aspects. *Processes, 10*(11). https://doi.org/10.3390/pr10112304

Zhang, S., Zhang, L., Xu, G., Li, F., & Li, X. (2022). A review on biodiesel production from microalgae: Influencing parameters and recent advanced technologies. In *Frontiers in microbiology* (Vol. 13). Frontiers Media S.A. https://doi.org/10.3389/fmicb.2022.970028

or non-biodegradable materials. Despite having these limitations, the above conventional methods are still widely employed. But it is not adequate for managing the vast and complex types of waste currently generated with changing consumption patterns.

Given these concerns, it is a growing need for developing sustainable waste management approaches that can complement or even replace existing practices (Das & Ghosh, 2022; Das et al., 2024b; Mishra et al., 2019). These novel solutions must be environmentally friendly, cost-effective, and capable of processing a wide range of waste materials (Mishra & Das, 2023; Mohanty et al., 2018). Microorganisms, particularly extremophiles, possess a huge potential to contribute to develop advanced waste treatment technologies that can handle current and future waste management challenges (Zhu et al., 2020). Extremophiles are microorganisms that thrive in extreme environmental conditions. They have remarkable abilities to degrade organic pollutants, break down hazardous chemicals, and even transform waste into valuable resources such as biofuels, bioplastics, and essential minerals (Das & Ghosh, 2022). The use of extremophiles in waste management represents a shift toward more sustainable, nature-based solutions that align with the principles of a circular economy. Here waste is considered not as a problem but as a resource. In this chapter, it explores the huge potential of extremophiles in sustainable waste management. It also highlights their advantages over conventional methods and discusses the emerging research and applications that could revolutionize future waste management approaches.

2 Extremophiles and Waste Management

Extremophiles

Extremophiles thrive under extreme environmental conditions that are generally considered inhospitable to other forms of life (Rampelotto, 2013). They have evolved specialized adaptations that enable them to survive and thrive under extreme temperatures, pH, salinity, pressure, and radiation (Table 1). Understanding the diversity of extremophiles and their physiological and molecular adaptations provides valuable insights into the resilience of life, and offers potential applications in biotechnology and waste management.

Different Types of Extremophiles for Waste Treatment Applications

Extremophiles, including thermophiles, psychrophiles, halophiles, acidophiles, alkaliphiles, and metalophiles can play crucial roles in waste treatment by thriving

Table 1 Different groups of extremophiles

Group of extremophile	Description	Optimal conditions	Examples	Natural habitats
Thermophiles	Thrive in high temperatures	Typically 45 °C to 80 °C or higher	*Thermus aquaticus*, *Pyrococcus furiosus*	Hot springs, hydrothermal vents, deep-sea volcanic environments
Psychrophiles	Thrive in very cold environments	Typically below 15 °C, some as low as -20 °C	*Pseudomonas syringae*, *Psychrobacter* species	Polar ice caps, deep ocean, high-altitude regions
Halophiles	Thrive in high-salt concentrations	Typically 3 M NaCl or more	*Halobacterium salinarum*, *Salinibacter ruber*	Salt lakes, hypersaline ponds, salt mines
Acidophiles	Thrive in acidic environments	pH < 3	*Ferroplasma acidarmanus*, *Acidithiobacillus ferrooxidans*	Acidic hot springs, acid mine drainage, sulfur springs
Alkaliphiles	Thrive in alkaline environments	pH > 9	*Bacillus alcalophilus*, *Alkalimonas* species	Alkaline lakes, soda lakes, soda soils
Piezophiles	Thrive under high pressure	Typically high-pressure environments	*Halomonas* species, *Shewanella benthica*	Deep-sea trenches, deep ocean floors
Radiophiles	Thrive in environments with high levels of radiation	High radiation exposure	*Deinococcus radiodurans*, *Clostridium species*	Nuclear waste sites, high-radiation environments
Xerophiles	Thrive in extremely dry environments	Low water availability	*Bacillus subtilis*, *Sporobolomyces* species	Deserts, dry caves, and other low-water environments
Metalophiles	Thrive in environments with high metal concentrations	High concentrations of metals like iron, copper, or mercury	*Cupriavidus metallidurans*, *Ralstonia metallidurans*	Metal-rich soils, mine tailings, industrial waste sites

in conditions that inhibit conventional microbial activity. Thermophiles accelerate organic waste degradation in high-temperature composting and anaerobic digestion with enhanced biogas production. Psychrophiles enable efficient biodegradation in cold environments, making them valuable for wastewater treatment in alpine and polar regions (Kanekar & Kanekar, 2022). Halophiles contribute to the remediation of saline wastewater from industries such as textiles and tanning (Castillo-Carvajal et al., 2014), whereas acidophiles contributes to the treatment of waste generated by acid mine drainage and bioleaching for metal recovery (Fashola et al., 2015; Zhao & Wang, 2019). Alkaliphiles facilitate the breakdown of pollutants in alkaline industrial effluents, such as those from the cement and paper industries (Mamo & Mattiasson, 2020). Metalophiles are crucial for the bioremediation of heavy metal-contaminated sites and e-waste recycling (Aishwarya et al., 2024). Other groups of extremophiles, such as piezophiles, radiophiles, and xerophiles are relatively understudied in terms of their abilities for waste treatment. However, recent studies indicate that these microorganisms possess unexploited potential for sustainable waste management applications (Chaturvedi, 2023; Sekiguchi et al., 2010).

Potential of Thermophiles in Waste Management

Thermophiles are organisms that thrive at high temperatures, typically between 45 and 122 °C. They have evolved heat-stable enzymes and proteins known as thermozymes, which maintain their structural integrity and catalytic activity at high temperatures (Mohammad et al., 2023; Wu et al., 2023). These adaptations include an increased number of ionic bonds and hydrophobic interactions, as well as more compact protein-folding patterns. It prevents denaturation of enzymes under thermal stress. Thermophiles also possess specialized chaperone proteins that assist in proper protein folding and refolding, ensuring that cellular functions remain uninterrupted by high temperatures. The very famous example for thermophiles is *Thermus aquaticus*, a bacterium that thrives at temperatures of around 70 °C. This bacterium was first isolated from hot springs in Yellowstone National Park, US (Williams, 2021). It has been instrumental in the development of polymerase chain reaction (PCR) technology because of the heat-stable Taq DNA polymerase (Amils, 2023). Thermophiles play a critical role in sustainable waste management. Their unique enzymatic properties and metabolic diversity make them ideal candidates for bioconversion of organic waste, biodegradation of pollutants, and bioenergy production.

Thermophiles in Organic Waste Decomposition

Organic waste such as agricultural residues, food waste, and municipal solid waste, can be efficiently converted into compost by thermophilic microorganisms. Thermophilic composting not only accelerates biodegradation to improve the nutrient availability but also minimizes contamination risks by suppressing the growth of pathogenic microorganisms (Zhang et al., 2024). Thermophilic bacteria such as *Bacillus*, *Geobacillus*, and *Thermus* species play a key role in the decomposition of cellulose, lignin, and hemicellulose, which are often recalcitrant under mesophilic conditions (Cotta et al., 2021; Finore et al., 2023). Furthermore, many thermophiles exhibit a high tolerance to other extreme conditions, including extreme pH and moisture content, making them robust candidates for large-scale organic waste decomposition.

Thermophiles in Anaerobic Digestion and Biogas Production

Thermophilic anaerobic digestion (TAD) is also an important approach to converting organic waste into bioenergy (Phan et al., 2023). Compared to mesophilic anaerobic digestion (MAD), TAD operates at temperatures between 50 and 60 °C, enhancing the hydrolysis and methanogenesis stages of the process (Kaur et al., 2024). It also shows higher degradation rates of organic matter, reduced retention times, and greater pathogen reduction compared to mesophilic processes. Thermophilic methanogens such as *Methanothermobacter* and *Methanosaeta* species are important in converting complex organic compounds into methane, increasing biogas yield (Sun et al., 2015). Another important advantage is that higher temperatures facilitate the breakdown of fats, oils, and grease-rich waste, of which the degradation is often inhibited under mesophilic conditions. The application of thermophiles in biogas production contributes to sustainable waste-to-energy conversion.

Thermophiles in Industrial and Hazardous Waste Degradation

Thermophiles are also capable of degrading industrial and hazardous pollutants such as hydrocarbons, persistent organic pollutants, and toxic heavy metals in contaminated environments (Bhattacharya & Gupta, 2022). Certain thermophilic archaea and bacteria produce thermostable enzymes that catalyze the degradation of xenobiotic compounds, including polychlorinated biphenyls (PCBs) and petroleum hydrocarbons (Nzila, 2018). Furthermore, thermophilic microorganisms have shown promising results in removing phenolic compounds and dyes from textile and chemical industry effluents using thermophilic peroxidases and laccases. The robustness of thermophiles in extreme environments makes them valuable for applications in hazardous waste bioremediation.

Thermophilic Enzymes for Waste Valorization

One of the most promising aspects of thermophile-based waste management is the utilization of their thermostable enzymes for the valorization of waste into valuable bioproducts (Tanruean et al., 2021). Thermophilic proteases, cellulases, amylases, and lipases have been extensively used in bioconversion processes, facilitating the production of biofuels, biofertilizers, and biodegradable plastics. For example, thermophilic cellulases and hemicellulases enhance the degradation of lignocellulosic biomass, enabling the production of bioethanol and other biofuels from agricultural waste. Similarly, thermophilic lipases contribute to biodiesel production by catalyzing the transesterification of waste oils and fats. These enzymatic processes offer environmentally friendly alternatives to traditional chemical-based waste treatment, reducing energy consumption and minimizing harmful byproducts (Najar et al., 2024).

Potential of Psychrophiles in Waste Management

Psychrophiles are organisms that adapted to cold environments, with optimal growth occurring at temperatures of 15 °C or lower (Kanekar & Kanekar, 2022; Morita, 1975). These organisms have developed enzymes that remain flexible and functional at low temperatures, often by reducing the number of stabilizing interactions within their protein structures. Their cell membranes contain a higher proportion of unsaturated fatty acids that help them to maintain their fluidity under cold conditions. Psychrophiles also produce antifreeze proteins that prevent ice crystal formation within the cells.

Psychrophiles also demonstrate a huge potential in sustainable waste management. Their ability to function in extreme cold conditions makes them valuable for waste degradation in cold environments where conventional mesophilic and thermophilic microbes are ineffective. Psychrophiles offer a unique solution for managing waste in cold climates and refrigerated environments. The cold-adaptive enzymes of these microorganisms play a crucial role in these biodegradation processes, enabling resource recovery and pollution degradation in low-temperature ecosystems.

Psychrophiles in Organic Waste Decomposition

Cold environments pose significant challenges for microbial waste degradation due to reduced enzymatic activity and slowed metabolic processes. However, psychrophilic microorganisms have evolved unique biochemical adaptations, such as increased membrane fluidity and cold-active enzymes. Those modifications allow them to remain active in low-temperature waste treatment systems. These organisms, including *Psychrobacter*, *Polaromonas*, and *Flavobacterium* species, contribute to the decomposition of organic waste, such as food residues, agricultural byproducts, and municipal solid waste at low-temperature conditions (Dhole et al., 2023). Psychrophilic composting and landfill bioremediation have been proposed as viable strategies for sustainable waste management in polar regions and high-altitude areas where conventional biodegradation processes are hindered by low temperatures. The metabolic efficiency of psychrophiles in breaking down lignocellulosic biomass and other recalcitrant organic matter ensures the sustained activity of microbial communities in cold waste environments.

Psychrophiles in Cold-Adapted Anaerobic Digestion and Biogas Production

Anaerobic digestion (AD) is widely used approach to convert organic waste into biogas. However, its efficiency is often compromised in cold environments. Psychrophilic methanogens, such as *Methanococcoides*, *Methanosarcina*, and *Methanogenium* species, have adapted to produce methane at low temperatures, ensuring efficient operation of anaerobic digesters in cold environments without the need for artificial heating (Akindolire et al., 2022). This psychrophilic AD approach reduces energy requirements and makes biogas production feasible in cold environments. Cold-adapted hydrolytic and fermentative bacteria play a crucial role in breaking down complex organic matter into volatile fatty acids (VFAs), which are subsequently converted into methane by psychrophilic methanogens. It is demonstrated that psychrophilic anaerobic digestion processes can achieve comparable methane yields to mesophilic digestion, given sufficient adaptation time and microbial community optimization (Liu et al., 2023). The implementation of psychrophilic AD systems in waste treatment facilities, particularly in northern regions and high-altitude environments, offers a promising avenue for bioenergy production with minimal environmental impact.

Psychrophiles for Bioremediation of Contaminated Sites

Cold environments, such as Arctic and Antarctic regions, pose a unique challenge for the bioremediation of pollutants, as conventional microbial degradation is significantly reduced. Psychrophilic bacteria and fungi have demonstrated exceptional capabilities in breaking down hydrocarbons, persistent organic pollutants (POPs), and heavy metals in contaminated sites. Certain psychrophiles, including *Pseudomonas*, *Rhodococcus*, and *Sphingomonas* species, possess cold-active enzymes that facilitate the degradation of petroleum hydrocarbons in polar regions. In addition to hydrocarbon degradation, psychrophiles also contribute to the breakdown of synthetic dyes, pharmaceuticals, and pesticide residues in cold wastewater treatment systems. Cold-adapted peroxidases, laccases, and oxygenases play a critical role in detoxifying industrial effluents, ensuring efficient pollutant removal at low temperatures. The use of psychrophilic bioremediation strategies offers a cost-effective and environmentally friendly alternative to traditional chemical treatments in low-temperature environments.

Psychrophiles for Waste Valorization

Psychrophilic microorganisms produce a wide range of cold-active enzymes, including proteases, cellulases, lipases, and amylases, which have significant applications in sustainable waste management. These enzymes remain active at low temperatures, reducing the energy requirement for enzymatic waste degradation processes. For example, cold-adapted cellulases facilitate the degradation of plant biomass into fermentable sugars, enabling biofuel production from lignocellulosic waste. Furthermore, psychrophilic lipases have been utilized in biodiesel production from waste oils and fats, as their high catalytic efficiency at low temperatures allows for more efficient transesterification reactions. The bioconversion of organic waste into biodegradable plastics, biofertilizers, and other high-value bioproducts using psychrophilic enzymes presents a novel approach to waste valorization, reducing reliance on fossil-derived materials and contributing to circular economy initiatives.

Potential of Halophiles in Waste Management

Halophiles thrive in high-salinity environments such as salt lakes and saline industrial effluents. Halophiles accumulate compatible solutes such as potassium ions and specific organic compounds to balance the external salt concentration. Their cellular machinery, including enzymes and structural proteins, is adapted to function optimally under high-salt conditions. The cell walls and membranes of halophiles are uniquely structured to prevent salt-induced denaturation and retain water, ensuring cellular integrity in hypertonic environments. *Halobacterium salinarum* is a well-studied halophile that grows in saturated salt solutions. Halophiles have emerged as a promising solution in sustainable waste management. Metabolic diversity of halophiles allows them to degrade organic pollutants, detoxify heavy metals, and contribute to bioenergy production in saline waste streams where conventional microbes are unable to survive.

Halophiles for Biodegradation of Organic Pollutants

Saline environments pose unique challenges for microbial-mediated waste degradation due to osmotic stress, ionic toxicity, and reduced microbial diversity. However, halophilic microorganisms have evolved specialized adaptations, including the production of compatible solutes, salt-stable enzymes, and ion transport mechanisms, enabling them to efficiently break down organic pollutants. Halophilic bacteria such as *Halomonas*, *Marinobacter*, and *Salinivibrio* species play a crucial role in the degradation of hydrocarbons, phenols, and complex organic compounds found in industrial effluents.

Halophilic archaea, particularly *Haloferax* and *Halobacterium* species, exhibit remarkable metabolic flexibility, utilizing a wide range of carbon sources under extreme conditions. These organisms have been successfully employed in the bioremediation of petroleum-contaminated sites, where they contribute to breakdown crude oil and polycyclic aromatic hydrocarbons (PAHs). Their ability to function in hypersaline environments makes them particularly useful for treating wastewater from oil refineries and coastal pollution.

Additionally, halophilic fungi such as *Wallemia* and *Aspergillus* species produce extracellular enzymes capable of degrading recalcitrant organic pollutants, including synthetic dyes and agricultural pesticides. These fungi are particularly useful in saline textile wastewater treatment, where they facilitate the breakdown of azo dyes and other persistent organic pollutants.

Halophilic Microorganisms in Heavy Metal Bioremediation

Industrial processes such as mining, electroplating, and desalination generate saline wastewater contaminated with toxic heavy metals, posing significant environmental and health risks. Halophilic bacteria and archaea have developed unique mechanisms for metal detoxification, including biosorption, bioaccumulation, enzymatic reduction, and precipitation. Species such as *Halomonas elongata* and *Chromohalobacter salexigens* have been shown to effectively remove heavy metals like arsenic, lead, and cadmium from saline effluents by binding them to their cell walls or intracellular compartments. Some halophiles produce extracellular polysaccharides (EPS) that act as natural biosorbents, sequestering heavy metals from contaminated water. Moreover, halophilic sulfate-reducing bacteria (SRB), including *Desulfohalobium* and *Desulfosalsimonas* species, play a crucial role in the bioremediation of heavy metal-contaminated wastewater by converting toxic metal ions into insoluble sulfides, which can be easily removed from the environment. These biological processes provide an efficient and cost-effective alternative to conventional chemical treatments for heavy metal removal.

Halophiles for Converting Waste to Energy

Halophilic microorganisms have also been explored for their potential in bioenergy production, particularly in microbial fuel cells (MFCs) and anaerobic digestion. Their ability to function in hypersaline environments makes them ideal for generating bioelectricity and biogas from saline wastewater.

(a) **Microbial Fuel Cells (MFCs)**

Microbial fuel cells utilize electroactive bacteria to convert organic waste into electricity. Halophilic electroactive bacteria, including *Shewanella* and *Geobacter* species, have demonstrated significant potential in saline MFCs, where they transfer electrons to an electrode while oxidizing organic matter. These systems can be employed for the simultaneous treatment of saline wastewater and generation of renewable energy.

(b) **Anaerobic Digestion for Biogas Production**

Anaerobic digestion (AD) is a widely used process for converting organic waste into methane-rich biogas. However, high salinity can inhibit microbial activity, reducing the efficiency of the process. Halophilic methanogens, such as *Methanohalophilus* and *Methanosalsum* species, have been identified as key players in saline AD systems, where they produce methane even under extreme salt conditions. These organisms enable biogas production from high-salinity waste streams, including food industry effluents and hypersaline lagoons.

Potential of Acidophiles and Alkaliphiles in Waste Management

Acidophiles and alkaliphiles are adapted to thrive in environments with extreme pH levels. Acidophiles that thrive in highly acidic conditions (pH < 3) have evolved mechanisms to pump out excess protons, maintaining neutral intracellular pH. Their cell membranes are less permeable to protons, which reduces the influx of acidity into cells. On the other hand, alkaliphiles that thrive in highly alkaline environments (pH > 9) possess adaptations that allow them to efficiently import protons to counteract the high external pH. This ensures that their internal environment is conducive to cellular processes. Both acidophiles and alkaliphiles have enzymes that are optimized to function at their respective external pH extremes and show remarkable biochemical diversity. Therefore, both groups have gained significant attention in sustainable waste management due to their ability to function in extreme pH conditions. Their application in environmental biotechnology offers promising solutions for managing acidic and alkaline waste streams generated from mining, tannery, and chemical industries.

Role of Acidophiles in Waste Management

Acidophiles in Heavy Metal Bioremediation and Bioleaching

Acidophilic microorganisms play a crucial role in the bioremediation of heavy metal-contaminated environments, particularly in acid mine drainage (AMD) and industrial effluents. AMD results from the oxidation of sulfide minerals, producing acidic wastewater enriched with toxic metals such as iron, arsenic, and lead. Conventional treatment methods involve costly chemical neutralization, but acidophiles offer a biological alternative through bioleaching and metal precipitation.

Species such as *Acidithiobacillus ferrooxidans* and *Leptospirillum ferrooxidans* oxidize ferrous iron and sulfide minerals, facilitating the dissolution and recovery of valuable metals from waste materials (Das et al., 2011; Prabhakar et al., 2019; Priyadarsini & Das, 2025; Raj & Das, 2023). This microbial-driven bioleaching process is widely used in metal recovery from electronic waste and mine tailings (Das & Ghosh, 2018; Ghosh et al., 2023; Mohanty et al., 2016). Additionally, sulfate-reducing acidophiles, such as *Desulfosporosinus acidiphilus*, contribute to metal precipitation by reducing sulfate to sulfide, forming insoluble metal sulfides that can be easily removed from wastewater.

Acidophilic Microorganisms in Organic Waste Degradation

Acidophilic fungi and bacteria are effective in breaking down organic waste under low pH conditions. Composting processes involving acidic environments benefit from acid-tolerant fungi such as *Aspergillus* and *Penicillium* species, which degrade lignocellulosic materials into simpler compounds. Furthermore, acidophilic lactic acid bacteria (LAB), including *Lactobacillus acidophilus*, play a crucial role in acidic fermentation processes, helping with the stabilization of organic waste in landfill sites and anaerobic digesters.

Acidophiles for Industrial Effluent Treatment

Industries such as the pharmaceutical and food processing sectors generate acidic wastewater rich in complex organic pollutants. Acidophilic microbes, including *Acidiphilium* and *Ferroplasma* species, have been utilized for the degradation of phenolic compounds, dyes, and petroleum hydrocarbons under acidic conditions. These organisms secrete acid-stable enzymes that catalyze the breakdown of recalcitrant compounds, enhancing the efficiency of wastewater treatment systems.

Role of Alkaliphiles in Waste Management

Alkaliphiles in Textile and Paper Industry Wastewater Treatment

The textile and paper industries discharge large volumes of alkaline wastewater containing dyes, lignin, and other organic pollutants. Alkaliphilic bacteria, such as *Pseudomonas* and *Bacillus* species, produce alkaline enzymes that degrade synthetic dyes and lignocellulosic compounds. Furthermore, alkaliphilic fungi, such as *Sporotrichum thermophile*, contribute to the biodegradation of textile dyes, reducing environmental contamination.

Bioremediation of Alkaline Industrial Waste

Cement, detergent, and leather processing industries generate alkaline waste streams that inhibit conventional microbial activity. Alkaliphilic microorganisms have evolved specialized mechanisms to survive and function in these conditions. For example, *Bacillus alcalophilus* and *Halomonas campisalis* are known for their ability to degrade hydrocarbons and phenols under alkaline conditions, making them suitable for cleaning up oil spills and industrial effluents.

Alkaliphiles in Organic Waste Conversion and Bioenergy Production

Alkaliphiles also contribute to waste-to-energy applications, particularly in biogas production and microbial fuel cells (MFCs). Alkaliphilic methanogens, such as *Methanobacterium alcaliphilum*, efficiently convert organic waste into methane under high pH conditions, providing an alternative for biogas production from alkaline waste streams. Furthermore, alkaliphilic bacteria involved in microbial fuel cells facilitate electron transfer processes that generate bioelectricity while simultaneously degrading organic pollutants.

Potential of Metalophiles in Waste Management

Metalophiles are a specialized group of extremophiles that have evolved to survive and thrive in environments with high concentrations of heavy metals that are typically toxic to most life forms. These unique organisms have developed a range of physiological and molecular adaptations that enable them to withstand metal-induced stress, making them useful in bioremediation and environmental detoxification, promoting circular economy principles.

Metalophiles exhibit a variety of adaptations that allow them to survive and function in metal-rich environments. These adaptations include biosorption, bioaccumulation, enzymatic detoxification, and biomineralization.

- **Biosorption and Bioaccumulation**: Many metalophiles possess cell wall components that can adsorb metal ions from the environment, facilitating their removal from contaminated sites. For example, species of *Bacillus* and *Pseudomonas* have been found to effectively accumulate lead, cadmium, and arsenic.
- **Enzymatic Detoxification**: Some metalophiles produce metallothioneins and metal-chelating proteins that neutralize toxic metal ions, preventing cellular damage. For example, *Cupriavidus metallidurans* is known for its ability to transform toxic metals into less harmful forms through enzymatic processes.
- **Biomineralization and Metal Precipitation**: Certain extremophiles can convert soluble metal ions into insoluble mineral forms, reducing their mobility and bioavailability. For example, *Desulfovibrio* species facilitate sulfate reduction processes that lead to metal sulfide precipitation, aiding in the stabilization of heavy metal contaminants.

Bioremediation of Heavy Metal-Contaminated Sites

Metalophiles are particularly valuable for cleaning up environments contaminated with heavy metals from industrial activities such as mining, smelting, and electronic waste disposal. Through bioaccumulation and biomineralization, they can reduce metal toxicity and prevent further environmental degradation.

- **Mining Waste Treatment**: Acid mine drainage (AMD) is a major environmental issue caused by the oxidation of sulfide minerals, leading to highly acidic and

metal-laden effluents. Metalophilic bacteria such as *Acidithiobacillus ferrooxidans* and *Leptospirillum ferrooxidans* play a crucial role in AMD bioremediation by promoting metal precipitation and neutralizing acidity.
- **Industrial Effluent Treatment**: Heavy metal-contaminated wastewater can be effectively treated with metalophiles to remove toxic elements such as mercury, lead, and chromium. Microbial strains such as *Geobacter* and *Shewanella* have been shown potential to reduce hexavalent chromium, a highly toxic pollutant, into less toxic trivalent chromium.

E-Waste Processing and Metal Recovery

Electronic waste (e-waste) contains significant amounts of precious and rare metals, including gold, silver, and palladium. Metalophilic microbes are increasingly being explored for bioleaching and biohydrometallurgy applications to recover these valuable resources sustainably.

- **Bioleaching of Valuable Metals**: Certain metalophiles, such as *Chromobacterium violaceum*, secrete organic acids and other metabolites that solubilize metal ions from e-waste components, enabling efficient metal extraction without the use of harsh chemicals.
- **Biogenic Nanoparticle Synthesis**: Some metalophiles are capable of reducing metal ions into nanoparticles with unique physicochemical properties. For example, *Pseudomonas aeruginosa* has been reported to facilitate silver nanoparticle synthesis, which has potential applications in medical and industrial fields.

Stabilization of Toxic Waste in Landfills

Toxic metal-containing waste often ends up in landfills, posing long-term environmental and health risks. Metalophiles can contribute to the stabilization and sequestration of these hazardous elements, reducing their leaching potential.

- **Formation of Metal Sulfides and Oxides**: Sulfate-reducing bacteria such as *Desulfovibrio desulfuricans* can convert soluble heavy metals into stable sulfide precipitates, minimizing groundwater contamination risks.
- **Metal-Immobilizing Biofilms**: Some extremophilic bacteria form biofilms that can immobilize heavy metals in waste dumps, preventing their spread into surrounding ecosystems.

Piezophiles and Their Role in Waste Degradation

Piezophiles, also known as barophiles, thrive at high hydrostatic pressures, such as those found in deep-ocean trenches. They have been adapted by developing uniquely structured enzymes and cellular membranes that remain functional under intense pressure. These adaptations often involve changes in protein structure to prevent compression and maintain flexibility, as well as alterations in lipid composition to preserve membrane fluidity. These unique adaptations enable piezophiles to contribute to bioremediation processes in high-pressure ecosystems and subsurface locations.

- **Deep-Sea Waste Decomposition**: Industrial waste, particularly microplastics, and hydrocarbons, tends to settle in deep oceanic trenches where conventional microbial activity is limited. Certain piezophilic bacteria, such as *Shewanella violacea* and *Colwellia psychrerythraea*, exhibit enzymatic capabilities suited for the breakdown of these pollutants under extreme pressures.
- **Subsurface Bioremediation**: Contaminated deep aquifers and underground waste repositories pose significant environmental challenges. Piezophiles have the potential to enhance the degradation of hazardous substances in deep geological storage settings.

Despite their promising attributes, piezophilic applications in waste treatment remain in the developmental phase. Advances in microbial biotechnology and synthetic biology could enable the design of high-pressure microbial systems optimized for waste breakdown in deep-sea and subsurface ecosystems.

Radiophiles as Candidates for Nuclear Waste Management

Radiophiles are organisms that can withstand high levels of ionizing radiation. They possess efficient DNA repair mechanisms that can reconstruct damaged genetic material as well as protective cellular compounds that neutralize reactive oxygen species generated by radiation exposure. *Deinococcus radiodurans* is well known for its exceptional resistance to ionizing radiation, which is attributed to its efficient DNA repair systems that can reconstruct shattered genomes. These unique traits make them highly valuable in addressing contamination issues linked to radioactive waste disposal and nuclear site rehabilitation.

- **Bioremediation of Radioactive Waste:** Some radiation-resistant bacteria, including *D. radiodurans* and *Rhodococcus* spp., exhibit the ability to metabolize organic pollutants and heavy metals in radiation-exposed environments. Their deployment in nuclear waste facilities could help neutralize hazardous compounds that conventional microbes cannot tolerate.
- **Heavy Metal Sequestration**: Certain radiophilic species are capable of bioaccumulating and reducing metals such as uranium and cesium, making them potential tools for radioactive waste sequestration. Genetically modified strains of *Deinococcus* have been investigated for their ability to remediate nuclear waste effluents by immobilizing toxic radionuclide.

Xerophiles and Their Potential in Arid Waste Treatment Systems

Xerophiles are a specialized group of extremophiles that thrive in environments with minimal water availability. These organisms have evolved unique physiological, biochemical, and genetic adaptations to survive in arid and desiccated conditions such as deserts, high-altitude plateaus, and hypersaline environments. Xerophiles include a diverse range of microorganisms, fungi, and some plants that have evolved to survive extreme desiccation. These adaptations make them particularly relevant waste treatment in arid and semi-arid regions where standard microbial degradation is often inefficient.

- **Waste Biodegradation in Dry Climates**: In desert and semi-arid regions, industrial waste and petroleum-based contaminants accumulate with limited biological breakdown. Xerophilic bacteria and fungi, such as *Arthrobacter* spp. and *Aspergillus niger*, have demonstrated the ability to metabolize hydrocarbons and persistent organic pollutants under dry conditions.
- **Composting and Organic Waste Stabilization**: Conventional composting methods require adequate moisture levels for microbial activity, but xerophilic microorganisms offer potential solutions for organic waste decomposition in water-scarce environments. They could enhance stabilization processes in landfills and composting units specifically designed for dry climates.
- **Plastic and Synthetic Polymer Degradation**: Some xerophilic fungi, such as *Aspergillus versicolor* and *Penicillium chrysogenum*, have been found to facilitate the breakdown of synthetic polymers under desiccated conditions. Their ability to degrade plastic waste in arid environments suggests potential applications in sustainable waste treatment strategies where water availability is a limiting factor.

Despite their limited industrial integration, xerophiles provide a viable approach for organic and plastic waste degradation in regions experiencing chronic water scarcity.

3 Challenges in Implementing Extremophile-Based Solutions for Waste Treatment Applications

Despite the promising potential of extremophiles for waste treatment, several challenges hinder their widespread implementation in real-world applications. These challenges span across biological, technological, economic, and regulatory domains, making the integration of extremophile-based solutions into current waste management systems complex and multifaceted. Understanding these barriers is crucial for the future success of utilizing extremophiles in sustainable waste management.

Cultivation and Maintenance of Extremophiles

One of the primary challenges in applying extremophiles to waste treatment is the difficulty in cultivating and maintaining these organisms under controlled conditions. Many extremophiles thrive in extreme environments such as high-temperature hydrothermal vents or highly acidic lakes, conditions that are difficult to replicate in industrial-scale bioreactors. The maintenance of these microorganisms in laboratory or industrial settings often requires specialized equipment, and the costs of maintaining the optimal conditions for their growth can be prohibitive. Additionally, some extremophiles have slow growth rates, which can limit their scalability and effectiveness in large-scale waste treatment processes.

Limited Knowledge on Metabolic Pathways of Extremophiles

Another challenge lies in the incomplete understanding of the metabolic pathways and biochemical processes of many extremophiles. While these microorganisms have been shown to possess unique enzymatic capabilities for degrading a wide range of pollutants, such as heavy metals, hydrocarbons, and organic toxins, much remains unknown about how these processes operate at a molecular level. For example, understanding how extremophiles interact with complex waste matrices in real-world conditions is still under investigation. Without a comprehensive understanding of their metabolic mechanisms, it is difficult to predict how they will behave in diverse waste treatment environments, leading to potential inefficiencies in their application.

Environmental and Operational Stability

Extreme environments often provide a stable environment for extremophiles to thrive, but when transferred to more variable conditions, their performance may fluctuate. Industrial waste treatment settings, which often deal with fluctuating temperatures, pH levels, and the presence of diverse pollutants, can be challenging for these microorganisms to adapt to. Extremophiles may not always perform optimally outside their native habitats, and their ability to withstand sudden environmental changes or toxic conditions in waste streams remains a concern. The operational stability of these microorganisms needs to be carefully assessed, especially when scaling up to treat large volumes of waste.

Regulatory and Safety Concerns

The use of genetically modified or non-native microorganisms in industrial applications raises a series of regulatory and safety concerns. While extremophiles are naturally occurring organisms, the manipulation of their genetic material to enhance their waste-degrading capabilities may lead to unintended ecological consequences if these organisms are inadvertently released into the environment. Regulatory bodies in many regions have stringent guidelines for the use of microorganisms in industrial applications, particularly those that may pose a risk to human health or the environment. Ensuring that extremophile-based solutions are safe, both in terms of their effectiveness and their potential environmental impact, is a key challenge for their widespread adoption.

Cost-Effectiveness and Economic Viability

The economic feasibility of implementing extremophile-based waste treatment systems remains uncertain. The initial costs of setting up bioreactors capable of maintaining extreme conditions, along with the expenses involved in the large-scale cultivation and monitoring of extremophiles, can be significant. Additionally, the time and resources required to optimize the waste treatment processes and integrate them into existing infrastructure may deter industries from adopting these technologies. While extremophiles have the potential to lower operational costs in the long term by reducing the need for chemical treatments and energy-intensive processes, the upfront investment required for research, development, and infrastructure may pose a barrier to their widespread use.

Public Perception and Acceptance

The use of microorganisms, especially genetically modified organisms (GMOs), in waste treatment applications may face public resistance. Concerns over the safety of genetically engineered organisms and their potential impact on human health or ecosystems can hinder the acceptance of extremophile-based technologies. Public perception plays a critical role in the success of any new technology, and without effective communication about the benefits and safety of these microorganisms, their adoption may be slow. Ensuring transparency in research and providing clear evidence of the safety and efficacy of extremophile-based solutions will be key to overcoming this hurdle.

4 Future Directions of Extremophile-Based Waste Treatment Applications

The potential of extremophiles to revolutionize waste management systems has been increasingly recognized, yet the full integration of these microorganisms into mainstream waste treatment remains a work in progress. As scientific understanding advances, several future directions emerge for the application of extremophiles in waste treatment, ranging from improved bioreactor designs to novel biotechnological advancements. Harnessing the capabilities of extremophiles for large-scale, sustainable waste management systems will require overcoming current limitations and innovating new methodologies. Here, we explore several promising avenues for the future of extremophile-based waste treatment applications.

Synthetic Biology and Genetic Engineering

The field of synthetic biology holds tremendous potential for enhancing the capabilities of extremophiles in waste treatment. Advances in genetic engineering techniques, such as CRISPR-Cas9, enable precise manipulation of extremophiles' genomes, allowing researchers to enhance their metabolic pathways for more efficient degradation of complex waste materials. Furthermore, designing microorganisms that can function optimally across a range of environmental conditions, such as temperature fluctuations or varying pH levels, will significantly enhance their utility in diverse waste streams. The ongoing progress in synthetic biology will likely lead to the development of tailored extremophile strains capable of addressing specific contaminants in waste management systems.

Advanced Bioreactor Designs

To effectively implement extremophiles on a large scale, innovative bioreactor designs are essential. Traditional bioreactor systems may not provide the optimal conditions required for extremophiles, which often necessitate specialized temperature, pressure, or pH environments. Recent developments in bioreactor technology, such as membrane bioreactors, biofilm reactors, and modular systems, are providing new avenues for improving the performance of extremophiles in waste treatment applications. These systems offer better control over environmental conditions and can be scaled up to treat large volumes of waste. Future bioreactors may incorporate advanced monitoring systems that allow real-time adjustments to environmental parameters, thus optimizing the activity of extremophiles and improving the overall efficiency of waste treatment processes.

Integration with Waste-to-Energy Technologies

The integration of extremophiles with waste-to-energy technologies represents a promising direction for sustainable waste management. Many extremophiles are capable of converting organic waste into valuable biofuels, such as biogas, methane, or ethanol, through anaerobic digestion or other metabolic pathways. By combining extremophile-based waste treatment with existing anaerobic digestion or thermochemical conversion technologies, it may be possible to not only detoxify hazardous waste but also recover valuable energy from waste streams. This approach could significantly reduce dependence on fossil fuels while contributing to the circular economy. Future research will likely focus on optimizing the production of biofuels by extremophiles and developing hybrid systems that combine bioremediation with energy recovery.

Bioremediation of Emerging Contaminants

A growing concern in environmental science is the presence of emerging contaminants in wastewater, such as microplastics, pharmaceuticals, and personal care products. These pollutants are not effectively removed by conventional wastewater treatment methods, leading to their accumulation in ecosystems and posing long-term ecological risks (Mishra & Das, 2021, 2022; Mishra et al., 2022, 2024). Extremophiles, with their remarkable ability to degrade a wide range of organic and inorganic compounds, offer a potential solution to this challenge. Future research will likely focus on identifying extremophiles capable of breaking down or transforming these emerging contaminants. Additionally, bioaugmentation strategies where selected strains of extremophiles are introduced into treatment systems could provide a targeted approach for addressing specific pollutants, enhancing the overall efficacy of wastewater treatment processes.

Monitoring and Automation in Extremophile-Based Systems

The complexity of extremophile-based waste treatment systems calls for advanced monitoring and automation techniques to ensure their stability and performance. Traditional waste treatment plants often rely on human oversight to manage the various stages of the treatment process. However, for large-scale implementation of extremophile-based systems, automation of environmental control and system monitoring is essential. Recent advancements in sensor technologies, artificial intelligence, and machine learning have opened up new possibilities for real-time monitoring of microbial activity, pollutant degradation, and environmental parameters in extremophile-based reactors. By integrating these technologies into bioreactor systems, it will be possible to optimize extremophile activity, improve system reliability, and reduce operational costs. Furthermore, automated systems could enable predictive modeling, allowing for more efficient management of waste treatment processes and better resource utilization.

Climate Change Adaptation and Environmental Impact

As climate change alters environmental conditions, the role of extremophiles in waste management may become even more crucial. Extremophiles are naturally adapted to survive under extreme conditions, and some species may prove resilient to the changing environmental stressors associated with global warming. For example, increasing temperatures, shifts in water availability, and the occurrence of extreme weather events may affect the stability and efficiency of traditional waste treatment methods. Extremophiles, with their inherent tolerance to temperature fluctuations, salinity changes, and other stresses, could be better suited to adapt to these evolving challenges. Research into climate-resilient extremophile strains will likely be an important area of focus, enabling waste treatment systems to remain effective and sustainable even as environmental conditions continue to change.

Cross-Sector Collaboration and Global Implementation

The successful integration of extremophiles into waste treatment applications will require collaboration across multiple sectors, including biotechnology, environmental engineering, and policymaking. Governments, industries, and academic institutions must work together to address the challenges of scaling up extremophile-based technologies. International collaboration will be crucial in ensuring the safety, regulatory compliance, and effectiveness of these technologies across different regions and waste types. Additionally, public–private partnerships may play a key role in driving innovation, providing funding for research and development, and facilitating the transition to sustainable waste management solutions. By fostering a collaborative approach, the global implementation of extremophile-based waste treatment systems can be accelerated, paving the way for more sustainable and circular waste management practices worldwide.

References

Aishwarya, C. S., Prusty, T., Mashakhetri, K. D., & Bast, F. (2024). Role of extremophiles in the removal of heavy metal and e-waste. In *Trends in biotechnology of polyextremophiles* (pp. 361–375).

Akindolire, M. A., Rama, H., & Roopnarain, A. (2022). Psychrophilic anaerobic digestion: A critical evaluation of microorganisms and enzymes to drive the process. *Renewable and Sustainable Energy Reviews, 161*, Article 112394.

Amils, R. (2023). Taq polymerase. In *Encyclopedia of astrobiology* (pp. 3001–3002). Springer Berlin Heidelberg.

Bhattacharya, A., & Gupta, A. (2022). Current trends in applicability of thermophiles and thermozymes in bioremediation of environmental pollutants. In *Microbial extremozymes* (pp. 161–176). Academic Press.

Castillo-Carvajal, L. C., Sanz-Martín, J. L., & Barragán-Huerta, B. E. (2014). Biodegradation of organic pollutants in saline wastewater by halophilic microorganisms: A review. *Environmental Science and Pollution Research, 21*, 9578–9588.

Chaturvedi, R. (2023). Extremophiles: Biofactories for bioremediation. In *Integrative approaches to biotechnology* (pp. 171–185). CRC Press.

Choudhary, M., Singh, D., Parihar, M., Choudhary, K. B., Nogia, M., Samal, S. K., & Mishra, R. (2024). Impact of municipal solid waste on the environment, soil, and human health. In *Waste management for sustainable and restored agricultural soil* (pp. 33–58). Academic Press.

Cotta, S. P. M., Marins, M. S., Marriel, I. E., Lana, U. D. P., Gomes, E. A., Figueiredo, J. E. F., & Oliveira-Paiva, C. A. (2021). Thermo-resistant enzyme-producing microorganisms isolated from composting. *Brazilian Journal of Biology, 83*, Article e244205.

Das, A. P., Ghosh, S., Mohanty, S., & Sukla, L. B. (2015). Advances in manganese pollution and its bioremediation. In L. B. Sukla, N. Pradhan, S. Panda, & B. K. Mishra (Eds.), *Environmental microbe biotechnology* (1st ed., pp. 313–328). Springer. https://doi.org/10.1007/978-3-319-19018-1_16

Das, A. P., & Ghosh, S. (2023). *Manganese mining microorganisms.*

Das, A. P., & Ghosh, S. (2018). Bioleaching of manganese from mining waste materials. *Materials Today: Proceedings, 5*(1), 2381–2390. https://doi.org/10.1016/J.MATPR.2017.11.459

Das, A., & Ghosh, S. (2022). Role of microorganisms in extenuation of mining and industrial wastes. *Geomicrobiology Journal, 39*, 173–175. https://doi.org/10.1080/01490451.2022.2038953

Das, A. P., Nayak, B., & Behera, I. D. (2024a). Applications of microorganisms in advanced geomicrobiological and biotechnological exploration. *Geomicrobiology Journal, 41*(4), 309–311. https://doi.org/10.1080/01490451.2024.2327059/ASSET//CMS/ASSET/DE2A1F67-08C1-475A-ABC7-2B3F4C9E8BCF/01490451.2024.2327059.FP.PNG

Das, A., Ray, S. K., Mohanty, M., Mohanty, J., Dey, S., & Das, A. P. (2024b). Ecotoxicity of microplastic wastes and their sustainable management: A review. *Environmental Chemistry and Ecotoxicology, 6*, 144–152. https://doi.org/10.1016/J.ENCECO.2024.05.003

Das, A. P., Sukla, L. B., Pradhan, N., & Nayak, S. (2011). Manganese biomining: A review. *Bioresource Technology, 102*(16), 7381–7387. https://doi.org/10.1016/J.BIORTECH.2011.05.018

Dey, S., Das, A., Mallick, K., Sahu, A., & Das, A. P. (2023). In *Environmental petroleum waste: Pollution, toxicity, sustainable remediation* (pp. 159–175). https://doi.org/10.1007/978-3-031-48220-5_7

Dhole, N. P., Dar, M. A., Ullah, K., Xie, R., Sun, J., Shahnawaz, M., & Pandit, R. S. (2023). A review on the bioprospection of psychrophilic bacteria and their potential applications for sustainability. In *Soil microbiome of the cold habitats* (pp. 127–146).

Fashola, M. O., Ngole-Jeme, V. M., & Babalola, O. O. (2015). Diversity of acidophilic bacteria and archaea and their roles in bioremediation of acid mine drainage. *British Microbiol Res J, 8*(3), 443–456.

Finore, I., Feola, A., Russo, L., Cattaneo, A., Di Donato, P., Nicolaus, B., Poli, A., & Romano, I. (2023). Thermophilic bacteria and their thermozymes in composting processes: A review. *Chemical and Biological Technologies in Agriculture, 10*(1), 7.

Ghosh, S., Tripathy, B., Dey, S., & Das, A. P. (2023). Comparative investigation of fungal and bacterial manganese biomining mechanisms. In *Manganese mining microorganisms* (pp. 99–115). https://doi.org/10.1016/b978-0-12-822147-1.00016-9

Jones, A. M., & Harrison, R. M. (2016). Emission of ultrafine particles from the incineration of municipal solid waste: A review. *Atmospheric Environment, 140*, 519–528.

Kanekar, P. P., & Kanekar, S. P. (2022). Psychrophilic, psychrotrophic, and psychrotolerant microorganisms. In *Diversity and biotechnology of extremophilic microorganisms from India* (pp. 215–249). Springer Nature Singapore.

Kaur, G., Basak, N., & Kumar, S. (2024). State-of-the-art techniques to enhance biomethane/biogas production in thermophilic anaerobic digestion. *Process Safety and Environmental Protection.*

Kaza, S., Yao, L., Bhada-Tata, P., & Van Woerden, F. (2018). *What a waste 2.0: A global snapshot of solid waste management to 2050.* World Bank Publications.

Liu, Y. C., Ramiro-Garcia, J., Paulo, L. M., Braguglia, C. M., Gagliano, M. C., & O'Flaherty, V. (2023). Psychrophilic and mesophilic anaerobic treatment of synthetic dairy wastewater with long chain fatty acids: Process performances and microbial community dynamics. *Bioresource technology, 380*, 129124.

Mamo, G., & Mattiasson, B. (2020). Alkaliphiles: The versatile tools in biotechnology. In *Alkaliphiles in biotechnology* (pp. 1–51).

Mishra, P., Singh, N., Shrama, C., & Pathak, A. K. (2020). Landfill emissions and their impact on the environment. *International Journal of Engineering Research & Technology, 8.*

Mishra, S., & Das, A. P. (2021). Current treatment technologies for removal of microplastic and microfiber pollutants from wastewater. In *Wastewater treatment: cutting-edge molecular tools, techniques and applied aspects* (pp. 237–251). https://doi.org/10.1016/B978-0-12-821881-5.00011-8

Mishra, S., & Das, A. P. (2022). Treatment of the wastewater polluted with synthetic microfiber released from washing machine. In B. B. Das, H. Hettiarachchi, P. K. Sahu, & S. Nanda (Eds.), *Lecture notes in civil engineering* (Vol. 207, pp. 109–117). Springer Science and Business Media Deutschland GmbH. https://doi.org/10.1007/978-981-16-7509-6_9/COVER

Mishra, S., & Das, A. P. (2023). Microbial remediation of synthetic microfiber contaminated wastewater. In M. P. Shah (Ed.), *Microbial technologies in industrial wastewater treatment* (pp. 337–350). Springer. https://doi.org/10.1007/978-981-99-2435-6_17

Mishra, S., Dash, D., & Das, A. P. (2024). Isolation and molecular characterization of polyester-degrading bacterial strains from municipal wastewater and their application in microplastic degradation in contaminated wastewater. *Geomicrobiology Journal.* https://doi.org/10.1080/01490451.2024.2376101

Mishra, S., Ghosh, S., van Hullebusch, E. D., Singh, S., & Das, A. P. (2023). A critical review on the recovery of base and critical elements from electronic waste-contaminated streams using microbial biotechnology. *Applied Biochemistry and Biotechnology, 2023*, 1–30. https://doi.org/10.1007/S12010-023-04440-X

Mishra, S., Rout, P. K., & Das, A. P. (2019). Solar photovoltaic panels as next generation waste: a review. *Biointerface Research in Applied Chemistry, 9*(6).

Mishra, S., Singh, R. P., Rout, P. K., & Das, A. P. (2022). Membrane bioreactor (MBR) as an advanced wastewater treatment technology for removal of synthetic microplastics. In *Development in wastewater treatment research and processes* (pp. 45–60). https://doi.org/10.1016/B978-0-323-85583-9.00022-3

Mohammad, I. E., Syed, S., Darukamalli, M. R., & Alapati, K. S. (2023). Review on thermozymes produced by thermophilic fungi: A gold mine for industrial applications. *European Journal of Biology and Biotechnology, 4*(1), 1–5.

Mohanty, S., Ghosh, S., Bal, B., & Das, A. P. (2018). A review of biotechnology processes applied for manganese recovery from wastes. *Reviews in Environmental Science and Bio/Technology, 17*(4), 791–811. https://doi.org/10.1007/S11157-018-9482-1

Mohanty, S., Ghosh, S., Nayak, S., & Das, A. P. (2016). Isolation, identification and screening of manganese solubilizing fungi from low-grade manganese ore deposits. *Geomicrobiological Journal, 34*(4), 309–316. https://doi.org/10.1080/01490451.2016.1189016

Morita, R. Y. (1975). Psychrophilic bacteria. *Bacteriological Reviews, 39*(2), 144–167.

Nag, M., Lahiri, D., Ghosh, S., Sarkar, T., Pati, S., Das, A. P., Ram, D. K., Bhattacharya, D., & Ray, R. R. (2023). Application of microorganisms in biotransformation and bioremediation of environmental contaminant: A review. *Geomicrobiology Journal*. https://doi.org/10.1080/01490451.2023.2261443

Najar, I. N., Sharma, P., Das, R., Tamang, S., Mondal, K., Thakur, N., Gandhi, S. G., & Kumar, V. (2024). From waste management to circular economy: Leveraging thermophiles for sustainable growth and global resource optimization. *Journal of Environmental Management, 360*, 121136.

Nanda, S., & Berruti, F. (2021). Municipal solid waste management and landfilling technologies: A review. *Environmental Chemistry Letters, 19*(2), 1433–1456.

Nzila, A. (2018). Current status of the degradation of aliphatic and aromatic petroleum hydrocarbons by thermophilic microbes and future perspectives. *International Journal of Environmental Research and Public Health, 15*(12), 2782.

Phan, K. H., Nguyen, A. T., Le, S. L., Ngo, T. M., Cao, T. N. D., Nguyen, T. T., Le, L. T., Mai, T. P., & Bui, X. T. (2023). Thermophilic anaerobic digestion and emerging methods for organic waste treatment: A review. *Bioresource Technology Reports, 22*, 101402.

Prabhakar, A., Mishra, S., & Das, A. P. (2019). Isolation and identification of lead (Pb) solubilizing bacteria from automobile waste and its potential for recovery of lead from end of life waste batteries. *Geomicrobiological Journal, 36*(10), 894–903. https://doi.org/10.1080/01490451.2019.1654044

Priyadarsini, S., & Das, A. P. (2024). Analytical and structural characterization of waste lithium-ion batteries for their effective recycling strategy. *Environmental Chemistry and Ecotoxicology*. https://doi.org/10.1016/j.enceco.2024.12.004

Priyadarsini, S., & Das, A. P. (2025). Lithium bioleaching: A review on microbial-assisted sustainable technology for lithium bio-circularity. *Journal of Water Process Engineering, 69*, Article 106744. https://doi.org/10.1016/j.jwpe.2024.106744

Raj, K., & Das, A. P. (2023). Lead pollution: Impact on environment and human health and approach for a sustainable solution. *Environmental Chemistry and Ecotoxicology, 5*, 79–85. https://doi.org/10.1016/J.ENCECO.2023.02.001

Rampelotto, P. H. (2013). Extremophiles and extreme environments. *Life, 3*(3), 482–485.

Sekiguchi, T., Sato, T., Enoki, M., Kanehiro, H., & Kato, C. (2010). Procedure for isolation of the plastic degrading piezophilic bacteria from deep-sea environments. *極限環境微生物学会誌, 9*(1), 25–30.

Sharma, R., Sharma, M., Sharma, R., & Sharma, V. (2013). The impact of incinerators on human health and environment. *Reviews on Environmental Health, 28*(1), 67–72.

Sun, W., Yu, G., Louie, T., Liu, T., Zhu, C., Xue, G., & Gao, P. (2015). From mesophilic to thermophilic digestion: The transitions of anaerobic bacterial, archaeal, and fungal community structures in sludge and manure samples. *Applied Microbiology and Biotechnology, 99*, 10271–10282.

Tanruean, K., Penkhrue, W., Kumla, J., Suwannarach, N., & Lumyong, S. (2021). Valorization of lignocellulosic wastes to produce phytase and cellulolytic enzymes from a thermophilic fungus, Thermoascus aurantiacus SL16W, under semi-solid state fermentation. *Journal of Fungi, 7*(4), 286.

Williams, R. A. D. (2021). The genus thermus. In *Thermophilic bacteria* (pp. 51–62). CRC Press.

Wu, H., Chen, Q., Zhang, W., & Mu, W. (2023). Overview of strategies for developing high thermostability industrial enzymes: Discovery, mechanism, modification and challenges. *Critical Reviews in Food Science and Nutrition, 63*(14), 2057–2073.

Zhang, C., Xu, T., Feng, H., & Chen, S. (2019). Greenhouse gas emissions from landfills: A review and bibliometric analysis. *Sustainability, 11*(8), 2282.

Zhang, J., Kong, Y., Yang, Y., Ma, R., Li, G., Wang, J., Cui, Z., & Yuan, J. (2024). Effects of thermophilic bacteria inoculation on maturity, gaseous emission and bacterial community succession in hyperthermophilic composting. *Science of the Total Environment*, *927*, 172304.
Zhao, F., & Wang, S. (2019). Bioleaching of electronic waste using extreme acidophiles. In *Electronic waste management and treatment technology* (pp. 153–174). Butterworth-Heinemann.
Zhu, D., Adebisi, W. A., Ahmad, F., Sethupathy, S., Danso, B., & Sun, J. (2020). Recent development of extremophilic bacteria and their application in biorefinery. *Frontiers in Bioengineering and Biotechnology, 8*, 483.

Challenges and Innovations in Biomedical Waste Management (BMW): A Focus on Sustainable Technologies and Zero-Waste Approaches

G. Koteswara Reddy, S. K. Ayaaz Basha, A. Ramya Sri, Kurra Shrinidhi, Rushitha Deekshitula, Jeeshitha Kudithipudi, Sailaja Priyadarsini, and Alok Prasad Das

1 Introduction

The concept of a circular bioeconomy, however, offers a promising vision of agricultural waste not as a problem, but as a resource to drive sustainability and innovation (Priyadarsini & Das, 2025). By applying circular principles, agricultural residues can be upcycled into high-value products such as biofuels, bioplastics, metals, bio-based chemicals, and even food and feed supplements (Priyadarsini & Das, 2024). This transformation not only addresses the twin challenges of waste management and resource scarcity but also presents an economic opportunity to reduce waste, minimize pollution, and drive innovation (Paco et al., 2022). Waste can be classified into several types based on its origin and nature, such as plastic waste (Dey et al., 2024; Priyadarshini et al., 2024; Singh Jadaun et al., 2022), which is non-biodegradable such as bottles and packaging. Electronic waste, or e-waste, consists of discarded devices like phones and computers, many of which contain toxic materials (Mishra et al., 2023; Prabhakar et al., 2019; Raj & Das, 2023), mining waste containing many heavy metals (Das & Ghosh, 2018, 2022; Das et al., 2023; Dey et al., 2023; Ghosh & Das, 2017), and biomedical waste. The extensive generation of agricultural waste worldwide poses significant environmental challenges. Traditional disposal

G. K. Reddy (✉) · S. K. Ayaaz Basha · A. Ramya Sri · K. Shrinidhi · R. Deekshitula · J. Kudithipudi
Department of Biotechnology, Koneru Lakshmaiah Education Foundation (Deemed to be University), Vaddeswaram, India
e-mail: koteswarareddy@kluniversity.in

S. Priyadarsini · A. P. Das
Department of Life Sciences, Rama Devi Women's University, Bhubaneswar, Odisha, India

A. P. Das et al. (eds.), *Sustainable Municipal Waste Management with a Zero Waste Approach*, Sustainable Environmental Waste Management Strategies,
https://doi.org/10.1007/978-3-032-05950-5_6

Table 1 Biomedical waste and corresponding generation of plastic components

Biomedical waste	Plastic component
Masks, face shield, eye shield	Polypropylene (PP), polycarbonate (PC), polystyrene (PS)
Gowns	Low-density polyethylene (LDPE)
Hand sanitizer bottles, aprons, gloves, goggles	Polyvinyl chloride (PVC)
Packaging materials	PP, PS, LDPE, high-density polyethylene (HDPE), polyethylene terephthalate (PET)
Plastic bottles	PET, PP

Parashar and Hait (2021)

methods, such as crop burning, contribute to severe air pollution and ecological degradation (Adejumo, 2020). In the recent decade, coronaviruses have caused large-scale pandemics, namely severe acute respiratory syndrome coronavirus-1 (SARS-CoV-1) and the Middle East respiratory syndrome (MERS). A new outbreak in this family is added in November–December of 2019 as the novel coronavirus disease 2019 (COVID-19) caused by a large group of highly diverse, enveloped, positive-sense, and single-stranded RNA viruses, namely SARS-CoV-2. The drastic increase in COVID waste from 40 tons/day to 240 tons/day in Wuhan and the estimated increase from 5 million tons/year to 2.5 million tons/month in the USA is scary if not treated properly. Since the pandemic outbreak, South Korea has generated about 2000 tons of COVID waste until the start of May 2020 (Cesaro, 2020). It also includes discarded surgical gloves and instruments, needles, lancets, cultures, stocks, and swabs used to inoculate cultures and remove body organs. The healthcare waste consists primarily of pathological, infectious, chemical, pharmaceutical, and domestic wastes as well as sharps that have been contaminated with blood, infectious agents, tissues, organs, and many more (Kumar & Singh, 2020). Bioleaching is the process where microorganisms can be used to leach metals and other precious materials from various waste (Das & Ghosh, 2023a, 2023b; Ghosh et al., 2016; Mohanty et al., 2017).

Remediation of Biomedical Waste

See Tables 1 and 2.

2 BMW Generation in India During Pandemic

According to the CPCB reports from the year 2007 to 2019, there was a steady increase in the amount of BMW generation per day, as shown in Fig. 1 (CPCB, 2020). In 2008, the BMW generation rate was only 410 tons per day. In the next ten years, the value increased up to 619 tons per day. India has set up different BMW treatment facilities in its states and union territories with a treatment capacity of 800 tons per day (Dehal et al., 2021). Figure 1 shows that the daily generation of BMWs was much

Table 2 Worldwide production and increased demand for biomedical waste during the COVID-19 pandemic

Continent	Country	Production and increased demand	Time	Reference
Asia	Bangladesh	Produced 1.63–1.99 kg per day per bed in the hospital in Dhaka city	Peak pandemic	Rahman et al. (2020)
	China	Produced 116 million facemasks per day	February 2020	Yudell et al. (2020)
	China	12.5% demand for medical gloves is increased	During pandemic	Yudell et al. (2020)
	China (Wuhan)	200 tons of medical waste produced per day	February 2020	Saadat et al. (2020)
	Indonesia	30% demand for medical waste is increased	During pandemic	Nurhati and Cordova (2020)
	India	600 metric ton per day	May 2020	Ramteke and Sahu (2020)
	Malaysia	Increase in 27% medical waste	During pandemic	Yudell et al. (2020)
	Philippine	280 tons per day produced in Manila	Peak pandemic	Ramteke and Sahu (2020)
	Thailand	Demand for medical waste is 62% increased	January to April 2020	Ramteke and Sahu (2020)
	Jakarta	Produce 212 tons of medical waste per day	Peak pandemic	Ramteke and Sahu (2020)
	Japan	600 million face masks a day	April 2020	Ramteke and Sahu (2020)
	China (Hubei)	600% increase in hazardous waste	During pandemic	ADB (2020)
	Southeast Asia	1000 tons of medical waste produced per day	During pandemic	Bridges (2020)
	Jordan	650 kg medical waste, which is tenfold higher	During pandemic	Abu-Qdais et al. (2020)
Europe	Western Europe	27% increase in medical waste	During pandemic	Parashar and Hait (2021)
	UK	Plastic medical kit needs 7.5–12 million daily, 39,500 face masks, 1500 surgical gloves and gowns	February 2020	Duer (2020)
	France	40 million surgical masks per week	During pandemic	Parashar and Hait (2021)
	Catalonia and Spain	350% medical waste increased	During pandemic	Klemeš et al. (2020)
North America	USA	312% increase in face shield growth rate	During pandemic	Nzediegwu and Chang (2020)

(continued)

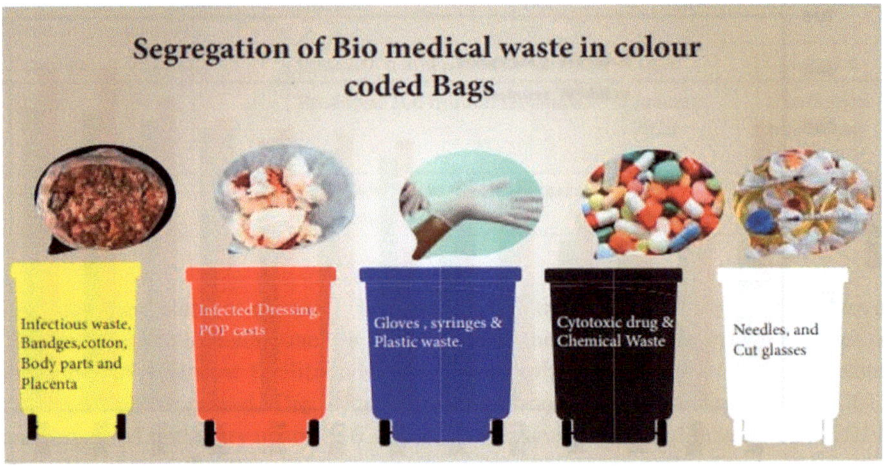

Fig. 3 Segregation of Biomedical waste in color coded bags

features of the renewed 2016 guidelines proposed by the Indian Ministry of Environment, Forest, and Climate change department. The CPCB of India had revised the recommendations for the fourth time and had released renewed recommendations for BMW management concerning COVID-19 Central Pollution Control Board (2020a, 2020b). Considering the magnitude of problem associated with BMW management during pandemic, Central Pollution Control Board (CPCB) has initially formulated guidelines (Guidelines for Handling, Treatment and Disposal of Waste Generated during Treatment/Diagnosis/Quarantine of COVID-19 Patients) based on current knowledge on COVID-19 and from other contagious diseases like HIV and H1N1, and existing practices in managing infectious and hazardous wastes. CPCB has been issuing specific guidelines about segregation, collection, storing, transportation, and disposal of BMW generating from the COVID-19 treatment facilities. These newer guidelines have been suggested after collecting and evaluating additional current information on the disease. The procedures for solid BMW collection, segregation, treatment, and disposal regarding the BMW were generated at healthcare facilities, quarantine centers, home care facilities, sample collection centers, diagnostic laboratories, and CBWTFs. Also, the revised guidelines stress the disposal of PPE. Some important guidelines included mandatory labeling of BMW as COVID-19 waste, registration of COVID-19-related facility (quarantine facility, sample collection centers, etc.) with the CPCB, the SPCB, and informing it to the CBWTFs, the use of foot-operated bins, use of double-layered plastic bags, sodium hypochlorite disinfection sprays on the COVID-19 waste bins, among others. Central Pollution Control Board (2020a, 2020b). *Guidelines for management of biomedical waste generated during COVID-19 pandemic*. Ministry of Environment, Forest and Climate Change, Government of India.

4 Biomedical Waste Management

Traditionally, the biomedical waste was processed in the hospital and laboratory premises by using incinerators. Incineration is the process of burning the BMWs at an extremely high temperature (> 1000 °C) (Gautam et al., 2010a, 2010b). This process although was effective in controlling and preventing the spread of infections, there is an imminent possibility of the generation of toxic substances like the products of incomplete combustion including dioxins, furans, and other metal-related toxic remnants. Also, the ash produced from the incineration of BMWs could mix with the environment and pose both public and animal health hazards. The use of incinerators with high-efficiency particulate air filters (HEPA) that can filter out the hazardous and toxic wastes before being disposed of in the form of landfills was also previously practiced.

5 Biomedical Waste Classification

Waste that Is not Dangerous

This is around 85% of the waste generated in the majority of healthcare environments. This includes waste materials such as organic food waste, fruit peels, wastewater, paper cartons, and packaging materials.

Dangerous Waste

Waste that Could Be Infectious

Multiple terms have been employed in scientific literature, regulations, guidance manuals, and standards to refer to infectious waste throughout the years. These examples include infectious, infective, medical, biomedical, hazardous, red bag, contaminated, medical infectious, regulated, and regulated medical waste. These expressions all describe the same kind of waste, although the terminology used in regulations is typically more specific. It constitutes 10% of the overall waste, encompassing.

Contaminated dressings and swabs containing blood, pus, or body fluids, along with laboratory waste, specifically culture stocks of pathogenic agents, are considered hazardous. Additionally, infectious tissues, such as surgically removed tumors, organs, expelled placenta, extracted teeth, and similar materials, pose significant health risks (Khatun et al., 2023; Sadasivuni et al., 2019). Animals used in clinical and research operations that may be infected also fall into this category. Sharps, which include needles, syringes, and blades, are another key concern due to their potential for injury and infection. Furthermore, hematological substances and other substances derived from blood require careful handling due to their infectious nature. These materials all fall under the category of biohazardous waste and must be managed with the utmost caution.

Potentially Hazardous Garbage

Radioactive waste refers to materials contaminated with radionuclides, which can exist in solid, liquid, or gaseous forms. These materials are often derived from laboratory analysis of bodily fluids and tissues, medical imaging, and therapeutic procedures performed outside the body. Chemical waste consists of various substances that are no longer needed and could pose a risk to human health or the environment. Examples include disinfectants such as hypochlorite, glutaraldehyde, iodophors, phenolic derivatives, and alcohol-based preparations, as well as X-ray processing solutions, monomers, associated reagents, and base metal debris like dental amalgam found in extracted teeth. Pharmaceutical waste includes substances such as anesthetics, sedatives, antibiotics, analgesics, and other pharmaceutical products, all of which require careful disposal due to their potential hazards Zubair and Khan (2020).

6 Segregation of Biomedical Management

Color coding is the first step of BMW management. Different wastes are classified into different types, and therefore, they must be handled and disposed of according to their classification. The bins used for waste disposal in all healthcare facilities worldwide are always color-coded. Based on the rule of universality, bins are assigned a specific color, according to which the waste is segregated. This step helps avoid the chaos that occurs when all types of waste are jumbled, which can lead to improper handling and disposal and further result in the contraction of several diseases. The different kinds of categories of waste include sharp waste such as scalpels, blades, needles, and objects that can cause a puncture wound, anatomical waste, recyclable contaminated waste, chemicals, and laboratory waste such as specimens, blood bags, vaccines, and medicines that are discarded. All the above-mentioned wastes are segregated in different colored bins and sent for treatment. Yellow bins collect anatomical waste, infectious waste, chemical waste, laboratory waste, and pharmaceutical waste, covering almost all types of BMW. Different bins and various types of sterilization methods are used depending on how hazardous the waste is. The best tools for sterilization are autoclaves. Red bins collect recyclable contaminated wastes, and non-chlorinated plastic bags are used for BMW collection. Blue containers collect hospital glassware waste such as vials and ampoules. White bins are translucent where discarded and contaminated sharps are disposed of. Sharp wastes must always be disposed of in puncture-proof containers to avoid accidents leading to handlers contracting diseases.

Dry Heat Treatment

The dry heat technique is used for sterilization. It uses very high temperature to kill bacterial spores and microbes (Parida et al., 2026). It is generally used for handling a small volume of BMWs. It is used on dry items like glasswares, metal equipment, paper-wrapped items, powders, and oil. In this technique, moisture-free high-temperature air is used over a lengthened period for the sterilization of BMWs. It is an eco-friendly as well as an inexpensive process. It is noncorrosive treatment process suitable for metals and sharp objects; however, it cannot be used for plastic or rubber items as there is a possibility of damaging these items due to exposure to heat over a prolonged period. Therefore, both temperature and time should be optimized to minimize the volatile organic carbon released from the plastic BMWs during the dry heat treatment, particularly during the pandemic period when the daily BMW generation exceeds the capacity of treatment facilities. During the pandemic, many countries have used this ad hoc technique for the emergency management of the huge COVID-19-related BMWs. Although encapsulation is simple to execute, it is considered an outdated technique.

Incineration

Incineration is a common practice and widely used process for the treatment of BMWs. All hospitals have mandatory set up of an incinerator. It is conducted between 800 and 1200 °C in the presence of sufficient air for the complete obliteration of pathogens and organic wastes, which reduces the BMW volume and mass up to 90 and 75%, respectively (Gautam et al., 2010b). The aerated burning process generates flue gas containing different toxins like furan and dioxins. Therefore, it needs a secondary treatment plant or a high-quality flue gas cleaning system in order to mitigate air pollution. Also, the ash generated from the incinerators should be properly characterized before its disposal following the statutory regulations. Incineration is a high energy-intensive treatment process, and combined with the secondary treatment plant for the toxins, the overall process becomes expensive. However, its high efficiency for the BMW treatment makes it the most suitable and widely adopted process by different countries. According to the CPCB reports, there are 232 incinerator plants available in India for the treatment of BMWs (CPCB, 2019). (CPCB, 2019. Information of common biomedical waste treatment facilities. https://cpcb.nic.in/uploads/Projects/BioMedicalWaste/CBWTF_Status_2019.pdf.)

Sanitary Landfill

Sanitary landfill is a simple and inexpensive disposal method for disposal of the BMWs. It can be conducted in the existing municipal waste management system. It is the traditional landfill technique and completely banned because of its environmental concern which includes the decomposition of waste resulting in the production of landfills leachate and greenhouse gases. This method can only be adopted if there are no acceptable means of treating the waste before disposal. Few things are considered before the landfill that the site should be geographically isolated and away from the water stream, the operation is managed by a competent authority, and there is an acceptable limit to the quantity that can be landfilled at a location Manzoor and Sharma (2019).

Shredding

In this process, a shredder is used to cut the pretreated BMWs into small chunks. The shredders should be operated by a competent staff. Shredders are gene Manzoor and Sharma (2019). Impact of biomedical waste on environment and human health. *Environmental Science and Pollution Research*, 26(12), 12045–12054. https://doi.org/10.1007/s11356-019-04567-5rally built-in devices to the integrated chemical or thermal disinfection system (Gautam et al., 2010a, 2010b). (Gautam, V., Thapar, R., Sharma, M., 2010. Biomedical waste management: incineration versus environmental safety. Indian J. Med. Microbiol. 28(3), 191–192. https://doi.org/10.4103/0255-0857.66465.) This technique facilitates the recycling of plastics and significantly reduces the volume of BMWs. Metallic wastes should be avoided in the shredders as they can damage the metal blade on them.

Chemical Treatment

Chemical treatment is a disinfection technique used for the pretreatment of COVID-19-related BMWs. Chemical disinfectants like chlorine (Cl_2), sodium hypochlorite (NaOCl), chlorine dioxide (ClO_2), ozone (O_3), and UV irradiation are generally used in this process (Wang et al., 2020). This process is conducted in a covered system for a fixed period so that all pathogens and organic substances are effectively inactivated. The possibility of chemical aerosol formation during the process is minimized by using the high-efficiency particulate absolute filters. The major disadvantage of the process is that it releases large-scale Cl_2 gas into the environment. Both residue and effluent of the process should be analyzed and disposed of with proper regulation. Also, it generates toxic effluent which requires further treatment resulting in the high expenditure of the overall process. Combining the process with a mechanical shredder is used for the complete disposal of BMWs. It is a simple disinfection technique, but the BMWs should be properly shredded before they undergo the chemical disinfection (Datta et al., 2018; Wang et al., 2020).

Advanced Technologies Involved in BMW

Management of general waste from quarantine facilities National Green Tribunal, India, expressed the need for further revision of the guidelines so that all aspects of scientific disposal of liquid and solid waste management are taken care of not only at institution level but also at individual levels, such as manner of disposal of used personal protection equipment (PPE), without the same getting mixed with other municipal solid waste causing contamination (Samantaray et al., 2026). Administrations and the management of the service providers are required to provide adequate PPEs for the human resources handling BMW as prescribed in the guidelines. Training of the staff on proper provided disposal of waste material arising from COVID-19 care facilities is another key precautionary measure to be conducted to prevent exposure risks to contaminated substances. Subsequent revisions (CPCB, 2020a) of the guidelines incorporated guidance on: segregation of general solid waste and BMW, the safety of waste handlers and sanitation workers providing training to waste handlers about infection prevention measures, electronic tracking of BMW through an App "COVID19BMW,"and disposal of used PPEs. India's utmost effort in combating this pandemic has significantly increased the quantity of waste generated from healthcare facilities, quarantine facilities, and laboratories. Technology-enabled monitoring and tracking of BMW from generation to safe disposal became essential for efficiency optimization. CPCB developed COVID-19 waste tracking software named "COVID19BMW," which can track BMW at the time of generation, collection, and disposal. This application was designed for waste generators, CBWTF Operators, State Pollution Control Boards/Pollution Control Committees, and Urban Local Bodies.

Biological Ways for Sustainable BMW

Homemade face masks made of household items should only be used as a final prophylactic measure against large respiratory droplets. Homemade masks are typically manufactured without quality control using a rudimentary fabric base, mostly made of cotton and similar types of fibers. Homemade masks were recorded to be woven (cross-thread), knitted (fibers with interconnecting loops), and felted (disoriented fibers in compressed form) MacIntyre and Chughtai (2020). No well-defined protocol was observed adequately describing mask design, layer count, filtering efficiency, material to be used, or breathability rate. Various bio-based filtering media, which include chitosan, gelatin, polylactic acid, and cellulose, have recently been commercialized as biodegradable facemasks. Additionally, other biopolymer combinations, such as polylactic acid/polyhydroxy butyrate, sericin/polyvinyl alcohol/clay (Cloisite 30B), cellulose/polyethyleneimine, and Ag nanoparticles, are also used Kaur and Kaur (2021).

These materials feature high filtering efficiencies and provide filtration of bacteria, and other foreign particles act as a barrier against respiratory droplets carrying pathogens, particulate, and bacterial filtration and provide additional fit and comfort (2) Li, J., Zhang et al. Beyond these, future facemasks need to be antiviral, and the development of biocidal masks may significantly reduce the chances of transmission and contamination of the pathogens. To achieve so, antiviral or antibacterial compounds can be added to the surface or into the polymeric fibers. Due to their inherent broad-spectrum antibacterial and antiviral activity, durability, and capacity to be active at much lower doses, synthetic nanoparticles based on metal derivatives such as metal–organic frameworks (MOFs), silver (Ag), copper (Cu), or titanium dioxide (TiO_2) nanoparticles have been recommended over the years as chemical disinfectants (Armentano et al., 2021). Several investigations have recently been conducted to increase the effectiveness of respirators and masks in combating diseases—metal organic frameworks (MOFs) are used as filtration materials in personal devices instead of conventional textiles; they are used to synthesize PM2.5-protective face masks by developing a versatile hybrid fiber substrate. Li et al. (2020)

7 Recycling

The usage of face masks and plastic gloves has increased dramatically during the coronavirus (COVID-19) outbreak. It is believed that over 700 million Africans utilize face masks on a, while more than 2.2 billion Asians do so. As a result, masks are getting dumped in garbage bins or landfills. Consequently, wind and rains may easily carry the masks into city streets, rivers, and seas due to their lightweight nature. Not only that, other medical plastics are frequently thought to be contagious and hence cannot be disposed of with regular municipal garbage. The waste plastic must be decontaminated in this circumstance by breaking it down into little bits and then heating it at a high temperature (Priyadarsini et al., 2025; Priyadarsini & Das, 2026). This may then be disposed of with municipal plastic garbage. This is done by classifying things that may be expired but have not been opened and are not polluted. All these processes can help to decrease waste production. Some nations, including Korea, utilize steam sterilization before reuse, which involves heating supplies to 121 °C at 1 atmospheric pressure for 30 min. If all these things are done in a well-organized and methodical manner, the quantity of garbage produced can be minimized before it is recycled (Bahl & Bahl, 2021).

In the development of reusable, efficient, high-filtration, thermally controlled, and antiviral face masks, there is still more to be discovered. However, in the event of an unanticipated pandemic, such as COVID-19, surgical masks and respirators will always be in high demand and in limited supply. This results in a greater reliance on fabric masks, despite their limited efficiency, particularly in low-income nations. As a result, evaluating the filtration efficiencies of existing materials that potentially provide considerable pathogen protection is critical, and other attributes, such as thermal stability management, reusability, and self-sanitizing capability, were also examined. Yang et al. (2017) developed a unique nanofiber/nanoporous polyethylene system with excellent adhesion characteristics for particulate matter and efficiency of capture. Thermal management seeks to decrease radiative dissipation in a low temperature environment. Due to the increased rate of transmission dispersion in the body heat of humans, this material has a good radiative cooling effect. More recently, researchers from Australia's Queensland University of Technology have developed a cellulose nanofiber facemask that can filter particles as small as 100 nm and is breathable with the use of a disposable filter cartridge. In addition, LIGC Applications Ltd. of the USA has developed a respirator made of microporous, reusable, conductive graphene foam. The mask can effectively entrap bacterial matter and sterilize itself by the passage of an electrical charge. Among many other adversities, the COVID-19 pandemic has deteriorated the environmental problem related to plastic pollution. Over time, environmental factors such as waves, rain, turbulence, wind, and sun cause plastics to break into smaller fragments, called microplastics (< 5 mm; MPs). Biomedical waste and other plastic wastes degraded by mechanical and photodegradation processes can be a source of MPs in the environment. Bioaccumulation is the net uptake of a pollutant such as MPs from the environment through viable routes (e.g., respiration, touch, and ingestion) from water, air, and soil (Sharma & Chatterjee, 2017).

8 Conclusion

Effective management of biomedical waste is crucial for safeguarding public health and the environment, particularly during health emergencies like the COVID-19 pandemic. The surge in medical waste highlights the importance of proper segregation, treatment, and disposal at the source. Employing green technologies and adopting zero-waste strategies can reduce environmental harm, encourage resource recovery, and support sustainable practices. Strengthening infrastructure, training healthcare personnel, and following standardized guidelines are vital for safe waste handling. By fostering innovation and responsibility within healthcare systems, it is possible to protect both human health and ecological well-being while moving toward sustainable biomedical waste management.

References

Abu-Qdais, H. A., Al-Ghazo, M. A., & Al-Ghazo, E. M. (2020). Statistical analysis and characteristics of hospital medical waste under novel coronavirus outbreak. *Global Journal of Environmental Science and Management, 6*(4), 389–396. https://doi.org/10.22034/gjesm.2020.04.01

Adejumo, I. O. (2020). Agricultural solid wastes: Causes, effects, and management strategies. In *Agricultural solid wastes: Causes, effects, and management strategies*. IntechOpen.

Armentano, I., Dottori, M., & Puglia, D. (2021). Nanomaterials for the development of antiviral and antibacterial face masks. *Materials Today Chemistry, 20*, 100426. https://doi.org/10.1016/j.mtchem.2021.100426

Asian Development Bank. (2020). Managing infectious medical waste during the COVID-19 pandemic. https://www.adb.org/publications/managing-medical-waste-covid19

Bahl, P., & Bahl, S. (2021). The impact of COVID-19 on medical waste management: A review. *Environmental Science and Pollution Research, 28*(15), 18879–18890. https://doi.org/10.1007/s11356-021-12948-x

Bridges, T. (2020). Innovating for impact: Dealing with a mountain of medical waste. LinkedIn. https://www.linkedin.com/pulse/innovating-impact-dealing-mountain-medical-waste-trista-bridges

Central Pollution Control Board. (2020a). *Guidelines for management of biomedical waste during COVID-19 pandemic*. Ministry of Environment, Forest and Climate Change, Government of India.

Central Pollution Control Board. (2020b). *Management of biomedical waste in India: Status report 2020*. Ministry of Environment, Forest and Climate Change, Government of India.

Cesaro, A. (2020). About the effects of Covid-19 on solid waste management. *Tema, 1*(1), 1–10.

Chand, S., Kumar, S., & Singh, R. (2021). Impact of COVID-19 on biomedical waste management in India: A review. *Environmental Science and Pollution Research, 28*(15), 18857–18867. https://doi.org/10.1007/s11356-021-12949-w

Das, A., & Ghosh, S. (2022). Role of microorganisms in extenuation of mining and industrial wastes. *Geomicrobiology Journal, 39*, 173–175. https://doi.org/10.1080/01490451.2022.2038953

Das, A. P., & Ghosh, S. (2018). Bioleaching of manganese from mining waste materials. *Materials Today: Proceedings, 5*(1), 2381–2390. https://doi.org/10.1016/J.MATPR.2017.11.459

Das, A. P., & Ghosh, S. (2023a). Mixed culture bioleaching: An insight into manganese biomining process and efficacy. *Manganese Mining Microorganisms*, 129–141. https://doi.org/10.1016/B978-0-12-822147-1.00001-7

Das, A. P., & Ghosh, S. (2023b). Role of metabolites and metabolomics for understanding of manganese bioleaching. *Manganese Mining Microorganisms*, 117–127. https://doi.org/10.1016/B978-0-12-822147-1.00011-X

Das, A. P., Mallavarapu, M., & Ghosh, S. (2023). Ecotoxicity of mining pollutants on the environment and their remediation. *Environmental Chemistry and Ecotoxicology, 5*, 165–167. https://doi.org/10.1016/J.ENCECO.2023.08.002

Datta, P., Mohi, G. K., & Chander, J. (2018). Biomedical waste management in India: Critical appraisal. *Journal of Laboratory Physicians, 10*(1), 6–14. https://doi.org/10.4103/JLP.JLP_36_17

Dehal, A., Kumar, S., & Singh, R. (2021). Biomedical waste management in India: Current status and future challenges. *Environmental Science and Pollution Research, 28*(15), 18845–18856. https://doi.org/10.1007/s11356-021-12950-1

Dey, S., Dash, N., Ray, S. K., Priyadarsini, S., Chhotaray, C., & Das, A. P. (2024). Occurrence and quantification of tire wear plastics pollutants from highway city dust of Bhubaneswar City of Odisha. In *Sustainable microbial technology for synthetic and cellulosic microfiber bioremediation* (pp. 77–89). https://doi.org/10.1007/978-3-031-62600-5_5

Dey, S., Tripathy, B., Kumar, M. S., & Das, A. P. (2023). Ecotoxicological consequences of manganese mining pollutants and their biological remediation. *Environmental Chemistry and Ecotoxicology*. https://doi.org/10.1016/J.ENCECO.2023.01.001

Duer, J. (2020, October 5). The plastic pandemic: COVID-19 trashed the recycling dream. Reuters. https://www.reuters.com/investigates/special-report/health-coronavirus-plastic-recycling/

Gautam, P., Kumar, S., & Singh, R. (2010a). Biomedical waste management: A review of the current practices and challenges in India. *Waste Management, 30*(12), 2460–2467. https://doi.org/10.1016/j.wasman.2010.06.002

Gautam, V., Thapar, R., & Sharma, M. (2010b). Biomedical waste management: Incineration versus environmental safety. *Indian Journal of Medical Microbiology*, 28(3), 191–192. https://doi.org/10.4103/0255-0857.66465

Ghosh, S., & Das, A. P. (2017). Bioleaching of manganese from mining waste residues using *Acinetobacter* sp. *Geology, Ecology, and Landscapes, 1*(2), 77–83. https://doi.org/10.1080/24749508.2017.1332847

Ghosh, S., Mohanty, S., Akcil, A., Sukla, L. B., & Das, A. P. (2016). A greener approach for resource recycling: Manganese bioleaching. *Chemosphere, 154*, 628–639. https://doi.org/10.1016/J.CHEMOSPHERE.2016.04.028

Goswami, A., Kumar, S., & Singh, R. (2021). Biomedical waste management during the COVID-19 pandemic: A review of the challenges and solutions. *Environmental Science and Pollution Research, 28*(15), 18868–18878. https://doi.org/10.1007/s11356-021-12951-0

Kaur, S., & Kaur, R. (2021). Biodegradable face masks: A review on the recent advancements in biopolymer-based materials for face masks. *Journal of Polymers and the Environment, 29*(1), 1–15. https://doi.org/10.1007/s10924-020-01873-5

Khatun, R., Singh, S., Dubey, N. K., & Prasad Das, A. (2023). A review on marine-based phytochemicals and their application in biomedical research. In *Recent frontiers of phytochemicals: Applications in food, pharmacy, cosmetics, and biotechnology* (pp. 383–395). https://doi.org/10.1016/B978-0-443-19143-5.00013-X

Klemeš, J. J., Van Fan, Y., Tan, R. R., & Jiang, P. (2020). Minimising the present and future plastic waste, energy and environmental footprints due to COVID-19. *Renewable and Sustainable Energy Reviews, 127*, 109883. https://doi.org/10.1016/j.rser.2020.109883

Kumar, S., & Singh, R. (2020). Management of healthcare waste during COVID-19 pandemic: A review. *Environmental Science and Pollution Research, 27*(38), 47700–47710. https://doi.org/10.1007/s11356-020-10383-5

Li, J., Zhang, Y., & Wang, Y. (2020). Development of advanced face masks with high filtration efficiency and comfort. *Journal of Hazardous Materials, 392*, Article 122267. https://doi.org/10.1016/j.jhazmat.2020.122267

MacIntyre, C. R., & Chughtai, A. A. (2020). A rapid systematic review of the efficacy of face masks and respirators against coronavirus. *Medical Journal of Australia, 213*(4), 210–215. https://doi.org/10.5694/mja2.50729

Manzoor, J., & Sharma, M. (2019). Impact of biomedical waste on environment and human health. *Environmental Science and Pollution Research, 26*(12), 12045–12054. https://doi.org/10.1007/s11356-019-04567-5

Mishra, S., Ghosh, S., van Hullebusch, E. D., Singh, S., & Das, A. P. (2023). A critical review on the recovery of base and critical elements from electronic waste-contaminated streams using microbial biotechnology. *Applied Biochemistry and Biotechnology, 2023*, 1–30. https://doi.org/10.1007/S12010-023-04440-X

Mohanty, S., Ghosh, S., Nayak, S., & Das, A. P. (2017). Bioleaching of manganese by *Aspergillus* sp. isolated from mining deposits. *Chemosphere, 172*, 302–309. https://doi.org/10.1016/J.CHEMOSPHERE.2016.12.136

Nurhati, I. S., & Cordova, M. R. (2020). Marine plastic debris in Indonesia: Baseline estimates (2010–2019) and monitoring strategies (2021–2025*). Marine Research in Indonesia, 45*(2), 97–102.https://doi.org/10.1016/J.CHEMOSPHERE.2016.12.136

Nzediegwu, C., & Chang, S. X. (2020). Improper solid waste management increases potential for COVID-19 spread in developing countries. *Resources, Conservation and Recycling, 161*, 104947. https://doi.org/10.1016/j.resconrec.2020.104947.

Paço, A., Duarte, K., & Ramos, M. M. (2022). The circular bioeconomy in agriculture: Valorization of waste and sustainability challenges. *Journal of Cleaner Production, 365*, Article 132812. https://doi.org/10.1016/j.jclepro.2021.132812

Parashar, N., & Hait, S. (2021). Plastics in the time of COVID-19 pandemic: Protector or polluter? *Science of the Total Environment, 759*, 144274.

Prabhakar, A., Mishra, S., & Das, A. P. (2019). Isolation and identification of lead (Pb) solubilizing bacteria from automobile waste and its potential for recovery of lead from end of life waste batteries. *Geomicrobiological Journal, 36*(10), 894–903. https://doi.org/10.1080/01490451.2019.1654044

Priyadarshini, B., Santhosh Kumar, M., & Das, A. P. (2024). Microbial degradation of synthetic plastics released from COVID-19 generated single use disposable face masks. *Environmental Science and Engineering, Part F2827*, 1–16. https://doi.org/10.1007/978-3-031-50840-0_1

Priyadarsini, S., & Das, A. P. (2024). Analytical and structural characterization of waste lithium-ion batteries for their effective recycling strategy. *Environmental Chemistry and Ecotoxicology*. https://doi.org/10.1016/j.enceco.2024.12.004

Priyadarsini, S., & Das, A. P. (2025). Lithium bioleaching: A review on microbial-assisted sustainable technology for lithium bio-circularity. *Journal of Water Process Engineering, 69*, 106744. https://doi.org/10.1016/j.jwpe.2024.106744

Parida, S., Samal, D. P. K., Sukla, L. B., & Das, A. P. (2026). Green technology for the treatment of wastewater using microbes. *Decarbonization of Wastewater Pollutants as a Sustainable Solution*, 181–210. https://doi.org/10.1016/B978-0-443-33062-9.00023-0

Priyadarsini, S., Alok, Das, P., Journal, G., & Das, A. P. (2025). Isolation and identification of native lithium-resistant bacteria from electronic waste contaminated sites: A focused investigation on tolerance assessment with potential lithium bio-recovery application. *Geomicrobiology Journal*. https://doi.org/10.1080/01490451.2025.2539509

Priyadarsini, S., & Das, A. P. (2026). Lithium in waste water: Impacts on pollution and human health. *Decarbonization of Wastewater Pollutants as a Sustainable Solution*, 353–364. https://doi.org/10.1016/B978-0-443-33062-9.00010-2

hazardous chemicals and toxic metals that, if not handled in an appropriate manner, are dangerous to human health and the environment. With the exception of some regions of the world where the management of electronic waste has been inadequate, there is generally low awareness among public members about the rules, and government officials have been unable to properly implement them (Odili et al., 2018).

Decisions about purchasing and disposal are heavily influenced by factors that drive upstream, such as consumer volume and attitudes. The rise of cheap, low quality new goods is an example of consumer behavior that creates a cycle of short-term gains to the detriment of long-term environmental damage. These products have negative effects that manifest themselves later, affecting both the environment and those who are involved in the disposal. Notably, electronic products are often rendered disposable before they hit the official end-of-life (EoL), and very few EoL electronics are processed in conventional recycling facilities. Much of these electronic devices eventually either end up in a landfill or are burned in waste-to-energy incineration.

Electronic waste also makes countries like India, China, and several African countries more susceptible to electronic waste risks. This is mainly attributed to the export of vast amounts of electronic and electrical equipment (EEE) from developed nations to these areas. Often this waste is managed under the wrong practices like inappropriate safety (in its handling), disposal into landfills, or via incineration without consideration of the possible effect of contamination on water and air quality. This problem affects countries at all stages of development; however, in developing nations, it vanishes as a result of inadequate policy mechanisms and a lack of government support, combined with a deficiency of funding which is necessary for the creation of developed garbage dumps (Kiddee et al., 2014).

Many developing countries, such as Nigeria, have a long way to go when it comes to handling electronic waste, since there is little recycling infrastructure in place. In response, the household sector relies on informal sectors of the economy practicing rudimentary dismantling and backyard recycling at the source, for example by recovering copper from wires through open burning and disposing of e-waste in dumpsites. Participation in such practices leads to human contact with harmful chemicals (Kaya, 2019). Also, a big problem with electronic waste concern is its recycling, recovery, and disposal process. Currently, only about 20–30% of global waste gets recycled. Attempts to recover metals such as gold, silver, platinum, and copper have been made, but they are often carried out through activities that can be extremely harmful to human health. There is still insufficient documentation and research about e-waste disposal facilities (Bakare et al., 2009; Balde et al., 2017). As a result, large volumes of e-waste end up in landfills or are burned at high temperatures, both highly detrimental to the environment. A number of routine pretreatment procedures, including component removal, manual disassembling, crushing, and size reduction, can result in substantial pollution on a local level (Khanna et al., 2020a, 2020b).

Insufficient e-waste management and destruction encourages hazardous substances into the human food chain right to the environment mainly via the contamination of water, air, and soil (Deng et al., 2006; Leung et al., 2006). As shown in the figure above, the different environmental compartments have been interlinked by contaminants emitted from e-waste of living organisms. Pollutants from e-waste include heavy metals, polyaromatic hydrocarbons (PAHs), polychlorinated biphenyls (PCBs), polybrominated diphenyl ethers (PBDEs), and other pollutants such as polyvinyl chloride phthalates (Wong et al., 2007), and these substances have already damages to ecosystems. These contaminants remain in soil for long time and causes many negative impacts on soil microorganisms and higher organisms health. These drivers also lead to changes in soil microbiology, with respect to both its phylogenetic structure and functionality. Here, we review the worldwide generation of e-waste and its components, as well as their effects on crucial ecosystem components such as soil, plants, microbes, and animals, including humans. From this study, a lot of methods for recovering valuable metals are reviewed. The results reveal a strong need to shift toward innovative, high-level, and environmentally friendly approaches for managing e-waste.

2 Methodology

This overview covers diverse literature published from 1996 to 2021, with emphasis on the journal articles, book chapters, books, conference papers, and technical reports published from 2011 to 2021. A systematic literature review was performed to collect relevant information for electronic waste. This method was performed with the related academic databases such as Google, Google scholar, Web of Science, Research Gate, and Scopus. During the search, a variety of keywords were used, including, electrical and electronic waste, e-waste management, unorganized e-waste, e-waste leachates, e-scrap, recycling, environmental pollution caused due to e-waste, leaching and landfilling of e-waste, recovery of useful metals, phyto-management of e-waste and its leachates, incineration, and adverse ecological effects of e-waste.

Focus was given to certain journals such as Science of the Total Environment, Journal of Hazardous Materials, Waste Management, Resources, Conservation and Recycling, Environmental Technology and Innovation, Journal of Waste Management, Ecotoxicology and Environmental Safety and Ecological Engineering. These sources were systematically searched through for relevant literature until June 2021. Then, the inclusion of quotative and duplicate publications were excluded via an individual investigation of each publication. From the literature documents retrieved, 268 articles were selected and referenced. Result: A total of 234 journal articles, 11 book chapters, 4 book, 11 technical reports, and eight conference papers were identified. The plurality (about 60%) of the documents cited were from between 2015 and 2021.

3 Global and National Status of E-waste Generation

The world's electronic waste output rises and falls with the economy and technology changes. In fact, electronic waste produced in any country is not only dependent on the number of computers, but it is also influenced by the gross domestic product (GDP) of that very nation. Such connection emphasizes the need of electronic hardware in order to work properly for other techniques (Kiddee et al., 2014). According to recent studies, the world produced approximately 53.6 million tons of electronics waste, which equals about 7.3 kg per inhabitant (Forti et al., 2020). That number has grown by nine million tons since 2014, and estimates predict it will reach 74.7 million tons in 2030—nine kilograms per resident.

Asia generates the most electronic waste, 24.9 million tons or 5.6 kg per capita. This is followed by the Americas, which emit 13.1 million tons or 13.3 kilos per inhabitant. Europe accounts for 12 million tons or 16.2 kg per inhabitant. The per capita generation per inhabitant, by continent, is as follows: Africa (2.9 million tons, 2.5 kg per inhabitant) and Oceania (0.7 million tons, 16.1 kg per inhabitant). In terms of sub-regions, central Asia is the first in the ranking of e-waste generation with a total of 13.7 million tons and 8.6 kg per inhabitant. Next comes Southern Asia, producing 4.8 million tons or 2.6 kg per person. Southeastern Asia has 3.5 million tons or 5.4 kg per inhabitant; Western Asia, 2.6 million tons or 9.6 kg per inhabitant. Finally, Central Asia: 0.2 million tons with 7.1 kg per capita. In Northern America waste generation statistics, 7.7 million tons is equal to 20.9 kg per person. Next is South America, with 3.9 million tons or 9.1 kg per person. In Central America, it corresponds to 1.5 million tons, which is equivalent to 8.3 kg per inhabitant, and in the Caribbean, 0.1 million tons, equivalent to 7.8 kg per inhabitant. In Western Europe, the calculation is 4 million tons or 20.3 kg per inhabitant. Next, there is Eastern Europe, with 3.2 million tons for 11 kg per habitant, followed by Southern Europe with about 2.5 million tons (16.7 kg/inhabitant) and Northern Europe with 2.4 million tons (22.4 kg/inhabitant) (Forti et al., 2020).

$$E(\text{kg/year}) = N \times M/L \tag{1}$$

The components of E, including M (kg), N, L (yr), refer to the mass of the product, the total amount of objects in use, and the average life of the product, respectively. Equation (1) can be used to obtain the calculation. Robinson

$$(1)E = MN/L$$

Computers usually last about three years, generating more e-waste in that period than electric cookers and refrigerators in ten years (Betts, 2008). Of this, 10.19 metric kilotons (kt) constitute electronic waste generation, the highest in China. It is followed by India at 3.230 kt, Japan at 2.569 kt, and Indonesia at 1.618 kt (Forti et al., 2020). India is also the second-largest market after China. This is due to a large consumer market for electronics and electrical equipment, as well as a high growth economy representing a large market for these products. According to a new study, with economic growth and living standards on the rise in the country, India has become the world's largest home appliance market for television sets, refrigerators, air conditioners, washing machines, and ovens. These findings were presented by Sabha back in 2011, and Ramakrishnan et al. further expanded on these results. The consumerization of e-governance, in turn, had a transformative effect on the information technology sectors of the economy and resulted in a conspicuous demand for personal computers (PCs) and information technology (IT) equipment.

The western region of India generates the highest amount of electronic waste (EW), contributing approximately 35% of the total, followed by the southern region, around 30%, while the northern and the eastern regions contribute, respectively, about 21 and 14%. More than 60% of electronic waste in India has its source in the sixty-five biggest cities as shown in research. Additionally, it has been found that only 10 states are responsible for 70% of this waste: Andhra Pradesh, Maharashtra, Delhi, Tamil Nadu, Uttar Pradesh, Karnataka, Gujarat, Madhya Pradesh, West Bengal, and Punjab. Together, these states contribute to 70% of the total e-waste of the country.

E-waste behavior in countries like India differs in scope due to the massive presence of the informal segment to handle e-waste. Instead of doing away with such informal sectors, it would serve the government better to harness their extensive collection networks. An incentive-based system that helps connect the formal with the informal recycling sector is essential. This means that informal recyclers would be motivated to take more amount of e-waste to formal recycling plants (Kumar et al., 2017). The absence of robust regulations and implementable management practices and uncontrolled recycling, especially at the lower levels, have resulted in the release of heavy metals into the environment (Kyere et al., 2016).

4 Sources and Types of E-waste

Electronic waste may originate from different sources including the government and commercial sectors, institutional and research organizations, homes and households, and manufacturing industries. These fields have grown considerably and, more importantly, act as a redeeming factor when the machine becomes obsolete. Though big institutions and government bodies are usually responsible for proper waste disposal, individual household waste often goes to local rag pickers. Such practices would often lead to improper handling that can affect the people involved in the process of recycling.

Leading contributors include home appliances, government entities, information technology and telecommunications equipment, hospital and commercial medical devices, and industrial machinery and equipment. Around 30% of electronic waste consists of common household appliances such as washing machines, dryers, vacuum cleaners, air conditioners, coffee machines, and irons. Computers, telephones, printers, CDs, DVDs, fax machines, hi-fi sets, et cetera, make up a further 30%, while refrigeration accounts for 20%. For instance, televisions and monitors account for 10% and 10%, respectively (Kaya, 2016). The largest component of e-waste was small equipment, at 17.4 million tons—half of all e-waste produced worldwide—in 2019. Then came the large equipment at a total of 13.1 million tons, followed by temperature exchange equipment at 10.8 million tons (Forti et al., 2020).

5 Composition of E-Waste

The electronic waste generally consists of several metals such as copper (Cu), aluminum (Al), nickel (Ni), lead (Pb), tin (Sn), manganese (Mn), and iron (Fe) from which we can extract the metals (Akcin, 2021; Akdeniz & Özmen, 2011; Akram et al., 2019; Althaf et al., 2021). Moreover, it consists of various types of plastics, glass fiber, metal oxides, and ceramics (Ilankoon et al., 2018). For instance, an old personal computer with a cathode-ray tube (CRT) monitor can weigh around 25 kg. Depending on the electronic components, the composition of this device is formed by 43.7% metals, 17.3% electronic components, 15% glass materials, and 23.3% plastics (Berkhout & Hertin, 2004).

Almost all e-waste contains some amounts of valuable metals, with copper being the major one. Characteristics in Metal Composition of Different Electrical and Electronics Products: A recent trend has been identified based on the type of electronic waste being generated, implying that, generally, heavier e-appliances exhibit higher concentrations of potential contaminants than lightweight e-waste. Refrigerators, washing machines, etc., are mainly made of steel and a small dose of your environmental impurities. To the contrary, the presence of heavy metals and fire retardants is more pronounced in devices like computer, laptop, and mobile phones.

Due to their exceptional oxidation resistance, wear resistance, and other beneficial properties, platinum group metals are used in electrically conductive materials. The concentrations of valuable metals contained in waste printed circuit boards (PCBs) are several orders of magnitude, approximately ten times, greater than that found in minerals through conventional mining techniques (Ilankoon et al., 2018). Nonetheless, the composition of e-waste is highly variable based on the type, origin, age, and manufacturer of the device (Bigum et al., 2017). For example, processing one million mobile phones can produce about 24 kg of gold, 9 kg of palladium, 250 kg of silver, and 9000 kg of copper (Ilankoon et al., 2018). Some e-waste may also include uncommon and complex mixtures of hazardous pollutants which are unique to that waste stream. Lithium, lanthanum, beryllium, cerium, antimony, indium, and gallium are used in batteries, studio lighting, conductive materials, fire retardants, LCD monitors, and silicon chips (Ladou & Lovegrove, 2008).

6 Leachates from E-Waste

Leachate is a high concentration, toxic type of by-product that is produced from landfills, composting facilities, e-waste processing centers, solid waste disposal sites, and incineration plants. Leachates are characterized by a higher complexity than residential sewage and usually consist of a mixture of toxic organo-pollutants, heavy metals, ammonia nitrogen compounds, and other aromatic compounds (Youcai, 2018). It is a strongly varied and diverse composition. Electronic waste disposal sites produce more toxic and harmful leachate than other sources.

Leachate composition is a function of the location, season, landfill lifespan and design, waste type, storage cell retention time in incineration plants, and waste source (Youcai, 2018) (negative excerpts). The volume of leachate depends on the site, moisture condition, time, and treatment type (Armstrong & Rowe, 1999).

Persistent organic pollutants (POPs) and polychlorinated biphenyls (PCBs) can be released into various ecosystems when e-waste is disposed of in landfills. PCBs are made up of ~ 200 chemical compounds known as congeners. Persistent organic pollutants (POPs) can be found in remote locations and in areas where they have never been used; they can be transported over 1600 km by oceanic or atmospheric currents (Castro-Jiménez et al., 2007).

The main contributors of leachate from landfills include rainfall, infiltration, surface leachate infiltration, groundwater infiltration, leachate from the coverage materials, free water contained within the waste material, and leachate resulting from organic decay (Youcai, 2018).

Properties of E-Waste Leachates

E-waste are complex materials requiring specialized treatment and cannot be dumped in open landfills. Electronic wastes are composed of toxic materials such as polychlorinated biphenyls (PCBs), mercury, polybrominated biphenyls (PBBs), brominated flame retardants (BFRs), polybrominated diphenyl ethers (PBDEs), and lead along with precious materials like steel, iron, aluminum, copper, silver, gold, platinum, plastics, and palladium (Chancerel et al., 2009). Other toxic substances present in e-waste can include antimony, arsenic, barium, beryllium, TBBPA (tetrabromobisphenol A), chlorofluorocarbons (CFCs), hexavalent chromium (Cr VI), nickel, and polyvinyl chloride (PVC) (Kiddee et al., 2020).

The characteristics of leachate depend on waste composition, moisture content, source, and conditions of environment. One of the factors that influencing the characteristics of landfill leachate is precipitation. Rainwater pulls the soluble waste from a solid to liquid as it percolates through the dump sites. Leachates are a highly concentrated outflows of both organic and inorganic contaminants that pose hazards.

When e-waste is disposed of in landfills with industrial and municipal solid waste, it becomes extremely poisonous and hazardous, and the leachate content obtained through the leaching of contaminants from the e-waste becomes complicated with time. Studies show that leachate generated from open sites of e-waste dumping contains deleterious agents such as aneugenic and clastogenic agents to living organisms (Bakare et al., 2009). In addition, studies at laboratory scale reveal that leachates generated from shredded e-waste display significantly higher concentrations of metals like barium, beryllium, aluminum, cadmium, chromium, copper, cobalt, lead, nickel, and vanadium than those found in leachates resulting from other waste classes (Kiddee et al., 2013, 2014).

7 Toxicity of E-Waste

Poor electronics waste management and handling not only leads to the release of toxic elements to the water, air, and soil (Deng et al., 2006; Leung et al., 2006) but also introduces these pollutants in the human food chain and environment. These can include dangerous metals, polyaromatic hydrocarbons (PAHs), polychlorinated biphenyls (PCBs), polybrominated diphenyl ethers (PBDEs), and polyvinyl chloride phthalates and other contaminants (Wong et al., 2007). These pollutants persist for long periods in soil due to their settlement after long-range transport significantly impacting soil microorganisms and higher species health. They also generate changes in soil microbiology phylogenetically and functionally. According to research done by Salam and Varma in 2019, e-waste contamination caused drastic reductions in bacterial diversity and was established by denaturing gradient gel electrophoresis (DGGE) study of bacterial 16s-RNA. The pollution led to a decline of important bacterial taxa including Firmicutes and Proteobacteria.

Many studies have been conducted on the effects of metals (cadmium, lead, chromium, beryllium, mercury) and organic pollutants (PAH, PCB, etc.), which are incorporated into the soil during the process of e-waste materials disposal, on soil enzymatic activity, soil ecosystems, and the physicochemical properties of the soil (Chaperon & Sauve, 2007; Chaperon & Sauv, 2008; Kurek & Bollag, 2004).

In addition, heavy-metals-laced leachate can easily contaminate water bodies, soil, and air and enter the food chain through primary producers, affecting living beings directly. Most metals and compounds have an ionic nature and therefore can easily be taken up by plants and wildlife.

Many studies have been conducted to investigate the leachability of e-waste constituents in real and artificial landfill as well as laboratory settings. Studies have shown that mean concentrations of cadmium, antimony, chromium, aluminum, barium, beryllium, cobalt, copper, nickel, lead, and vanadium in leachates from fragmented e-waste column were found to be significantly higher in the columns with e-waste than in columns without e-waste (Kiddee et al., 2013). Many studies have concluded that e-waste bearing landfills had higher levels of toxic chemicals than landfills that did not contain e-waste.

Of all examples of e-waste, used-up mobile phones constitute the fastest-growing batch of obsolete devices. Brominated flame retardants (BFRs) and toxic metals are common toxicants that are often incorporated into plastics used in electrical and electronic equipment. Singh et al. investigated toxicity trends for 19 chemicals in basic and cellphones from 2001 to 2015 and found that the amount of harmful substances increased in relation to non-harmful compounds over the 15-year period. Plastic waste from e-waste (Awasthi et al., 2019; Berkhout & Hertin, 2004) is growing to a historic level globally in a very short time; it is a major issue.

Toxic Effects on Human Health

It is harmful to humans in a variety of ways; exposure can disrupt thyroid function, cause bad outcomes in newborns, modify cellular function and expression, cause psychological changes to behavior and temperament, and cause lung dysfunction. Populations from e-waste recycling towns have significantly more DNA damage than those in control towns (Grant et al., 2013). Leachate has been studied in both plant and bacterial cells to investigate the effects of leachate on genetic material; in addition, leachate has been shown to act as a genotoxic agent in mammalian cells (Youcai, 2018).

Cell absorption of leachates causes pH variations inside and outside the cell and subsequently modifies enzyme activity and DNA conformation (Youcai, 2018). Alternatively, free radical formation into the cell leads to lipid peroxidation of the cell membrane in tissues, which results in DNA fragmentation in human tissues. In addition, the other contribution of free radicals is base substitution by modifying components of nucleic acids, such as purine and pyrimidine, with consequent fragmentation of the DNA strands and inducing the downstream mutations (Youcai, 2018).

Hair, placentas, and human milk show high levels of dioxins and other e-waste toxic contaminants, indicating their body infiltration via food, water, and air in enough quantities to pose significant health risks (Chan et al., 2007). Electronic waste PCBs leach into surface water and groundwater, alimentary soils, rice, fish, eggs, and meat, impacting humans. Zhao et al. set Paylan and colleagues (2008) conducted further studies supporting the presence of PBDEs, PCBs, and PBBs in human hair. They found concentrations of 30.182 and 58 ng/g in the hair of persons working in e-waste disassembly areas.

Many biomonitoring studies have reported that women and babies in the womb are considerably exposed to a multitude of chemicals emitting from these reprocessing factories in developing countries (Cong et al., 2018; Dai et al., 2017; Jedrychowski et al., 2015; Li et al., 2017). While this exposure to toxic hazards affects all groups, young and old, the consequences manifest differently in adult and pediatric bodies.

From the beginning, children are exposed to contaminants through breastfeeding by their mothers (Li et al., 2017). In addition, a fetus can also be exposed during gestation (Li et al., 2017). Studies have shown that children show significantly higher levels of lead and cadmium, have lower cognitive skills in comparison with non-e-waste affected populations, and also have higher incidences of respiratory problems such as coughing (Li et al., 2008). A highly concentrated level of lead in children's blood from e-waste recycling areas is possibly associated with immunoglobulin E (IgE) production and is widely accepted in causing asthma in humans and impaired immunity, resulting in the human body unable to resist hepatitis (Xu et al., 2015b; Zeng et al., 2016).

Long-term ingestion or accumulation of toxins in the human body will disrupt the body system and cause damage to the nervous system, physical and muscle, leading to diseases such as Parkinson's disease, Alzheimer's disease, multiple sclerosis, and muscular dystrophy. The major threats are posed by metal-contaminated soils that can accurately be evaluated by the determination of the transfer of metals and metalloids into the human food chain.

Eguchi et al. (2014) collected 131 e-waste site serum samples from Northern Vietnam and reported that e-waste-derived biopersistent pollutants (such as perchlorate (ClO_4^-) and thiocyanate (SCN^-)) may affect iodide (I^-) absorption through the iodide/sodium cotransporter and, hence, the synthesis of thyroid hormone (TH) in the thyroid gland. Perchlorate (ClO_4^-) is another significant contaminant, which generally comes from explosives, fireworks, rocket propellant, and the polyvinyl chloride (PVC) industry (Eguchi et al., 2014).

Heavy metals are stable, and they do not disintegrate, thus allowing them to persist in the environment for a very long time. Upon entry into the body through various routes, metals (e.g., antimony, arsenic, chromium, and cadmium) adversely affect human health (Li et al., 2014). These metals then accumulate in plants and subsequently in animals and humans through the food chain. Different metals also have their unique harmful effects when in contact with humans; for example, lead can affect behavior and cognitive functions, copper can affect hepatic function (Danzeisen et al., 2007), and increased levels of cadmium can result in lung cancer and renal damage (Ebrahimi et al., 2020). Electronic waste (e-waste) can negatively affect the health of people who produce/manufacture and recycle it (Wang et al., 2009). The lead can float in the air for 10 days, exploding with wind and then raining down to contaminate surrounding soil and water. Lead is one of the most well-established carcinogens (Garcia-Leston et al., 2010; Grover et al., 2010).

Toxic Effects on Plants

Much less research has been done on the negative effects of e-waste on vegetation; however, its constituents, particularly the toxicants (cadmium, manganese, nickel, chromium, zinc, and palladium) in leachates, are known Bauddh and Singh (2011). Bakare et al. (2012) found that the phytotoxicity of e-waste components was able to induce cytotoxicity and somatic mutations in Allium cepa. The cytogenetic effects and root length were determined after 48 and 72 h on exposure to five different concentrations of e-waste leachate (1, 5, 10, 25, and 50% v/v leachate/tap water). The greatest root growth was at a concentration of 1%, while the lower was at a concentration of 50% (Bakare et al., 2012).

Chromosomes disease like sticky, anaphase bridge, and bi-nucleate cells were identified in e-waste leachate. Identified metals in Allium cepa that have been indicated as responsible for such anomalies include, but are not limited to, iron, nickel, chromium, arsenic, zinc, lead, and cadmium. These abnormalities show that e-waste leachates can cause somatic mutations and cytotoxic effects.

The adverse impact of e-waste leachates on chlorophyll production and the primary production of key microalga *Pseudokirchneriella subcapitata* has been shown to stem from reduction in the number of active reaction centers of the photosystem II (PSII), indicating that e-waste leachates have a negative influence on photosynthetic performance. PBDEs and metals can cause oxidative stress and inhibit the activity of antioxidative enzymes like ascorbate peroxidase (APX), chloramphenicol acetyltransferase (CAT), glutathione S-transferases (GST), and superoxide dismutase (SOD) in algae.

Another study reported that PAHs inhibit PSII and reduce cell growth (Kummerová et al., 2008). Generally, values of e-waste leachates significantly exceed limits with copper, whose physiological effects on plants have been extensively researched. Copper is a known strong inhibitor of PSII electron transport activity linked to the photolysis of water. Copper can affect energy storage in algae and cause a decrease of chlorophyll a during photosynthesis. According to previous studies, plant growth in a medium containing 100 μg/L of copper affects the rate of growth and the biomass (Bossuyt & Janssen, 2004).

The Allium assays have been used to evaluate DNA damage from e-wastes in the forms of chromosomal defects and mitotic cycle interruptions, as these tests are sensitive to multiple mutagenic agents and have a short cycle time (Akdeniz & Özmen, 2011; Evseeva et al., 2003). The laboratory report results could indicate the presence of certain mutagenic, genotoxic, or cytotoxic compounds in the environment, which can pose both direct and indirect risks to all living organisms.

Heavy metals in large amounts are harmful to plants and uptake in leaves or the roots. These metals block plants from absorbing other essential nutrients (Djingova & Kuleff, 2000). By substituting critical nutrient cations at the cation exchange site of the plant, the metals inhibit the functioning of cytoplasmic enzymes of the plant cell and cause the cell to be injured by oxidative stress (Jadia & Fulekar, 2009).

Depending on the relatively heavy metals, the reaction of plant activity can be either positive, negative, or neutral; most of the heavy metals such as lead, cadmium, mercury, chromium, arsenic, and beryllium are harmful, while copper, iron, and aluminum are beneficial to plant growth. Kibra (2008) reported significant inhibition of tiller and panicle formation in rice plants grown at a mercury concentration of 1 mg Hg/kg. For instance, Nicholls and Mal studied the combined effect of lead and copper at doses of 500 mg/kg and 1000 mg/kg, respectively, and found that leaves of the plant were dying more in *Lythrum salicaria*.

Toxic Effect on Soil and Soil Microbial Community

Healthy soil systems are necessary for functioning ecosystems—natural and cultivated alike. Soil is considered one of the two major ultimate sinks of chemical pollutants. Soil and its biota as well as dynamic processes are affected by many contaminants, especially the more toxic and persistent ones. Microorganisms make significant contributions to the functionality of soil and are known to be indicative of soil quality based on their dependability on the contaminants present in the soil (Abd Elsalam et al., 2006; Andreoni et al., 2004).

Microorganisms are essential to maintaining soil fertility and primary production through organic matter degradation and nutrient cycling. E-waste contains hazardous elements and compounds that affect soil quality and the microbes living in the soil. Even higher concentrations of toxic metal and chemicals are known to have effects on soil microbial communities and their functions. It is well known that accumulated heavy metals can decline soil microbial biomass (de Quadros et al., 2016) and diverse enzyme activities. This leads to a decrease in functional diversity of soil microorganisms (Abdu et al., 2017) and a change in microbial community structure (Beattie et al., 2018). In fact, studies show chronic exposure to metals could lead to the development of metal tolerance in microbiota (Ellis et al., 2003).

Microorganisms can play a major role in the reclamation of contaminated soils. Liu et al. (2015) found that toxicants such as copper and decabromodiphenyl ether (BDE-209) explained about 70% of the variation in microbial assembly in a canonical correspondence analysis among several environmental factors. The results suggest that poor handling of electronic waste can significantly affect microbial population diversity (Liu et al., 2015). Microorganisms are key catalysts of biological cycles of Earth and are essential for the functioning of nearly all ecosystems. Microbial metabolic processes are sensitive to small changes in the environment.

Many species (e.g., *Pseudomonas* spp., *Acinetobacter* spp. and *Sphingomonas* spp.) that show tolerance and resistance of organisms to heavy metals such as cadmium, zinc, and copper have been isolated and studied (Gotebiewski et al., 2014). The latest studies focus on the isolation of pure cultures of bacterial communities of soils that are either polluted with hazardous organic pollutants or are able to successfully degrade PCBs, PBDEs, and PAHs (Golyshin et al., 2003; Liang et al., 2012). A better understanding of these bacteria will, however, assist to extend their degradative potential for bioremediation of polluted biosphere (Field & Sierra-Alvarez, 2008; Haritash & Kaushik, 2009). Some studies have attempted to elucidate the individual and collective effects of certain metals or chemicals on soil microorganism assemblages in the laboratory.

Toxic Effects on Aquatic Organisms

By leaching into the water bodies, the pollutants from the e-waste easily enter and impact the aquatic species in various ways such as growth and reproduction. There is a negative correlation between e-waste pollution and the amount of fungi, suggesting a strong relationship between pollution and organisms. A significant change in biochemical oxygen demand (BOD) and pH of water samples was observed due to pollution. There have been reports of fish and other aquatic creatures being exposed to multiple heavy metals. Javed and Usmani (2011) these metals are accumulated in their tissues by biomagnification.

A study by Luo et al. at China's Nanyang River revealed carp fish bioaccumulation of PBDEs as high as 766 ng/g (fresh weight). However, based on a follow-up study, the levels of PBDEs detected in the sediments were also high, even as high as 16,000 ng/g in the same polluted e-waste river. And an apex predator water snake showed an average concentration of 1091 ng/g PBDEs and 16,512 ng/g PCBs (in fresh weight) near an e-waste recycling site (Wu et al., 2008). In addition, other wild aquatic species at these study sites, including mud carp, crucian carp, northern snakehead, Chinese mystery snail, and prawns, showed elevated concentrations of PBDEs (830, 316, 490, 67.5, and 254 ng/g) and PCBs (12,390, 8338, 7052, 62.8, and 3503 ng/g), respectively. Ambient water concentrations of PCBs and PBDEs were 204 and 24.4 ng/L, respectively. High levels of PBDEs and PCBs were found in extraterrestrial aquatic avifauna, including waterfowl, in the downstream sector of the Pearl River Delta of China. Decabromodiphenyl ethane, 1,2-bis(2,4,6-tribromophenoxy) ethane and TBBPA bis(2,3-dibromopropyl) brominated flame retardants were also detected in several aquatics from the same river near e-waste recycling villages.

8 E-Waste and Groundwater Pollution

Groundwater is an important resource, increasingly threatened by pollution and contamination. One major contributor to this issue is e-waste. The lack of knowledge regarding how to dispose of the e-waste correctly leads toward a drastic environmental and health problems arising mainly in developing countries such as India and to some extent in developed countries (Gupta & Nath, 2020).

Additionally, heavy metals like cyanide and mercury as well as auxiliary residues have the potential to leach into the soil and form organic and inorganic complexes in the soil matrix. The dumping sites of e-waste are very near, and thus, the quality of groundwater is highly depleted. In a similar study, Panwar and Ahmed (2018) found significant correlations between copper and cadmium and between lead and chromium in soils, indicating that the release of one metal was dependent on the presence of the other. The photographs show a strong correlation between copper and cadmium and nickel and chromium in the analyzed water samples. The observed increase in cadmium concentration indicates that there was a positive correlation with elevated levels of copper recorded from both water and soil samples. The high correlation between copper and cadmium and lead and chromium in soil samples indicates that there may have been a common source of these metals (Panwar & Ahmed, 2018).

Toxic elements in e-waste have been found to react with groundwater, changing its pH, total dissolved solids (TDS), turbidity, chloride content, and conductivity. Such changes are associated with potential health risks in both children and adults (Dharini et al., 2017). Many industries are based next to water sources; toxic electronic waste seeps into the soil and permeates the deep aquifers above it. This water is not safe to drink or use for the other purposes listed. Studies have found elevated levels of pollutants near the region of landfills, which decrease as the aquifers are farther away from the landfills (Gupta & Nath, 2020).

9 Management of E-Waste

Proper e-waste management techniques can help combat the alarming rise in electronic waste. The presence of such valuable metals highlights the need for efficient waste management practices that can recover both base and precious metals, including metals like copper and gold, from electronic waste. IoT technology has also been applied to the collection of e-waste, thus providing unique and intelligent solutions (Kang et al., 2020).

This is highly relevant to several of the UN Sustainable Development Goals (SDGs) which are linked to the need for proper management of e-waste. It particularly plays a role in responsible production and consumption (Goal 12), building sustainable cities and communities (Goal 11), conserving life below water (Goal 14), and ensuring clean water and sanitation (Goal 6) toward 2030 (Yong et al., 2019). In the recycling of such wastes and the recovery of valuable components, some nations have successfully implemented e-waste management strategies and improved recycling technologies (for instance, integrated hydrometallurgical and pyrometallurgical smelters) (Yong et al., 2019).

The scope of bioprospecting of e-waste shows a great promise and has major benefits in terms of environmental sustainability, public health, and economic development (Awasthi & Li, 2019). Even electronic waste management and disposal has emerged as a considerable challenge in the wake of expansive population increase and industrialization, which implies reinforced dependence on electronic devices and equipment. E-waste has become an important part of the solid waste. Currently, we are left with a handful of ways to deal with this issue, including landfilling and incineration. Each of these methods comes with its own pros and cons.

Landfilling

Note that landfilling remains one of the widely used electronic waste disposal methods. This method consists of digging trenches in the ground, where e-waste is placed and then covered with a thick layer of soil to reduce air exposure. However, such a method is not suitable for the leachate, the liquid that emerges from the runoff, which pollutes the surrounding soil and water sources. To solve this problem, improved landfill technology like secure landfills includes an impervious layer (barrier) of plastic or clay to prevent leachates from entering the soil and water. At the basin level, collecting leachates and transferring them to wastewater treatment plants may also be part of the solution to leachate disposal problems.

In e-waste recycling, barrier technology has received great acceptance as a suitable alternative for the e-waste challenges, mainly owing to its cost-effective nature (Gupta & Nath, 2020). As noted by Kui and Xiangping (2019), the classification of this technology includes three categories: horizontal barrier technology, vertical barrier technology, and cover barrier technology. Horizontal barriers may play a major role in controlling the downward migration of metals contaminants and can function as under lining tints. These barriers can work vertically and prevent contaminated water from traveling through deep aquifers. It should have properties such as impermeability, high tensile strength, high flexibility, and flexibility to stop the motion of leachate. In addition, the material must have a black color to resist UV light (Gupta & Nath, 2020).

However, the techniques used in landfilling do not solve various problems such as the leaching of the landfills, chemical reactions, vaporization, and uncontrolled fires. Moreover, the landfilling does not find applications as rightful treatment for volatile and non-biologically degradable substances such as cadmium, mercury, and CFCs, nor for persistent compounds like PCBs and BFRs (Annamalai, 2015).

Incineration

Incineration is an alternative waste disposal process where waste materials are combusted at high temperature (900–1000 °C) in specially designed incinerators. These include reducing the amount of electronic waste and taking advantage of the energy content of combustible materials. The main drawbacks of e-waste incineration are release of flue gases and large volumes of by-products, both from the combustion process and from the cleaning of the gas. Incineration processes used in e-waste facilities make a significant contribution to the annual emission of mercury and cadmium into the environment. Therefore, unless timely and appropriate remediation methods are undertaken, the emission flue gas and other by-products are expected to increase during the incineration of electronic waste (Annamalai, 2015).

One of the biggest disadvantages of incinerating toxic waste is the generation of harmful emissions throughout the process, including furans and dioxins. Pyrolysis of e-waste converts polybrominated diphenyl ethers (PBDEs) into other more hazardous compounds, including polychlorinated dibenzo-p-dioxin (PBDD) and polychlorinated dibenzofuran (PBDF) (Fu et al., 2016). These are synthetic poisons that remain in the environment, leading to chronic diseases in people with genetic disorders and other ill effects, including breastmilk contamination (Devika, 2010). As persistent lipophilic compounds, PBDD/Fs bioaccumulate along the food chain and thus in the human body.

E-Waste Recycling and Management

E-waste recycling refers to developing, building, purchasing, or marketing new electronics that are generated from used materials. Several factors shape the journey of a fully functioning electronic device to electronic waste, such as the increasing market presence in the global south, the rapid turnover in the global north, and the skyrocketing rates of obsolescence (Garlapati, 2016). Apart from this rapid change and development in technology, changing trends and social value, irresistible offers on electronic devices, and the short lifespan of electronic devices contribute to the huge amount of e-waste generation (Borthakur & Singh, 2021).

As Shaikh et al. explain, one of the benefits of electronic waste recycling is mining critical materials from obsolete electronics.

Research at present is receiving considerable interest in the fabrication of innovative equipment through the use of the recycled materials. We are seeing a positive side, which not only resolves the crisis of waste disposal but also paves the path toward sustainability for diverse industries. Using recycled resources, you can design and manufacture equipment or machinery according to the needs of society and with minimal harm to the environment.

Benefits include:

Conservation of finite natural resources,
Reducing the effects of pollution, and
Enabling the generation of job opportunities.

It is critical to recycle electronic waste since it has a way to help us reduce a great danger toward important natural resource. Today, there are a set of sequential steps to the procedure because e-waste is an emerging issue that requires that practical considerations are engaged in response. The solution lies in creating a competitive system of recycling and reuse.

There are plenty of compelling reasons that strongly support the need for new systems and methodologies to be established for the collection, reuse, and recycling of e-waste. Our data are relevant only till October 2023, and hence, the strategies used to dismantle electronic waste must focus on reusable components and devices, enhancing the recyclability of materials from electronic waste, and a societal setting that strikes the right balance between the rapid advancement of technology and its responsible disposal. This model stands in stark contrast to the current system, in which electronic waste ends up dumped in a landfill once no longer in use. High-level frameworks must also be informed by the cycles of materials that flow through electrical products, for example, by integrating collectors, recyclers, and reusers into the overall system.

Another fascinating proposal regarding the problem of e-waste in the USA is the creation of an e-market for returned deposit. A cyber-infrastructure-enabled e-market (Dzombak et al., 2019) addresses e-waste management. This model is a consumer depositing with a seller that lets some of the deposit be refunded to the consumer when the product is returned at the end of its life cycle. Its minimum deposit shall be sufficient to cover the costs of the product's recycling and transport. This brings up a very interesting concept, as it addresses the possibility of reusing. In that case, consumers may receive a price higher than the amount paid to the reseller, especially when a working computer or the like is still valuable to the reuse market. This system can be implemented by attaching a radiofrequency identification device (RFID) to the product to track the flows of both economy and materials (Condemi et al., 2019; Kahhat et al., 2008).

Printed circuit board recycling is mostly performed using pyrometallurgical methods, which have considerable environmental impact and implication on several environmental problems (Kaya, 2019). Hydrometallurgical methods are currently under investigation for extracting precious and base metals from waste printed circuit boards. The predictability, controllability, and precision of these methods are well recognized (Kaya, 2019). Percentage of magnetic fraction, size distribution, leachability, and metal content were determined through atomic absorption spectroscopy (AAS), volumetric analysis, and X-ray fluorescent spectrometry (XRF). As such, hydrometallurgical techniques provide a wider range than classical methods (Kamberović et al., 2018).

The electronic waste is typically recycled through three methods, depending on the structure/ingredients of the waste materials: chemical recycling, mechanical recycling, and thermal recycling (Kaya, 2016). A range of electronic accessories such as CRT (cathode-ray tubes) and monitors, telephone boards, hard and floppy drives, modems, keyboards, compact disks, mobile phones, fax machines, laptops, printers, memory chips, CPUs, cables, and connecting wires can all be recycled (Annamalai, 2015). The main process of recycling is disassembly, which is the methodical removal of incremental elements of e-waste that would produce hazardous pollution such as PCB and mercury. This also forms a part of the removal of PVC-containing plastics from the back of the device (CRT extraction) along with the collection of valuable metals like copper, lead, silver, and gold (from the waste stream).

Recycling is not only efficient [up to 95% of valuable materials can be extracted from discarded computers (Ladou & Lovegrove, 2008)] but also high yielding [between 45 and 95% of CRT monitors can be recycled (Ladou & Lovegrove, 2008)]. There are many other modern methods which can successfully recover high-lead glass from discarded CRTs while protecting the environment (Andreola et al., 2007). Recycling has proven to be the safest option, and, compared to incineration of e-waste, or landfilling, construction accounts the lowest adverse environmental impact (Hischier et al., 2005; Kaya, 2016). Although recycling has many benefits, it is important to address the impacts of the potential hazards of electronic waste recycling. One important fear is the liberation of toxic elements that can appear as fine airborne particulates. Some deposits can be close to the emission site, and others can be transported to large distances that can cause large areas to be polluted (Khanna et al., 2020a, 2020b).

Reuse

To prevent retirement that happens in a short span of time, the practice of eLogs reuse comes handy. Это или некое вторичное народное использование или адаптация с незначительным редактированием оригинала. These can include computers and even cell phones—the latter being a particularly useful example, as one or more of these devices can frequently be put to work with inexperienced users who cannot afford to buy new kits. Not only are people able to maximize how they use their electronic devices, but it also helps to reduce the amount of electronic waste that is created. Reuse reduces the creation of hazardous waste through the extended lifetimes of electronic and electrical products. Reusing gadgets is also where you have a positive impact on the environment and are preventing the depletion of precious natural resources. This strategy also temporarily alleviates the need for either recycling or disposal (Forti et al., 2020).

Recycling includes recovering reusable materials separate from production waste or other scrap material—but despite the significance of e-waste recycling, consumers remain unaware about it. Additionally, there are few options available for recycling. The infrastructure to recycle and reuse has fragmented and roots of inconsistency which subsequently end in costly and inconvenient solutions for consumers. One major impediment to software reuse is the software licensing agreements (Cairns, 2005). New approaches are being studied for the recovery of electronic waste such as applying use of low-cost biomaterials to recover/purify metals. This method focuses on eco-efficient technology which provides a high degree of purity and improves the reusability of the recovered copper. These methods are defined under "greening the waste".

10 E-Waste as Secondary Metal Resource

Moreover, e-waste is not only hazardous, but also it is actually an advantage since it contains a secondary metal resources, especially precious and rare metals (Baniasadi et al., 2019). In other words, the concentration of metals like copper, gold, silver, and tin in discarded printed circuit boards (PCBs) can be up to ten times more pure than that in the natural minerals used in mining. This property makes the waste PCBs a rich secondary source for these valuable metals. Metals are almost 60% of the electronic waste which also includes some rare metals and precious metals. Due to the expected increase in the global demand of the metals within the upcoming years, metal recovery from e-waste is urgent (Islam et al., 2020).

Beyond metals, electronic waste—especially high-volume streams like printed circuit boards (PCBs)—includes toxic compounds (for example, dioxins, dibenzofurans) and toxic metals (for example, cadmium, mercury, lead, chromium). This highlights their recycling for environmental reasons (Guo et al., 2019; Wang et al., 2017). Gold, silver, lead, palladium, zinc, cobalt, copper, manganese, and nickel are some of the major metals recognized for recovery from electronic waste. The Tokyo 2020 medal concept is based on the recycling of 78,985 tons of e-waste aggregated across Japan to produce medals for the both Olympics and Paralympics Games. Out of the collected e-waste, 32 kg gold, 3500 kg silver, and 2200 kg bronze were recovered.

Recycling was largely informal, and informal recycling involved manual disassembly and burning in the open. Not only were these processes lower in efficiency, but they also had significant negative consequences on the environment. Collecting materials, followed by pretreatment processes and metal recovery, is just one part of the structured recycling process of electronic waste. Pretreatment involves critical steps such as physical disassembly, de-soldering, size reduction, and fractionation of metals and non-metals (Ashiq et al., 2019; Ilankoon et al., 2018). The metallic fraction is being recovered using different methods such as pyrometallurgical methods like pyrolysis, smelting, burning, roasting, heating, and electrometallurgy. Hydrometallurgical techniques, inclusive pretreatment, leaching and metal purification processes (Arya & Kumar, 2020a), and bio-hydrometallurgical methods are also utilized.

Hydrometallurgical Recovery of Metals Resources from E-Waste

This technique uses aqueous solutions to selectively leach metals, enabling their recovery and recycling in the process. This approach also has the potential to reduce the environmental burden, as electronic waste is managed in the context of resource efficiency.

Hydrometallurgy in the context of e-waste recycling is the process of dissolving soluble metals (such as Au, Ag, Pd, Pt, Cu, and others) from a solid state into a liquid solvent, which is required for its efficient extraction. The established and traditional approach of extracting gold includes cyanide leaching (Lee et al., 2018). Compared to pyrometallurgy, hydrometallurgical processes for the extraction of metals are more predictable and controllable, enabling better control of the process. However, this method also makes use of toxic solvents like cyanides and chlorides. For leaching, cementation, and solvent extraction, highly acid and caustic aqueous solutions can be used to dissolve and precipitate metals (Debnath et al., 2018).

The hydrometallurgical recovery of gold and silver is well established, but metals extraction from the leachate is a tedious task. Various techniques have been used to accomplish this goal, such as ion-exchange, adsorption, membrane filtration, solvent extraction, and cementation (Lahtinen et al., 2017). The successful recovery of gold with a nanoporous anion colorant sensor and captors was reported by Hussein et al. (2020). Researchers fabricated synthetic mesoporous silica nanospheres designed for ion-selective detection and extraction of Au(III) from the leach liquor of electronic waste.

Gold was recovered from ore using cyanide in very large numbers for the first time for over a century because of the high selectivity of cyanide for gold and the ability of cyanide to form the dicyanoaurate complex. What you describe that specifically relates to the dissolution of gold and silver from surfaces. There have also been documented cases of contamination of groundwater and rivers from cyanide leaching. As a result, there has been increasing focus on non-cyanide processes, such as thiourea, thiosulfate, and halide leaching methods. Chlorine and chlorides are used in the industrial recovery of gold, as well as sodium and ammonium thiosulfates, which are commonly used to recover gold from electronic waste. Several inorganic- and organic-based acids, such as sulfuric, hydrochloric, nitric, aqua regia, citric, ascorbic, malic, and oxalic acids, are commonly used to extract metals from e-waste (Abdelbasir et al., 2018; Li et al., 2018).

Pyrometallurgical Method

The main processes of TM process consist of collecting crushed electronic waste scrap into a molten bath by eliminating plastic and refractory oxides [310]. This approach generates toxic by-products like polybrominated dibenzofurans, polybrominated dibenzo-p-dioxins, hexabromocyclododecane, and dibromopropyldian, while metal recovery is incomplete. While it is far and away more sustainable than other methods that are based on physical or chemical processing, thermal consolidation for recovering valuable metals from electronic waste is still both high in energy and carbon because (due to its own nature) this part requires crack-caching. This approach can help reduce human and environmental threats posed by emerging, uncontrolled common pollutants.

Innovative approaches to address this environmental challenge have demonstrated the removal of metals from electronic waste (EW) via incineration, roasting, sintering, smelting, and melting (Khaliq et al., 2014). The pyrolysis implementation on waste printed circuit boards (WPCBs) has proven to be environmentally sustainable, with such distinctive characteristics as high energy utilization efficiency and reduced pollutant emissions (Gao et al., 2019). The metal-rich pyrolysis residues are generally used for the recovery of pure metals from them by hydrometallurgical methods (Chen et al., 2020).

Anhydrous Thermal Degradation of E-Waste Polymers to Produce Oil and Combustible Gases Panda et al. investigated the anhydrous thermal decomposition of printed circuit boards (PCBs) under anoxic conditions, examining the conversion of e-waste polymers to oil and combustible gases. Recovery of the metals was achieved by heating the metals-enriched residue from pyrolysis mixed with ammonium chloride. The results demonstrated that under the optimized roasting conditions, about 93% of copper was recovered, along with 100% of nickel, zinc, and lead. These results stand as further testament to the efficacy of the current approach as a sustainable method for the recovery of base and precious metals from e-waste.

To extract metal values, Liu et al. adopted microwave pyrolysis (60 min, 700 °C) to recover copper, tin, and lead from waste printed circuit boards (WPCBs). The results suggest that 91% copper, 96% tin, and 94% lead were successfully recovered after microwave pyrolysis. This work proves that the assessed method has real potential to be highly efficient while significantly lowering the environmental pollutants generated in the recovery of value-added metals from waste printed circuit boards (WPCBs).

*The study by Jadhao et al. (2020) details the use of ultrasonication, an environmentally friendly method, combined with low-temperature pyrolysis at 400 °C for extracting precious metals from waste printed circuit boards (WPCB). The study confirmed nearly total recovery of the following alloys of precious metals (gold, silver, palladium, and platinum) from the metallic fraction of WPCB. It was revealed from this study that the coupling of pyrolysis with ultrasonication is highly effective for treating electronic waste, recovering both base and precious metals, and generating value-added gases that may be used as an energy source.

The study by Sahle Demessie et al. (2021) evaluated the recovery of valuable metals (between others, indium, cobalt, lithium, gold, silver, and platinum) and energy from different electronic waste components by means of pyrolysis.

Ball Milling Process for Recovery of Valuable Metals

The ball milling process plays a major role in the extraction of beneficial metals. This method can wash off the metal particles from platinum ore from the ground ores by mechanical grinding. Indeed, it constitutes an energy-critical component of extracting and recycling value.

Hydrometallurgical methods raise environmental issues owing to the use of diluted and strong acids. Moreover, these processes are quasimonetary and inherently inefficient for simultaneous recovery of precious metals. Surprisingly, a large number of acidic or hazardous reagents like nitric acid, aqua regia, cyanide, and halides are used, causing the formation of much toxic and acid vapor or waste solution.

The ball milling process serves as a significant method for the recovery of valuable metals. This technique involves the mechanical grinding of materials, which facilitates the liberation of metals from their ores. Through this process, the efficiency of metal recovery can be enhanced, making it a crucial step in the extraction and recycling of valuable resources.

Hydrometallurgical methods present significant environmental concerns due to their reliance on strong and diluted acids. Additionally, these processes are often time-consuming and demonstrate inefficiency when it comes to the simultaneous recovery of precious metals. Remarkably, a significant volume of acidic or harmful reagents, such as nitric acid, aqua regia, cyanide, and halides, is employed, leading to the production of substantial quantities of toxic and acidic fumes or waste solutions. Consequently, identifying innovative approaches for the recovery of valuable metal resources from electronic waste is crucial, particularly those that are both economically viable and environmentally sustainable.

Zhang et al. (2013) describe that during mechanochemical (MC) treatment, the particles involved experience frequent breakage as they collide while e-wastes are milled with reactive chemicals in a ball mill. Zhang and colleagues introduced a comprehensive MC technology aimed at the recovery of valuable metals from electronic waste. The treatment of e-waste scrap through ball milling resulted in the formation of acid-soluble metallic compounds, which subsequently enhanced the leaching process. Remarkably, it is possible to circumvent the use of nitric acid or its combinations with hydrochloric or sulfuric acid, typically utilized in the extraction of copper, silver, and palladium, thereby preventing the emission of nitrogen dioxide or nitric oxide.

This research indicated that the bioleaching efficiencies for copper, nickel, and zinc were approximately 20% greater than those observed in unactivated WPCBs. The findings of this study indicate that the ball milling process resulted in a reduction of particle size, alterations in the electrode potential of metals, and the exposure of encapsulated metals. These changes contributed to an improved bioleaching efficiency of the metals present in waste printed circuit boards (WPCBs).

In their 2021 study, Qin et al. investigated the recovery of indium and tin through an acid leaching process that was enhanced by rapid milling techniques. The integrated approach demonstrates enhanced energy efficiency, reduced time requirements, and a lower environmental impact when juxtaposed with conventional methods. This research indicates that the dimensions of the LCD were decreased from 3 mm to 0.075 mm through the application of ball milling techniques. In addition, the recoveries of indium and tin from the particle size range of 0.075 to 0.035 mm in liquid crystal display (LCD) materials were found to be 86.3% and 76.1%, respectively, after a milling duration of 10 min using 3 M H_2SO_4 at a temperature of 85 °C. Notably, when the milling time was extended to 90 min, the recovery rates improved significantly, reaching 96.9% for indium and 85.6% for tin.

Bioleaching of Metals from E-Waste

This method utilizes microorganisms to facilitate the dissolution of metals, thereby enabling their recovery from complex waste matrices. The efficiency and effectiveness of bioleaching can vary based on several factors, including the type of microorganisms employed, the composition of the e-waste, and the environmental conditions under which the bioleaching occurs. As electronic waste continues to pose significant, both pyrometallurgy and hydrometallurgy present several drawbacks. These methods are associated with high costs, necessitate elevated temperatures, and emit toxic gases such as dioxins and furans. Additionally, they utilize potent toxic and hazardous solvents for the leaching process, which further contributes to their complexity. Moreover, these techniques demand a significant energy input (Brar et al., 2020; Harumi & Santosa, 2019). To address these challenges, research is being conducted on an ecological and economic approach for the recovery of metals from electronic waste. In recent years, researchers have been exploring a biological method that utilizes microorganisms for the extraction of metals from electronic waste. This approach is noted for its environmental efficiency and cost-effectiveness when compared to traditional methods previously examined.

Microorganisms engage in essential activities that involve the absorption and bioaccumulation of metals. This biosorption process occurs through biochemical oxidation and reduction mechanisms. The process of bioleaching in electronic waste involves the utilization of various microorganisms, including acidophilic bacteria, chemolithoautotrophic bacteria, cyanogenic bacteria, and fungi that produce organic acids. Bioleaching is a process that involves the transformation of metals from their insoluble solid forms into water-soluble states, facilitated by the activity of microorganisms. The metabolism by-products and the enzyme activities they produce play a specific role (Pant & Dhiman, 2020).

Microorganisms utilized for bioleaching are typically sourced from their natural habitats and can be obtained from leachates as well as acid mine drainage. The presence of iron-oxidizing bacteria in sulfidic concentrates has been extensively documented in relation to the bioleaching process. In recent times, there has been a growing emphasis on employing microbial techniques for the extraction of metals from electronic waste. The concept of biological recovery encompasses various processes, one of which is biosorption. This process involves the binding of diverse metals to biomass and metabolites derived from aqueous solutions. The critical factors influencing biosorption include selectivity, binding capacity with metals, material stability, cost efficiency, and renewability. The utilization of fungi, bacteria, and agricultural waste biomass for the purpose of heavy metal biosorption has been thoroughly documented in the literature (Awasthi et al., 2019).

11 Phyto-Management of E-Waste Leachates

There are two different treatment approaches for leachate treatment including (i) biological treatment methods and (ii) physicochemical treatment methods. Each treatment approach has its own benefits and downsides and various abilities for contaminant elimination. Leachate, for instance, produced over a five-year period is typically characterized by the presence of many organic compounds, especially low-molecular-weight fatty acids. Biological treatment provides a good solution in these circumstances (Youcai, 2018). On the other hand, leachate age exceeding five years results in decreased organic matter content and low COD/BOD5 ratio, situating the leachate in the order of biodegradability, as increasing humic substances negatively

(EPR) system, and these have developed into other key components of its approach to tackling this issue (Chung & Zhang, 2011).

The summary of regulations regarding effective e-waste management noted above demonstrates that there is no comprehensive approach behind these legal frameworks. The different aspects of e-waste management have been piecemeal consolidated into distinct rules, which piles up several issues on effective enforcement of these rules. In addition, China must put in place strong measures to manage one of the world's biggest informal economies involved in dismantling and recycling electronic waste. Sadly, unlike Japan, there are no legal requirements for consumers in these cases to properly dispose of waste.

Prior to the notification of "E-Waste (Management and Handling) Rules, 2011," e-waste issues in India were governed by the Hazardous Waste (Management and Handling) Rules, 2008. India introduced its first E-waste (Management and Handling) Rule 2011 in 2011 with the possibility of EPR.

The erstwhile regulation has been accompanied by a number of challenges and issues and thus was finally succeeded by the E-waste (Management) Rules, 2016. This marked the introduction of a new framework under which the responsibility of producers was expanded and other stakeholders were brought on board, including manufacturers, collection centers, consumers, bulk consumers, dealers, re-retailers, dismantlers, and recyclers, whose responsibilities were more narrowly defined under the e-waste rule of 2011 (CPCB, 2011). It was also mentioned the word EPR, but the term was never properly defined in connection to the e-waste rule framed in 2011. India passed the e-waste management rule in 2016 which builds on the previous regulations and has elaborated the concept of extended producer responsibility (EPR) in more detail. In July 2016, the regulations created additional stakeholders with the introduction of refurbishers, dealers, and product responsibility organizations (PROs) to the system. This absence of mind on the part of the government to engage with the producers is now being made up for in this evolution of regulations and would make sense in holding some producers responsible for the collection and management of e-waste generated at the end of a product's life cycle. Under this rule, a feature that has been added is called the "Deposit Refund Scheme." The 2016 regulation has simplified the authorization and registration processes (CPCB, 2011). Although adequate legislation is available for electronic waste management, there are number of loopholes in these rules which can be enlisted as follows:

The regulations are silent on how best to bring informal sectors, which are critical to e-waste management, into the fold of law and order alongside the assigned and registered sectors which are seen as formal. As much of e-waste is processed through informal sectors, it is essential for policymakers to address and design frameworks that lead to the formalization of informal systems for e-waste disposal. Also, it is important to provide the beneficiaries with incentive, training, and skill development access to get their livelihoods strengthened in these sectors (Arya & Kumar, 2020b).

Current legislation holds consumers and bulk consumers partially responsible for the appropriate disposal of electronic waste. Nevertheless, this framework seems insufficient for genuinely stimulating their engagement. With this in mind, it is possible to introduce extended consumer responsibility (ECR), similar to the concept of extended producer responsibility (EPR) that actively fosters and facilitates participation of consumers and bulk consumers in e-waste management.

The lack of clarity regarding responsibilities of various stakeholders was also noted. The rules must clearly define the roles of all the actors involved so as to leave no doubt as to who does what.

Under these regulations, State Pollution Control Boards (SPCB) have been made responsible for inventorizing e-waste in India. Nonetheless, the country still faces a major challenge of unreliable data due to disparate use of data sources. Without accurate and consistent data, applying rules and policies becomes a daunting task.

An established system within urban local bodies (ULBs) to do door-to-door collection of municipal solid waste, as envisaged in the solid waste management rules, 2016 can prove as a robust mechanism that can further be developed for carrying out e-waste management in the country. This approach can ensure that e-waste from households is collected effectively by engaging ULBs as key stakeholders.

Although the 2011 and 2016 e-waste management rules have scores of good provisions for ensuring management of electrical and electronic waste, the major challenge to this initiative is the inability of these rules to be exercised at the grassroots level. Therefore, more awareness-raising initiatives aimed at the general public, informal sectors, and different stakeholders need to be put in place. Moreover, one needs to develop competencies for deconstruction, reuse, and recovery of electric and electronic equipment. Das Land wollte sich jetzt jedoch auch nicht dazu verstiegen, die Behörden zur Bekämpfung von illegalem Recycling von Elektro-Abfällen im Land nicht schon jetzt nicht schon zu weit an die Leine zu lasten.

Ultimately, after analyzing the varying degrees of existing country laws in many countries around the world, in relation to the illegal importation and transmigration from the developed world, it is obvious that the existing limits are limited or specifically designed to be less regulated to aid a smoother access to third world nations. Putting a few of these laws in place is recognized as possibly having an impact, but there is serious worry about how its enactment might hurt countless poor folks in developing countries like India and China. Above all, it is essential that physical and financial responsibilities be equitably distributed among the various stakeholders. Moreover, common and coherent legislation with slight modifications adapted to local needs should be imposed to all countries.

Proposals to combat electronic waste alongside with multiple legislative frameworks:

E-Waste Import Restrictions

How to properly dispose of your home waste

Combining recycling initiatives with product take-back programs

There is a powerful potential to invest profitably in this sector.

Formal and informal sectors: Integrating the operations

Creating their well-formed models for a process

Retraining for effective and sustainable recycling management technologies.

Precautionary and polluter pays principles

Highlighting the need for processing on Kenyan soil

Accordingly, the units for recycling and testing have been promoted to facilitate their disposal and encourage the generator to follow responsible disposal.

Training generators on how to manage electronic waste

Initiatives that are focused on educating people on the need to recycle

Defining roles and responsibilities and accountability among recyclers

Offering tax incentives to scrap dealers.

Chaturvedi et al. (2010) have discussed how the carrot and stick approach can be adopted for performance and non-compliance issue dealing with e-waste management.

Informal handling of e-waste is an issue in this arena, especially in developing countries, which needs addressing. Roughly 95% of recycling is done in the informal sector, which consists of old-school, badly polluting backyard operations. Such practices are mostly risk-indifferent, where ambient air, water, and soil are getting contaminated, and the workers using these practices prone to several health hazards. Various policy statements by the government of India have highlighted this importance of informal sector. (4) National Environment Policy: The NEP asserts that adequate laws should be made to provide legal recognition and support to the informal sector systems for collecting and recycling different materials, emphasizing the criticality of strengthening the systems of the informal sector involved in the collection and recycling of different materials. "In particular, enhance access to institutional finance and relevant technologies". Connecting the informal sector to the formal sector could help overcome many of the problems that beset the informal sector in developing countries (Chaturvedi et al., 2010).

13 Conclusion and Future Prospects

Nearly every nation is facing the challenge of a rapidly growing volume of electronic waste. The sustainable management of electronic waste has emerged as a critical necessity in contemporary society. The adverse effects linked to electronic waste are numerous and varied. The implications are not limited to endangering human health; they also extend to the health of the environment. In contrast to the management of e-waste, which is often overlooked, the handling of municipal solid waste remains the foremost priority for numerous countries, thereby neglecting the escalating issue posed by electronic waste. In many developing economies, a significant portion of electronic waste is handled and processed through informal and unauthorized channels.

The presence of inefficient technology, coupled with a lackadaisical approach from the administration, significantly hinders the establishment of formal e-waste recycling setups. Additionally, recyclers face challenges in obtaining industry certification for the responsible recycling of e-waste. A significant turning point in this context could arise from the evolution of informal participants into small business owners. The implementation of this approach has the potential to contribute significantly to the attainment of global sustainable development goals set for 2030. It aims to reduce health complications and lower mortality rates associated with toxic and hazardous substances found in leachates.

Awareness campaigns have the potential to yield positive outcomes for both suppliers and users. Addressing the current disparities between informal and formal recycling sectors through a sustainable framework is essential for effectively tackling the issue of e-waste. The disposal of electrical and electronic equipment that cannot be repaired, as well as replaced, outdated, broken, or discarded components, often occurs alongside other municipal solid waste. This practice typically involves placing these items in informal dumpsites and unlined landfills. A significant quantity of waste and remnants from repairs often ends up in various locations, including backyards, workshops, open fields, roadsides, irrigation canals, rivers, ponds, and along riverbanks.

The infiltration of rainwater through landfills, open dumps, and dumpsites leads to the leaching of toxic substances into the soil, which subsequently results in the contamination of nearby water bodies. The use of such water for domestic purposes can present significant environmental risks and pose serious threats to human health (Fent, 2004). A number of chemical assessments have investigated the potential hazards associated with the leaching of substances from electronic waste. The complexity of toxicants produced from electronic waste presents significant challenges, and toxicity assays could serve as an effective method for evaluating potential environmental risks (Dagan et al., 2007). Individuals residing in the surrounding area are susceptible to respiratory and neurological issues. To mitigate the emission of potentially toxic elements during the processing of electronic waste, it is essential to minimize the presence of elements characterized by high vapor pressure and low melting points, such as tin, lead, and zinc (Khanna et al., 2020a, 2020b). The transition from conventional traditional methods to innovative, environmentally sustainable practices for the management of e-waste is increasingly evident. It is important to recognize that a singular solution does not exist to address all issues comprehensively. The removal of various heavy metals necessitates distinct processes tailored to each specific metal. Furthermore, addressing the issue of electronic waste requires a collaborative approach. It is essential for both government entities and citizens to fulfill their respective responsibilities to effectively mitigate issues and maintain societal stability.

The extraction of valuable metals, including both base and precious metals, from electronic waste via bio-metallurgy presents a method that is not only economically viable but also efficient and environmentally sustainable when contrasted with traditional hydrometallurgical and pyrometallurgical processes. The development of sustainable and integrated technologies is essential for the recovery of energy and valuable materials from electronic waste.

References

Abd Elsalam, H., Shameseldin, A., & Hafez, E. E. (2006). PAH degradation by two native Egyptian strains *Flavobacterium* sp. and *Pseudomonas putida*. *Journal of Applied Sciences Research, 2*, 1092–1098.

Abdelbasir, S. M., Hassan, S. S., Kamel, A. H., & El-Nasr, R. S. (2018). Status of electronic waste recycling techniques: A review. *Environmental Science and Pollution Research, 25*(17), 16533–16547.

Abdu, N., Abdullahi, A. A., & Abdulkadir, A. (2017). Heavy metals and soil microbes. *Environmental Chemistry Letters, 15*(1), 65–84.

Akcin, A. (2021). The effects of fulvic acid on physiological and anatomical characteristics of bread wheat (*Triticum aestivum* L.) cv. Flamura 85 exposed to chromium stress. *Soil and Sediment Contamination: An International Journal*, 1–20.

Akdeniz, D., & Özmen, A. (2011). Antimitotic effects of the biopesticide oxymatrine. *Caryologia, 64*(1), 117–120.

Akram, R., Fahad, S., Hashmi, M. Z., Wahid, A., Adnan, M., Mubeen, M., & Nasim, W. (2019). Trends of electronic waste pollution and its impact on the global environment and ecosystem. *Environmental Science and Pollution Research, 26*(17), 16923–16938.

Althaf, S., Babbitt, C. W., & Chen, R. (2021). The evolution of consumer electronic waste in the United States. *Journal of Industrial Ecology, 25*(3), 693–706.

Andreola, F., Barbieri, L., Corradi, A., Ferrari, A. M., Lancellotti, I., & Neri, P. (2007). Recycling of EOL CRT glass into ceramic glaze formulations and its environmental impact by LCA approach. *International Journal of Life Cycle Assessment, 12*(6), 448–454.

Andreoni, V., Cavalca, L., Rao, M. A., Nocerino, G., Bernasconi, S., Dell Amico, E., Colombo, M., & Gianfreda, L. (2004). Bacterial communities and enzyme activities of PAHs polluted soils. *Chemosphere, 57*, 401–412.

Annamalai, J. (2015). Occupational health hazards related to informal recycling of E-waste in India: An overview. *Indian Journal of Occupational and Environmental Medicine, 19*(1), 61.

Armstrong, M. D., & Rowe, R. K. (1999). Effect of landfill operations on the quality of municipal solid waste leachate. In *Proceedings of the Third International Landfill Symposium, Cagliari* (pp. 81–88).

Arya, S., & Kumar, S. (2020a). Bioleaching: Urban mining option to curb the menace of E-waste challenge. *Bioengineered, 11*(1), 640–660.

Arya, S., & Kumar, S. (2020b). E-waste in India at a glance: Current trends, regulations, challenges and management strategies. *Journal of Cleaner Production, 271*, Article 122707.

Ashiq, A., Kulkarni, J., & Vithanage, M. (2019). Hydrometallurgical recovery of metals from E-waste. In *Electronic waste management and treatment technology* (pp. 225–246). Butterworth-Heinemann.

Awasthi, A. K., Hasan, M., Mishra, Y. K., Pandey, A. K., Tiwary, B. N., Kuhad, R. C., Gupta, V. C., & Thakur, V. K. (2019). Environmentally sound system for E-waste: Biotechnological perspectives. *Current Research in Biotechnology, 1*, 58–64.

Awasthi, A. K., & Li, J. (2019). Sustainable bioprospecting of electronic waste. *Trends in Biotechnology, 37*(7), 677–680.

Bakare, A. A., Adeyemi, A. O., Adeyemi, A., Alabi, O. A., & Osibanjo, O. (2012). Cytogenotoxic effects of electronic waste leachate in *Allium cepa*. *Caryologia, 65*(2), 94–100.

Bakare, A. A., Okunola, A. A., Olusanmi, A. A., & Jenmi, H. B. (2009). Genotoxicity assessment of a pharmaceutical effluent using four bioassays. *Genetics and Molecular Biology, 32*(2), 373–381.

Balde, C. P., Wang, F., Kuehr, R., & Huisman, J. (2017). *The global E-waste monitor. United Nations University (UNU), International Telecommunication Union (ITU) & International Solid Waste Association (ISWA), Bonn/Geneva/Vienna* (pp. 1–109).

Baniasadi, M., Vakilchap, F., Bahaloo-Horeh, N., Mousavi, S. M., & Farnaud, S. (2019). Advances in bioleaching as a sustainable method for metal recovery from e-waste: A review. *Journal of Industrial and Engineering Chemistry, 76*, 75–90.

Bauddh, K., & Singh, R. P. (2011). Differential toxicity of cadmium to mustard (*Brassica juncia* L.) genotypes under higher metal levels. *Journal of Environmental Biology, 32*, 355–362.

Beattie, R. E., Henke, W., Campa, M. F., Hazen, T. C., McAliley, L. R., & Campbell, J. H. (2018). Variation in microbial community structure correlates with heavy-metal contamination in soils decades after mining ceased. *Soil Biology & Biochemistry, 126*, 57–63.

Berkhout, F., & Hertin, J. (2004). De-materialising and re-materialising: Digital technologies and the environment. *Futures, 36*(8), 903–920.

Betts, K. (2008). Producing usable materials from e-waste. *Environmental Science and Technology, 42*(18), 6782–6783.

Bigum, M., Damgaard, A., Scheutz, C., & Christensen, T. H. (2017). Environmental impacts and resource losses of incinerating misplaced household special wastes (WEEE, batteries, ink cartridges and cables). *Resources, Conservation and Recycling, 122*, 251–260.

Borthakur, A., & Singh, P. (2021). The journey from products to waste: A pilot study on perception and discarding of electronic waste in contemporary urban India. *Environmental Science and Pollution Research, 28*(19), 24511–24520.

Bossuyt, B. T., & Janssen, C. R. (2004). Long-term acclimation of *Pseudokirchneriella subcapitata* (Korshikov) Hindak to different copper concentrations: Changes in tolerance and physiology. *Aquatic Toxicology, 68*(1), 61–74.

Brar, K. K., Magdouli, S., Etteieb, S., Zolfaghari, M., Fathollahzadeh, H., Calugaru, L., & Brar, S. K. (2020). Integrated bioleaching-electrometallurgy for copper recovery—A critical review. *Journal of Cleaner Production*, 125257.

Cairns, C. N. (2005). E-waste and the consumer: Improving options to reduce, reuse and recycle. In: *Proceedings of the 2005 IEEE International Symposium on Electronics and the Environment* (pp. 237–242). IEEE.

Castro-Jiménez, J., Eisenreich, S. J., & Vives, I. (2007). *Persistent Organic Pollutants (POPs) in the European atmosphere: An updated overview*. European Commission Joint Research Centre.

Chan, J. K., Xing, G. H., Xu, Y., Liang, Y., Chen, L. X., Wu, S. C., & Wong, M. H. (2007). Body loadings and health risk assessment of polychlorinated dibenzo-p-dioxins and dibenzofurans at an intensive electronic waste recycling site in China. *Environmental Science and Technology, 41*(22), 7668–7674.

Chancerel, P., Meskers, C. E., Hagelüken, C., & Potter, V. S. (2009). Assessment of precious metal flows during preprocessing of waste electrical and electronic equipment. *Journal of Industrial Ecology, 13*(5), 791–810.

Chaperon, S., & Sauv, S. (2008). Toxicity interactions of Cd, Cu, and lead on soil urease and dehydrogenase activity in relation to chemical partitioning. *Ecotoxicology and Environmental Safety, 70*, 1–9.

Chaperon, S., & Sauve, S. (2007). Toxicity interaction of metals (Ag, Cu, Hg, Zn) to urease and dehydrogenase activities in soils. *Soil Biology & Biochemistry, 39*, 2329–2338.

Chaturvedi, A., Arora, R., & Ahmed, S. (2010). *Mainstreaming the informal sector in E-waste management* (pp. 1–5). Urban, Industrial and Hospital Waste Management Ahmedabad.

Chen, B., He, J., Sun, X., Zhao, J., Jiang, H., & Zhang, L. (2020). Separating and recycling metal mixture of pyrolyzed waste printed circuit boards by a combined method. *Waste Management, 107*, 113–120.

Chung, S. S., & Zhang, C. (2011). An evaluation of legislative measures on electrical and electronic waste in the People's Republic of China. *Waste Management, 31*(12), 2638–2646.

Chung, S. W., & Murakami-Suzuki, R. (2008). *A comparative study of E-waste recycling systems in Japan, South Korea and Taiwan from the EPR perspective: Implications for developing countries* (Vol. 21). Kojima Chiba.

Condemi, A., Cucchiella, F., & Schettini, D. (2019). Circular economy and E-waste: An opportunity from RFID TAGs. *Applied Sciences, 9*(16), 3422.

Cong, X., Xu, X., Xu, L., Li, M., Xu, C., Qin, Q., & Huo, X. (2018). Elevatedbiomarkers of sympatho-adrenomedullary activity linked to e-waste air pollutant exposure in preschool children. *Environment International, 115*, 117–126.

Central Pollution Control Board. (2011). *Implementation of E-waste rules 2011 guidelines*. Central Pollution Control Board.

Dagan, R., Dubey, B., Bitton, G., & Townsend, T. (2007). Aquatic toxicity of leachates generated from electronic devices. *Archives of Environmental Contamination and Toxicology, 53*, 168–173.

Dai, Y., Huo, X., Zhang, Y., Yang, T., Li, M., & Xu, X. (2017). Elevated lead levels and changes in blood morphology and erythrocyte CR1 in preschool children from an e-waste area. *Science of the Total Environment, 592*, 51–59.

Danzeisen, R., Araya, M., Harrison, B., Keen, C., Solioz, M., Thiele, D., & McArdle, H. J. (2007). How reliable and robust are current biomarkers for copper status? *British Journal of Nutrition, 98*, 676–683.

de Quadros, P. D., Zhalnina, K., Davis-Richardson, A. G., Drew, J. C., Menezes, F. B., Flávio, A. D. O., & Triplett, E. W. (2016). Coal mining practices reduce the microbial biomass, richness and diversity of soil. *Applied Soil Ecology, 98*, 195–203.

Debnath, B., Chowdhury, R., & Ghosh, S. K. (2018). Sustainability of metal recovery from E-waste. *Frontiers of Environmental Science & Engineering, 12*(6), 1–12.

Deng, W. J., Louie, P. K. K., Liu, W. K., Bi, X. H., Fu, J. M., & Wong, M. H. (2006). Atmospheric levels and cytotoxicity of PAHs and heavy metals in TSP and PM 2.5 at an electronic waste recycling site in southeast China. *Atmospheric Environment, 40*, 6945–6955.

Devika, S. (2010). Environmental impact of improper disposal of electronic waste. In: *Paper presented at the recent advances in space technology services and climate change 2010* (RSTS & CC-2010).

Dharini, K., Cynthia, J. B., Kamalambikai, B., Celestina, J. A. S., & Muthu, D. (2017). Hazardous E-waste and its impact on soil structure. *IOP Conference Series: Environmental and Earth Science, 80*(1), 012057.

Djingova, R., & Kuleff, I. (2000). In Vernet, J. P. (Ed.), *Instrumental techniques for trace analysis in trace elements: Their distribution and effects in the environment*. Elsevier.

Dzombak, R., Antonopoulos, C., & Dillon, H. E. (2019). Balancing technological innovation with waste burden minimization: An examination of the global lighting industry. *Waste Management, 92*, 68–74.

Ebrahimi, M., Khalili, N., Razi, S., Keshavarz-Fathi, M., Khalili, N., & Rezaei, N. (2020). Effects of lead and cadmium on the immune system and cancer progression. *Journal of Environmental Health Science and Engineering, 18*(1), 335–343.

Eguchi, A., Kunisue, T., Wu, Q., Trang, P. T. K., Viet, P. H., Kannan, K., & Tanabe, S. (2014). Occurrence of perchlorate and thiocyanate in human serum from e-waste recycling and reference sites in Vietnam: Association with thyroid hormone and iodide levels. *Archives of Environmental Contamination and Toxicology, 67*(1), 29–41.

Ellis, R. J., Morgan, P., Weightman, A. J., & Fry, J. C. (2003). Cultivation dependant and independent approaches for determining bacterial diversity in heavy-metal contaminated soil. *Applied and Environment Microbiology, 69*(6), 3223–3230.

Evseeva, T. I., Geras'kin, S. A., & Shuktomova, I. I. (2003). Genotoxicity and toxicity assay of water sampled from a radium production industry storage cell territory by means of Allium-test. *Journal of Environmental Radioactivity, 68*, 235–248.

Fent, K. (2004). Ecotoxicological effects at contaminated sites. *Toxicology, 20*, 223–240.

Field, J. A., & Sierra-Alvarez, R. (2008). Microbial transformation and degradation of polychlorinated biphenyls. *Environmental Pollution, 155*(1), 1–12.

Forti, V., Balde, C. P., Kuehr, R., & Bel, G. (2020). The global E-waste monitor 2020: Quantities, flows and the circular economy potential.

Fu, Z., Wang, Y., Chen, J., Wang, Z., & Wang, X. (2016). How PBDEs are transformed into dihydroxylated and dioxin metabolites catalyzed by the active center of cytochrome P450s: A DFT study. *Environmental Science and Technology, 50*(15), 8155–8163.

Gao, R., Liu, B., & Xu, Z. (2019). Fabrication of magnetic zeolite coated with carbon fiber using pyrolysis products from waste printed circuit boards. *Journal of Cleaner Production, 231*, 1149–1157.

Garcia-Leston, J., Mendez, J., Pasaro, E., & Laffon, B. (2010). Genotoxic effects of lead: An updated review. *Environment International, 36*(6), 623–636.

Garlapati, V. K. (2016). E-waste in India and developed countries: Management, recycling, business and biotechnological initiatives. *Renewable and Sustainable Energy Reviews, 54*, 874–881.

Ghimire, H., & Ariya, P. A. (2020). E-wastes: Bridging the knowledge gaps in global production budgets, composition, recycling and sustainability implications. *Sustainable Chemistry, 1*(2), 154–182.

Golyshin, P. N., Martins Dos Santos, V. A., Kaiser, O., Ferrer, M., Sabirova, Y. S., Lünsdorf, H., Chernikova, T. N., Golyshina, O. V., Yakimov, M. M., Pühler, A., & Timmis, K. N. (2003). Genome sequence completed of alcanivorax borkumensis, a hydrocarbon-degrading bacterium that plays a global role in oil removal from marine systems. *Journal of Biotechnology, 106*(2–3), 215–220.

Gotebiewski, M., Deja-Sikora, E., Cichosz, M., Tretyn, A., & Wróbel, B. (2014). 16S rDNA pyrosequencing analysis of bacterial community in heavy metals polluted soils. *Microbial Ecology, 67*(3), 635–647.

Grant, K., Goldizen, F. C., Sly, P. D., Brune, M. N., Neira, M., van den Berg, M., & Norman, R. E. (2013). Health consequences of exposure to E-waste: A systematic review. *The Lancet Global Health, 1*(6), e350–e361.

Grover, P., Rekhadevi, P. V., Danadevi, K., Vuyyuri, S. B., Mahboob, M., & Rahman, M. F. (2010). Genotoxicity evaluation in workers occupationally exposed to lead. *International Journal of Hygiene and Environmental Health, 213*(2), 99–106.

Guo, X., Qin, H., Tian, Q., & Li, D. (2019). Recovery of metals from waste printed circuit boards by selective leaching combined with cyclone electrowinning process. *Journal of Hazardous Materials, 384*, Article 121355.

Gupta, N., & Nath, M. (2020). Groundwater contamination by E-waste and its remedial measure-a literature review. *Journal of Physics: Conference Series, 1531*(012023, 1).

Haritash, A. K., & Kaushik, C. P. (2009). Biodegradation aspects of polycyclic aromatic hydrocarbons (PAHs): A review. *Journal of Hazardous Materials, 169*(1–3), 1–15.

Harumi, M., & Santosa, S. J. (2019). Recovery of Au (III) from printed circuit board waste by chitosan/SiO_2 coated on iron sand magnetic material. *Materials Today, 19*, 1101–1110.

Hischier, R., Wäger, P., & Gauglhofer, J. (2005). Does WEEE recycling make sense from an environmental perspective?: The environmental impacts of the SWISS take-back and recycling systems for waste electrical and electronic equipment (WEEE). *Environmental Impact Assessment, 25*(5), 525–539.

Hussein, M. A., Alamry, K. A., El Shishtawy, R. M., Elshehy, E. A., & El-Said, W. A. (2020). Nanoporous colorant sensors and captors for simultaneous recognition and recovery of gold from E-wastes. *Waste Management, 116*, 166–178.

Ilankoon, I. M. S. K., Ghorbani, Y., Chong, M. N., Herath, G., Moyo, T., & Petersen, J. (2018). E-waste in the international context—A review of trade flows, regulations, hazards, waste management strategies and technologies for value recovery. *Waste Management, 82*, 258–275.

Islam, A., Ahmed, T., Awual, M. R., Rahman, A., Sultana, M., Abd Aziz, A., & Hasan, M. (2020). Advances in sustainable approaches to recover metals from e-waste—A review. *Journal of Cleaner Production, 244*, Article 118815.

Jadhao, P. R., Ahmad, E., Pant, K. K., & Nigam, K. D. P. (2020). Environmentally friendly approach for the recovery of metallic fraction from waste printed circuit boards using pyrolysis and ultrasonication. *Waste Management, 118*, 150–160.

Jadia, C. D., & Fulekar, M. H. (2009). Phytoremediation of heavy metals: Recent techniques. *African Journal of Biotechnology, 8*, 921–928.

Javed, M., & Usmani, N. (2011). Accumulation of heavy metals in fishes: A human health concern. *International Journal of Environmental Sciences, 2.*

Jedrychowski, W. A., Perera, F. P., Majewska, R., Mrozek-Budzyn, D., Mroz, E., Roen, E. L., & Jacek, R. (2015). Depressed height gain of children associated with intrauterine exposure to polycyclic aromatic hydrocarbons (PAH) and heavy metals: The cohort prospective study. *Environmental Research, 136*, 141–147.

Kahhat, R., Kim, J., Xu, M., Allenby, B., Williams, E., & Zhang, P. (2008). Exploring e-waste management systems in the United States. *Resources, Conservation and Recycling, 52*(7), 955–964.

Kamberović, Ž., Korać, M., Ivšić, D., Nikolić, V., & Ranitović, M. (2018). Hydrometallurgical process for extraction of metals from electronic waste-part I: Material characterization and process option selection. *Metallurgical and Materials Engineering.*

Kang, K. D., Kang, H., Ilankoon, I. M. S. K., & Chong, C. Y. (2020). Electronic waste collection systems using Internet of Things (IoT): Household electronic waste management in Malaysia. *Journal of Cleaner Production, 252*, Article 119801.

Kaya, M. (2016). Recovery of metals and nonmetals from electronic waste by physical and chemical recycling processes. *Waste Management, 57*, 64–90.

Kaya, M. (2019). *Electronic waste and printed circuit board recycling technologies*. Springer.

Khaliq, A., Rhamdhani, M., Brooks, G., & Masood, S. (2014). Metal extraction processes for electronic waste and existing industrial routes: A review and Australian perspective. *Resources, 3*, 152–179.

Khanna, R., Mukherjee, P. S., & Park, M. (2020a). A critical assessment on resource recovery from electronic waste: Impact of mechanical pre-treatment. *Journal of Cleaner Production*, 122319.

Khanna, R., Saini, R., Park, M., Ellamparuthy, G., Biswal, S. K., & Mukherjee, P. S. (2020b). Factors influencing the release of potentially toxic elements (PTEs) during thermal processing of electronic waste. *Waste Management, 105*, 414–424.

Kibra, M. G. (2008). Effects of mercury on some growth parameters of rice (*Oryza sativa* L.). *Soil Environment, 27*, 23–28.

Kiddee, P., Naidu, R., & Wong, M. H. (2013). Metals and polybrominated diphenyl ethers leaching from electronic waste in simulated landfills. *Journal of Hazardous Materials, 252*, 243–249.

Kiddee, P., Naidu, R., Wong, M. H., Hearn, L., & Muller, J. F. (2014). Field investigation of the quality of fresh and aged leachates from selected landfills receiving e-waste in an arid climate. *Waste Management, 34*(11), 2292–2304.

Kiddee, P., Pradhan, J. K., Mandal, S., Biswas, J. K., & Sarkar, B. (2020). An overview of treatment technologies of E-waste. In *Handbook of electronic waste management* (pp. 1–18).

Kui, H., & Xiangping, L. (2019). Application status of vertical barrier technology in site contamination remediation. *IOP Conference Series: Earth and Environmental Science, 242*(5), 052029.

Kumar, A., Holuszko, M., & Espinosa, D. C. R. (2017). E-waste: An overview ongeneration, collection, legislation and recycling practices. *Resources, Conservation and Recycling, 122*, 32–42.

Kummerová, M., Váňová, L., Krulová, J., & Zezulka, Š. (2008). The use of physiological characteristics for comparison of organic compounds phytotoxicity. *Chemosphere, 71*(11), 2050–2059.

Kurek, E., & Bollag, J. M. (2004). Microbial immobilization of Cd released from CdO in the soil. *Biogeochemistry, 69*, 227–239.

Kyere, V. N., Greve, K., & Atiemo, S. M. (2016). Spatial assessment of soil contamination by heavy metals from informal electronic waste recycling in Agbogbloshie, Ghana. *Environmental Health and Toxicology, 31*.

Ladou, J., & Lovegrove, S. (2008). Export of electronics equipment waste. *International Journal of Occupational and Environmental Health, 14*(1), 1–10.

Lahtinen, E., Kivijärvi, L., Tatikonda, R., Väisänen, A., Rissanen, K., & Haukka, M. (2017). Selective recovery of gold from electronic waste using 3D-printed scavenger. *ACS Omega, 2*(10), 7299–7304.

Lee, H., Molstad, E., & Mishra, B. (2018). Recovery of gold and silver from secondary sources of electronic waste processing by thiourea leaching. *Journal of the Minerals Metals and Materials Society, 70*, 1616–1621.

Leung, A., Cai, Z. W., & Wong, M. H. (2006). Environmental contamination from electronic waste recycling at Guiyu, southeast China. *Journal of Material Cycles and Waste Management, 8*, 21–33.

Li, H., Eksteen, J., & Oraby, E. (2018). Hydrometallurgical recovery of metals from waste printed circuit boards (WPCBs): Current status and perspectives—A review. *Resources, Conservation and Recycling, 139*, 122–139.

Li, J., Wei, Y., Zhao, L., Zhang, J., Shangguan, Y., Li, F., & Hou, H. (2014). Bioaccessibility of antimony and arsenic in highly polluted soils of the mine area and health risk assessment associated with oral ingestion exposure. *Ecotoxicology and Environmental Safety, 110*, 308–315.

Li, X., Tian, Y., Zhang, Y., Ben, Y., & Lv, Q. (2017). Accumulation of polybrominated diphenyl ethers in breast milk of women from an e-waste recycling centre in China. *Journal of Environmental Sciences, 52*, 305–313.

Li, Y., Xu, X., Wu, K., Chen, G., Liu, J., Chen, S., & Huo, X. (2008). Monitoring of lead load and its effect on neonatal behavioural neurological assessment scores in Guiyu, an electronic waste recycling town in China. *Journal of Environmental Monitoring, 10*(10), 1233–1238.

Liang, R. B., Liu, H., Tao, F., Liu, Y., Ma, C., Liu, X. P., & Liu, J. H. (2012). Genome sequence of Pseudomonas putida Strain SJTE-1, a bacterium capable of degrading estrogens and persistent organic pollutants. *Journal of Bacteriology, 194*(17), 4781–4782.

Liu, J., He, X. X., Lin, X. R., Chen, W. C., Zhou, Q. X., Shu, W. S., & Huang, L. N. (2015). Ecological effects of combined pollution associated with e-waste recycling on the composition and diversity of soil microbial communities. *Environmental Science and Technology, 49*(11), 6438–6447.

Nnorom, I. C., & Osibanjo, O. (2008). Overview of electronic waste (e-waste) management practices and legislations, and their poor applications in the developing countries. *Resources, Conservation and Recycling, 52*(6), 843–858.

Odili, J. B., Noraziah, A., Ambar, R., Wahab, M. H., & Fakheraldin, M. A. I. (2018). E-waste generation, awareness and management in third world countries: Prospects and challenges. *The Eurasia Proceedings of Educational and Social Sciences, 9*, 349–353.

Pant, D., & Dhiman, V. (2020). An overview on environmental pollution caused by heavy metals released from e-waste and their bioleaching.

Panwar, R. M., & Ahmed, S. (2018). Assessment of contamination of soil and groundwater due to e-waste handling. *Current Science, 114*(1), 166.

Wang, H., Zhang, S., Li, B., Pan, D. A., Wu, Y., & Zuo, T. (2017). Recovery of waste printed circuit boards through pyrometallurgical processing: A review. *Resources, Conservation and Recycling, 126*, 209–218.

Wang, T., Fu, J., Wang, Y., Liao, C., Tao, Y., & Jiang, G. (2009). Use of scalp hair as indicator of human exposure to heavy metals in an electronic waste recycling area. *Environmental Pollution, 157*(8–9), 2445–2451.

Wiszniowski, J., Robert, D., Surmacz-Gorska, J., Miksch, K., & Weber, J. V. (2006). Landfill leachate treatment methods: A review. *Environmental Chemistry Letters, 4*(1), 51–61.

Wong, M. H., Wu, S. C., Deng, W. J., Yu, X. Z., Luo, Q., Leung, A. O., Wong, C. S., Luksemburg, W. J., & Wong, A. S. (2007). Export of toxic chemicals—A review of the case of uncontrolled electronic-waste recycling. *Environmental Pollution, 149*(2), 131–140.

Wu, J. P., Luo, X. J., Zhang, Y., Luo, Y., Chen, S. J., Mai, B. X., & Yang, Z. Y. (2008). Bioaccumulation of polybrominated diphenyl ethers (PBDEs) and polychlorinated biphenyls (PCBs) in wild aquatic species from an electronic waste (e-waste) recycling site in South China. *Environment International, 34*(8), 1109–1113.

Xu, X., Chen, X., Zhang, J., Guo, P., Fu, T., Dai, Y., & Huo, X. (2015a). Decreased blood hepatitis b surface antibody levels linked to e-waste lead exposure in preschool children. *Journal of Hazardous Materials, 298*, 122–128.

Yong, Y. S., Lim, Y. A., & Ilankoon, I. M. S. K. (2019). An analysis of electronic waste management strategies and recycling operations in Malaysia: Challenges and future prospects. *Journal of Cleaner Production, 224*, 151–166.

Youcai, Z. (2018). Leachate generation and characteristics. In Y. Zhao (Ed.), *Pollution control technology for leachate from municipal solid waste* (pp. 1–30). Elsevier.

Zeng, X., Xu, X., Zheng, X., Reponen, T., Chen, A., & Huo, X. (2016). Heavy metals in PM2.5 and in blood, and children's respiratory symptoms and asthma from an e-waste recycling area. *Environmental Pollution, 210*, 346–353.

Zhang, K. L., Huang, J., Yu, G., Zhang, Q. W., Deng, S. B., & Wang, B. (2013). Destruction of perfluorooctane sulfonate (PFOS) and perfluorooctanoic acid (PFOA) by ball milling. *Environmental Science and Technology, 47*, 6471–6477.

Zhang, W. L., Du, Y., Zhai, M. M., & Shang, Q. (2014). Cadmium exposure and its health effects: A 19-year follow-up study of a polluted area in China. *Science of the Total Environment, 1*, 470–471.

Smart and Innovative Technologies for Waste Disposal: Tools for Sustainable Environmental Management

Ankita Mohanty, Aditya Kishore Dash, Akshaya Kumar Verma, and Alok Prasad Das

1 Introduction

Lifestyle change, increasing commercial and industrial growth in many countries throughout the world during the last few decades has resulted in a rapid increase in both industrial and municipal solid wastes (Renou et al., 2008). Improper management of solid waste results in serious environmental, ecological, and health issues (Priyadarsini & Das, 2024). It is reported that up to 95% of the MSW collected was disposed of through landfill methods worldwide (Kurniawan et al., 2006). Landfill generates the polluted liquid called leachate (Parida et al., 2023). Waste management sustainability is found to be more complex due to rapid urbanization and industrialization, and due to changes in waste generation and composition, particularly in developing nations (Agamuthu et al., 2007). Waste collection rate is mostly linked with the area, socioeconomic level, and the region, and waste management presents a greater challenge because of the complexity in waste collection and disposal (Dey et al., 2023; Mishra et al., 2019b, 2023). This ultimately is influenced by the political, legal, environmental as well and economic factors of the area.

The rapid increase in waste generation has posed a great challenge for the management of waste through traditional methods. Hence, there is a need to assess and

A. Mohanty · A. K. Dash (✉)
Chemistry Department, ITER, Siksha 'O' Anusandhan (Deemed to be University), Bhubaneswar, India
e-mail: adityadash@soa.ac.in

A. K. Verma
Department of Civil Engineering, ITER, Siksha 'O' Anusandhan (Deemed to be University), Bhubaneswar, India

A. P. Das
Department of Life Sciences, Rama Devi Women's University, Bhubaneswar, India

A. P. Das et al. (eds.), *Sustainable Municipal Waste Management with a Zero Waste Approach*, Sustainable Environmental Waste Management Strategies,
https://doi.org/10.1007/978-3-032-05950-5_8

explore the present advancements in different computational tools for the management of waste. In this context, the cutting-edge waste management technologies like IoT, artificial intelligence, big data analysis, etc., can be adopted. This can help governments and municipalities to optimize waste collection, waste recycling, and disposal methods for enhancing resource efficiency, protecting the environment, and thereby improving the quality of urban lifestyle (Abdallah et al., 2020). The different stages for waste management with innovative computational tools are waste generation, waste accumulation, waste collection, and waste disposal (Das & Ghosh, 2022; Das et al., 2024). The various components of waste generation are source classification, waste reduction, prediction for waste generation and smart charging. Smart bins, wireless communication, capacity design are the components of waste accumulation. Waste collection involves route optimization and dynamic routing. Similarly, final waste disposal includes smart industry technology, separation mechanisms, circular economy, and waste-to-energy.

Impact of Waste Disposal on the Environment

Waste disposal plays a critical role in environmental sustainability, particularly in mitigating the adverse effects of industrial production (Mishra & Das, 2023; Mohanty et al., 2018; Tripathy et al., 2024). Without effective waste management, excessive waste accumulation causes significant environmental degradation, including soil contamination, water pollution, air quality deterioration, and adverse effects on biodiversity, as well as human and animal health (Choudhary et al. 2024; Mishra & Das, 2021; Mishra et al., 2022, 2024). According to the United Nations Environment Programme (UNEP, 2024), global municipal solid waste generation exceeds 1.3 billion tons annually, contributing 25% of greenhouse gas emissions. Only a small fraction of this waste is recycled, while the remainder is landfilled or incinerated, exacerbating pollution. Recycling conserves energy, and composting reduces water demand, yet waste treatment methods vary in effectiveness and environmental impact (Parida et al., 2023). Improper waste disposal refers to the careless and unsystematic discharge of waste, including municipal, recyclable, and hazardous materials, which can result in significant environmental and public health hazards (Prabhakar et al., 2019; Priyadarsini & Das, 2025; Raj & Das, 2023). Such practice contaminates air, soil, and water, posing serious health risks to humans and numerous wildlife species (Fig. 1). As reported by Siddiqua et al. (2022), the collection of non-biodegradable and poisonous materials further intensifies the already existing pollution, accelerates the loss of biodiversity, and elevates emissions of greenhouse gases. These factors also pose health hazards to the inhabitants in the vicinity of polluted groundwater and gas emissions with carcinogenic and non-carcinogenic effects. These impacts highlight the need for proper planning and strategies focused on the efficient disposal of waste to diminish the damage inflicted on nature and promote healthier industrial activities.

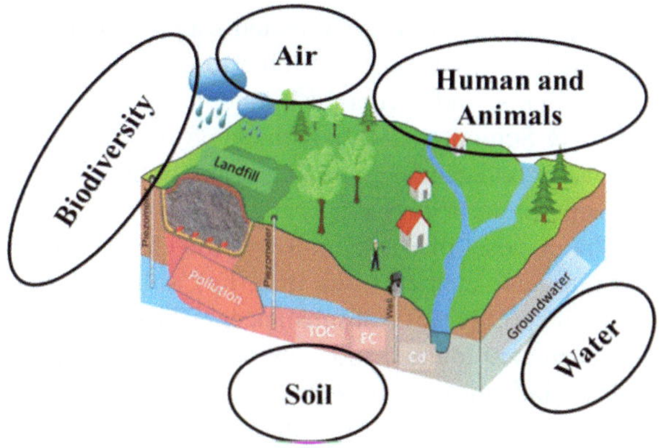

Fig. 1 Various environmental effects related to improper waste disposal (Siddiqua et al., 2022)

Environmental Pollution and Climate Change

Atmospheric pollution is a problem affecting everyone, and careless waste management practices contribute heavily to the accumulation of pollutants in the environment. The biological breakdown of waste in landfills and open dumping sites emits greenhouse gases like methane, which worsens climate change and environmental degradation (Bala et al., 2024a; Kumar et al., 2023; Mishra & Das, 2022). Landfills account for around 5% of the total emissions of greenhouse gases, primarily due to methane emissions from the anaerobic decomposition of organic wastes (Zhang et al., 2019). According to reports, the amount of carbon dioxide and other pollutants created by burning garbage frequently exceeds the rates of greenhouse gas emissions from several conventional energy sources per unit of electricity generated (Tangri, 2023). These practices put both human and animal health at risk while deteriorating air quality due to pollution. Waste that is not adequately handled contributes to climate change by speeding up sea level rise and ice cap melting, putting ecosystems and human infrastructure at risk (Ghosh & Das, 2017; Pradhan et al., 2023; Ray et al., 2024; Sahu et al., 2024). Additionally, it makes extreme weather events like hurricanes and floods more frequent. To reduce the negative effects, sustainable waste management techniques must be put into place, such as recycling, trash minimization, and enhanced treatment technologies (Abubakar et al., 2022).

The harmful waste, like plastics and chemicals, can pollute the water bodies and there would be a serious impact on the ecosystem as well as on the human population (Bala et al., 2024a; Das et al., 2025; Priyadarshini et al., 2024a; Singh Jadaun et al., 2022). The decomposition of these pollutants generates toxic chemicals, raises water toxicity levels, and endangers biodiversity (Dey et al., 2024; John & Das, 2024; Reddy Koteswara et al., 2025; Sahoo et al., 2022). Contaminant accumulation in freshwater systems makes them unsuitable for drinking and recreational uses, which poses significant ecological and societal health threats (Jambeck et al., 2015). Landfills may ooze hazardous chemicals (leachate) to the ground's surface, and can carry through soil strata, causing contamination of subterranean water reserves

(Parida et al., 2023). Contaminated water, which includes organic waste, pathogens, and nutrient runoff, all contribute to degradation of water quality, eutrophication, algal blooms, and increased bio-depletion processes leading to anoxic conditions (Smith & Schindler, 2009). Effluents from industries are often discharged undiluted into water bodies which leading to the pollution and destruction of water resources and ecosystems (Singh et al., 2023).

The disposal of waste increases the concentration of dissolved organic material and inorganic elements like iron (Fe), sulphate (SO_4^{2-}), chloride (Cl^-) as well as other potentially harmful substances (PTEs) (Cittadino et al., 2020). Soil pollution is also attributed to the disposal of toxic elements, polycyclic aromatic hydrocarbons, and microplastics. While well-designed sanitary landfills solve these problems, other uncontrolled dumpsites without proper liners tend to allow leachate and gas migration, which contaminates soil via infiltration and diffusion (Rouhani & Hejcman, 2024). Open burning of waste leads to the formation of ash with highly concentrated deposits of potentially toxic elements (PTEs). These undergo oxidation and the subsequent mobilization by rainwater percolation aids in encapsulating the soil's vegetation, thus entering the food chain (Kanmani & Gandhimathi, 2013).

Biodiversity, Human and Animal Health

Improper waste disposal practices cause irreparable harm to ecological systems and biodiversity by disrupting the balance of the ecosystem and damaging natural habitats (Mohanty et al., 2024a; Prabina et al., 2024; Satapathy et al., 2024). Poor waste disposal practices cause habitat destruction by polluting air, water, and soil while posing risks to animals in terms of drowning or choking (Mishra et al., 2019a; Priyadarshini et al., 2024b; Singh et al., 2020). Harmful food chain pollutants change the composition of species that comprise food chains and, consequently, jeopardize the balance in the ecosystem (Dubey et al., 2024). With the accumulation of hazardous waste and plastic pollution, both marine and terrestrial ecosystems suffer the loss of biodiversity (Bala et al., 2024b; Das et al., 2015; Mohanty et al., 2024b; Paul et al., 2023). For example, Fig. 2 shows plastic marine pollution as a multifaceted problem impacting coastal and marine resources far beyond ecology, leading to socioeconomic decline in tourism and fisheries reliant communities. Plastic debris causes loss of biodiversity and the extinction of many species alters the trophic structure, biological productivity, and population maintenance of various organisms resulting in the disruption of marine ecosystem functions such as entwined, ingestion death, reproductive sterility, and biologically loss of species. Eco-system services crucial for climate moderation, like carbon storage, are further endangered due to this decline (Vergara et al., 2025). Addressing these issues requires more comprehensive approaches such as better education about public policy enforcement, public promotion, effective legislation, and improvement of recycling programs. Incorporating

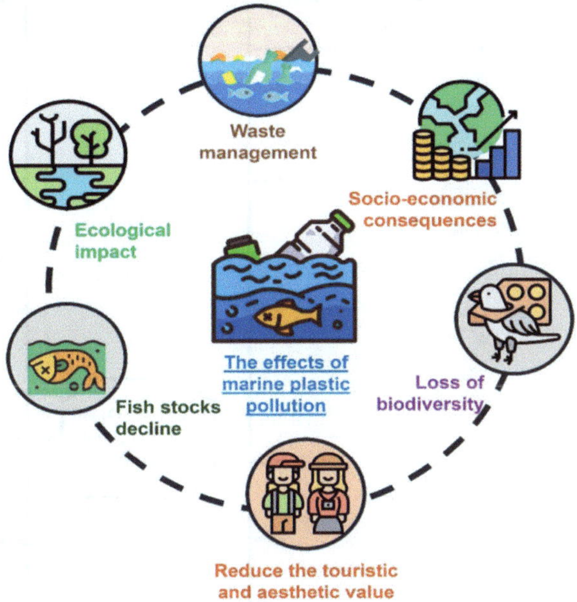

Fig. 2 Multifaceted impacts of marine plastic pollution (Vergara et al., 2025)

sustainable methods of waste management is critical in order to prevent further deterioration of the environment and protect biodiversity for future generations. Figure 2 depicts the multifaceted impacts of plastics on marine environment.

The breakdown of municipal solid waste in landfills leads to the emission of bioaerosols rich in bacteria, fungi, and enteric pathogens, which can significantly impede occupational safety and public health. They are easily inhaled and remain airborne during waste transferring, and prolonged exposure to this increases the risk of respiratory diseases such as chronic obstructive pulmonary disease (Nair, 2021). A study by Ithnin et al. (2013) observed high lead concentrations and challenging respiratory symptoms among students living near an abandoned landfill, exemplifying how pollution affects waste as well as air quality. Adana, uncontrolled urbanization along with the progressive increase in population has raised serious problems related to the collection and disposition of municipal solid waste (MSW) (Unal et al., 2019). The uncontrolled placement of MSW facilities has led to worsening health and humanitarian issues (Unal et al., 2019). Some of the world's most densely populated areas in Africa, particularly Nigeria, Kenya, and Sub-Saharan Africa, face simultaneous increases in resource consumption and waste production, as well as extremely deficient waste collection, treatment, and disposal systems (Kumar & Agrawal, 2020). Poor waste management practices aggravate pollution and increases the risk of various respiratory diseases. Attaining these objectives requires the adoption of environmentally responsible waste management techniques. Various sources of wastes to garbage open dumping and burning with possible environmental and health hazards is depicted in Fig. 3.

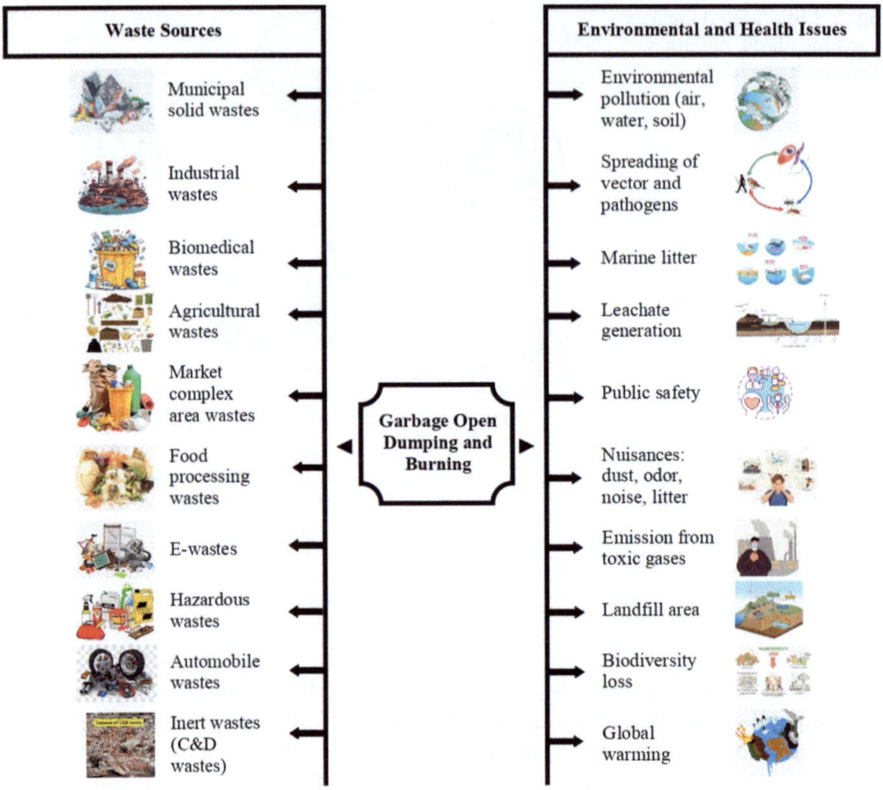

Fig. 3 Sources of waste to garbage open dumping and burning with possible environmental and health hazards

Innovative Solutions for Waste Management for a Clean and Green Future

Traditional waste disposal methods were unscientific, due to which it leads to environmental degradation, causing harm to the ecosystem and human health. Innovative solutions for waste management promote a clean and green future. In the process of waste-to-energy technology, electricity is generated from the waste. Advanced recycling techniques help to promote environmental purification. Other innovative techniques for better management of wastes are smart waste bins and sensors, a data analysis system for smart waste management, waste level sensors, AI recycling robots, garbage truck weighing machines, pneumatic waste pipes, solar-powered trash compactors, and e-waste kiosks.

Waste-to-Energy Conversion

By converting waste into usable energy, primarily electricity, through processes like incineration, waste-to-Energy (WtE) plays a major role in Municipal Solid Waste (MSW) management. This approach not only reduces waste volume sent to landfills but also provides an alternative energy source, thus serving as a dual-purpose solution for waste disposal and energy generation. For example, India has the potential to generate 5690 MW of electricity from municipal solid waste and industrial waste, but as of May 2023, only 556 MW of this capacity had been realized. WtE offers several advantages in MSW management. These facilities significantly reduce the amount of solid waste that requires land filling through incinerating waste in a controlled environment, thus conserving space and potentially lowering landfill gas emissions. Additionally, this offers a renewable energy alternative as the heat produced from waste combustion generates steam, drives turbines to produce electricity. When properly managed, by reducing greenhouse gas emissions compared to traditional land filling, WtE can also contribute to environmental benefits as it captures and utilizes the energy within waste rather than allowing it to decompose and release methane. WtE systems also come with challenges. To minimize air pollution from WtE plants, emission control measures must be implemented. Waste also needs to be sorted and pre-processed to remove recyclable materials before being utilized for energy generation. Apart from this, the high costs associated with building and operating WtE facilities is a barrier for its widespread adoption. There are different technologies for WtE conversion. Incineration, the most common method, directly burns waste to produce heat for electricity generation. Gasification heats waste materials in the absence of oxygen, producing a combustible gas (syngas) that is used for power generation. Pyrolysis, a similar process conducted at higher temperatures, produces a liquid fuel (bio-oil) in addition to gas. One of the most successful examples of WtE implementation is Sweden's waste-to-energy program. It has developed an advanced system where over 50% of its MSW is incinerated to produce electricity and district heating. As a result, less than 1% of Sweden's waste ends up in landfills, and the nation even imports waste from other countries to utilize in its WtE plants. Overall, WtE can be a valuable component of an integrated MSW management strategy. However, to achieve optimal sustainability, it should be used alongside other waste reduction measures such as recycling and composting, ensuring a comprehensive approach to waste management. Figure 4 shows the different pathways through which the municipal solid wastes can be converted in to valuable products, the step towards conversion of waste-to-energy.

In this process, the organic wastes are converted into biogas through anaerobic digestion. It provides renewable energy source. The anaerobic bacteria breakdown organic matter in an oxygen free environment releasing a mixture of gases such as CH_4 and CO_2. In this process the organic wastes such as food scraps, animal manure, and agricultural residues are placed inside a digester chamber. In the presence of anaerobic bacteria, these organic wastes are digested to release biogas. In waste to energy conversion, there are three vortices at the waste triangle, i.e., plastics,

of molecules that make up plastics. Polyethylene terephthalate, polyethylene, and polypropylene are degraded by enzymes. There are two naturally occurring enzymes, such as leaf-branch compost cutinases and metagenome-derived polyester hydrolase, utilized for the recycling of waste.

Chemical recycling: Chemical recycling is an economically and environmentally possible way to convert waste plastics into value-added products. In the process, liquid oils with high heating values, different carbon materials, hydrogen, syngas, and other chemicals is recovered from plastic waste, by applying techniques such as pyrolysis, reforming, gasification, and chemolysis. The plastic waste streams that are not recycled mechanically can be recycled by chemical recycling. In this process, the plastic waste streams can be converted back into feedstock for the chemical industry and allocated to the products manufactured in BASF's integrated chemical production network through a mass balance approach. In the process of chemical recycling, the plastic wastes are broken down by heat or chemicals. Depending on the type of plastic and desired product, the reaction is carried out involving heat, pressure, solvents or catalysts. The end goal is to decompose long, stubborn polymer chains into simpler and versatile monomers. Then it can be purified and re-polymerized into new, high-quality plastics.

Mobile recycling units: Mobile recycling units are convenient for people who do not have access to a recycling centre and cannot transport their recyclables. In this service, the unit comes to the home to collect recyclable materials like e-waste. Types of mobile recycling units are:

Trashpresso: It is a solar-powered and AI-powered recycling machine that converts plastic waste into construction tiles. It involves three steps that are: size reduction, reshaping, and purification. It protects the environmental discharge from the process with air and water filtration.

Mobile C&D waste recycling plant: it is a small, self-contained system mint for waste material processing on the spot, which minimizes the transportation cost. It has several units such as crushing unit, screening unit, and dust control system. They help for conversion of waste into valuable products.

Hook-type mobile recycling machines: They can be loaded and transported on a truck and are easy to handle. These units can be moved to different locations to respond to the waste generation. Many mobile machines are equipped with hybrid technology.

Product distribution service: Product distribution service for waste management describes to the logistical process of collecting, transporting, and delivering different types of waste materials. It involves in following such as waste collection, waste segregation, transportation, & disposal. It is also made from natural materials that are grown without chemicals and fertilizers. It is also more sustainable because they do not produce harmful toxins.

Zero waste initiatives: It is a holistic step for addressing the issues of unsustainable resource flow. Zero waste aims to reduce waste, promote reuse, and keep materials out of landfills. Adopting zero waste policy can significantly minimize the environmental footprint of a business which can lower the emission of green house gases, reduces pollution load and also conserves natural resources. It includes policies, programs, and every day practices. Zero waste initiatives have several aspects like:

Policy development: So many policies are developed to create strategies to reduce waste, promote recycling, and reuse.
Closed-loop system: The systems are designed to keep product and materials in use for as long as possible.
Reduce consumption: To create zero waste, we have to reduce consumption and avoiding single use plastics.

3 Smart Waste Bin and Sensor

Utilizing advanced sensors and a data analysis system is a clever and sophisticated method of keeping rubbish in the bins. This is to control the bins' capacity to fill, in which the sensor determines the waste kind and filling level. When the bin filled up to its upper limit than the sensor sends the alert to the respective data analysing system. It is very easy to manage the waste and ultimately it improves the waste management system by managing real-time data of type of waste and level or quantity of waste in the bin.

The smart waste bin sensors are various types i.e. ultrasonic sensor for measure filled level, infrared sensors for distinguish the type of waste, weight sensors for weighing of waste, capacitive sensors for detection of changes in waste composition and RFID sensors for tracking different types of bins and waste types. It is used to first detect the type of waste i.e. e-waste or domestic waste or biomedical waste or industrial waste, etc. and then the waste automatically stored in the separate containers which are already under the conveyor belt que. After that, when the container filled up to the upper level then it sends the alert to the system to stop the filling process of container. The sensors data is transmitted through wirelessly by the help of bluetooth or Wi-Fi. Because of this, it prevents from overflow and reducing environmental impact from the hazardous wastes.

4 Data Analysing System for Smart Waste Management

Data analytics or analysing system is used to store and manage the complete data of various types of waste, how many times it filled in a day, amount of waste filled in the containers, how many containers are in the que, waste disposal habits like reducing, recycling, reusing, composting, the sensor's real-time signals or alerts to the system which is a centralized control panel. The smart waste bins data can be easily accessed from the control panel and due to the stored data, the control panel easily understand the waste trends and habits of the waste. The various types of analysers used are image-based analysers, weight-based analysers, sensor-based analysers (including near infrared, Raman spectroscopy), and combined systems.

5 Waste Level Sensor

Businesses centres and houses over the nation depend on the routine waste collection service to discard of their garbage. Although weekly services have existed for many years, they are not always the best choice. Waste level sensor also called as fill level sensor or ultrasonic sensors, tracks the amount of waste in bins and containers. It can be water-resistant and can accurately transfer data in different conditions. It is most beneficial for waste management and can help cities and businesses centres to understand the waste production and optimize collection routes. This device is most efficient among other devices. It can helps in maximizing efficiencies by monitoring waste capacity demands. These are different types of waste level sensors like:

Recycle bin ultra eye sensor: It can be mounted on bin lids, poles, or other customized arrangements.

Sensoneo smart sensor: It is an ultrasonic sensor for monitoring the fill levels in different waste containers.

Smart bin sensor: Sensor transmits data in real-time to a smart waste management platform.

6 AI Recycling Robots

Recycling equipment are necessary to lowering the quantity of waste that reaches landfills and waterways annually. Many facilities are finding it difficult to meet demand as a result of a decreased workforce during the COVID-19 pandemic. Tool like artificial intelligence (AI) powered recycling robots can help and make up part of the difference. Designed to precisely find and sort recyclable items, these robots boost efficiency and lower the need for human labour. Long-term financial savings for recycling facilities are also achieved, as materials that would otherwise wind up in landfills are diverted.

7 Garbage Truck Weighing Mechanisms

The waste and recycling industry is dependent on careful measurement for monitoring the volume transported through trucks. The scales and weighing system, is also called 'on board weighing system', is a device installed on a garbage tank that measures the weight of the collected waste in real time. By measuring and documenting garbage containers, they then use the data to project fill levels over time. By adopting this technology, cities may more precisely project how often they need to send their trucks and reduce their annual collecting costs.

8 Pneumatic Waste Pipes

These are underground pipes that use air pressure to transport waste to a central collection point. It is an alternative to the traditional waste collection systems. In this process the fans generate air that creates negative pressure in the pipes. Then the air enters into the pipes at atmospheric pressure and the solid wastes are trapped into it. The waste is transported to collection stations and into sealed containers. It improves sanitation and environmental conditions. These pipes are used in residential, roadside, and other locations. The pipes carry trash to a waste collection facility, where it is either sorted or removed. This system lowers energy expenses, boosts overall efficiency, and does away with the need for conventional waste collection.

9 Solar Powered Trash Compactors

A solar powered trash compactor can hold up to five times as much as conventional trash cans. It includes solar panel, battery storage, a compaction machine, a fill level sensor and a control unit trash. The bin can hold significantly more waste before needing to be emptied, minimizing collecting trips. It also helps to reduce environmental pollution and it is a most cost saving device.

10 E-waste Kiosks

An e-waste kiosk is a designated drop-off point usually present in public places like shopping malls, community centres, etc. The e-waste is disposed properly and it undergoes recycling, minimizing the harmful substances properly. These kiosks collect items like old phones, computers, TVs, and small appliances. These devices reduce the amount of hazardous materials entering landfills. It also helps for extraction of valuable metals like copper, gold and silver by the extraction process. It provides a user-friendly way for people to dispose e-waste without the need to actively search for recycling facilities. The important references that are reviewed in this article under different heads like route optimization, application of computational intelligence with respect to waste classification and identification, role and functioning of smart bins, waste management applying IoT and communications are reflected in Table 1 (Nesmachnow et al., 2025). The table also includes the study concept with their key limitations and the corresponding references under each head.

Table 1 Study concept and their key limitations for different aspects of waste management like smart waste management, computational intelligence for waste classification and waste identification, IoT system, and communications for waste management and smart bins

Study concept	Key limitations	References
Smart waste management with respect to rout optimization techniques		
IoTbased WMS for optimizing the schedule and routes for waste collection	There is no comparison with similar methods	Idwan et al. (2020)
Real time bin status for improving collection efficiency of the waste and to reduce cost of collection and emission	Did not present any real evaluation	Hannan et al. (2020)
Metaheuristic optimization for e-waste vehicle routing	No report on validation over actual scenarios	Nowakowski et al. (2020)
Metaheuristic for optimizing routs and schedule for vehicles to collect from waste bins	Real validation in the area was reported	Bin et al. (2021)
Used real data for waste collection routing	There is no description on real implementation or validation	Burduk et al. (2022)
Sustainable waste collection routes identification framework	There is no real implementation	Hu et al. (2023)
Autonomous emptying process approach and to predict the filling levels of the bin	No methods for route optimization have been studied for efficient time and energy for the robots	Pollak et al. (2024)
Computational intelligence applied for waste classification and waste identification		
IoT for detection of human activity and classification of waste management	No implication or validation	Alqahtani et al. (2020)
ANN for waste classification and identification	No real evaluation, accurate baseline methods	Alsubaei et al. (2022)
Automatic visual warning and identification for illegal waste disposal	Performed with less dataset, real implementation was not presented	Liu et al. (2023)
Waste classification by using IoT, edge/cloud computing, and deep learning	No real implementation/ validation	Cheema et al. (2022)
Waste classification applying deep learning	No state of the art classification, only adhoc database	Kumar et al. (2023)
IoT system and communications for management of wastes		
IoT application as a multilevel architecture for smart city waste management	Communication metrics evaluation only	Marques et al. (2019)
Real time smart bin monitoring applying IoT and AI	No real evaluation	Bano et al. (2020)
Smart bins	Chances of theft and damage	Jhaveri et al. (2021)

(continued)

Table 1 (continued)

Study concept	Key limitations	References
Smart waste removal system based on rapid exploration and random trees	Validated only in few case studies	Zhang et al. (2020)
Smart bin for communal WMS	No deployment or real evaluation	Azyze et al. (2022)
GA- fuzzy method for smart waste management	No real scenario, only done by simulation	Thaseen et al. (2023)
IoT based household WMS	No real scenarios, simple evaluation	Wong et al. (2022)
Smart bins		
Household e-waste collection smart bin	No real validation/ implementation	Kang et al. (2020)
Underground structure for landfill for waste collection and separation	Proposed system has no real evaluation	Majidi (2021)
Waste disposal prediction and monitoring with smart bin using IoT	Real implementation was not reported	John et al. (2021)
Deep learning and IoT for improving collection rate and minimizing cost	Need few training data, GPU, lifetime of the batteries	Sallang et al. (2021)
Smart cities self-powered WMS, smart bin using IoT sensors	No real implementation	Thamarai and Naresh (2023)

11 Conclusion

Effective Municipal Solid Waste management has become more critical than ever with rapid urbanization, population growth, and increasing waste generation. In order to minimize environmental impact, the future of waste management must focus on circular economy principles, sustainability, and technological advancements. Integration of new technologies which convert low-calorific waste into energy at a reasonable cost is essential, as they can aid in the cleanup of ageing landfill sites while simultaneously generating energy. Waste-to-energy technologies have redefined waste as a resource rather than a burden. However, the composition of waste is quite crucial in energy recovery, as the existing WtE systems function best with high-calorific-value waste. Garbage-to-energy solutions are emerging as a viable alternative to traditional disposal methods in India, as increasing amounts of high-calorie waste are ending up in landfill site. The future in MSW management involves a multi-pronged approach that integrates technology, policy, community participation, and sustainability principles. In order to establish a cleaner, greener, and more livable environments, cities must prioritize resource recovery, waste minimization, and eco-friendly disposal techniques. Municipalities can enhance waste management efficiency while reducing their environmental footprint through adoption of strategies such as improved waste segregation, circular economy-based recycling programs, and innovative waste processing methods in addition to the emerging technologies like IoT, artificial intelligence, and big data analysis.

References

Abdallah, M., Talib, M., Feroz, S., Nasir, Q., Abdalla, H., & Mahfood, B. (2020). Artificial intelligence applications in solid waste management: A systematic research review. *Waste Management, 109*, 231–246.

Abubakar, I. R., Maniruzzaman, K. M., Dano, U. L., AlShihri, F. S., AlShammari, M. S., Ahmed, S. M. S., Al-Gehlani, W. A. G., & Alrawaf, T. I. (2022). Environmental sustainability impacts of solid waste management practices in the Global South. *International Journal of Environmental Research and Public Health, 19*(19), 12717. https://doi.org/10.3390/ijerph191912717

Agamuthu, P., Fauziah, S. H., Khidzir, K. M., & Noorazamimah Aiza, A. (2007). Sustainable waste management—Asian perspectives. In *Proceedings of the international conference on sustainable solid waste management* (pp. 15–26).

Alqahtani, F., Al-Makhadmeh, Z., Tolba, A., & Said, W. (2020). Internet of Things-based urban waste management system for smart cities using a cuckoo search algorithm. *Cluster Computing, 23*, 1769–1780.

Alsubaei, F., Al-Wesabi, F., & Hilal, A. (2022). Deep learning-based small object detection and classification model for garbage waste management in smart cities and IoT environment. *Applied Sciences, 12*, 2281.

Azyze, N., Isa, I., & Chin, T. (2022). IoT-based communal garbage monitoring system for smart cities. *Indonesian Journal of Electrical Engineering and Computer Science, 27*, 37.

Bala, K., Anshul, Verma, S., Dogra, A., Das, A. P., Kumar, R., & Nandini. (2024a). Sustainable management for urban plastic waste generation. In *Sustainable management of urban plastic waste through circular economic approaches* (pp. 211–236). https://doi.org/10.1201/9781003465447-13

Bala, K., Dogra, A., Lata, P., & Das, A. P. (2024b). Discussing sources and biological-chemical remediation approaches for microplastic pollution. In *Sustainable microbial technology for synthetic and cellulosic microfiber bioremediation* (pp. 37–59). https://doi.org/10.1007/978-3-031-62600-5_3

Bano, A., Ud Din, I., & Al-Huqail, A. (2020). A IoT-based smart bin for real-time monitoring and management of solid waste. *Scientific Programming, 2020*, Article 613263.

Bin, C., Xinghan, C., Zhihan, L., Ruichang, L., & Shanshan, F. (2021). Optimization of classified municipal waste collection based on the internet of connected vehicles. *IEEE Transactions on Intelligent Transportation Systems, 22*, 5364–5373.

Burduk, A., Łapczyn´ska, D., Kochan´ska, J., Musiał, K., Wie˛cek, D., & Kuric, I. (2022). Waste management with the use of heuristic algorithms and Internet of Things technology. *Sensors,22*, 8786.

Cheema, S., Hannan, A., & Pires, I. (2022). Smart waste management and classification systems using cutting edge approach. *Sustainability., 14*, 10226.

Choudhary, M., Singh, D., Parihar, M., Choudhary, K. B., Nogia, M., Samal, S. K., & Mishra, R. (2024): Impact of municipal solid waste on the environment, soil, and human health. In *Agricultural soil degradation and restoration. Waste management for sustainable and restored agricultural soil* (pp. 33–58). Academic Press. ISBN 9780443184864.

Cittadino, A., Ocello, N., Majul, M. V., Ajhuacho, R., Dietrich, P., & Igarzabal, M. A. (2020). Heavy metal pollution and health risk assessment of soils from open dumps in the Metropolitan Area of Buenos Aires, Argentina. *Environmental Monitoring and Assessment, 192*(5), 291. https://doi.org/10.1007/s10661-020-8246-x

Das, A. P., Ghosh, S., Mohanty, S., & Sukla, L. B. (2015). Advances in manganese pollution and its bioremediation. In L. B. Sukla, N. Pradhan, S. Panda, & B. K. Mishra (Eds.), *Environmental microbe biotechnology* (1st ed., pp. 313–328). Springer. https://doi.org/10.1007/978-3-319-19018-1_16

Das, A., Dey, S., Gacem, A., Yadav, K. K., Nath, H., Ram, D. K., & Das, A. P. (2025). Revolutionizing solid hydrocarbon sourcing: Plastic's journey from waste to upcycled treasure. *Environmental Science and Engineering, Part, F218*, 331–351. https://doi.org/10.1007/978-3-031-81615-4_14

Das, A., & Ghosh, S. (2022). Role of microorganisms in extenuation of mining and industrial wastes. *Geomicrobiology Journal, 39*, 173–175. https://doi.org/10.1080/01490451.2022.2038953

Das, A., Ray, S. K., Mohanty, M., Mohanty, J., Dey, S., & Das, A. P. (2024). Ecotoxicity of microplastic wastes and their sustainable management: A review. *Environmental Chemistry and Ecotoxicology, 6*, 144–152. https://doi.org/10.1016/J.ENCECO.2024.05.003

Dey, S., Das, A., Mallick, K., Sahu, A., & Das, A. P. (2023). *Environmental petroleum waste: Pollution, toxicity, sustainable remediation* (pp. 159–175). https://doi.org/10.1007/978-3-031-48220-5_7

Dey, S., Dash, N., Ray, S. K., Priyadarsini, S., Chhotaray, C., & Das, A. P. (2024). Occurrence and quantification of tire wear plastics pollutants from highway city dust of Bhubaneswar city of Odisha. In *Sustainable microbial technology for synthetic and cellulosic microfiber bioremediation* (pp. 77–89). https://doi.org/10.1007/978-3-031-62600-5_5

Dubey, D. K., Agarwal, S., Yadav, M. P., Goswami, A., & Ravat, A. (2024). The study of the effects of improper hazardous waste disposal on ecosystems. *Indian Journal of Scientific Research, 4*(4), 130–136. https://www.researchgate.net/publication/385241984

Fan, Z., Zhang, Y., Pan, L., Ouyang, J., & Zhang, Q. (2021). Recent developments in flexible thermoelectric: From materials to devices. *Renewable and Sustainable Energy Reviews, 137*, Article 110448. https://doi.org/10.1016/j.rser.2020.110448

Ghosh, S., & Das, A. P. (2017). Bioleaching of manganese from mining waste residues using Acinetobacter sp. *Geology, Ecology, and Landscapes, 1*(2), 77–83. https://doi.org/10.1080/24749508.2017.1332847

Hannan, M., Begum, R., Al-Shetwi, A., Ker, P., Al Mamun, M., Hussain, A., Basri, H., & Mahlia, T. (2020). Waste collection route optimisation model for linking cost saving and emission reduction to achieve sustainable development goals. *Sustainable Cities and Society, 62*, Article 102393.

Ho, H.-J., Iizuka, A. and Shibata, E. (2021): Chemical recycling and use of various types of concrete waste: A review. *Journal of Cleaner Production*, 124785. https://doi.org/10.1016/j.jclepro.2020.124785

Hu, S., An, L., & Shen, L. (2023). A multi-objective modeling and optimization approach to municipal solid waste collection for classified treatment in China towards sustainable development. *Sustainable Cities and Society, 98*, Article 104846.

Idwan, S., Mahmood, I., Zubairi, J., & Matar, I. (2020). Optimal management of solid waste in smart cities using Internet of Things. *Wireless Personal Communications, 110*, 485–501.

Ithnin, A., Abd Rahman, M. S., Awang, N., Mohd Yusuf, N., Abdullah, R., & Ariffin, F. D. (2013). Study on air quality in a school located near a former landfill site and its influences on students' respiratory health. *Middle-East Journal of Scientific Research, 14*, 371–374.

Jambeck, J. R., Geyer, R., Wilcox, C., Siegler, T. R., Perryman, M., Andrady, A., & Law, K. L. (2015). Plastic waste inputs from land into the ocean. *Science, 347*(6223), 768–771.

Jhaveri, R., Shah, P., Patel, N., Patel, D., Thanki, S., Bhoi, A., & Bhatia, J. (2021). An ICT-based solid waste management system for smart cities: A case of municipality in India. *International Journal of Ad Hoc and Ubiquitous Computing, 38*, 158.

John, A., & Das, A. P. (2024). Government strategies for achieving circular economy for plastic waste management in Southeast Asia. In *Sustainable management of urban plastic waste through circular economic approaches* (pp. 72–82). https://doi.org/10.1201/9781003465447-6/GOVERNMENT-STRATEGIES-ACHIEVING-CIRCULAR-ECONOMY-PLASTIC-WASTE-MANAGEMENT-SOUTHEAST-ASIA-AKBAR-JOHN-ALOK-PRASAD-DAS

John, J., Varkey, M., Podder, R., Sensarma, N., Selvi, M., Santhosh, S., & Kannan, A. (2021). Smart prediction and monitoring of waste disposal system using IoT and cloud for IoT based smart cities. *Wireless Personal Communications, 122*, 243–275.

Kang, K., Kang, H., Ilankoon, I., & Chong, C. (2020). Electronic waste collection systems using Internet of Things (IoT): Household electronic waste management in Malaysia. *Journal of Cleaner Production, 252*, Article 119801.

Kanmani, S., & Gandhimathi, R. (2013). Assessment of heavy metal contamination in soil due to leachate migration from an open dumping site. *Applied Water Science, 3*, 193–205. https://doi.org/10.1007/s13201-012-0072-z

Kumar, A., & Agrawal, A. (2020). Recent trends in solid waste management status, challenges, and potential for the future Indian cities–A review. *Current Research in Environment Sustainability, 2*, Article 100011. https://doi.org/10.1016/j.crsust.2020.100011

Kumar, M., Pandit, S., Patel, V., Khanna, N., Nag, M., Lahiri, D., Ray, R. R., Das, A., & Das, D. (2023). Maximization of energy recovery from starch processing wastewater by thermophilic dark fermentation coupled with microbial fuel cell technology. *Geomicrobiology Journal*. https://doi.org/10.1080/01490451.2023.2209555

Kurniawan, T. A., Lo, W. H., & Chan, G. Y. S. (2006). Physicochemical treatments for removal of recalcitrant contaminants from landfill leachate. *Journal of Hazardous Materials, 129*(1–3), 80–100.

Liu, J., Prabuwono, A., Abulfaraj, A., Miniaoui, S., & Taheri, N. (2023). Cognitive cloud framework for waste dumping analysis using deep learning vision computing in healthy environment. *Computers & Electrical Engineering, 110*, Article 108814.

Majidi, A. (2021). City with smart waste management: Introducing waste bin technologies. *Nexon Revista Científica, 34*, 1032–1038.

Marques, P., Manfroi, D., Deitos, E., Cegoni, J., Castilhos, R., Rochol, J., Pignaton, E., & Kunst, R. (2019). An IoT-based smart cities infrastructure architecture applied to a waste management scenario. *Ad Hoc Networks, 87*, 200–208.

Mishra, S., & Das, A. P. (2021). Current treatment technologies for removal of microplastic and microfiber pollutants from wastewater. In *Wastewater treatment: Cutting-edge molecular tools, techniques and applied aspects* (pp. 237–251). https://doi.org/10.1016/B978-0-12-821881-5.00011-8

Mishra, S., & Das, A. P. (2022). Treatment of the wastewater polluted with synthetic microfiber released from washing machine. In B. B. Das, H. Hettiarachchi, P. K. Sahu, & S. Nanda (Eds.), *Lecture notes in civil engineering* (Vol. 207, pp. 109–117). Springer Science and Business Media Deutschland GmbH. https://doi.org/10.1007/978-981-16-7509-6_9/COVER

Mishra, S., & Das, A. P. (2023). Microbial remediation of synthetic microfiber contaminated wastewater. In M. P. Shah (Ed.), *Microbial technologies in industrial wastewater treatment* (pp. 337–350). Springer. https://doi.org/10.1007/978-981-99-2435-6_17

Mishra, S., Dash, D., & Das, A. P. (2024). Isolation and molecular characterization of polyester-degrading bacterial strains from municipal wastewater and their application in microplastic degradation in contaminated wastewater. *Geomicrobiology Journal*. https://doi.org/10.1080/01490451.2024.2376101

Mishra, S., Ghosh, S., van Hullebusch, E. D., Singh, S., & Das, A. P. (2023). A critical review on the recovery of base and critical elements from electronic waste-contaminated streams using microbial biotechnology. In *Applied biochemistry and biotechnology 2023* (pp. 1–30). https://doi.org/10.1007/S12010-023-04440-X

Mishra, S., Rath, C. C., & Das, A. P. (2019a). Marine microfiber pollution: A review on present status and future challenges. *Marine Pollution Bulletin, 140*, 188–197. https://doi.org/10.1016/J.MARPOLBUL.2019.01.039

Mishra, S., Rout, P. K., & Das, A. P. (2019b). Solar photovoltaic panels as next generation waste: A review. *Biointerface Research in Applied Chemistry, 9*(6).

Mishra, S., Singh, R. P., Rout, P. K., & Das, A. P. (2022). Membrane bioreactor (MBR) as an advanced wastewater treatment technology for removal of synthetic microplastics. In *Development in wastewater treatment research and processes* (pp. 45–60). https://doi.org/10.1016/B978-0-323-85583-9.00022-3

Mohanty, J., Dey, S., Bala, K., & Das, A. P. (2024a). Recent advances in microplastic pollution for its sustainable management. In *Sustainable management of urban plastic waste through circular economic approaches* (pp. 27–48). https://doi.org/10.1201/9781003465447-3

Mohanty, M., Tripathy, B., Koteswara Reddy, G., & Das, A. P. (2024b). Greener technologies for establishing restraint over microplastic pollution. In *Sustainable microbial technology for synthetic and cellulosic microfiber bioremediation* (pp. 239–258). https://doi.org/10.1007/978-3-031-62600-5_13

Mohanty, S., Ghosh, S., Bal, B., & Das, A. P. (2018). A review of biotechnology processes applied for manganese recovery from wastes. *Reviews in Environmental Science and Bio/Technology, 17*(4), 791–811. https://doi.org/10.1007/S11157-018-9482-1

Nair, A. T. (2021). Bioaerosols in the landfill environment: An overview of microbial diversity and potential health hazards. *Aerobiologia, 37*, 185–203.

Nesmachnow, S., Rossit, D., & Moreno-Bernal, P. (2025). A literature review of recent advances on innovative computational tools for waste management in smart cities. *Urban Science, 9*(16), 1–61. https://doi.org/10.3390/urbansci9010016

Nowakowski, P., Szwarc, K., & Boryczka, U. (2020). Combining an artificial intelligence algorithm and a novel vehicle for sustainable e-waste collection. *Science of the Total Environment, 730*, Article 138726.

Parida, B., Dash, A. K., & Verma, A. K. (2023). Technologies for treatment of landfill leachate: A brief review. *Recent Innovations in Chemical Engineering, 16*(4), 260–284. https://doi.org/10.2174/0124055204276847231013040437

Paul, A., Dey, S., Ram, D. K., & Das, A. P. (2023). Hexavalent chromium pollution and its sustainable management through bioremediation. *Geomicrobiology Journal*, 1–11. https://doi.org/10.1080/01490451.2023.2218377

Pollak, A., Gupta, A., & Göhlich, D. (2024). Optimized operation management with predicted filling levels of the litter bins for a fleet of autonomous urban Service Robots. *IEEE Access, 12*, 7689–7703.

Prabhakar, A., Mishra, S., & Das, A. P. (2019). Isolation and identification of lead (Pb) solubilizing bacteria from automobile waste and its potential for recovery of lead from end of life waste batteries. *Geomicrobiological Journal, 36*(10), 894–903. https://doi.org/10.1080/01490451.2019.1654044

Prabina, K. E., Benya, A., Mohanty, S., Van Hullebusch, E. D., Das, A. P., & Singh, S. (2024). Recovery of multi-functional carbon fiber from COVID-19 generated waste for an effective management through circular economy. *Environmental Science and Engineering, Part, F2827*, 83–97. https://doi.org/10.1007/978-3-031-50840-0_6

Pradhan, G., Tripathy, B., Ram, D. K., Digal, A. K., & Das, A. P. (2023). Bauxite mining waste pollution and its sustainable management through bioremediation. *Geomicrobiology Journal*, 1–10. https://doi.org/10.1080/01490451.2023.2235353

Priyadarshini, B., Santhosh Kumar, M., & Das, A. P. (2024a). Microbial degradation of synthetic plastics released from COVID-19 generated single use disposable face masks. *Environmental Science and Engineering, Part, F2827*, 1–16. https://doi.org/10.1007/978-3-031-50840-0_1

Priyadarshini, S., Jagatee, S., & Das, A. P. (2024b). Synthetic fabrics and microfiber pollution–An assessment of their global impact. *Environmental Science and Engineering, Part, F2333*, 137–157. https://doi.org/10.1007/978-3-031-51792-1_8

Priyadarsini, S., & Das, A. P. (2024). Analytical and structural characterization of waste lithium-ion batteries for their effective recycling strategy. *Environmental Chemistry and Ecotoxicology*. https://doi.org/10.1016/j.enceco.2024.12.004

Priyadarsini, S., & Das, A. P. (2025). Lithium bioleaching: A review on microbial-assisted sustainable technology for lithium bio-circularity. *Journal of Water Process Engineering, 69*, Article 106744. https://doi.org/10.1016/j.jwpe.2024.106744

Raj, K., & Das, A. P. (2023). Lead pollution: Impact on environment and human health and approach for a sustainable solution. *Environmental Chemistry and Ecotoxicology, 5*, 79–85. https://doi.org/10.1016/J.ENCECO.2023.02.001

Ray, S. K., Dey, S., Das, A. P., & John, A. (2024). Microplastic waste management through greener technologies: A review. In *Sustainable management of urban plastic waste through circular economic approaches* (pp. 237–261). https://doi.org/10.1201/9781003465447-14

Reddy Koteswara, G., Bhargavi, S., Yamuna, D., Anjani Devi, M., Jeevana Lakshmi Kumar, G., Manas, T., Adarsh, S., Bhavya Sai, V., Nath, H., Alreshidi, M. A., Togun, H., Ram, D. K., & Das, A. P. (2025). Recent advancements in the remediation of marine hydrocarbon and plastic pollutants. *Environmental Science and Engineering, Part F218*, 127–142. https://doi.org/10.1007/978-3-031-81615-4_7

Renou, S., Givaudan, J. G., Poulain, S., Dirassouyan, F., & Moulin, P. (2008). Landfill leachate treatment: Review and opportunity. *Journal of Hazardous Materials, 150*(3), 468–493. https://doi.org/10.1016/j.jhazmat.2007.09.077

Rouhani, A., & Hejcman, M. (2024). A review of soil pollution around municipal solid waste landfills in Iran and comparable instances from other parts of the world. *International Journal of Environmental Science and Technology*. https://doi.org/10.1007/s13762-024-05728-z

Sahoo, P. P., Singh, S., Rout, P. K., Mishra, S., & Das, A. P. (2022). Microbial remediation of plastic pollutants generated from discarded and abandoned marine fishing nets. *Biotechnology and Genetic Engineering Reviews*. https://doi.org/10.1080/02648725.2022.2152629

Sahu, A., Dey, S., Mallick, K., Priyadarsini, S., Behera, I. D., & Das, A. P. (2024). A review on advanced technology for sustainable management of synthetic microplastic waste. In *Sustainable management of urban plastic waste through circular economic approaches* (pp. 262–279). https://doi.org/10.1201/9781003465447-15

Sallang, N., Islam, M., & Arshad, H. (2021). A CNN-based smart waste management system using tensorflow lite and LoRa-GPS shield in Internet of Things environment. *IEEE Access, 9*, 153560–153574.

Satapathy, S., Kar, D., Priyadarsini, S., Das, A. P., & Kuanar, A. (2024). Application of microorganisms in emerging waste management strategies and remediation of synthetic and cellulosic microfibers. In *Sustainable microbial technology for synthetic and cellulosic microfiber bioremediation* (pp. 183–206). https://doi.org/10.1007/978-3-031-62600-5_10

Siddiqua, A., Hahladakis, J. N., & Al-Attiya, W. A. K. A. (2022). An overview of the environmental pollution and health effects associated with waste landfilling and open dumping. *Environmental Science and Pollution Research, 29*, 58514–58536. https://doi.org/10.1007/s11356-022-21578-z

Singh Jadaun, J., Bansal, S., Sonthalia, A., Rai, A. K., & Singh, S. P. (2022). Biodegradation of plastics for sustainable environment. *Bioresource Technology, 347*, Article 126697. https://doi.org/10.1016/J.BIORTECH.2022.126697

Singh, B. J., Chakraborty, A., & Sehgal, R. (2023). A systematic review of industrial wastewater management: Evaluating challenges and enablers. *Journal of Environment Management, 348*, Article 119230. https://doi.org/10.1016/j.jenvman.2023.119230

Singh, R. P., Mishra, S., & Das, A. P. (2020). Synthetic microfibers: Pollution toxicity and remediation. *Chemosphere, 257*, Article 127199. https://doi.org/10.1016/j.chemosphere.2020.127199

Smith, V. H., & Schindler, D. W. (2009). Eutrophication science: Where do we go from here? *Trends in Ecology & Evolution, 24*(4), 201–207. https://doi.org/10.1016/j.tree.2008.11.009

Tangri, N. (2023). Waste incinerators undermine clean energy goals. *PLOS Climate, 2*(6), Article e0000100. https://doi.org/10.1371/journal.pclm.0000100

Thamarai, M., & Naresh, V. (2023). Smart self-power generating garbage management system using deep learning for smart cities. *Microprocessors and Microsystems, 98*, Article 104816.

Thaseen, S., Mohanraj, V., Ramachandran, S., & Balakrishnan, A. (2023). An intelligent waste management application using IoT and a genetic algorithm–fuzzy inference system. *Applied Sciences, 13*, 3943.

Tripathy, B., Dey, S., Dubey, N. K., Lo, W.-C., & Das, A. P. (2024). *An omics-based approach for managing the risk of COVID-19-generated wastes* (pp. 51–66). https://doi.org/10.1007/978-3-031-50840-0_4

Unal, M., Cilek, A., & Guner, E. D. (2019). Implementation of fuzzy, Simos and strengths, weaknesses, opportunities and threats analysis for municipal solid waste landfill site selection: Adana City case study. *Waste Management Research, 38*, 45–64.

United Nations Environment Programme. (2024). Global waste management outlook 2024. https://www.unep.org/resources/global-waste-management-outlook-2024. Accessed April 3, 2025.

Vergara, D., de la Hoz-M, J., Ariza-Echeverri, E. A., Fernández-Arias, P., & Antón-Sancho, Á. (2025). Evaluating solutions to marine plastic pollution. *Environments, 12*(3), 86. https://doi.org/10.3390/environments12030086

Wong, S., Han, H., Cheng, K., Koo, A., & Yussof, S. (2022). ESS-IoT: The smart waste management system for general household. *Pertanika Journal of Science and Technology, 31*, 311–325.

Zhang, C., Xu, T., Feng, H., & Chen, S. (2019). Greenhouse gas emissions from landfills: A review and bibliometric analysis. *Sustainability., 11*(8), 2282. https://doi.org/10.3390/su11082282

Zhang, Q., Li, H., Wan, X., Skitmore, M., & Sun, H. (2020). An intelligent waste removal system for smarter communities. *Sustainability, 12*, 6829.

Microbial Processes for Sustainable Municipal Waste Management

Chathuri Yapa Bandara and Kosala Sirisena

1 Introduction

Municipal Waste Management Challenges

The rapid rate of urbanization, industrialization, and changes in consumption patterns hugely triggered the generation of municipal solid waste (MSW) worldwide (Chen et al., 2020). The United Nations estimates that the annual MSW production will be 3.4 billion metric tons by 2050, leading to serious environmental and public health issues (Küfeoğlu, 2024). Furthermore, the composition of municipal waste has also become more complex with the emergence of recently discovered contaminants such as pharmaceuticals and microplastics. Conventional waste disposal methods, including open dumping, landfilling, and incineration, cause substantial environmental hazards, leading to groundwater contamination, air pollution, and greenhouse gas emissions. For example, landfilling contributes to approximately 11% of global emissions of methane, which is a potent greenhouse gas contributing to climate change. Additionally, the leaching of toxic substances from poorly managed waste sites creates severe threats to soil and water quality, eventually causing serious damage to both ecosystems and human populations (Fig. 1). The growing urgency for sustainable waste management strategies has led to the exploration of innovative eco-friendly methods that help reduce environmental degradation while maximizing resource recovery.

C. Y. Bandara
Arcadia Naturae Foundation, Colombo, Sri Lanka

K. Sirisena (✉)
Department of Environmental Technology, Faculty of Technology, University of Colombo, Colombo, Sri Lanka
e-mail: kosala@et.cmb.ac.lk

A. P. Das et al. (eds.), *Sustainable Municipal Waste Management with a Zero Waste Approach*, Sustainable Environmental Waste Management Strategies,
https://doi.org/10.1007/978-3-032-05950-5_9

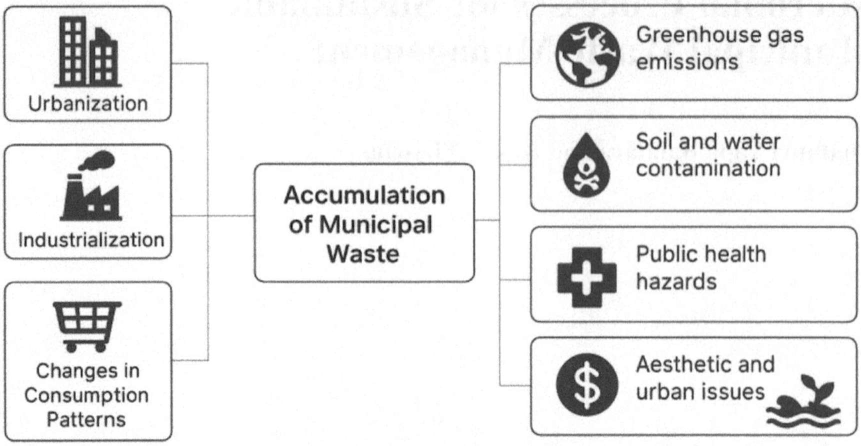

Fig. 1 Illustration of major reasons for exponential accumulation of municipal waste and its' impacts on the environment and human health

The Role of Microorganisms in Waste Treatment

Microbial processes play a crucial role in municipal waste treatment as it provide a green and sustainable approach to the degradation of waste and resource recovery (Das & Ghosh, 2023a; Das et al., 2012; Mishra & Das, 2023). Microorganisms are capable of degrading literally any organic waste, removing pollutants, and stabilizing waste streams (Adebayo & Obiekezie, 2018). Aerobic and anaerobic microbial communities contribute to waste treatment processes through different approaches such as composting, anaerobic digestion, and wastewater bioremediation (2024b; Das et al., 2015; Paul et al., 2023; Pradhan et al., 2023). In the composting process, microbial communities, including Actinobacteria and fungi such as *Aspergillus* and *Penicillium*, decompose organic matter into nutrient-rich compost that can be used as a soil amendment (Das et al., 2023; Dey et al., 2024; Nag et al., 2023a). Anaerobic digestion is mainly conducted by hydrolytic, acidogenic, acetogenic, and methanogenic bacteria that convert organic waste into biogas, primarily methane and carbon dioxide. This process provides a renewable energy source while reducing landfill dependency. Furthermore, microbial bioremediation is governed by specialized bacteria such as *Pseudomonas* and *Bacillus* species to degrade hydrocarbons, heavy metals, and other toxic contaminants in waste streams. This process improves the quality of treated waste, enabling safe disposal or reuse. The metabolic diversity of microorganisms thus provides a robust foundation for developing innovative waste treatment technologies that align with environmental sustainability (Das & Ghosh, 2022; Das et al., 2024a; Ghosh et al., 2023; Nag et al., 2023b).

Sustainable Waste Management and the Circular Economy

Sustainable waste management practices should be able to reduce ecological destruction, enhance effective resource utilization, and promote long-term environmental stability (Bala et al., 2024; Dey et al., 2023; Mohanty et al., 2024; Singh Jadaun et al., 2022). Traditional linear waste management practices focus on the "take-make-dispose" approach, which ultimately results in resource depletion and environmental pollution. However, the circular economy approach promotes waste valorization, where waste materials are converted into useful products through recovery, recycling, and reuse, ultimately reducing the negative impacts on natural ecosystems (Fig. 2) (Priyadarsini & Das, 2024, 2025). Microbial waste treatment technologies follow circular economy principles as those methodologies basically transform waste into valuable by-products. For example, microbial fermentation processes convert renewable resources, such as agricultural by-products and waste materials, into biodegradable polymers like polyhydroxyalkanoates (PHAs) (Samuel et al., 2024). This approach not only reduces reliance on fossil fuels but also addresses environmental concerns associated with plastic pollution (Priyadarshini et al., 2024). Similarly, microbial electrochemical systems, such as microbial fuel cells, utilize the metabolic activity of electrogenic bacteria to generate bioelectricity from organic waste components (Das & Mishra, 2010; Kumar & Das, 2016; Mishra & Das, 2008). Therefore, microbial processes have significant potential for sustainable waste management by enhancing waste degradation and pollutant removal. It also supports the sustainable recovery of energy and materials, resulting in a more resilient and circular waste management system (Behera et al., 2021; Mishra et al., 2024).

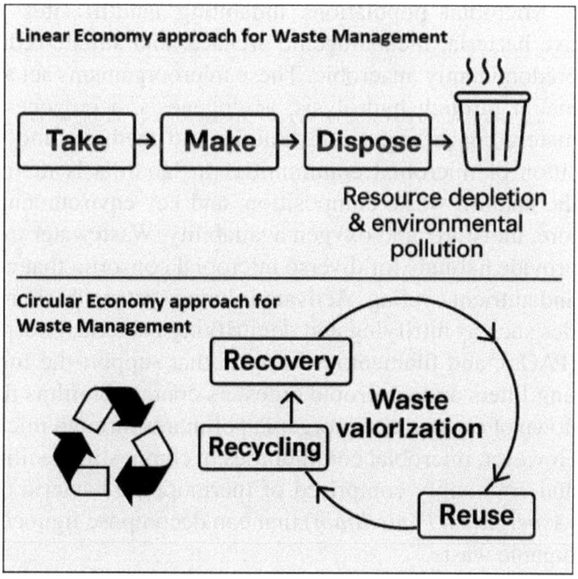

Fig. 2 Comparison between the linear economy and circular economy approaches for waste management

2 Microbial Ecology in Waste Management

Microbial Diversity in Municipal Waste Environments

Microorganisms are the most diverse and abundant life forms on Earth. They occupy every ecological niche, from deep-sea hydrothermal vents to arid deserts and glacial ice (Das & Ghosh, 2023b, 2023c). Microorganisms have been exposed to various environmental conditions, including extreme temperatures, pH levels, salinity, and nutrient availability, and have adapted to those conditions over billions of years of evolution, resulted in this enormous diversity. Due to the huge genetic variability within microbial populations, their ability for horizontal gene transfer and rapid reproduction rates, microbes are capable of exploiting diverse metabolic pathways and survive in rapidly changing environments. Furthermore, the complexity of microbial communities is further driven by their interactions with other organisms, environmental pressures, and their ability to form resilient biofilms that enhance survival and functionality in diverse ecosystems.

Microorganisms act as principal decomposers that transform waste materials found in municipal waste environments. Due to the heterogeneous nature of waste, fluctuating environmental conditions, and selective pressures from pollutants and nutrient availability, municipal waste environments also possess a huge microbial diversity. These microbial groups, including bacteria, archaea, fungi, and protists are well adapted to these specific ecological niches and evolved to metabolize diverse substrates, ranging from simple organic compounds to complex polymers and xenobiotic contaminants. That makes them crucial for sustainable waste management applications.

Microbial populations inhabiting landfill sites mainly consist of fermentative bacteria, methanogenic archaea, and sulfate-reducing bacteria (SRB) and are predominantly anaerobic. These microorganisms act as consortia to degrade organic matter through hydrolysis, acidogenesis, acetogenesis, and methanogenesis, ultimately converting waste mainly into methane and carbon dioxide. The composition of microbial communities in landfills is mainly determined by the age of the landfill, waste composition, and key environmental factors including temperature, moisture, and oxygen availability. Wastewater treatment plants (WWTPs) also provide habitats for diverse microbial consortia that contribute to pollutant removal and nutrient cycling. Activated sludge systems harbor complex microbial communities such as nitrifying and denitrifying bacteria, phosphate-accumulating organisms (PAOs), and filamentous bacteria that support the formation of sludge floc. Trickling filters and anaerobic digesters contain biofilms that further increase the breakdown of organic and inorganic pollutants through microbial synergistic interactions. However, microbial communities in composting facilities promote aerobic digestion and are mainly comprised of thermophilic bacteria (*Bacillus*, *Thermus*) and fungi (*Aspergillus*, *Penicillium*) that can decompose lignocellulosic material and stabilize organic waste.

Functional Roles of Microbial Communities in Decomposition

Microbial degradation of complex waste materials into simple, bioavailable forms is driven by microbial enzymatic processes. These processes play crucial roles in converting waste into useful products, such as compost, biogas, and treated effluent. Organic waste decomposition is mainly conducted through the activity of hydrolytic enzymes such as cellulases, proteases, amylases, and lipases. These enzymes degrade macromolecules into simpler sugars, amino acids, and fatty acids. Then the fermentative bacteria will metabolize these intermediate products into volatile fatty acids (VFAs) and hydrogen, which act as precursors for methanogenesis. Ammonia-oxidizing bacteria (AOB) and nitrite-oxidizing bacteria (NOB) play important roles by contributing to nitrification: converting ammonia into nitrite and then nitrite into nitrate in wastewater treatment. Then the denitrifying bacteria convert nitrates into nitrogen gas. The removal of excess amounts of nitrate helps mitigate eutrophication risks in aquatic systems. Some bacterial species including *Pseudomonas* and *Rhodococcus* are capable of bioremediating petroleum-contaminated waste streams. On the other hand, inorganic waste decomposition is governed by specialized microbial metabolic pathways. For example, sulfate-reducing bacteria (SRB) and metal-reducing bacteria (MRB) eliminate toxic metals by precipitating them as sulfides or binding them onto exopolysaccharides.

Adaptations of Microorganisms to Waste Processing Conditions

Waste environments can be broadly characterized by low nutrient availability, high toxicity, and oxygen limitations. These are generally extreme conditions and microorganisms that inhabit such environments require various physiological and genetic adaptations to enhance their resilience and metabolic capabilities. Microorganisms live in nutrient-poor conditions adapted to those conditions by utilizing alternative energy sources or following oligotrophic lifestyles. Some bacteria, like *Acidithiobacillus* thrive in acidic waste environments using special membrane transport systems to gather essential nutrients very efficiently. Other microbes form symbiotic partnerships, working together to help each other survive. For example, in methanogenic microbial groups, certain bacteria and archaea form symbiotic associations to break down complex organic waste materials.

Some microbes can also withstand toxic environments contaminated with heavy metals and persistent organic pollutants. They achieve this by using special methods such as metal efflux pumps, enzymatic detoxification, and the production of metal-chelating exopolysaccharides. Some extremophiles like halophilic bacteria produce special substances called osmoprotectants and employ special ion transport mechanisms to help them survive in saline waste environments. In oxygen-limited environments, like landfills and anaerobic digesters, microbes rely on anaerobic respiration and fermentation to obtain energy. Facultative anaerobes, such as *Escherichia coli* and *Shewanella*, have the ability to switch between aerobic and anaerobic respiration for energy, allowing them to adapt to changing oxygen levels. In contrast, obligate anaerobes, like *Clostridium* and *Methanobacterium*, thrive in oxygen-limited waste environments by using alternative electron acceptors like sulfate and carbon dioxide to produce energy.

3 Microbial Processes in Municipal Waste Treatment

Microbes are vital in municipal waste treatment worldwide. They help in breaking down organic matter, degrading pollutants, and retrieving valuable resources through their metabolic activities. Therefore, microbial approach to waste treatment significantly improved the waste management efficiency, reduced environmental pollution, and promoted sustainable practices in urban areas.

Aerobic and Anaerobic Biodegradation

Biodegradation is a fundamental process in municipal waste treatment. This process is conducted by microorganisms that break down organic matter into simpler compounds. The efficiency and end products of microbial degradation are largely dependent on the presence or absence of oxygen. This leads to two distinct metabolic pathways: aerobic biodegradation and anaerobic biodegradation. Microorganisms show distinct metabolisms in oxygen-rich (aerobic) and oxygen-limited (anaerobic) environments. The main differences include how they generate energy, electron acceptors, and the types of metabolic by-products they produce.

Aerobic metabolism involves microorganisms using oxygen to produce energy. Oxygen acts as the terminal electron acceptor in the electron transport chain (ETC), allowing these organisms to produce energy efficiently in the form of adenosine triphosphate (ATP). The main steps involved in this process are glycolysis, the tricarboxylic acid (TCA) cycle, and oxidative phosphorylation. When microorganisms completely break down organic substances like glucose, they produce carbon dioxide (CO_2) and water as end products. Some of the key microbial groups involved in aerobic degradation include *Pseudomonas*, *Bacillus*, and *Aspergillus* species.

Anaerobic metabolism takes place in the absence of oxygen. In these conditions, microorganisms utilize alternative electron acceptors, like nitrate, sulfate, iron, or carbon dioxide. Anaerobic respiration produces less energy (ATP) compared to aerobic respiration, because alternative electron acceptors yield lower energy. Instead of complete oxidation, organic matter undergoes fermentation, producing organic acids, hydrogen, methane, and alcohols. The key metabolic pathways involved here are glycolysis, fermentation pathways, and different forms of anaerobic respiration. In anaerobic biodegradation, different microbial groups play important roles in sequential manner: hydrolytic bacteria break down large molecules; acidogenic and acetogenic bacteria convert those products into acids and acetates; finally, methanogenic archaea convert those products into methane. They all work in a sequence to help break down materials in anaerobic digestion.

Both aerobic and anaerobic pathways are used in various waste treatment approaches, like composting, activated sludge, and anaerobic digestion. These processes benefit from microbial metabolic activities to decompose organic waste, remove pollutants, and recover useful resources.

Composting: Aerobic Microbial Decomposition

Composting is an aerobic waste treatment process, where organic waste is turned into nutrient-rich compost. The process takes place with the synergistic action of bacteria, fungi, and actinomycetes to break down organic matter under oxygen-rich conditions.

Composting process progresses through three phases:

- **Mesophilic phase (25–40 °C)**: This is the initial decomposition, which is carried out by mesophilic bacteria (*Pseudomonas*, *Bacillus*). They break down simple compounds like sugars and amino acids.
- **Thermophilic phase (45–70 °C)**: In this phase, mainly thermophiles (*Bacillus*, *Aspergillus*) degrade complex organic compounds like lignin and cellulose. Due to higher microbial activity, the temperature is increased. That further enhances the activity of thermophiles, while eliminating pathogenic organisms and weed seeds.
- **Maturation phase**: In this phase, microbial activity slows down, stabilizing the compost and enriching it with humic substances, which are essential for soil health.

Composting is widely used for municipal solid waste (MSW), agricultural waste, and sewage sludge. It is a green alternative to landfilling. This process minimizes methane emissions and produces a valuable soil conditioner.

Activated Sludge: Aerobic Wastewater Treatment

The activated sludge process is a commonly used aerobic wastewater treatment method. Here, microbial communities are used to degrade organic pollutants in sewage. In this system, wastewater is aerated in a reactor, encouraging aerobic bacteria to form microbial flocs. These flocs are important in efficiently removing organic matter and nutrients.

Microbial metabolism in activated sludge mainly involves:

- **Oxidation of organic matter**: Bacteria such as *Pseudomonas* and *Flavobacterium* consume organic pollutants and convert them into CO_2, water, and microbial biomass.
- **Nitrification**: Autotrophic bacteria such as *Nitrosomonas* and *Nitrobacter* oxidize ammonia to nitrate. This is a crucial step in removing excess nitrogen in sewage.
- **Phosphorus uptake**: In this step, Polyphosphate-accumulating organisms (PAOs), such as *Acinetobacter*, store phosphorus intracellularly, which helps to decrease the amount of phosphorus released into the treated water.

After these processes, the microbial biomass is separated from treated water. However, a fraction of the sludge is recirculated to maintain the microbial activity. The activated sludge system is highly effective at removing organic matter and nutrients from wastewater.

Anaerobic Digestion

Anaerobic digestion (AD) is a commonly used process that transforms waste into energy through microbial approaches. It is mainly applied to treat sewage sludge, organic waste, and waste accumulated in landfill bioreactors. The process takes place in four sequential steps, and each step is driven by specialized microbial groups:

- **Hydrolysis**: In this step, hydrolytic bacteria such as *Clostridium* and *Bacteroides* break down complex polymers (carbohydrates, proteins, lipids) into simpler monomers (sugars, amino acids, and fatty acids).
- **Acidogenesis**: Here, acidogenic bacteria like *Enterobacter* and *Lactobacillus* convert those simple monomers into volatile fatty acids (VFAs), alcohols, hydrogen (H_2), and carbon dioxide (CO_2).
- **Acetogenesis**: In this stage, acetogenic bacteria such as *Syntrophomonas*, *Pelotomaculum,* and *Syntrophobacter* convert the intermediate products, especially VFAs and alcohols, into acetic acid, CO_2, and H_2.
- **Methanogenesis**: Methanogenic archaea such as *Methanosarcina* and *Methanobacterium* utilize acetate, H_2, and CO_2 to produce methane (CH_4). There are two main pathways for producing methane: (1) Acetolastic methanogenesis, where CH_4 is produced from acetic acid; and (2) Hydrogenotrophic methanogenesis, in which CH_4 is synthesized from CO_2, and H_2.

Biogas generated from anaerobic digestion (AD) typically consists of 60–70% methane (CH_4), along with trace amounts of carbon dioxide (CO_2), hydrogen (H_2), and hydrogen sulfide (H_2S). This process makes anaerobic digestion an excellent, sustainable method for managing organic waste and producing renewable energy. It helps reduce our reliance on fossil fuels and cuts down greenhouse gas emissions. Furthermore, the remaining sludge can be reused as a nutrient-rich fertilizer for agriculture. Table 1. Shows a comparative overview of aerobic and anaerobic processes in waste management.

Table 1 Comparative overview of aerobic and anaerobic processes in waste management

Parameter	Aerobic biodegradation (composting, activated sludge)	Anaerobic biodegradation (anaerobic digestion)
Oxygen requirement	Requires oxygen	Operates in oxygen-deprived conditions
Microbial energy yield	High ATP yield via oxidative phosphorylation	Lower ATP yield via fermentation/anaerobic respiration
End products	CO_2, H_2O, microbial biomass	CH_4, CO_2, organic acids, digestate
Decomposition rate	Faster due to efficient metabolism	Slower due to sequential microbial action
Waste treatment applications	Composting, activated sludge, bioremediation	Anaerobic digestion, landfill bioreactors
Environmental impact	Reduces organic waste, mitigates pollution, produces compost	Generates biogas, reduces methane emissions, recovers nutrients

Microbial Processes for Leachate and Nutrient Treatment

Landfill leachate is a complex, heavily contaminated liquid produced when rainwater seeps through solid waste in landfills, picking up organic and inorganic contaminants along the way. Leachate may contain high levels of ammonia (NH_4^+), nitrate (NO_3^-), phosphate (PO_4^{3-}), organic matter (biodegradable or refractory), heavy metals, chlorinated organic compounds, and inorganic salts, which can cause serious environmental issues if not properly treated (Li et al., 2010). Additionally, wastewater from municipal and industrial sources also contains significant amounts of nutrients that can cause eutrophication, which in turn causes oxygen depletion and loss of biodiversity in aquatic ecosystems. Microbial processes are crucial for treating nutrients in leachate and wastewater. These processes include:

- **Nitrification–denitrification**

  Here, the nitrification process is coupled with the denitrification process to remove excess amounts of nitrogen.

  Nitrification: This is a two-step process where Ammonia-oxidizing bacteria (AOB) such as *Nitrosomonas* first convert ammonia into nitrite, and then nitrite-oxidizing bacteria (NOB) such as *Nitrobacter* converts nitrite into nitrate through aerobic respiration (Chisholm et al., 2024; Costa et al., 2006). It was recently discovered that comammox *Nitrospira,* a group of bacteria traditionally thought to be responsible for nitrite oxidation, could undertake both steps of nitrification (Daims et al., 2015; Van Kessel et al., 2015).

  Denitrification: Conventional denitrifying bacteria such as *Paracoccus denitrificans, Pseudomonas stutzeri* reduce NO_3^- to nitrogen gas (N_2), which is released into the atmosphere, preventing nitrate contamination. However, these conventional denitrification microorganisms can only denitrify under anoxic conditions, which restrict their applications in nitrogen removal treatment (Luo et al., 2016).

However, it was discovered that heterotrophic nitrification and aerobic denitrification microorganisms (HNADMs) could simultaneously achieve nitrification and denitrification under aerobic conditions. With sufficient organic carbon, HNADMs, can achieve faster growth and effectively undertake the nitrification and denitrification processes (Chen & Ni, 2012; Chen et al., 2014; Song et al., 2021).

- **Phosphorus removal through Enhanced Biological Phosphorus Removal (EBPR)**

Phosphorus is an element for the growth and metabolism of living organisms, but it can create serious environmental problems when it accumulates in excess amounts. The basis of the Enhanced Biological Phosphorus Removal (EBPR) technology is the enrichment of polyphosphate-accumulating organisms (PAOs) such as *Accumulibacter* in activated sludge to take up large quantities of polyphosphate inside their cell under changing aerobic and anaerobic conditions. This technology offers an efficient mechanism for phosphorus removal and recovery while preventing eutrophication in water bodies (Izadi et al., 2020).

- **Microbial degradation of organic pollutants in leachate**

Various organic pollutants such as phthalates, polycyclic aromatic hydrocarbons (PAH), phenols, and nitrogen-containing pollutants can be present in landfill leachate. These pollutants may initially be present in the landfilled waste or produced during the waste degradation processes (Marttinen et al., 2003). Microbial processes, mainly aerobic processes, provide promising tools for leachate treatment. Aerobic processes are relatively cheap and particularly effective in degrading not only organic pollutants contained in young leachate but also nitrogen compounds through nitrification/denitrification processes (Bove et al., 2015). Microbial communities comprised of some fungi (*Trametes versicolor*) and bacteria (*Pseudomonas, Bacillus*) could effectively break down humic substances, phenols, and volatile organic compounds (VOCs), reducing the toxicity of landfill leachate.

These microbial processes play a key role in landfill leachate treatment. We can effectively optimize these processes to clean up the leachate and minimize the impacts on the environment and human health (Remmas et al., 2023).

Bioremediation of Hazardous Waste

Microbial processes can be effectively used to treat hazardous waste in municipal waste systems with a minimal environmental impact. These processes degrade, detoxify, or transform hazardous pollutants into less harmful forms, enabling safe disposal or reuse. Microbial enzymatic oxidation–reduction reactions facilitate the degradation of persistent contaminants such as heavy metals, pharmaceuticals, and plastics. These microbial methods are more sustainable, economical, and eco-friendly compared to conventional chemical treatments.

Certain bacteria and fungi are capable of immobilizing or transforming heavy metals through different processes: biosorption, bioaccumulation, bioprecipitation, and enzymatic reduction. Biosorption is the metabolism-independent adsorption of toxic compounds to the surface of the microbial cell wall. Some bacteria, fungi, and algae have shown the ability to adsorb various metals, including Ag, Cd, Cu, Hg, Ni, and Pb (Kuippers et al., 2021; Torres, 2020). For example, *Pseudomonas* sp. 375 has shown promising biosorbent properties to remove Cd^{2+} from contaminated wastewater (Xu et al., 2020). In bioprecipitation, microorganisms are employed to precipitate soluble harmful metals/metaloids, which enables the removal of these contaminants from wastewater (Sahinkaya et al., 2017). Sulfate-reducing bacteria such as *Desulfovibrio desulfuricans* have shown the capability of converting soluble lead and mercury into insoluble sulfides, enabling the removal of those toxic metals from the system (Zhang et al., 2016). Furthermore, *Zobellella denitrificans*, *Pseudomonas putida,* and *Shewanella oneidensis* have shown the ability to reduce toxic chromium (Cr^{6+}) into the less toxic Cr^{3+} form (Balamurugan et al., 2014; He et al., 2016; Hu et al., 2024). Pharmaceuticals such as antibiotics, hormones, and painkillers could also cause serious threats to aquatic ecosystems. Bacteria such as *Mycobacterium* and *Sphingomonas* have shown the potential to degrade pharmaceutical compounds (Aulestia et al., 2021; Carneiro, 2023; Lara-Moreno et al., 2024; Poupin et al., 1999). Furthermore, *Ideonella sakaiensis* has shown the capability of degrading hazardous plastics such as polyethylene terephthalate (PET), offering potential solutions to plastic waste accumulation (Sevilla et al., 2023; Son et al., 2019; Yoshida et al., 2021).

4 Bioproducts from Waste: Harnessing Microbial Processes for Sustainable Resource Recovery

Waste valorization through microbial processes is currently gaining huge attention as it is a promising approach to mitigate greenhouse gas emission, reduce landfill dependency, and move towards the circular economy (Moustakas et al., 2020; Varjani et al., 2021). Two key microbial-driven waste valorization processes: microbial conversion of organic waste into biofertilizers and microbial synthesis of biopolymers and bioplastics are briefly discussed below.

Microbial Conversion of Organic Waste into Biofertilizers

Composting is an aerobic microbial process that converts organic waste into biofertilizers. This process improves the soil fertility and microbial activity. Compost adds important nutrients such as nitrogen (N), phosphorus (P), and potassium (K) into soil and improves microbial diversity and promotes plant health. Vermicomposting is a special type of composting that integrates microbial decomposition with the activity of earthworms (*Eisenia fetida, Eudrilus eugeniae*) to speed up the bioconversion of organic waste into biofertilizers (Acquah et al., 2021; Gebrehana et al., 2023). Vermicompost is similar to compost, but it is well-aerated with an excellent structure, permeability, and moisture retention capacity that promotes plant growth. Earthworm gut microbiota especially contribute to these beneficial features in vermicompost. Microbial-driven composting and vermicomposting processes significantly contribute to a closed-loop nutrient cycle. It also reduces the dependency on synthetic fertilizers and promotes sustainable agriculture (Kavya et al., 2024; Mendoza-Hernández et al., 2014; Varjani et al., 2021).

Microbial Synthesis of Biopolymers and Bioplastics

Plastics have become an integral part of modern society with a wide range of applications. However, non-biodegradable plastic pollution creates severe environmental concerns and hazards to animal and human health. This leads to exploring alternatives to petroleum-based plastics and bioplastics received a significant interest as a sustainable alternative (de Souza & Gupta, 2024; Nanda et al., 2022). Microbial biopolymers and bioplastics such as polyhydroxyalkanoates (PHAs) and polylactic acid (PLA), provide eco-friendly solutions to plastic waste problem (Anitha & Srivastava, 2021; Varghese et al., 2022). Polyhydroxyalkanoates (PHAs) are biodegradable plastics produced by bacteria as intracellular storage compounds under nutrient-limiting conditions. Bacteria such as *Cupriavidus necator*, *P. putida*, and *Bacillus megaterium* demonstrate excellent mechanisms to produce PHAs from organic waste substrates, such as food waste, wastewater sludge, and agricultural residues (Ho et al., 2024; Joyline & Aruna, 2019; Ronďošová et al., 2022). Polylactic Acid (PLA) is another biopolymer derived from microbial fermentation of lactic acid by *Lactobacillus* species (Ahmad et al., 2024; Nicolescu et al., 2023). Biopolymer production can be more economical by utilizing waste-derived carbon sources. Agricultural waste can be effectively used as feedstocks for lactic acid fermentation. Food waste such as oils, starch and dairy residues can be used as substrates for biosynthesis of PHAs. Advances in metabolic engineering and synthetic biology further enhance the production of cost-effective and eco-friendly bioplastics (Arora et al., 2023).

5 Microbial Biotechnology for Waste Management

The recent advances in microbial biotechnology have significantly improved waste management strategies by increasing efficiency in waste degradation, resource recovery, and bioenergy production (Ezejiofor et al., 2014; Haddadi et al., 2018). Emerging and ever-evolving biotechnological tools such as omics technologies, microbiome engineering, CRISPR/Cas9 gene-editing tools, and microbial fuel cells, have shown novel avenues to utilize microbial systems for sustainable waste treatment (Dixit et al., 2025; Lawson et al., 2019; Sharma et al., 2022). These advanced technologies significantly improve the efficiency of waste treatment process and reduce the environmental footprint by increasing the precision, efficiency, and adaptability in microbial applications.

Omics Technologies and Microbiome Engineering for Sustainable Waste Treatment

The term OMICS can be described as exploring and analyzing large amount of data pertaining to the structure and function of an entire makeup of a given biological system at a particular level. In parallel to the development of various other scientific disciplines such as DNA sequencing technologies, bioinformatics, and imaging technologies, various promising omics technologies have emerged over the last few decades (Zhang et al., 2010; Dai & Shen, 2022). With the emergence of the big four omics: genomics, transcriptomics, proteomics, and metabolomics, new lineages of omics technologies evolved, which are particularly useful in gaining insights into microbial systems. These are some of the key omics approaches that can be effectively used for microbial-driven waste treatment technologies:

- **Metagenomics**: Metagenomics is the analysis of all the genomes contained within an environmental sample. Conventional phylogenetic gene-based analyses only provide taxonomic information of the microbial community. But metagenomics provides information on the functional gene composition of microbial communities. Furthermore, it gives genetic information on potentially novel biocatalysts or enzymes, genomic linkages between function and phylogeny for uncultured organisms, and evolutionary profiles of community function and structure (Nam et al., 2023; Thomas et al., 2012). This insight is very useful to manipulate microbial communities to achieve sustainable waste management.
- **Metatranscriptomics**: Metatranscriptomics is defined as the analysis of a gene expression in a population or a whole community (Aransay & Trueba, 2016). It provides information on the functional capabilities of the community: gene expression levels, regulations, and pathways that are mainly related to environmental changes. Therefore, matatranscriptomics can be employed on various complex environmental samples including municipal solid waste to understand how we

The different regulatory and legislative frameworks operating in different parts of the world could also act as a barrier for large-scale implementation of microbial waste treatment technologies. Many countries have strict regulations on engineered microbes due to concerns about biosafety and environmental risks. Furthermore, the commercialization of bio-based products such as biofertilizers, bioplastics, and biogas derived from microbial waste treatment requires extensive safety testing to ensure that these products are safe for human consumption. This will take more time and cost to make these products available in the market. To overcome these barriers, it is important to provide financial incentives and government subsidies to promote microbial-driven waste treatment technologies.

References

Acquah, M. N., Essandoh, H. M. K., Oduro-Kwarteng, S., Appiah-Effah, E., & Owusu, P. A. (2021). Degradation and accumulation rates of fresh human excreta during vermicomposting by Eisenia fetida and Eudrilus eugeniae. *Journal of Environmental Management, 293*, Article 112817.

Adebayo, F. O., & Obiekezie, S. O. (2018). Microorganisms in waste management. *Research Journal of Science and Technology, 10*(1), 28–39.

Ahmad, A., Banat, F., Alsafar, H., & Hasan, S. W. (2024). An overview of biodegradable poly (lactic acid) production from fermentative lactic acid for biomedical and bioplastic applications. *Biomass Conversion and Biorefinery, 14*(3), 3057–3076.

Anitha, N. N. N., & Srivastava, R. K. (2021). Microbial synthesis of polyhydroxyalkanoates (PHAs) and their applications. In *Environmental and agricultural microbiology: Applications for sustainability* (pp. 151–181).

Aransay, A. M., & Trueba, J. L. L. (Eds.). (2016). *Field guidelines for genetic experimental designs in high-throughput sequencing*. Springer.

Arora, Y., Sharma, S., & Sharma, V. (2023). Microalgae in bioplastic production: A comprehensive review. *Arabian Journal for Science and Engineering, 48*(6), 7225–7241.

Aulestia, M., Flores, A., Mangas, E. L., Pérez-Pulido, A. J., Santero, E., & Camacho, E. M. (2021). Isolation and genomic characterization of the ibuprofen-degrading bacterium Sphingomonas strain MPO218. *Environmental Microbiology, 23*(1), 267–280.

Bala, K., Anshul, Verma, S., Dogra, A., Das, A. P., Kumar, R., & Nandini. (2024). Sustainable management for urban plastic waste generation. In *Sustainable management of urban plastic waste through circular economic approaches* (pp. 211–236). https://doi.org/10.1201/9781003465447-13

Balamurugan, D., Udayasooriyan, C., & Kamaladevi, B. (2014). Chromium (VI) reduction by Pseudomonas putida and Bacillus subtilis isolated from contaminated soils. *International Journal of Environmental Sciences, 5*(3), 522–529.

Behera, I. D., Nayak, M., Biswas, S., Meikap, B. C., & Sen, R. (2021). Enhanced biodegradation of total petroleum hydrocarbons by implementing a novel two-step bioaugmentation strategy using indigenous bacterial consortium. *Journal of Environmental Management, 292*, Article 112746. https://doi.org/10.1016/J.JENVMAN.2021.112746

Bove, D., Merello, S., Frumento, D., Arni, S. A., Aliakbarian, B., & Converti, A. (2015). A critical review of biological processes and technologies for landfill leachate treatment. *Chemical Engineering & Technology, 38*(12), 2115–2126.

Carneiro, R. B. (2023). Biodegradation assessment of pharmaceuticals and parabens in lab-made sewage driven by Brevundimonas sp., Burkolderia sp. and Sphingomonas sp. *Biodegradation, 603*, 3.

Chavez, M., Chen, X., Finn, P. B., & Qi, L. S. (2023). Advances in CRISPR therapeutics. *Nature Reviews Nephrology, 19*(1), 9–22.

Chen, D., Chen, D., Bodirsky, B. L., Krueger, T., Mishra, A., Mishra, A., & Popp, A. (2020). The world's growing municipal solid waste: Trends and impacts. *Environmental Research Letters.*

Chen, M., Wang, W., Feng, Y., Zhu, X., Zhou, H., Tan, Z., & Li, X. (2014). Impact resistance of different factors on ammonia removal by heterotrophic nitrification–aerobic denitrification bacterium Aeromonas sp. HN-02. *Bioresource Technology, 167*, 456–461.

Chen, Q., & Ni, J. (2012). Ammonium removal by Agrobacterium sp. LAD9 capable of heterotrophic nitrification–aerobic denitrification. *Journal of Bioscience and Bioengineering, 113*(5), 619–623.

Chisholm, C., Di, H., Cameron, K., Podolyan, A., Shen, J., Zhang, L., Sirisena, K., & Godsoe, W. (2024). Contrasting response of comammox Nitrospira, ammonia oxidising bacteria, and archaea to soil pH and nitrogen inputs. *Science of the Total Environment, 924*, 171627.

Costa, E., Pérez, J., & Kreft, J. U. (2006). Why is metabolic labour divided in nitrification? *Trends in Microbiology, 14*(5), 213–219.

Dai, X., & Shen, L. (2022). Advances and trends in omics technology development. *Frontiers in Medicine, 9*, Article 911861.

Daims, H., Lebedeva, E. V., Pjevac, P., Han, P., Herbold, C., Albertsen, M., Jehmlich, N., Palatinszky, M., Vierheilig, J., Bulaev, A., Kirkegaard, R. H., von Bergen, M., Rattei, T., Bendinger, B., Nielsen, P. H., & Wagner, M. (2015). Complete nitrification by Nitrospira bacteria. *Nature, 528*(7583), 504–509.

Das, A. P., & Ghosh, S. (2023a). Metagenomic insights into unculturable microbial diversity. In *Manganese mining microorganisms* (pp. 63–76). https://doi.org/10.1016/B978-0-12-822147-1.00002-9

Das, A. P., & Ghosh, S. (2023b). Mixed culture bioleaching: an insight into manganese biomining process and efficacy. In *Manganese mining microorganisms* (pp. 129–141). https://doi.org/10.1016/B978-0-12-822147-1.00001-7

Das, A. P., & Ghosh, S. (2023c). Role of metabolites and metabolomics for understanding of manganese bioleaching. In *Manganese mining microorganisms* (pp. 117–127). https://doi.org/10.1016/B978-0-12-822147-1.00011-X

Das, A. P., Ghosh, S., Mohanty, S., & Sukla, L. B. (2015). Advances in manganese pollution and its bioremediation. In L. B. Sukla, N. Pradhan, S. Panda, & B. K. Mishra (Eds.), *Environmental microbe biotechnology* (1st ed., pp. 313–328). Springer. https://doi.org/10.1007/978-3-319-19018-1_16

Das, A. P., Mohanty, A. K., van Hullebusch, E. D., Figueiredo, G., Das, A. P., Mohanty, A. K., van Hullebusch, E. D., & Figueiredo, G. (2023). Bioremediation as an emerging technology for the removal of synthetic microplastic pollutants from marine ecosystem. *MarPB, 194*, Article 115297. https://doi.org/10.1016/J.MARPOLBUL.2023.115297

Das, A. P., Nayak, B., & Behera, I. D. (2024a). Applications of microorganisms in advanced geomicrobiological and biotechnological exploration. *Geomicrobiology Journal, 41*(4), 309–311. https://doi.org/10.1080/01490451.2024.2327059/ASSET//CMS/ASSET/DE2A1F67-08C1-475A-ABC7-2B3F4C9E8BCF/01490451.2024.2327059.FP.PNG

Das, A. P., Sukla, L. B., & Pradhan, N. (2012). Microbial recovery of manganese using *Staphylococcus Epidermidis. International Journal of Nonferrous Metallurgy, 2012*(02), 9–12. https://doi.org/10.4236/IJNM.2012.12002

Das, A., & Ghosh, S. (2022). Role of microorganisms in extenuation of mining and industrial wastes. *Geomicrobiology Journal, 39*, 173–175. https://doi.org/10.1080/01490451.2022.2038953

Das, A., & Mishra, S. (2010). Biodegradation of the metallic carcinogen hexavalent chromium Cr(VI) by an indigenously isolated bacterial strain. *Journal of Carcinogenesis, 9*. https://doi.org/10.4103/1477-3163.63584

Das, A., Dey, S., Bala, K., Gnanasekaran, R., & Das, A. P. (2024). Microbial bioremediation technology for sustainable treatment and management of synthetic microfiber waste. In *Sustainable microbial technology for synthetic and cellulosic microfiber bioremediation* (pp. 221–238). https://doi.org/10.1007/978-3-031-62600-5_12

de Souza, F. M., & Gupta, R. K. (2024). Bacteria for bioplastics: Progress, applications, and challenges. *ACS Omega, 9*(8), 8666–8686.

Dey, S., Das, A., Mallick, K., Sahu, A., & Das, A. P. (2023). *Environmental petroleum waste: Pollution, toxicity, sustainable remediation* (pp. 159–175). https://doi.org/10.1007/978-3-031-48220-5_7

Dey, S., Dash, A., & Das, A. P. (2024). Bioremediation as a sustainable and circular solution for mining waste management and bioenergy generation. In *Sustainable management of mining waste and tailings: A circular economy approach* (pp. 107–119). https://doi.org/10.1201/978 1003442455-6/BIOREMEDIATION-SUSTAINABLE-CIRCULAR-SOLUTION-MINING-WASTE-MANAGEMENT-BIOENERGY-GENERATION-SUDESHNA-DEY-AKANSHYA-DASH-ALOK-PRASAD-DAS

Dixit, Y., Yadav, P., Asnani, H., & Sharma, A. K. (2025). CRISPR/Cas9-engineering for increased amylolytic potential of microbes for sustainable wastewater treatment: A review. *Current Microbiology, 82*(1), 44.

Ezejiofor, T. I. N., Enebaku, U. E., & Ogueke, C. (2014). Waste to wealth-value recovery from agro-food processing wastes using biotechnology: A review. *British Biotechnology Journal, 4*(4), 418.

Gao, D., Zhao, H., Wang, L., Li, Y., Tang, T., Bai, Y., & Liang, H. (2022). Current and emerging trends in bioaugmentation of organic contaminated soils: A review. *Journal of Environmental Management, 320*, Article 115799.

Gebrehana, Z. G., Gebremikael, M. T., Beyene, S., Sleutel, S., Wesemael, W. M., & De Neve, S. (2023). Organic residue valorization for Ethiopian agriculture through vermicomposting with native (Eudrilus eugeniae) and exotic (Eisenia fetida and Eisenia andrei) earthworms. *European Journal of Soil Biology, 116*, Article 103488.

Gentry, T., Rensing, C., & Pepper, I. A. N. (2004). New approaches for bioaugmentation as a remediation technology. *Critical Reviews in Environmental Science and Technology, 34*(5), 447–494.

Ghosh, S., Tripathy, B., Dey, S., & Das, A. P. (2023). Comparative investigation of fungal and bacterial manganese biomining mechanisms. In *Manganese mining microorganisms* (pp. 99–115). https://doi.org/10.1016/b978-0-12-822147-1.00016-9

Haddadi, M. H., Aiyelabegan, H. T., & Negahdari, B. (2018). Advanced biotechnology in biorefinery: A new insight into municipal waste management to the production of high-value products. *International Journal of Environmental Science and Technology, 15*, 675–686.

He, Z., Hu, Y., Yin, Z., Hu, Y., & Zhong, H. (2016). Microbial diversity of chromium-contaminated soils and characterization of six chromium-removing bacteria. *Environmental Management, 57*, 1319–1328.

Ho, L. Y. L., Pan, L., Meng, F., Ho, K. T. M., Liu, F., Wu, M. T., Lei, H. I., Bhachu, G., Wang, X., Dahlsten, O., Sun, Y., Lee, P.-H., & Tan, G. Y. A. (2024). Quantum modeling simulates nutrient effect of bioplastic polyhydroxyalkanoate (PHA) production in Pseudomonas putida. *Scientific Reports, 14*(1), 18255.

Hollywood, K., Brison, D. R., & Goodacre, R. (2006). Metabolomics: Current technologies and future trends. *Proteomics, 6*(17), 4716–4723.

Hu, H., Wang, M., Huang, Y., Xu, Z., Xu, P., Nie, Y., & Tang, H. (2022). Guided by the principles of microbiome engineering: Accomplishments and perspectives for environmental use. *Mlife, 1*(4), 382–398.

Hu, S., Wang, Q., Zhang, H., Yang, Y., Chen, G., Wang, S., Liu, C., & Liu, T. (2024). Reduction and transformation of Cr (VI)-associated ferrihydrite by Shewanella oneidensis MR-1: Kinetics and secondary minerals. *Geochimica et Cosmochimica Acta, 374*, 15–32.

Izadi, P., Izadi, P., & Eldyasti, A. (2020). Design, operation and technology configurations for enhanced biological phosphorus removal (EBPR) process: A review. *Reviews in Environmental Science and Bio/technology, 19*, 561–593.

Javaid, D., Ganie, S. Y., Hajam, Y. A., & Reshi, M. S. (2022). CRISPR/Cas9 system: A reliable and facile genome editing tool in modern biology. *Molecular Biology Reports, 49*(12), 12133–12150.

Jeevanandam, V., & Osborne, J. (2022). Understanding the fundamentals of microbial remediation with emphasize on metabolomics. *Preparative Biochemistry & Biotechnology, 52*(3), 351–363.

Joyline, M., & Aruna, K. (2019). Production and characterization of polyhydroxyalkanoates (PHA) by Bacillus megaterium strain JHA using inexpensive agro-industrial wastes. *International Journal of Recent Scientific Research, 10*(7), 33359–33374.

Kavya, Vashisht, M., Jain, B., & Shrivastava, S. (2024). Transforming waste into wealth: a review on microbial conversion of organic municipal wastes to value-added products. *Discover Environment*, *2*(1), 112.

Khaksar, G., Sirijan, M., Suntichaikamolkul, N., & Sirikantaramas, S. (2022). Metabolomics for agricultural waste valorization: Shifting toward a sustainable bioeconomy. *Frontiers in Plant Science, 13*, Article 938480.

Küfeoğlu, S. (2024). Waste management emissions. In *Net zero: Decarbonizing the global economies* (pp. 555–611). Springer Nature Switzerland.

Kuippers, G., Bassil, N. M., & Lloyd, J. R. (2021). Microbial colonization of cementitious geodisposal facilities, and potential "biobarrier" formation. In *The microbiology of nuclear waste disposal* (pp. 157–192). Elsevier.

Kumar, M. S., & Das, A. P. (2016). Molecular identification of multi drug resistant bacteria from urinary tract infected urine samples. *Microbial Pathogenesis, 98*, 37–44. https://doi.org/10.1016/J.MICPATH.2016.06.029

Lam, T. K., Yang, J. T., Lai, S. J., Liang, S. Y., & Wu, S. H. (2021). Meta-proteomics analysis of microbial ecosystem during the anaerobic digestion of chicken manure in biogas production farm. *Bioresource Technology Reports, 13*, Article 100643.

Lara-Moreno, A., El-Sayed, F., Cox, C. J., Costa, M. C., & Carlier, J. D. (2024). Batch studies on the biodegradation of paracetamol and 1, 4-hydroquinone by novel bacterial strains isolated from extreme environmental samples and the identification of candidate catabolic genes. *Applied Water Science, 14*(9), 198.

Lawson, C. E., Harcombe, W. R., Hatzenpichler, R., Lindemann, S. R., Löffler, F. E., O'Malley, M. A., Martin, H. G., Pfleger, B. F., Raskin, L.,Venturelli, O. S., Weissbrodt, D. G., Noguera, D. R., & McMahon, K. D. (2019). Common principles and best practices for engineering microbiomes. *Nature Reviews Microbiology, 17*(12), 725–741.

Li, W., Zhou, Q., & Hua, T. (2010). Removal of organic matter from landfill leachate by advanced oxidation processes: A review. *International Journal of Chemical Engineering, 2010*(1), Article 270532.

Luo, G., Xu, G., Tan, H., Gao, J., & Liu, W. (2016). Effect of dissolved oxygen on denitrification using polycaprolactone as both the organic carbon source and the biofilm carrier. *International Biodeterioration & Biodegradation, 110*, 155–162.

Lyu, X., Nuhu, M., Candry, P., Wolfanger, J., Betenbaugh, M., Saldivar, A., Zuniga, C., Wang, Y., & Shrestha, S. (2024). Top-down and bottom-up microbiome engineering approaches to enable biomanufacturing from waste biomass. *Journal of Industrial Microbiology and Biotechnology, 51*, kuae025.

Marttinen, S. K., Kettunen, R. H., & Rintala, J. A. (2003). Occurrence and removal of organic pollutants in sewages and landfill leachates. *Science of the Total Environment, 301*(1–3), 1–12.

Mendoza-Hernández, D., Fornes, F., & Belda, R. M. (2014). Compost and vermicompost of horticultural waste as substrates for cutting rooting and growth of rosemary. *Scientia Horticulturae, 178*, 192–202.

Mishra, S., & Das, A. P. (2008). Hexavalent chromium reduction and 16S rDNA identification of bacteria isolated from a Cr (VI) contaminated site. *The Internet Journal of Microbiology, 7*(1).

Mishra, S., & Das, A. P. (2023). Microbial remediation of synthetic microfiber contaminated wastewater. In M. P. Shah (Ed.), *Microbial technologies in industrial wastewater treatment* (pp. 337–350). Springer. https://doi.org/10.1007/978-981-99-2435-6_17

Mishra, S., Dash, D., & Das, A. P. (2024). Isolation and molecular characterization of polyester-degrading bacterial strains from municipal wastewater and their application in microplastic degradation in contaminated wastewater. *Geomicrobiology Journal*. https://doi.org/10.1080/01490451.2024.2376101

Mohanty, J., Dey, S., Bala, K., & Das, A. P. (2024). Recent advances in microplastic pollution for its sustainable management. In *Sustainable management of urban plastic waste through circular economic approaches* (pp. 27–48). https://doi.org/10.1201/9781003465447-3

Moustakas, K., Rehan, M., Loizidou, M., Nizami, A. S., & Naqvi, M. (2020). Energy and resource recovery through integrated sustainable waste management. *Applied Energy, 261*, Article 114372.

Nag, M., Lahiri, D., Ghosh, S., Sarkar, T., Pati, S., Das, A. P., Ram, D. K., Bhattacharya, D., & Ray, R. R. (2023). Application of microorganisms in biotransformation and bioremediation of environmental contaminant: A review. *Geomicrobiology Journal*, 1–18. https://doi.org/10.1080/01490451.2023.2261443

Nam, N. N., Do, H. D. K., Loan Trinh, K. T., & Lee, N. Y. (2023). Metagenomics: An effective approach for exploring microbial diversity and functions. *Foods, 12*(11), 2140.

Nanda, S., Patra, B. R., Patel, R., Bakos, J., & Dalai, A. K. (2022). Innovations in applications and prospects of bioplastics and biopolymers: A review. *Environmental Chemistry Letters, 20*(1), 379–395.

Nebauer, D. J., Pearson, L. A., & Neilan, B. A. (2024). Critical steps in an environmental metaproteomics workflow. *Environmental Microbiology, 26*(5), Article e16637.

Nicolescu, C. M., Bumbac, M., Buruleanu, C. L., Popescu, E. C., Stanescu, S. G., Georgescu, A. A., & Toma, S. M. (2023). Biopolymers produced by lactic acid Bacteria: Characterization and food application. *Polymers, 15*(6), 1539.

Paul, A., Dey, S., Ram, D. K., & Das, A. P. (2023). Hexavalent chromium pollution and its sustainable management through bioremediation. *Geomicrobiology Journal*. https://doi.org/10.1080/01490451.2023.2218377

Poupin, P., Godon, J. J., Zumstein, E., & Truffaut, N. (1999). Degradation of morpholine, piperidine, and pyrrolidine by mycobacteria: Evidences for the involvement of a cytochrome P450. *Canadian Journal of Microbiology, 45*(3), 209–216.

Pradhan, G., Tripathy, B., Ram, D. K., Digal, A. K., & Das, A. P. (2023). Bauxite mining waste pollution and its sustainable management through bioremediation. *Geomicrobiology Journal*. https://doi.org/10.1080/01490451.2023.2235353

Priyadarshini, B., Santhosh Kumar, M., & Das, A. P. (2024). Microbial degradation of synthetic plastics released from COVID-19 generated single use disposable face masks. *Environmental Science and Engineering, Part, F2827*, 1–16. https://doi.org/10.1007/978-3-031-50840-0_1

Priyadarsini, S., & Das, A. P. (2024). Analytical and structural characterization of waste lithium-ion batteries for their effective recycling strategy. *Environmental Chemistry and Ecotoxicology*. https://doi.org/10.1016/j.enceco.2024.12.004

Priyadarsini, S., & Das, A. P. (2025). Lithium bioleaching: A review on microbial-assisted sustainable technology for lithium bio-circularity. *Journal of Water Process Engineering, 69*, Article 106744. https://doi.org/10.1016/j.jwpe.2024.106744

Remmas, N., Manfe, N., Zerva, I., Melidis, P., Raga, R., & Ntougias, S. (2023). A critical review on the microbial ecology of landfill leachate treatment systems. *Sustainability, 15*(2), 949.

Rond'ošová, S., Legerská, B., Chmelová, D., Ondrejovič, M., & Miertuš, S. (2022). Optimization of growth conditions to enhance PHA production by Cupriavidus necator. *Fermentation, 8*(9), 451.

Sahinkaya, E., Uçar, D., & Kaksonen, A. H. (2017). Bioprecipitation of metals and metalloids. In *Sustainable heavy metal remediation: Volume 1: Principles and processes* (pp. 199–231).

Samuel, H. S., Ekpan, F. D. M., & Ori, M. O. (2024). Biodegradable, recyclable, and renewable polymers as alternatives to traditional petroleum-based plastics. *Asian Journal of Environmental Research, 1*(3), 152–165.

Sevilla, M. E., Garcia, M. D., Perez-Castillo, Y., Armijos-Jaramillo, V., Casado, S., Vizuete, K., Debut, A., & Cerda-Mejía, L. (2023). Degradation of PET bottles by an engineered Ideonella sakaiensis PETase. *Polymers, 15*(7), 1779.

Sharma, P., Singh, S. P., Iqbal, H. M., & Tong, Y. W. (2022). Omics approaches in bioremediation of environmental contaminants: An integrated approach for environmental safety and sustainability. *Environmental Research, 211*, Article 113102.

Sharuddin, S. S., Ramli, N., Yusoff, M. Z. M., Muhammad, N. A. N., Ho, L. S., & Maeda, T. (2022). Advancement of metatranscriptomics towards productive agriculture and sustainable environment: A review. *International Journal of Molecular Sciences, 23*(7), 3737.

Singh Jadaun, J., Bansal, S., Sonthalia, A., Rai, A. K., & Singh, S. P. (2022). Biodegradation of plastics for a sustainable environment. *Bioresource Technology, 347*, Article 126697. https://doi.org/10.1016/J.BIORTECH.2022.126697

Singha, L. P., & Shukla, P. (2023). Microbiome engineering for bioremediation of emerging pollutants. *Bioprocess and Biosystems Engineering, 46*(3), 323–339.

Son, H. F., Cho, I. J., Joo, S., Seo, H., Sagong, H. Y., Choi, S. Y., Lee, S. Y., & Kim, K. J. (2019). Rational protein engineering of thermo-stable PETase from Ideonella sakaiensis for highly efficient PET degradation. *ACS Catalysis, 9*(4), 3519–3526.

Song, T., Zhang, X., Li, J., Wu, X., Feng, H., & Dong, W. (2021). A review of research progress of heterotrophic nitrification and aerobic denitrification microorganisms (HNADMs). *Science of the Total Environment, 801*, Article 149319.

Thomas, T., Gilbert, J., & Meyer, F. (2012). Metagenomics-a guide from sampling to data analysis. *Microbial Informatics and Experimentation, 2*, 1–12.

Torres, E. (2020). Biosorption: A review of the latest advances. *Processes, 8*(12), 1584.

Van Kessel, M. A., Speth, D. R., Albertsen, M., Nielsen, P. H., Op den Camp, H. J., Kartal, B., Jetten, M. S., & Lücker, S. (2015). Complete nitrification by a single microorganism. *Nature, 528*(7583), 555–559.

Varghese, S., Dhanraj, N. D., Rebello, S., Sindhu, R., Binod, P., Pandey, A., Jisha, M. S., & Awasthi, M. K. (2022). Leads and hurdles to sustainable microbial bioplastic production. *Chemosphere, 305*, 135390.

Varjani, S., Shah, A. V., Vyas, S., & Srivastava, V. K. (2021). Processes and prospects on valorizing solid waste for the production of valuable products employing bio-routes: A systematic review. *Chemosphere, 282*, Article 130954.

Xu, S., Xing, Y., Liu, S., Hao, X., Chen, W., & Huang, Q. (2020). Characterization of Cd^{2+} biosorption by Pseudomonas sp. strain 375, a novel biosorbent isolated from soil polluted with heavy metals in Southern China. *Chemosphere, 240*, 124893.

Yoshida, S., Hiraga, K., Taniguchi, I., & Oda, K. (2021). Ideonella sakaiensis, PETase, and MHETase: From identification of microbial PET degradation to enzyme characterization. In *Methods in enzymology* (Vol. 648, pp. 187–205). Academic Press.

Zhang, L., Lin, X., Wang, J., Jiang, F., Wei, L., Chen, G., & Hao, X. (2016). Effects of lead and mercury on sulfate-reducing bacterial activity in a biological process for flue gas desulfurization wastewater treatment. *Scientific Reports, 6*(1), 30455.

Zhang, W., Li, F., & Nie, L. (2010). Integrating multiple 'omics' analysis for microbial biology: Application and methodologies. *Microbiology, 156*(2), 287–301.

Ecological Impacts of Synthetic Microfiber Pollutants in Aquatic Ecosystems

Prachi Saxena, Tanmoy Roy, Jyoti Varma, and Chahat Dogra

1 The Importance of Clean Water

Seventy-one percent of the Earth's surface is covered by water. Within that, the oceans hold about 96.5% of it. About two-thirds of the human body is made of water, which is required for many different functions such as transporting nutrients, regulating temperature, removing harmful toxins, etc. It is a crucial component to not just humans but every living organism (Bidaisee, 2018). Degradation in the quality of water has global impacts, including human health, aquatic life, food chains, biodiversity loss, etc.

Sources of Water Contamination

Water pollution is the release of harmful substances (e.g., solid waste, chemicals, heat, radioactive matter, minerals, organic waste, etc.) into the bodies of water, rendering it detrimental to the consumption and functioning of ecosystems (Mishra & Das, 2023; Mishra et al., 2024a). Water can be contaminated naturally or by humans. Also, the water pollutants can stem from point sources or dispersed sources. Point sources are outlets confined to a small space, such as a pipe or a channel, usually used for industrial or sewage waste discharge, whereas dispersed sources have no definite boundaries (e.g., runoff from an agricultural area) and are difficult to control (Mishra & Das, 2022; Mishra et al., 2022c).

Natural sources of contamination, including sediments in the form of silt, clay, etc., can be transported into the water bodies by surface runoff as a result of soil

P. Saxena (✉) · T. Roy · J. Varma · C. Dogra
Department of Chemistry, Lovely Professional University, Jalandhar, Punjab, India
e-mail: prachi9saxena@gmail.com

A. P. Das et al. (eds.), *Sustainable Municipal Waste Management with a Zero Waste Approach*, Sustainable Environmental Waste Management Strategies,
https://doi.org/10.1007/978-3-032-05950-5_10

erosion. These sediments pose a threat to disrupt the natural ecological balance of the water bodies by obstructing the entrance of sunlight into them. In addition, it can interfere with the reproductive cycles of aquatic life. Land pollution can promote water pollution if the waste is carried to the water body by animals, rainfall, or wind (Kumar et al., 2023; Mishra et al., 2022b).

Man-made sources of contamination include domestic sewage, which consists of used water from households and is the primary source of pathogens, as well as putrifiable organic matter (e.g., food waste), which can degrade the quality of water and deplete the dissolved oxygen content in it (Madhav et al., 2019). Domestic sewage is one of the major sources of plant nutrients such as phosphates and nitrates. When present in excess, it can lead to the acceleration of the aging of the water body, termed cultural eutrophication, which can cause early death of the water body. Solid wastes such as plastics, garbage, electronics, construction waste, etc., deplete the quality and equilibrium of the water bodies. Industrial waste consists of toxic chemicals like lead, mercury, cadmium, etc. The discharge of hot water also disrupts the temperature and, thereby, the aquatic ecosystem. Agricultural waste contributes to the entry of fertilizers and pesticides into the water bodies due to surface runoff (Chaudhry & Mallick, 2017; Mokarram et al., 2020). Textile wastewater contains high concentrations of suspended solids, metals (e.g., lead, copper, chromium, iron, etc.), chlorides, nitrates, synthetic microfibers, etc. (Pal, 2016).

2 Introduction to Microfibers

The production and consumption of textile materials have been rapidly increasing due to population growth and the trend toward fast fashion. The fiber fragments ranging in size from 1 μm to 5 mm released during garment production and domestic sewage during washing and drying have significantly added to the water pollution (Jagatee et al., 2024; Mishra et al., 2020; Singh et al., 2020a, 2020b). Global fiber production (both natural and synthetic) has almost doubled from 57 million metric tons (MMT) in 2000 to 111 MMT in 2020. This figure is expected to cross 145 MMT in just another decade. Synthetic microfibers are man-made fibers with a size range of 10–30 μm. These can enter the water bodies due to textile waste, domestic washing, fishing gear, single-use wipes, sanitary products, etc. (Periyasamy & Tehrani-Bagha, 2022). The major components of microfibers found in environmental samples mainly comprise polyethylene terephthalate (PET, polyester), polyethylene (PE), polyamide (PA, nylon), polypropylene (PP), polyacrylonitrile (PAN), etc. (Dey et al., 2024; Li et al., 2023; Mishra et al., 2024b; Paul et al., 2024).

Several researchers understand and highlight the problems and implications of microplastics in the present time but still, there are some research gaps where researchers need to work further. Here, in Table 1 given discusses some information that is essential to address key gaps in knowledge on textile microfiber loss.

Table 1 Key gaps in knowledge on textile microfiber loss

Knowledge gap	Requirement of research or details
1. Terminology	Elucidation of some important terms to assist data splitting and communication that goes perfectly with the agreement that includes definitions for- – Nanofiber and microfiber – Primary and secondary MPs – Synthetic and natural microfibers
2. Prevalence of Units for MF	When compared to the outcomes of research some common units like mass per cloth or per kg clothing vs number of microfibers per article of clothing/kg clothing; kg/number/ volume water/soil
3. Sampling protocols and measurement	– Protocols need to be upgraded for sampling, examination, and characterization of nanofibers and MFs – Protocols for indicating conviction in observation (instrument used, limit of fiber size) – Protocols for investigational trials for the loss of fiber during every rotation of washing
4. Scale of loss of MFs at every phase of textile existence	– Loss during the production, consumption, and disposal of microfiber in the atmosphere – Wastewater treatment plants (WWTPs) effluent release to aquatic media (marine, coastal, freshwater) – Ample amount of sewage sludge to soil (agricultural) – Fragments from textile to atmosphere and landfills
5. The fate of MFs in the environment	– Deposition rate from atmosphere – Transfer of MFs from the river to land to marine or coastal habitats – Fragmentation and degradation rate of MFs to nanofibers – Biodegradation rate according to the type of fiber (soil, marine media, pure freshwater)

(continued)

Table 1 (continued)

Knowledge gap	Requirement of research or details
6. The physical impact of MFs	– Amount of microfiber ingested at various trophic levels along with their occurrence – Intake of textile MFs via beverages and foods – Rate of inhalation of textile fiber by humans, type of fiber (micro- or nanofiber), and various health effects associated with it – Existence and place of MFs within the human system – Microfiber ejection rate from both humans and organisms – Impacts of microfibers when ingested on heath, overall reproduction, growth rate, and survival rate of organisms as per the type of fiber being ingested
7. Chemical impacts of MFs	– Leaching (a process of extracting a substance from a solid material that gets dissolved in a liquid) rate of additives from textile MFs – Amount of MF sorption (a chemical process that occurs between two substances) of water hating components (hydrophobic) and metals – Impact of absorbed chemicals on the overall health, reproduction, growth rate, and life expectancy of organisms
8. Mitigation strategies	– Exchange of techniques and information for better outputs to reduce microplastic pollution throughout the supply chain of the textile sector – Protocols for evaluating best practices and initiatives to prevent MF loss

Background of Microfibers

Microfibers evolved during the mid-twentieth century, when there was a huge development in textile technology. In the beginning, microfibers were used for industrial cleaning and automotive (industries related to the production and maintenance of motor vehicles) applications, where they showed good absorbency and capability to capture dirt effectively. Now, microfiber has gained admiration due to its feathery texture and flexibility.

Various Sources of Microfiber:

Sources of microfiber are broadly divided into natural microfiber and man-made microfiber (anthropogenic ones). Natural microfibers include plant fibers (cotton, hemp, flax, and jute), which easily get broken down into microfibers via weathering and degradation), animal fibers (like wool and silk, which can be fragmented into MFs over a period of time by physical or biological degradation), atmospheric dust like wind erosion of soils, certain volcanic activity, etc. (Mishra et al., 2021; Priyadarshini et al., 2024b; Satapathy et al., 2024).

Anthropogenic sources (man-made): these sources include synthetic textiles (polyester, nylon, acrylic, and polypropylene), household items (cleaning clothes, sponges), industrial processes (manufacturing of textiles and their production), single-use products (disposable wipes, sanitary products, and diapers), fishing gear (including fishing nets and lines that can shed microfibers into aquatic media), which ultimately pose a threat to marine ecosystems (Carr, 2017).

Types of Fibers

Synthetic fibers are formed from polyester or nylon. Such fibers are fabricated by a process called extrusion, wherein the melted polymer is strained via fine nozzles to generate thin filaments. Geographically, China and India hold the top positions as producers of synthetic fibers, contributing 66% and 8% of total production globally (Periyasamy & Tehrani-Bagha, 2022). Polyester and polyamide are most commonly used in the textile industry. Polyethylene fibers form a good basis for material for fishing gear such as lines, nets, and ropes. Polypropylene fibers are used in the production of medical nonwoven fabrics and filter cloths, while polyacrylonitrile is used in winterwear such as sweaters, gloves, carpets, etc. (Li et al., 2023). Recycled microfibers are generated from recycled materials that are non-natural, and this process involves discarding nylon or polyester materials like bottles containing plastic or fishing nets to form microfiber yarns (Das et al., 2024b; Mishra et al., 2019). Natural microfibers are less common and can be formed by using materials like cellulose, silk, or wood pulp. Natural materials have gone through certain specific treatments to convert them into strands of microfibers. Some microfibers are a blend of other fibers. Such a type of microfiber includes viscose or cotton so that their properties get improved or generate certain specific characteristics. However, it is also important to mention that microfiber itself has fine-diameter fibers, not some specific material; therefore, its composition and source differ based on the manufacturer and intended use (Liu et al., 2019b).

3 Sources of Synthetic Microfiber in Marine Media

Microfibers that are present in seawater possess a varied collection of fibers that are different in morphological appearance and may also vary in chemical composition. Marine pollution has its origin in man-made fibers that arise from heterogenous starting points that may be sea-based or land-based (Singh et al., 2020b).

Land-Based Origin

Apparently, almost all sorts of anthropogenic ventures are conducted on soil, so it is quite evident that most of the adulteration that is present in our environment has been done on land. It has been addressed that plastic or synthetic waste, which is around 13 million tons, either directly or indirectly, goes into oceans, and in some reports, it is shown that Asia alone faces 85% of such pollution. It has been found that around 98% of microfibers in marine ecosystems are land-based, and 2% of microfibers approach via sea-based pathways (Boucher & Friot, 2017). The in-depth details about the sources that are land-based on which microfibers get released are schematically represented in Fig. 1.

Synthetic Textiles

As there is a huge amount of expansion in household and commercial laundering of man-made textiles, the problem of marine pollution has spiked a lot. Synthetic MFs

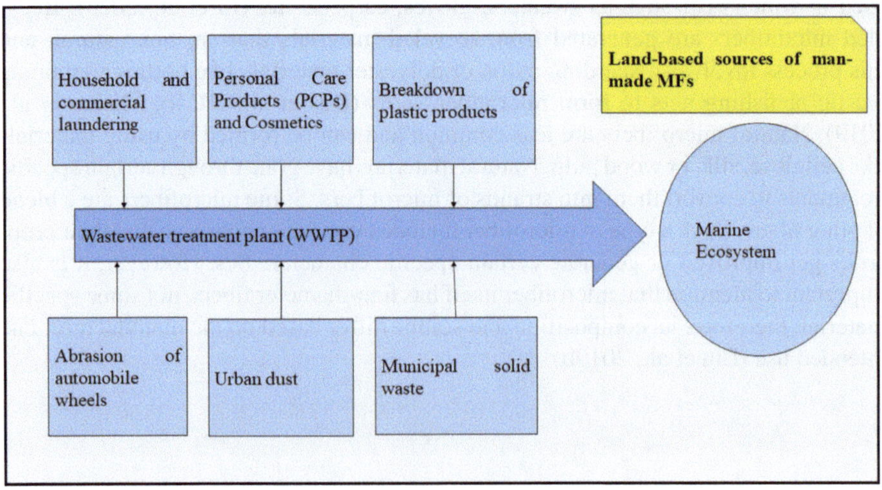

Fig. 1 Land-based microfibers

are allowed to emerge within the environment during the fabrication of textiles, use of personal care products, and dumping of waste (Das et al., 2024a). Microfibers that are used in the formation of garments without any type of further treatment are called primary microfibers, while secondary microfibers are such types of microfibers that are fragments of a huge number of primary microfibers (Mishra et al., 2022a; Mohanty et al., 2024b). Microfibers with sloppy composition include sweaters, blankets, socks, towels, yoga pants, etc. Such types of MFs, when liberated into the water, pose hazardous risks. Vassilenko et al. reported in 2019 that standard domestic wastewater in Canada and the USA emits around 533 million microfibers through laundering (Vassilenko et al., 2021). It is estimated that approximately 2000 microfibers are generated in one wash of a single fleece jacket. Likewise, clothes to be cleaned in a uniform laundry with 5–6 kg of synthetic garments may liberate around 65,000,000 microfibers. Discharge of synthetic fabrics might be dependent on the density factor, i.e., thick cloth sheds more microfiber than thin cloth, which has less shedding ability. It has been accounted that polyester allows more microfiber than nylon with a spike in density (Galvão et al., 2020).

Abrasion of Automobile Wheels

Currently, tires that are used in automobiles are constructed with several non-natural polymers, like styrene, polyester fiber, and butadiene. The tires of such automobiles are formed by using man-made polymers, i.e., around 60% of chemicals and additives were accumulated along with natural rubber. Processed microfibers are formed by some deterioration processes, and when it comes to the tires of automobiles, it becomes the second wide-reaching reason for microfiber pollution, the first being through the textile industry. Such types of fibers are present on land or roads when flow along with rainwater reaches the nearest resources of water through wastewater treatment plants (WWTPs). Moreover, in metropolitan cities, the spread of such microfibers is via rainfall or air to the closest water source or adjacent soils (Unice et al., 2019). Reports suggest that around 1500 million automobiles release around 4 million tons of MFs universally into the habitat, with a per capita discharge of 5 kg/year. A huge analysis has shown that the production of microfibers per capita ranges from 0.25 in India to 2.0 in Japan and around 5.0 kg/year in the USA. However, it has also been estimated that abrasion of tires leads to around 10% of the total comprehensive amount of plastic pollution that ends up being dumped into the oceans. The threads of automobile wheels are composed of styrene rubber, which is a precursor of Styrofoam, and a few supplementary compounds. As per certain estimates, around 503,600 ty^{-1} of microfibers are emitted from tire wear in the EU, and of them, around 52,000 to 136,000 ty^{-1} end up reaching water sources. Substances that help in the development of man-made polymers are polystyrene, polypropylene, and polyvinyl chloride, and it has been seen that the tires of cars are present in marine ecosystems in huge amounts. When weather conditions are not favorable, such trash tires get converted into small fragments and emit certain detrimental chemicals. Abrasion of automobile wheels is one of the major sources of microplastics (MPs) in our ecosystem, and innumerable components of tires may prove to be potentially toxic (Kole et al., 2017; Malachová et al., 2016).

Urban Dust

Particles of microfibers (MFs) that come from unused clothes and some fabrics that we throw away have ultimately been found in outdoor as well as indoor air samples. An investigation has been done by Dris et al., on the dimensions of MF concentration and has found that in 2017, in Paris, around 60 man-made microfibers were implicated as major sources of MF pollution both indoors and outdoors. Dust that is produced by urban cities is one of the vital origins of MF contaminants and it emerges from the spreading of microfibers during wiping or scrubbing synthetic objects, like cleaning the soles of synthetic footwear, household dust, coating buildings, using non-natural cooking materials, etc. If these above-mentioned actions cannot be controlled or managed properly, they will continue to produce more city dust, and the contribution of individuals to the control management system is of utmost importance. Currently, some studies have shown that human health is drastically impacted when exposed to such airborne microfibers (Carr, 2017; Mishra et al., 2019; Tiwari et al., 2019).

Aging and Fragments of Fibers in Aqueous Media (Water System)

Deterioration may have a huge effect on liberating microfibers when exposed to continued washing rotations. It has been reported that flees of microfibers when allowed to leave from houschold laundering of either new or older garments account for a remarkable growth of MFs of about 25%. Reports also show that the fraying of older jackets is one of the prime causes of the spike in microfiber pollution. As expected, there are several aging processes through which microfibers get exposed, like UV rays, chemical oxidation, biodegradation, and some mechanical abrasions (Liu et al., 2019a, 2019b). In aquatic media, when bigger fiber particles deteriorate, they are converted into minute fiber particles. Fragmentation is a monotonous and slow-moving operation that turns on various characteristics like heat, the intensity of light, oxygen concentration, and the properties of the material of that fiber. Due to the exposure of weather conditions for longer durations, like ultraviolet light and some mechanical scraping like tides (waves that move through oceans in response to the force exerted by the sun and the moon), fragments of microfibers can be fabricated. The quantity of microfibers released is based on various aspects like the type of garment, cleaning rotations (washing cycles), the type of detergent used, and the age of the cloth. In a report, it has been observed that the newer jacket will release around 2 g of microfiber, while older ones will release twice as much of the newer jacket in a single wash. The expected pathways of MFs entrance within the environment are via primary sources having the size criteria of 5 mm, and primary sources include domestic laundering and mass production of textiles, and shattered components of such MFs are regarded as secondary sources. When microfibers get released into certain habitats, their disintegration occurs at several intensities. For instance, when a fiber gets degraded in the presence of sunlight, bigger fibers are converted into small fragments. Such minute fiber particles remain in the habitat but cannot be located with ease. Recently, researchers' attention has been on evolving some modifications in enzymes that have a good ability to break polymer bits that are easy to assess (Austin et al., 2018). It is evident to observe the disintegration rates of microfibers to detect the spread of microfiber pollution in the coming times. However, it is important to note that, after disintegration, MFs remain consistent.

Microfibers via Electric Cloth Dryers

Electric cloth dryers serve as an underestimated cause of microfiber pollution. Microplastics, more specifically MFs, are present almost everywhere. Microplastics in the form of textiles easily enter our ecosystem via water treatment effluents or via washing machines, but very limited studies have been conducted on the emission of microfiber pollution through electric garment dryers. Their emission may vary based on the type of dryer used, lint trap capabilities, age, etc. Electric dryers produce microfiber pollution through a procedure known as abrasion. When clothes get dried in an electric dryer, friction occurs between the fabrics, resulting in the shedding of minute particles called microfibers. During the process of drying clothes via an electric dryer, thousands of microfibers get emitted into the air and run away through the dryer's exhaust vent, and in due course, these microfibers find their way to water sources like oceans and rivers. Considering that MFs are extremely small and cannot get filtered easily by most wastewater treatment plants, they therefore flow into aquatic media and adversely affect the whole food chain and marine ecosystem. It is important to mention that in 2019, the global market for electric dryers was around $7.3 billion (Kapp & Miller, 2020), and it will reach around $10.8 billion in 2023 and $13.4 billion in 2024 (Zhang, 2022). So, to mitigate this problem, a few practical steps need to be taken, like the use of a lint trap or filter to catch MFs before they emit into the environment or investing in a laundry bag with a built-in filter that attracts all the microfibers during the electric drying process.

Microfibers in Face Masks

The tools that have been worn by humans to control the COVID-19 pandemic may eventually appear to have an emerging origin in microfiber pollution. The use of face masks to control the COVID-19 epidemic can produce MFs because surgical face masks are composed of nonwoven materials comprised of polypropylene and polyethylene-based materials, consumed widely across the globe as a necessity (Mohanty et al., 2024a; Priyadarshini et al., 2024a; Sahu et al., 2024). It has been reported that in China, in February 2020, the production, consumption, and use of face masks were around 110 million, with a growth rate of 450% (Wu et al., 2022).

The demand for such face masks is still looking forward to an estimated annual spike of 20% from 2020 to 2025 during the postpandemic stretch (Singh et al., 2020a).

A substantial spike in the production of face masks and their use leads to the quick gathering of used personal protective equipment (PPE) in household solid waste (Priyadarshini & Das, 2024). If 1% of face masks get dumped improperly, then around 10 million face masks will be dumped into the environment monthly and comprise about 30–40,000 kg of microfibers (De-La-Torre et al., 2022). Therefore, nanofiber and microfiber that are released from the face masks will eventually disrupt Earth's overall balance if not controlled properly.

Microfibers in Wet Wipes and Tea Bags

Wet wipes that act as sanitizers for the purpose of disinfection usually contain alcohol within them. The use of wet wipes increased tremendously during COVID-19 and eventually became one of the sources of microfiber pollution (Shruti et al., 2021). These wet wipes, when disposed of, can degrade into MFs with the assistance of solar emission, mechanical abrasion, and microbial decomposition (Hu et al., 2021). Food packaging is also a primary source of microfiber contamination in food. The majority of food items manufactured in industries have plastic packaging due to its durability and price. Under certain conditions, the plastic can degrade into microfibers. A nylon derived tea bag, when dipped into water at 95 °C, can release up to 11.6 billion micron-sized particles along with 3.1 billion nanofiber particles. Upon ingestion, these microfibers can travel to various parts of the body in large quantities over a period of time (Li et al., 2023).

Microfibers in Cigarette Butts

Cigarette butts have come to light as one of the emerging contributors to microfiber pollution. They are mainly composed of cellulose acetate fibers, and a cigarette butt consists of about 15,000 cellulose acetate microfibers, or sometimes even more (Shen et al., 2021). It has been reported that around 300,000 tons of MFs get released from cigarette butts annually into aquatic media. Although the quantity of MFs released is in comparison, lower than other sources of microfiber pollution, the collective ecotoxicity of MFs and other toxic contaminants like carcinogenic tar, nicotine, and polycyclic HCs cannot be discarded. It is critical to know that the leachate (a liquid when passing through any matter, takes in the substances from it, hence making the liquid even more harmful or detrimental) of five smoked cigarettes if mixed with 1L of aqueous media will lead to 60% to 100% mortality in four different varieties of nonsaline invertebrates in not more than 5 days (Green et al., 2020). If small numbers of butts are present in the soil, it leads to a decline in soil-dwelling capacity.

Microfibers in Oceans

The most common type of plastic pollution is the microfibers present in oceans. The presence of MFs in the ocean includes acrylic, polyethene, polyamide, polyester, and polypropylene, and these fibers are primarily produced in the textile sector. It has been determined that fibers that are composed of polyester cause more than 50% of the combined MF specimens. In marine media, around 91% of the microplastics with a length of 0.1–1.5 mm are present in the surface water. During household washing, huge numbers of microfibers get released and therefore account for one of the utmost concerns for marine ecosystems. The presence of microfibers is on the surface of the water, subsurface, as well as deep down in the seawater. Consumption of microfibers is omnipresent among the nearby organisms present in marine ecosystems because MFs accumulate well in deepwater, but the consumption, assemblage, possible toxicity, and tropic transmission of MFs within marine environments are unsuccessful (Athey & Erdle, 2021). Moreover, it is crucial to know that MFs show identical characteristics to those of microplastics, which shows even more pessimistic consequences for those who ingest microfibers. The topmost surface of MFs has a good capability to adsorb several organic, inorganic, and miscellaneous contaminants (Zhao et al., 2022), including the presence of polycyclic hydrocarbons (which show carcinogenic effects) on the fiber. On top of that, it can be observed that around 10,000 dangerous additives, including synthetic or natural, antiviral or antibacterial, and UV-coated, have been extensively consumed in the course of fabricating yarns and other fibers. Apart from that, some dope additives and ultraviolet absorbers are used for the manufacturing and finishing of microfibers, which prove eco-toxic in the long run.

The total global fibers production in 2022 was 116 million metric tons. The contribution of various types of fibers is shown in the pie chart (Fig. 2). The global market is dominated by synthetic microfibers with 65% of total production, followed by plant fibers at 27%. Polyester constitutes more than 85% of the total synthetic fibers production, but less than 14% is generated from recycling (Textile Exchange Report, 2023).

Microfibers in Atmosphere

Microfibers that are liberated from several garment-producing industries like weaving, spinning, sewing, and cutting of microfibers during the shedding of clothes during laundry procedures are some prime sources of MF emission within the atmosphere. Although widely used microfibers are man-made, it is crucial to note that natural fibers are also a concerned source of microplastic that is supposed to transmit via rain and wind. Apart from that, there are certain studies that show that atmospheric pollution via MFs is influenced primarily by natural cellulosic fibers rather than non-natural ones. MFs that are present in the atmosphere have a direct impact

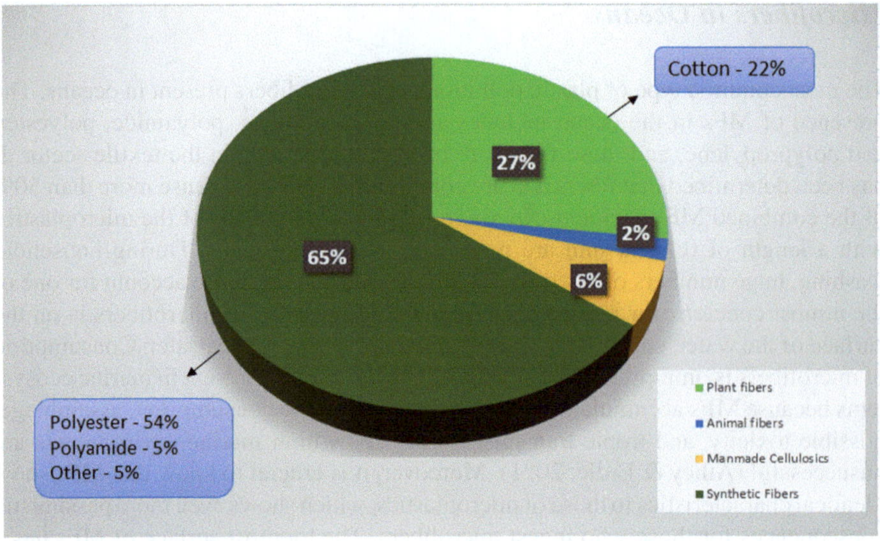

Fig. 2 Global fiber market 2022

on the respiratory tract, more specifically for those who work in the textile industry. Research that has been conducted recently shows that a minimum of 13,000 to 68,000 MFs get sniffed in from the textile sector every year (Catarino et al., 2018). When the sorption of such microfibers has been done, they (MFS) produce certain harmful chemicals that lead to textile dermatitis.

It has been studied that the lower density, average length, man-made or natural crimp, and substantial surface area to volume ratio of MFs, in contrast to monodisperse beads (plastic microbeads) and other particles, create ease for the movement of wind and transfer range from 10 to 1000 km, which shows how the atmosphere plays a significant role in the transfer of microfibers from one place to another. Studies also show that a larger range of transfer is shown by dry microfibers as compared to the regional habitats, which show various suspension deposition cycles for long periods. Rainstorms and winds at higher altitudes are observed as the principal forces behind the transfer and spread of atmospheric MFs across the globe, but eventually, the pathway of transfer of such microfibers is still not identified. Putting an application to some methods and concepts from geosciences helps to further stimulate our comprehension of the functioning of atmospheric contaminants of MF, the destiny of MFs within the atmosphere, and the effect of microfibers on air quality (Liu et al., 2023).

Microfibers in Groundwater

As is obvious, groundwater is one of the prime sources of freshwater on the planet, providing 2 billion individuals with pure water for household, agricultural, and drinking purposes both in developing and developed countries. Apart from that, in several rural regions, groundwater acts as a vital source of fresh water for major agricultural and household purposes. In some studies, it has been observed that around 38% of the irrigated portion depends on groundwater worldwide, which ultimately leads to a tenfold spike in groundwater usage for irrigation purposes in the past 50 years (Huang et al., 2021) (Fig. 3).

4 Methods for Quantifying and Detecting Microplastics

Development has proceeded on several methods for identifying microplastics, their extraction processes, and their detection, but still, there is no standardized agreement that is approved and used internationally. Microplastics can be found and identified using specific techniques, including scanning electron microscopy (SEM), optical microscopy, Raman spectroscopy, and Fourier transform infrared spectroscopy (FTIR). More refined methods for detecting microfibers at pertinent levels by using the reference of biological specimens are considered superior to giving upgraded estimates regarding their widespread presence. For instance, by using marine sedimetechnology, the presence of microplastics has been identified easily, but the validity of soil samples and its capability to find out the odds from other

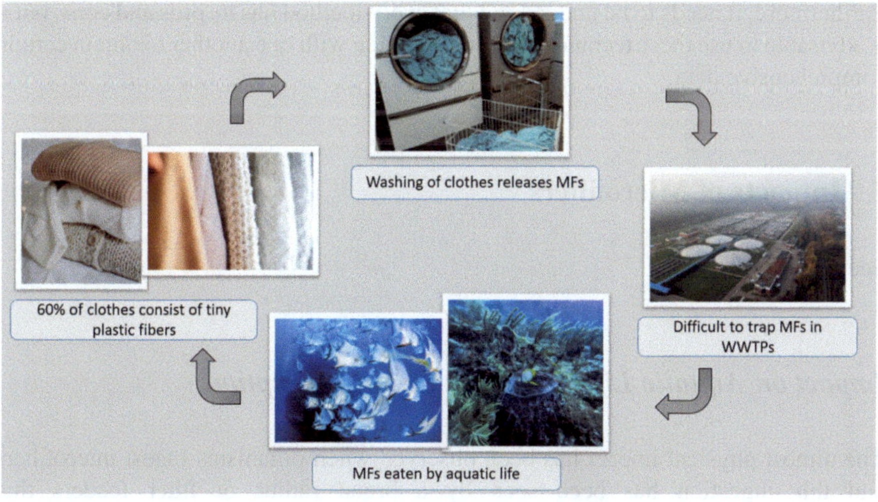

Fig. 3 The route of the microfiber entrance. *Image credit*—pexels.com

sources (compost, plastic pouches, sludge from sewage) are still unclear (Braun & Amelung, 2018). In some studies, it has been shown that analytical technologies are gaining progress, which specifies the presence of MFs of micro (< 100 μm) and nano (< 100 nm) particle sizes, but still, it requires some more definitive statistics to acknowledge the transfer speed routes and quantities. However, there ilittlell well-founded data that reveals how certain chemical properties like dyes, additives, and noninflammable materials affect the characterization of MFs. A few important methods for quantification and identification of microfibers are explained below.

Microscopy Technologies

Optical microscopy, like light microscopy and scanning electron microscopy (SEM), helps detect microfibers. Such techniques provide a higher-resolution view of discrete fibers, which ultimately helps for easier characterization based on outer morphology. The Fourier transform infrared spectroscopy (FTIR) technique helps identify the chemical structures of microfibers. Different MFs have different spectra that allow for the detection of polymer types that are present in the sample. Raman spectroscopy technique involves the use of laser light for the detection and characterization of the chemical structure or composition of MFs. It also helps to determine the types of polymers present within the microfiber. Microfiber extraction techniques include centrifugation and filtration, which can be used to trap microfibers from complex samples (sediments, biological tissues, and water). By embodying tracers into man-made fibers during the production of microfibers, it helps in aiding their detection and quantification. Such microplastic tracers can be identified by using certain specialized techniques that help in easy detection between natural fiber from synthetic ones. Furthermore, it needs to be understood that every method has its pros and cons, but it is advisable to use these techniques in combination with one another to obtain certain comprehensive data.

5 Impacts of Microfibers

See Fig. 4.

Impact on Aquatic Life and Food Chain Disruption

The utmost physical impact has been observed when organisms ingest microfibers and this ingestion has been seen by a broad radius of filter feeders that includes macroflora, species of phytoplankton, zooplankton, invertebrates, etc. The heterotrophic plankton take up the microfibers through phagocytosis and store them

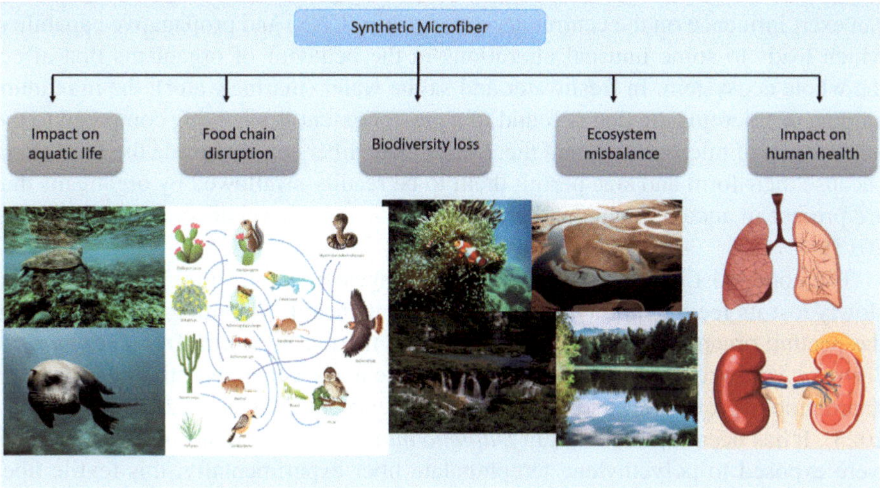

Fig. 4 Impact of synthetic microfiber pollution. *Image credit*— pexels.com, pixabay.com, free-stockimages.com

in their tissues. Taking aquatic vertebrates into account—sharks, whales, sea turtles, dolphins, etc.—their ingestion of MF can be achieved directly as filter feeders or indirectly by the intake of large amounts of prey that contains MFs. In a study, MFs were found in the intestines of sharks from the Sea of Cortez and whales from the Mediterranean Sea, while demersal sharks collected from the coastal parts of Chennai showed accumulation in the gills and gut. MFs were detected in the stomach and intestine areas of *Phoca vitulina.* In a recent report, they were found to have MFs of less than 300 mm in size in their intestinal cavities (Mao et al., 2018). When the ingested microfibers get transferred to certain tropic levels, they influence various aquatic organisms like salmon which is generally consumed by salmon, beings as a part of their food and birds like shearwaters (long-winged oceanic birds). MFs in sea birds can alter their feeding habits, reproductive health, oxidative stress, etc. Six different species of seabirds: *Pelecanoides garnotii, Pelecanoides urinatrix,, Spheniscus humboldti, Larus dominicus, and Phalacrocorax bougainvillii* showed the presence of MFs in their stomach areas. Among these, *Larus dominicanus*, which feeds on fishing nets and plastic waste, showed the maximum ability for MF ingestion. Ingestion of MF can cause the organism's body to treat it like a foreign particle, thus increasing the levels of oxidative stress in the gut tissue in species such as zebrafish (Thiel et al., 2018). This condition has the potential to cause genotoxicity, cell death, or even cancer. Therefore, the intake of microfibers as feed directly or indirectly leads to various gastrointestinal (GIT) problems. To give an example, microfiber particles having a uniform structure that is usual or regular might be easily ejected, but microfiber particles that are smaller than regular microplastic particles get easily absorbed, assemble, and remain within the important organs for a longer period of time. Nanoplastics (synthetic polymers with dimensions ranging from 1 nm to 1 μm) can exert influence on the central nervous system (CNS) and propagative capability, which leads to some unusual alterations in the behavior of organisms that affect the whole ecosystem. In freshwater and saline water (marine water), the maximum amount of microplastic that is found to a greater extent is synthetic compared to the other forms of microplastic, and these man-made fibers easily invade the food chain because their form and size permit them to be readily swallowed by organisms that are present in aquatic media and are highly prone to gastrointestinal issues (Rillig, 2018).

The copepod *Centropages typicus,* upon ingestion of 7.3 mm of microplastic, slowly lost its feeding ability, causing harm to its health. A similar effect is seen in the benthic organism *Hyalella azteca* and the marine lugworm *Arenicola marina. Tigropus japonicas* on exposure to surgical face mask-generated microfiber led to a retardation in growth rate and reproductive impairment (Mao et al., 2018; Priya et al., 2023). It has been reported that in *Daphnia magna, a* greater part of ingested fibers were exposed to polyethylene terephthalate fiber experimentally; this textile fiber is around 300 μm, and some bigger microfibers having 1400 μm were also found within the gut of *Daphnia magna,* but less information is available that tells about the impacts in terrestrial circumstances. Other feasible consequences, like changes in the bulk density of soil and carbon content, are inadequately known (Rillig, 2018). Additionally, microfiber debris was also detected in the stomach of Lantern fish species of the Mariana Islands of the Philippines Sea (Mishra et al., 2019) (Table 2).

Chemical Impacts on Aquatic Life

Potential mechanisms that come up with the chemical influence of microplastics in our ecosystems are discharge of plastic additives that include dyes or flame-resistant (De Souza Machado et al., 2018), antioxidants, stain-resistant finishes that can leach into the environment, transfer of components like (POPs) persistent organic pollutants. Microfibers possess a large ability to sorb various pollutants within solution due to their higher surface-to-volume ratio which leads to the possibility of intensifying the bioavailability of virulent components. The reaction of chemical impurity on the organism depends on the nature of pollutants that show hydrophobicity, amount of chemical contaminant, release rate of chemicals, area of bioaccumulation, and appearance of hazardous effects after ingesting such species. The use of higher concentrations and variables for experimental purposes may lead to irregularity in the outcomes that are issued in the literature, but still, there is some evidence regarding the potential effects of concentrations. However, when microfibers enter aquatic systems via washing machines or wastewater systems, emit certain hazardous chemicals into water bodies, and potentially harm aquatic organisms and whole aquatic media.

It has been accounted that, under environmentally relevant concentrations within sea water if an adult fish ate plastic trash is consumed then it disrupts the function of the endocrine system. Investigational procedures done on the exposure of polyethylene MPs (< 1 mm) and related chemicals, create disturbances in gene expression in

Table 2 Microfibers found in aquatic and terrestrial organisms

Sl. No.	Aquatic and terrestrial organisms	Source of MF or its accumulation	Impact on health	References
1	Phytoplankton	Intake of microfibers	Reduction of chlorophyll amount	Das et al. (2023)
2	Zooplankton	Entrapment in synthetic microfibers	Reduced rate of growth and reproduction	Das et al. (2023)
3	Microalgae	Uptake of plastic fragments	Decline in feeding rate	Mishra et al. (2019)
4	Mussels	Ingestion of synthetic microfibers from rope, bags, etc.	Blockage of the gut	Mishra et al. (2019)
5	Zebrafish	Ingestion of microfibers, Accumulation in Gut tissues	High oxidative stress levels, gut blockage, and liver damage	Das et al. (2023), Thiel et al. (2018)
6	Copepod *Centropages typicus*	Ingestion of microfiber	Reduced feeding ability	Priya et al. (2023)
7	*Tigropus japonicas*	Gut surface	Retardation in growth rate and reproductive impairment	Priya et al. (2023)
8	Fish	Ingestion of microfibers from fishing gear, clothes, etc	Damage to the intestine, liver, and reproductive system	Mishra et al. (2019)
9	Marine lugworms	Digestive tract	Decrease in ability to feed	Priya et al. (2023)
10	Snails	Uptake through soil	Reduced food intake, high oxidative stress, and GI tissue damage	Kwak et al. (2021)
11	Corals	Feed on plastic waste along the coast	Accumulation and blockage in the gut	Chatterjee and Sharma (2019)
12	Crustaceans	Ingestion of microfibers	Altered brachial function	Das et al. (2023)
13	Sea urchin	Engulfment of microfibers	Cytogenetic irregularity	Das et al. (2023)
14	Sea bird	Uptake of microfibers from fishing nets, plastic waste	Damage to reproductive health, altered feeding habits	Chatterjee and Sharma (2019), Thiel et al. (2018)
15	Seals	Entanglement innets and plastic waste and fishing nets	Effect on swimming ability and feeding habits	Chatterjee and Sharma (2019)

(continued)

Table 2 (continued)

Sl. No.	Aquatic and terrestrial organisms	Source of MF or its accumulation	Impact on health	References
16	Sea turtle	Uptake of microfibers through fishing nets, plastic straws, etc	Blockage of the digestive tract, increased risk of death	Chatterjee and Sharma (2019), Das et al. (2023)
17	Lobster	Engulfment of microfibers	Decreased rate of metamorphosis and survival of larvae	Das et al. (2023)
18	Sperm whale	Ingestion of plastic bags and bottles		Das et al. (2023)
19	Chlorella	Feed on microplastics	Reduction in photosynthetic ability	Das et al. (2023)
20	Humans	Uptake by air, seafood, water	Emphysema, renal disease, liver damage, cancer	Mishra et al. (2019), Tchounwou et al. (2012)

both female and male Japanese medaka. However, in terrestrial habitats, the possibility of soil biota (flora and fauna) when exposed to chemicals combined with MFs depends on the transport media, concentration, and selection of the size of the microparticle (Belzagui et al., 2020).

Biological Impacts (Reproductive and Overall Health)

Microplastics come up with new territory for microbial migration. Microplastic particles and microfibers via wastewater treatment plants (WWTPs) get supplemented with bacteria and microbe buildup enters freshwater via sludges. One of the major biological impacts is bioaccumulation i.e., here when MFs ingested organisms like plankton, fish, and mussels, these fibers can collectively mix within their bodies over a period and ultimately affect their health and whole food chain. When aquatic organisms ingest microfibers leads to physical blockages, reduces the efficiency of feeding, and impairs digestion in aquatic organisms. Microfibers can accumulate in various tissues and organs of crustaceans causing olfactory sense damage, endocrine system disruption, reproductive impairment, etc. (Acharya et al., 2021; Priya et al., 2023; Rathinamoorthy & Subramanian, 2021).

Diatoms were considered as the most varied groups having 14 genera, but the study detected fungi, bacteria, and cyanobacteria. In Asia–Pacific regions, evaluation has been done on the potential hazard of disease in 159 reefs, and 124,000 corals that show a huge spike that ranges from 4 to 9% when corals encounter microplastics (Lamb et al., 2018). Microplastics have a detrimental impact on corals leading to accumulation in the mesentrial tissue thereby reducing their ability to feed (Priya et al., 2023). However, there is a lack of confirmation at present to decide whether

MPs cause such deadly risk. Few microfibers contain additives within them like endocrine-disrupting compounds (EDCs) that eventually interfere with hormonal systems in organisms. This can lead to reproductive and developmental issues. The adverse effects of overexposure to microfibers include stunted growth, reduced reproductive output, shortening of lifespan, altered metabolic rates and gut microbiome, etc. (Mishra et al., 2019; Priya et al., 2023).

Impacts on Terrestrial Life (Soil and Humans)

Microplastics have been accounted for being extensively used in beverages and human food including seafood, salt, and sugar, drinking water, beer that can cause a threat to mankind through inhalation and ingestion (Periyasamy & Tehrani-Bagha, 2022). Acknowledgment regarding the risk factors due to adulterated beverages and food samples and the unreliability of some analytical technologies radiate a lack of certainty on microfiber ingestion with food. Even though, present-day study reveals that contamination of microfiber leads to some chronic toxicities at ongoing levels of exposure, absorption, and transformation of MFs and nanofibers through the GI tract or in the respiratory pathway. In addition, prolonged exposure to the same face masks on reuse increases the risk of inhalation of fiber-like particles which contain plastic additives with heavy metals like Ni, Cu and Cr which can in turn generate complexes with some of the reactive dyes. These dye molecules in turn have the potential to invade the cells and affect DNA synthesis by injecting themselves into the DNA and also bioaccumulate (Lellis et al., 2019). These chemicals containing heavy metals pose the risk of allergic reactions, and even more significant health issues such as cancer, emphysema, renal disease, and a potential threat to the unborn child during pregnancy (Tchounwou et al., 2012). Recent studies have also detected the ability of microfibers to cross the blood–brain barrier after absorption in the gut leading to neurodegeneration (Lusher et al., 2017).

Research on human beings is tough because experimental work on humans is not ethical and controversial at the same time. Apart from that, it is not feasible to find a controlled group of individuals who are not exposed to microfibers, and detecting the influence among vast populations is not only difficult but also expensive. Exposure to microfibers and nanofibers in humans leads to bioaccumulation. In vitro experiments have shown that resilience and the ability of a fiber to remain within the lungs for 180 days but the mechanism and nature of causing extreme toxicity by inhalation are unknown and utmost supervision is required to observe the consequences of textile nano- and microfibers. Direct contact with MFs can cause potential skin irritations, more specifically for people with sensitive skin. The friction between the skin and the fiber led to itching, redness, and rashes. Microfibers can also lead to some allergies by using synthetic textiles and the symptoms that trigger allergic reactions include watery eyes, sneezing, and irritation in the skin.

A limited number of studies have been performed to study the ecological impacts of synthetic microfibers on soil organisms. During a study performed artificially in the laboratory, it was seen that the presence of up to 1% of polyester microfibers demonstrated no harmful effects on earthworms after acute and chronic exposure, although the *mt* and *hsp70* gene expression associated with stress responses were affected. Similarly, on exposure of PET of up to 710 mg/kg of soil to snails, a reduction in the intake of food and rate of excretion along with heightened oxidative stress responses and damage to the GI tissues were observed after an exposure period of 28 days. In a study involving exposure of different sizes of polyester microfibers with different rates of food and soil conditions, the bioaccumulation of microfibers was recorded on the bodies of the soil fauna. In another study, 0.2% polyester increased the FDA hydrolase activities (indicators of microbial activity in the soil). The presence of acryl and nylon fibers altered the germination and chlorophyll content balance in the ryegrass species. Even though there is pointing evidence of the intake of microfibers by the soil organisms, there is still insufficient data on the degree of bioaccumulation (Kwak et al., 2021).

Impact on the Environment

One of the chief concerns with microfibers is the capacity to spread microfibers in the environment. When products containing microfibers are washed, minute microfibers get released into water which ultimately end up in oceans, rivers, lakes, and overall ecosystems. These microfibers are too small and cannot be trapped effectively by wastewater treatment plants (WWTPs), which ultimately leads to contamination in the aquatic environment.

6 Mitigation Strategies to Reduce or Prevent Microfiber Pollution

Different quantities of microfiber emissions from the masks and layers of masks have been observed by several researchers, so it is quite evident that the characteristics of masks trigger the release of microfibers (Wang et al., 2021). Therefore, textile industries need to analyze the production properties (method of manufacturing) and other properties like the type of fiber, thickness, weight per unit area, etc. Improvement of the above-mentioned parameters to produce masks helps to limit microfiber release. Studies have indicated that the release of microfibers is more when masks get exposed to extreme natural environmental conditions like water currents. Therefore, there is a critical need to develop certain waste management systems that stop throwing masks in such triggering environments. Apart from that, consumers should be educated properly regarding the responsible disposal of single-use surgical masks. When microfibers entered the aquatic media, it was nearly impossible to remove them to their minute size. Water pollution due to such particles is a growing problem and needs stringent actions to reduce microfiber contaminants. A few remediations include strict government policies, awareness among the masses regarding the harmful effects of microfiber pollution, and the development of technology so that microfibers get eliminated effectively. Well-established mitigation strategies for synthetic polymer garbage are crucial as MFs spiked at greater rates along with other man-made contaminants like heavy metals (Pivokonský et al., 2018). Products include Cora ball (revolutionary laundry ball that captures microplastics and visible fibers from clothes by tossing it in washing machines), lint filters, and guppy filter bag (effective, scientifically proven, and patented solution to stop or eliminate microplastic pollution) have used by this time to prevent microplastic pollution but still, these technologies are lacking to filter microparticles at a larger scale. However, some protocols are developed to stop microplastic pollution but fail to control extensive MF pollution. Micro-Fourier transform infrared technique (FTIR) imaging of MF particles when pretreated with hydrogen peroxide (H_2O_2) has proven an effective method for the accurate identification of pollutants in several aquatic samples. Pretreatment with hydrogen peroxide helps to eliminate various organic contaminants from surface samples. Fibers of polyesters can be easily observed and characterized by using certain spectroscopic techniques like the combination of Infrared technology with Raman spectroscopy. To examine the outer morphology of the individual particle that is present on the surface can be detected by using techniques like scanning electron microscopy (SEM) and stereo microscope. In a study, it has been shown that substantial amounts of MF particles emitted with washing machine effluents from a microfleece blanket and with the help of scanning electron microscopy (SEM) and stereo microscopy, researchers noticed the external morphological characteristics effortlessly. Microfibers can cause potentially hazardous risks for both wildlife and humans. Marine organisms like shellfish and fish mistakenly ingest microfibers. Apart from that synthetic fibers have been found within seafood when consumed by human beings raising concerns about their impact on the health of humans. Sewage water treatment can act as a great solution that helps to generate impact widely. Generally, wastewater treatment plants (WWTPs) work on minimizing chemical, biological, and physical pollutants in domestic sewage, wastewater coming from industries and runoff. For that purpose, researchers have formulated a filter plate that needs to be put in place in the extension pipe before the last outlet of the plant. The design of these plates comprises of macro- and microtextures apparently on the surface. Now, with the help of the micro tooth-like structure of filter plates, minute microfibers get trapped easily, thereby, preventing them from being entered into the water bodies. Of all the stages of textile consequences that include its manufacturing as well as its consumption, the process of laundering causes a huge disaster. As long as we know shedding means the spread of fibers and therefore, the procedure of laundering originates further harm. Therefore, there is a need to find a way to solve this problem of microfiber pollution (Le et al., 2022). Nowadays, several gadgets can be used to trap MFs present in household waste. The gadgets are referred to as in-drum appliances and exterior strainers or sieves. One of the laundry appliances for domestic motives is an In-drum gadget. These types of appliances can be used within the drum where clothes need to be cleaned and while washing this device must collect all the fiber particles that are shed from the outfit during the washing cycle.

Cora Balls

These are innovative balls outlined by the Rozalia Project to help minimize microfiber pollution in waterways and oceans. Cora balls are small and assembled using 100% recyclable materials. They are approximately the size of a tennis ball and hold a specific engineered fiber-trapping device. The creator of Cora ball stated that this device helps to reduce microfiber emissions by 35% during domestic washing sessions. Apart from that, it has also been outlined that the effectiveness of Cora balls goes on increasing with every spiked usage (Zhang et al., 2021), and their main feature is that they can be used in both settings of washing, i.e., top-loading and front-loading washing machines, in preference to spindles or without spindles. A study has also concluded that while collecting microfibers, cora balls also help to minimize MF shedding from textiles.

When the Cora ball is placed in the washing machine, it acts like a filter, capturing microfibers that get emitted from synthetic clothing during washing cycles. These minute strands of plastic, often measuring less than 5 mm, can escape traditional wastewater treatment plants (WWTPs) and end up contaminating water bodies. As we know, microfiber pollution is a noteworthy environmental concern. Synthetic fibers (nylon, polyester, and acrylic) from garments shed off easily during the cleaning process and move into lakes, rivers, and oceans. The focus of Cora balls is to smash this cycle by intercepting and keeping these microfibers, stopping them from entering water bodies and harming marine media. Cora balls have been proven to be extremely effective in trapping microfibers when put within washing machines during the garment cleaning process. Studies have shown that the use of these Cora balls drastically minimizes microfiber shedding during the garment cleaning process. However, it is important to mention that they cannot trap all microfibers, but they certainly make a positive impact. The usage of the Cora ball is simple. Just toss one or two balls into the washing machine along with a regular load of laundry. After the cleaning process has been done, just make sure that all the fibers that get accumulated during washing get removed and disposed of properly (De Falco et al., 2021).

Laundry Bags

Laundry bags are extremely useful and convenient tools during the cleaning process. There is another kind of bag, namely the guppy friend bag, which is composed of the filaments of nylon fabric having a fine pore size of about 50 $\mu\mu$m When MFs shed from the garments, they get collected in the hem of these bags and ultimately can't cross the threshold of wash effluent. In a study, it has been reported that there is a reduction of around 54 ± 4% in MFs by the usage of guppy friend bags (Yang et al., 2019). In another study, it was found that the consequences of using laundry bags composed of nylon with a 50 μm diameter resulted in a huge reduction in microfiber emissions of about 91% and 87% of MFs in the matter of fleece jackets and swimwear individually. Moreover, it is good to know that the MFs that get trapped by laundry bags are only 0.5–4% compared to when these bags are not used to trap microfibers. However, it cannot be ignored that using a laundry bag to trap microfibers can also expand the lifespan of synthetic clothing by reducing the wear and tear that occurs during washing cycles. This means that fewer microfibers are shed from the clothes over time, which ultimately leads to less pollution overall. It is basically a simple and effective way for individuals to take proactive initiatives to minimize microfiber pollution and therefore reduce their environmental impacts. Additionally, using a laundry bag also helps to prevent the clogging of washing machine filters. Overall, the consumption of laundry bags is a practical and eco-friendly solution to reduce microfiber pollution.

Furthermore, the utilization of the Guppy friend bag drastically minimizes the harm to the fabrics by creating a protective coating (behaves like a cushion for mechanical action and therefore lessens fiber deterioration, which automatically turns down the shedding action. A Guppy friend bag is basically a washing bag that prevents MFs from entering rivers and oceans. Working of guppy friend bag: This guppy friend washing bag drastically minimizes the breakage of fiber. The fibers that break off easily during the cleaning process get retained in the bag. Another researcher looked at the potency of guppy friend bags by cleaning newer (elastane and polyester) and older clothes (elastane/polyester/viscose) around five consecutive cycles and observed that in older garments there is a huge shedding of microfibers compared to the newer ones. There are several ways by which guppy friend bags prevent microfiber pollution. The bag's fine mesh material easily traps the microfibers that are shedding from synthetic clothes, preventing them from entering the aqueous media during washing cycles. By minimizing the wear and tear of clothes during washing cycles, Guppy Friend bags can help enhance or extend the life of man-made clothes, resulting in less mcrofiber shedding over time. By trapping MFs before entering the water systems, guppy friend bags help reduce pollution in rivers, oceans, and other aquatic environments, therefore helping to protect marine life. The use of guppy friend bags helps to prevent clogging of washing machine filters, therefore extending the life of machines and reducing the need for maintenance. The use of guppy friend bags helps maintain sustainable and eco-friendly laundry practices and promotes conscious consumer choices.

External Filters

The external filters can be connected to the washing machine's outlet pipe, which eventually filter the wastewater and capture fibers before they are released into the environment. Examples of external filters include LUV-R lint filters, filtrol

160™, XFiltra™, and planet care filters. LUV-R lint filters can stop the emission of microfibers via wash effluent because they are outlined to see if there is any failure in the septic system due to excessive buildup of lint. This mesh of metal can filter lint up to 285 μm and 175 μm. Researchers have reported that using LUV-R lint fibers, there is a huge reduction in terms of weight and count of microfibers of 80% and 87%, respectively (McIlwraith et al., 2019). To controvert this, another scientist accounted that the potency to remove microfibers from LUV-R lint is less, which is around 29 $\pm$ 15% (Napper et al., 2020). This might be due to some variations in the material type and methodology as well. Planet care filters are the type of microfiber filter designed to trap MPs and other detrimental bits that get emitted during washing cycles. These filters can be installed in washing machines so that microplastics can be minimized to a certain extent. The central portion of the system is grounded by electrically charged particles and membrane nanotechnology, which automatically self-cleans itself. The maker of planet care fibers claims that such filters with 200 μm can trap around 90% of MFs that are emitted from household washing effluent; therefore, by using these filters, individuals can make a huge positive difference in microplastic reduction. A study has been conducted on a small scale based on the reduction of microfiber by using planet care filters, and it has been found that 60-80% reduction. Installing a planet care filter is relatively effortless and can be done without any specific or special tools. It is one of the proactive steps that one can take to protect the environment and ensure cleaner water sources. Filtrol 160™ is a patented microfiber filter that can be used multiple times and relates to your washing machine to reduce MF emissions. This filter easily traps and prevents lint, hair, and debris from entering the septic system or wastewater treatment system. It helps minimize clogs, blockages in pipes, and potential damage to plumbing and septic tanks. This filter uses a unique filtration system to trap filter particles from the drainage system. Its innovative design allows for hassle-free installation and maintenance, ensuring optimal performance. This filter uses a canister (a round, cylindrical container) with a filter bag that is composed of a mesh with 100 microns that can trap minute MFs in every wash cycle. A study has shown that Filtrol 160™ helps minimize the fiber pollution by up to 89% by weight. Another study concluded that around 30–60% of the microfibers have been eliminated with every wash cycle by using Guppy Friends bags, Filtrol 160™, and Planet care filter. However, it is important to mention that, when extra detergents or fabric softeners are used in excess, will ultimately lead to clogged filters, and lessen the efficiency of clothes over a period (Herweyers et al., 2019). By capturing lint, and other particles, the Filtrol 160™ helps to protect the environment and maintain the efficiency of wastewater treatment systems. Overall, it is a great practical solution for homeowners to safeguard their septic systems and eliminate the risk of clogs and damage caused by lint from laundry. XFiltra™ filters are attached at the end terminal of the wastewater pipe of household laundry machines and the constructor of this filter states that these filters are 90% effective in eliminating microfiber pollution. Another research reported that among several filters including drum-in devices that claim to control MF pollution, X Filtra proves to be the most effective one (Dreillard et al., 2022). Overall, it can be concluded that all the external filters are more useful than in-drum devices to limit the pollution caused by microfibers. Moreover, looking forward to the combination of these above-mentioned filters will work even more effectively, but still, there is a lack of substantial research work in that aspect.

7 Regulatory Standards to Mitigate Microfiber Pollution

The lack of regulatory standards is one of the major shortcomings when it comes to the issue of microfiber pollution. Based on some scientific publications, it has been observed that governments have identified the hazardous impacts of MPs, more specifically microfiber pollution, but due to a deficiency of norms, standard assessment practices, and orders, consumers, and manufacturers are completely unaware of the detrimental effects of increasing microfiber pollution. Even so, there are no global standards that have been developed to reduce microfiber pollution yet, but various nations have formulated their policies on a national level and certain standards to minimize microfiber contamination. Regarding the USA, in 2018, a bill has passed by the New York Assembly that states that all the clothes that exceed 50% non-natural (synthetic material) fiber are categorized as microfiber trash and lead to MF pollution; therefore, such types of synthetic fiber need to be washed manually. Moreover, to date, there is no such study that scientifically proves that cleaning synthetic fiber is more beneficial by hand laundering compared to machine laundering. This act came into effect on January 1, 2021 (Plakantonaki et al., 2022). Currently, the same bill was also passed in California regarding the labeling of garments with more than 50% synthetic materials (AB 2379). The prime focus of this regulation is to intensify the knowledge of purchasers about the harm that has been caused by microfiber pollution and its hazardous effects on the environment. This norm has been implemented since January 2020. Two bills have been released by the General Assembly of Connecticut in 2018, Bill No. 5360 and Bill No. 341, to create awareness among buyers via some campaigns and programs. This bill also gives guidance regarding the need to label synthetic garments. With the help of this campaign, a law has been formulated that helps to create awareness among the masses regarding the microplastic and microfiber routes along with the possible influence on widespread individuals and industries and this regulation gets functional from December 2019 (Geyer et al., 2022). California state makes some strict laws regarding mitigation strategies by implementing a policy AB-129, which targets both private and public sectors by enhancing their responsiveness. This norm aims to make sure that microfiber pollution gets minimized in state boards along with public and private zones and the law became functional in January 2021 (AB 129 2019). In Europe, various worldwide laws and committees were enforced, but no usual standard or approach was put in place. The Sustainable Development Goals (SDG) address the issue of microfiber pollution, but the European Union (EU) countries give the consent to formulate certain standards and pledge to control the emissions of microfibers before 2030. As far as plastic pollution within the ocean is concerned, the United Nations Environment Program (UNDP) brought a law in 2017 considering all the pertinent regional, international, and sub-regional regulatory requirements. From the literature, it is understood that there are no such standards and regulations made by Asian nations that include China and India. The New Zealand government completely banned the use of plastic microbeads from 2018 which comes under the Waste Management Act. University of Colorado at Boulder proposed a few norms in their honors thesis to come up with awareness camps in the name of "blue standards" to preserve the environment from MF pollution.

8 Consumer Awareness and Education

Domestic laundering plays a key role in the discharge of synthetic microfibers into water bodies. Most people are not aware of the implications of fast fashion and the frequent washing and drying of clothes. Microfiber formation can be dependent on a number of factors, including the type of fiber, length, twist, dye, waterproofness, finishing, hairiness, and so forth. Some simple yet effective steps in the control of microfiber release from households can be to buy less, wash less and buy natural. Buying clothes only when essential not only reduces expense but also reduces energy for maintenance and hazards to the environment. Additionally, buying organic cotton is better since the natural microfibers present in it are readily able to biodegrade in water. This is unfortunately not the case when it comes to polyester microfibers. The use of a front-loading washing machine should be preferred over a top-loading one since the shedding of microfibers is reduced by seven times (Hartline et al., 2016). The use of cold water along with liquid detergent while washing also avoids additional shedding. Washing clothes with heavy items like backpacks can increase friction between the clothes and break more fibers. It is better to not mix light and heavy clothing for the same reasons. Natural air drying helps keep the clothes intact. Bleach is a major contributor and accelerator of microfiber shedding; hence, its use should be avoided. Using a microfiber filter in washing machines doesn't let them escape into the oceans (Miller et al., 2021). The clothes no longer in use can be donated to those in need instead of throwing them away in landfills.

9 Sustainable Management of Microfiber Pollutants

The growing prevalence of microfiber pollution, primarily from synthetic fibers like polyester, nylon, and acrylic, has raised significant environmental and health concerns. These nonbiodegradable microfibers, shed during textile production, washing, and daily use, persist in ecosystems, especially aquatic environments. Sustainable management strategies are vital to mitigating their impact and protecting the environment. Improvements in textile manufacturing, such as adopting eco-friendly production techniques and using biodegradable or recyclable materials like bio-based polymers, can reduce microfiber shedding (Adikari et al., 2023) Innovations in fabric design, including developing shedding-resistant fabrics, applying coatings to limit fiber release, and combining synthetic and natural fibers, further help address this issue. Upgraded laundry practices, such as promoting washing machine filters, using microfiber-catching devices, and encouraging lower-temperature washing cycles, can also limit microfiber pollution. Additionally, integrating advanced filtration systems in wastewater treatment plants can capture microfibers before they contaminate natural water bodies. Recycling and upcycling textile waste reduce the demand for virgin synthetic fibers, while regulatory frameworks enforce policies to limit the production and disposal of nonbiodegradable textiles. These regulations promote sustainable practices, foster eco-conscious consumer behavior, and support a shift toward a cleaner and healthier planet (Mohanty et al., 2024a, 2024b).

10 Conclusion

The increasing transition of microfiber dominance from plant-based to synthetic microfibers is causing a significant impact on the environment by various means. Within the environment, these microfibers are the most prevalent type of microplastic. The textile industry and domestic washing contribute significantly to their release into bodies of water. Together with the soil and groundwater, they have also been in the air. They can be detected by various methods, both quantitatively and qualitatively. These microfibers are consumed by aquatic animals and bioaccumulated in their bodies, which then enter the human body through their consumption and inhalation of air with high concentrations of microfibers. They have been known to cause a variety of health problems, alter feeding habits, and disrupt ecosystems as a result. Multistep treatment is required to decrease their growing prevalence by aiming for maximum reduction at the treatment and discharge stages. Government regulations play an important role to strictly adhere to the regulations and curb the prevalent MF levels in the environment.

References

Acharya, S., Rumi, S. S., Hu, Y., & Abidi, N. (2021). Microfibers from synthetic textiles as a major source of microplastics in the environment: A review. *Textile Research Journal*. https://doi.org/10.1177/0040517521991244

Adikari, M. U., Prasadi, N., & Jayasinghe, C. V. L. (2023). Microplastics in salt and drinking water. In *Microplastics in the ecosphere* (pp. 357–367). Wiley. https://doi.org/10.1002/9781119879534.ch21

Athey, S. N., & Erdle, L. M. (2021). Are we underestimating anthropogenic microfiber pollution? A critical review of occurrence, methods, and reporting. *Environmental Toxicology and Chemistry, 41*(4), 822–837. https://doi.org/10.1002/etc.5173

Austin, H. P., Allen, M. D., Donohoe, B. S., Rorrer, N. A., Kearns, F. L., Silveira, R. L., Pollard, B. C., Dominick, G., Duman, R., Omari, K. E., Mykhaylyk, V., Wagner, A., Michener, W. E., Amore, A., Skaf, M. S., Crowley, M. F., Thorne, A. W., Johnson, C. W., Woodcock, H. L.,

… Beckham, G. T. (2018). Characterization and engineering of a plastic-degrading aromatic polyesterase. *Proceedings of the National Academy of Sciences of the United States of America, 115*(19). https://doi.org/10.1073/pnas.1718804115
Belzagui, F., Gutiérrez-Bouzán, C., Álvarez-Sánchez, A., & Vilaseca, M. (2020). Textile microfibers reaching aquatic environments: A new estimation approach. *Environmental Pollution, 265*, Article 114889. https://doi.org/10.1016/j.envpol.2020.114889
Bidaisee, S. (2018). The importance of clean water. *Biomedical Journal of Scientific and Technical Research, 8*(5). https://doi.org/10.26717/bjstr.2018.08.001719
Boucher, J., & Friot, D. (2017). *Primary microplastics in the oceans: A global evaluation of sources*. https://doi.org/10.2305/iucn.ch.2017.01.en
Braun, M., & Amelung, W. (2018). Plastics in soil: Analytical methods and possible sources. *Science of the Total Environment, 612*, 422–435. https://doi.org/10.1016/j.scitotenv.2017.08.086
Carr, S. (2017). Sources and dispersive modes of micro-fibers in the environment. *Integrated Environmental Assessment and Management, 13*(3), 466–469. https://doi.org/10.1002/ieam.1916
Catarino, A. I., Macchia, V., Sanderson, W., Thompson, R. C., & Henry, T. B. (2018). Low levels of microplastics (MP) in wild mussels indicate that MP ingestion by humans is minimal compared to exposure via household fibres fallout during a meal. *Environmental Pollution, 237*, 675–684. https://doi.org/10.1016/j.envpol.2018.02.069
Chatterjee, S., & Sharma, S. (2019). Microplastics in our oceans and marine health. *Field Actions Science Reports. The Journal of Field Actions*, (Special Issue 19), 54–61. https://journals.openedition.org/factsreports/5257.
Chaudhry, F. N., Mallick, M. F. (2017). Factors affecting water pollution: A review. *Journal of Ecosystem & Ecography, 07*(01). https://doi.org/10.4172/2157-7625.1000225
Das, A. P., Dutta, K., Khatun, R., Behera, I. D., Singh, S., & Mishra, S. (2023). Microfiber pollution and its microbial mitigation: A review on current trends and future prospects. *Journal of the Taiwan Institute of Chemical Engineers*, 105104. https://doi.org/10.1016/j.jtice.2023.105104
Das, A., Dey, S., Bala, K., Gnanasekaran, R., & Das, A. P. (2024a). Microbial bioremediation technology for sustainable treatment and management of synthetic microfiber waste. In *Sustainable microbial technology for synthetic and cellulosic microfiber bioremediation* (pp. 221–238). https://doi.org/10.1007/978-3-031-62600-5_12
Das, A., Dey, S., & Das, A. P. (2024b). Ecotoxicological impacts of synthetic microfiber pollutants and development of sustainable mitigation strategies. *Environmental Chemistry and Ecotoxicology*. https://doi.org/10.1016/J.ENCECO.2024.12.003
De Falco, F., Di Pace, E., Avella, M., Gentile, G., Errico, M. E., Kržan, A., ElKhiar, H., Zupan, M., & Cocca, M. (2021). Development and performance evaluation of a filtration system for washing machines to reduce microfiber release in wastewater. *Water, Air, & Soil Pollution, 232*(10). https://doi.org/10.1007/s11270-021-05342-6
De Souza Machado, A. A., Kloas, W., Zarfl, C., Hempel, S., & Rillig, M. C. (2018). Microplastics as an emerging threat to terrestrial ecosystems. *Global Change Biology, 24*(4), 1405–1416. https://doi.org/10.1111/gcb.14020
De-La-Torre, G. E., Dioses-Salinas, D. C., Dobaradaran, S., Spitz, J., Keshtkar, M., Akhbarizadeh, R., Abedi, D., & Tavakolian, A. (2022). Physical and chemical degradation of littered personal protective equipment (PPE) under simulated environmental conditions. *Marine Pollution Bulletin, 178*, Article 113587. https://doi.org/10.1016/j.marpolbul.2022.113587
Dey, S., Bissoyi, A., Koteswara Reddy, G., & Das, A. P. (2024). Synthetic microfibers: An urban emerging pollutant demanding for sustainable environmental solutions. In *Sustainable microbial technology for synthetic and cellulosic microfiber bioremediation* (pp. 107–121). https://doi.org/10.1007/978-3-031-62600-5_7
Dreillard, M., Barros, C. D. F., Rouchon, V., Emonnot, C., Lefebvre, V., Moreaud, M., Guillaume, D., Rimbault, F., & Pagerey, F. (2022). Quantification and morphological characterization of microfibers emitted from textile washing. *Science of the Total Environment, 832*, Article 154973. https://doi.org/10.1016/j.scitotenv.2022.154973

Galvão, A., Aleixo, M., De Pablo, H., Lopes, C., & Raimundo, J. (2020). Microplastics in wastewater: Microfiber emissions from common household laundry. *Environmental Science and Pollution Research, 27*(21), 26643–26649. https://doi.org/10.1007/s11356-020-08765-6

Geyer, R., Gavigan, J., Jackson, A. M., Saccomanno, V. R., Suh, S., & Gleason, M. G. (2022). Quantity and fate of synthetic microfiber emissions from apparel washing in California and strategies for their reduction. *Environmental Pollution, 298*, Article 118835. https://doi.org/10.1016/j.envpol.2022.118835

Green, D. S., Kregting, L., & Boots, B. (2020). Smoked cigarette butt leachate impacts survival and behaviour of freshwater invertebrates. *Environmental Pollution, 266*, Article 115286. https://doi.org/10.1016/j.envpol.2020.115286

Hartline, N., Bruce, N. J., Karba, S. N., Ruff, E. O., Sonar, S. U., & Holden, P. A. (2016). Microfiber masses recovered from conventional machine washing of new or aged garments. *Environmental Science & Technology, 50*(21), 11532–11538. https://doi.org/10.1021/acs.est.6b03045

Herweyers, L., Catarci Carteny, C., Scheelen, L., Watts, R., & Du Bois, E. (2019). Consumers' perceptions and attitudes toward products preventing microfiber pollution in aquatic environments as a result of the domestic washing of synthetic clothes. *Sustainability, 12*(6), 2244. https://doi.org/10.3390/su12062244

Hu, T., Shen, M., & Tang, W. (2021). Wet wipes and disposable surgical masks are becoming new sources of fiber microplastic pollution during global COVID-19. *Environmental Science and Pollution Research, 29*(1), 284–292. https://doi.org/10.1007/s11356-021-17408-3

Huang, J., Chen, H., Zheng, Y., Yang, Y., Zhang, Y., & Gao, B. (2021). Microplastic pollution in soils and groundwater: Characteristics, analytical methods and impacts. *Chemical Engineering Journal, 425*, Article 131870. https://doi.org/10.1016/j.cej.2021.131870

Jagatee, S., Priyadarshini, S., Rath, C. C., & Das, A. P. (2024). Synthetic microfiber: an enduring environmental problem linked to sustainable development. *Environmental Science and Engineering, Part, F2333*, 93–112. https://doi.org/10.1007/978-3-031-51792-1_6

Kapp, K. J., & Miller, R. Z. (2020). Electric clothes dryers: An underestimated source of microfiber pollution. *PLoS ONE, 15*(10), Article e0239165. https://doi.org/10.1371/journal.pone.0239165

Kole, P. J., Löhr, A., Van Belleghem, F., & Ragas, A. (2017). Wear and tear of tyres: A stealthy source of microplastics in the environment. *International Journal of Environmental Research and Public Health, 14*(10), 1265. https://doi.org/10.3390/ijerph14101265

Kumar, M., Pandit, S., Patel, V., Khanna, N., Nag, M., Lahiri, D., Ray, R. R., Das, A., & Das, D. (2023). Maximization of energy recovery from starch processing wastewater by thermophilic dark fermentation coupled with microbial fuel cell technology. *Geomicrobiology Journal*. https://doi.org/10.1080/01490451.2023.2209555

Kwak, J. I., Liu, H., Wang, D., Lee, Y. H., Lee, J., & An, Y. (2021). Critical review of environmental impacts of microfibers in different environmental matrices. *Comparative Biochemistry and Physiology Part C: Toxicology & Pharmacology, 251*, Article 109196. https://doi.org/10.1016/j.cbpc.2021.109196

Lamb, J. B., Willis, B. L., Fiorenza, E. A., Couch, C. S., Howard, R., Rader, D. N., True, J. D., Kelly, L. A., Ahmad, A., Jompa, J., & Harvell, C. D. (2018). Plastic waste associated with disease on coral reefs. *Science, 359* (6374), 460–462. https://doi.org/10.1126/science.aar3320

Le, L., Nguyen, K. N., Nguyen, P., Duong, H. C., Bui, X., Hoang, N. B., & Nghiem, L. D. (2022). Microfibers in laundry wastewater: Problem and solution. *Science of the Total Environment, 852*, Article 158412. https://doi.org/10.1016/j.scitotenv.2022.158412

Lellis, B., Fávaro-Polonio, C. Z., Pamphile, J. A., & Polonio, J. C. (2019). Effects of textile dyes on health and the environment and bioremediation potential of living organisms. *Biotechnology Research and Innovation, 3*(2), 275–290. https://doi.org/10.1016/j.biori.2019.09.001

Li, Y., Lu, Q., Xing, Y., Liu, K., Ling, W., Yang, J., Yang, Q., Wu, T., Zhang, J., Pei, Z., Gao, Z., Li, X., Yang, F., Ma, H., Liu, K., & Zhao, D. (2023). Review of research on migration, distribution, biological effects, and analytical methods of microfibers in the environment. *Science of the Total Environment, 855*, Article 158922. https://doi.org/10.1016/j.scitotenv.2022.158922

Liu, G., Zhu, Z., Yang, Y., Sun, Y., Yu, F., & Ma, J. (2019a). Sorption behavior and mechanism of hydrophilic organic chemicals to virgin and aged microplastics in freshwater and seawater. *Environmental Pollution, 246*, 26–33. https://doi.org/10.1016/j.envpol.2018.11.100

Liu, J., Yue, Y., Ding, J., Zhu, B., & Gao, W. (2019b). Microfibers: A preliminary discussion on their definition and sources. *Environmental Science and Pollution Research, 26*(28), 29497–29501. https://doi.org/10.1007/s11356-019-06265-w

Liu, J., Zhu, B., An, L., Ding, J., & Xu, Y. (2023). Atmospheric microfibers dominated by natural and regenerated cellulosic fibers: Explanations from the textile engineering perspective. *Environmental Pollution, 317*, Article 120771. https://doi.org/10.1016/j.envpol.2022.120771

Lusher, A., Welden, N., Sobral, P., & Cole, M. (2017). Sampling, isolating and identifying microplastics ingested by fish and invertebrates. *Analytical Methods, 9*(9), 1346–1360. https://doi.org/10.1039/c6ay02415g

Madhav, S., Ahamad, A., Singh, A. K., Kushawaha, J., Chauhan, J. S., Sharma, S., & Singh, P. (2019). Water pollutants: sources and impact on the environment and human health. In *Advanced functional materials and sensors* (pp. 43–62). https://doi.org/10.1007/978-981-15-0671-0_4

Malachová, K., Kukutschová, J., Rybková, Z., Sezimová, H., Plachá, D., Čabanová, K., & Filip, P. (2016). Toxicity and mutagenicity of low-metallic automotive brake pad materials. *Ecotoxicology and Environmental Safety, 131*, 37–44. https://doi.org/10.1016/j.ecoenv.2016.05.003

Mao, Y., Ai, H., Chen, Y., Zhang, Z., Zeng, P., Kang, L., Li, W., Gu, W., He, Q., & Li, H. (2018). Phytoplankton response to polystyrene microplastics: Perspective from an entire growth period. *Chemosphere, 208*, 59–68. https://doi.org/10.1016/j.chemosphere.2018.05.170

McIlwraith, H. K., Lin, J. H., Erdle, L. M., Mallos, N. J., Diamond, M. L., & Rochman, C. M. (2019). Capturing microfibers – marketed technologies reduce microfiber emissions from washing machines. *Marine Pollution Bulletin, 139*, 40–45. https://doi.org/10.1016/j.marpolbul.2018.12.012

Miller, E., Sedlak, M., Lin, D., Box, C., Holleman, C., Rochman, C. M., & Sutton, R. (2021). Recommended best practices for collecting, analyzing, and reporting microplastics in environmental media: Lessons learned from comprehensive monitoring of San Francisco Bay. *Journal of Hazardous Materials, 409*, Article 124770. https://doi.org/10.1016/j.jhazmat.2020.124770

Mishra, S., & Das, A. P. (2022). Treatment of the wastewater polluted with synthetic microfiber released from washing machine. In B. B. Das, H. Hettiarachchi, P. K. Sahu, & S. Nanda (Eds.), *Lecture notes in civil engineering* (Vol. 207, pp. 109–117). Springer Science and Business Media Deutschland GmbH. https://doi.org/10.1007/978-981-16-7509-6_9/COVER

Mishra, S., & Das, A. P. (2023). Microbial remediation of synthetic microfiber contaminated wastewater. In M. P. Shah (Ed.), *Microbial technologies in industrial wastewater treatment* (pp. 337–350). Springer. https://doi.org/10.1007/978-981-99-2435-6_17

Mishra, S., Dash, D., Al-Tawaha, A. R. M. S., & Das, A. P. (2022a). A review on heavy metal ion adsorption on synthetic microfiber surface in aquatic environments. *Applied Biochemistry and Biotechnology, 194*(10), 4639–4654. https://doi.org/10.1007/S12010-022-04029-W/METRICS

Mishra, S., Dash, D., & Das, A. P. (2022b). Detection, characterization and possible biofragmentation of synthetic microfibers released from domestic laundering wastewater as an emerging source of marine pollution. *Marine Pollution Bulletin, 185*, Article 114254. https://doi.org/10.1016/J.MARPOLBUL.2022.114254

Mishra, S., Dash, D., & Das, A. P. (2024a). Isolation and molecular characterization of polyester-degrading bacterial strains from municipal wastewater and their application in microplastic degradation in contaminated wastewater. *Geomicrobiology Journal*. https://doi.org/10.1080/01490451.2024.2376101

Mishra, S., Dash, D., & Das, A. P. (2024b). Microbial degradation of polyester microfibers using indigenously isolated bacterial strain Exiguobacterium Sp. *CLEAN—Soil, Air, Water, 52*(12), e202300343. https://doi.org/10.1002/CLEN.202300343

Mishra, S., Rath, C. C., & Das, A. P. (2019). Marine microfiber pollution: A review on present status and future challenges. *Marine Pollution Bulletin, 140*, 188–197. https://doi.org/10.1016/j.marpolbul.2019.01.039

Mishra, S., Rout, P. K., & Das, A. P. (2021). Emerging microfiber pollution and its remediation. In R. Prasad (Ed.), *Environmental and microbial biotechnology* (pp. 247–266). Springer. https://doi.org/10.1007/978-981-15-5499-5_9

Mishra, S., Singh, R. P., Rath, C. C., & Das, A. P. (2020). Synthetic microfibers: Source, transport and their remediation. *Journal of Water Process Engineering, 38*, Article 101612. https://doi.org/10.1016/J.JWPE.2020.101612

Mishra, S., Singh, R. P., Rout, P. K., & Das, A. P. (2022c). Membrane bioreactor (MBR) as an advanced wastewater treatment technology for removal of synthetic microplastics. In *Development in wastewater treatment research and processes* (pp. 45–60). https://doi.org/10.1016/B978-0-323-85583-9.00022-3

Mohanty, M., Mohanty, J., Dey, S., Dutta, K., Shah, M. P., & Das, A. P. (2024a). The face mask: A tale from protection to pollution and demanding sustainable solution. *Emerging Contaminants, 10*(2), Article 100298. https://doi.org/10.1016/J.EMCON.2023.100298

Mohanty, M., Tripathy, B., Koteswara Reddy, G., & Das, A. P. (2024b). Greener technologies for establishing restraint over microplastic pollution. In *Sustainable microbial technology for synthetic and cellulosic microfiber bioremediation* (pp. 239–258). https://doi.org/10.1007/978-3-031-62600-5_13

Mokarram, M., Saber, A., & Sheykhi, V. (2020). Effects of heavy metal contamination on river water quality due to release of industrial effluents. *Journal of Cleaner Production, 277*, Article 123380. https://doi.org/10.1016/j.jclepro.2020.123380

Napper, I. E., Barrett, A. C., & Thompson, R. C. (2020). The efficiency of devices intended to reduce microfibre release during clothes washing. *Science of the Total Environment, 738*, Article 140412. https://doi.org/10.1016/j.scitotenv.2020.140412

Pal, P. (2016). Industry-specific water treatment: Case studies. In *Industrial water treatment process technology* (pp. 243–511). https://doi.org/10.1016/B978-0-12-810391-3.00006-0

Paul, A., Priyadarsini, S., Karakuş, S., Behera, J. K., Gnanasekaran, R., & Das, A. P. (2024). Fate and accumulation of synthetic microfibers in the environment. In *Sustainable microbial technology for synthetic and cellulosic microfiber bioremediation* (pp. 61–76). https://doi.org/10.1007/978-3-031-62600-5_4

Periyasamy, A. P., & Tehrani-Bagha, A. (2022). A review on microplastic emission from textile materials and its reduction techniques. *Polymer Degradation and Stability, 199*, Article 109901. https://doi.org/10.1016/j.polymdegradstab.2022.109901

Pivokonský, M., Čermáková, L., Novotná, K., Peer, P., Cajthaml, T., & Janda, V. (2018). Occurrence of microplastics in raw and treated drinking water. *Science of the Total Environment, 643*, 1644–1651. https://doi.org/10.1016/j.scitotenv.2018.08.102

Plakantonaki, S., Kiskira, K., Zacharopoulos, N., Chronis, I., Coelho, F., Togiani, A., Kalkanis, K., & Priniotakis, G. (2022). A review of sustainability standards and ecolabeling in the textile industry. *Sustainability, 15*(15), 11589. https://doi.org/10.3390/su151511589

Priya, K., Thilagam, H., Muthukumar, T., Gopalakrishnan, S., & Govarthanan, M. (2023). Impact of microfiber pollution on aquatic biota: A critical analysis of effects and preventive measures. *Science of the Total Environment, 887*, Article 163984. https://doi.org/10.1016/j.scitotenv.2023.163984

Priyadarshini, B., & Das, A. P. (2024). The adverse impact of synthetic microplastics released from disposable face masks pollutants on aquatic microbiota, aquatic species and humans: A review. *Environmental Science and Engineering, Part, F2827*, 67–81. https://doi.org/10.1007/978-3-031-50840-0_5

Priyadarshini, B., Santhosh Kumar, M., & Das, A. P. (2024a). Microbial degradation of synthetic plastics released from COVID-19 generated single use disposable face masks. *Environmental Science and Engineering, Part, F2827*, 1–16. https://doi.org/10.1007/978-3-031-50840-0_1

Priyadarshini, S., Jagatee, S., & Das, A. P. (2024b). Synthetic fabrics and microfiber pollution–An assessment of their global impact. *Environmental Science and Engineering, Part, F2333*, 137–157. https://doi.org/10.1007/978-3-031-51792-1_8

Rathinamoorthy, R., & Subramanian, R. B. (2021). Synthetic textile and microfiber pollution: A review on mitigation strategies. *Environmental Science and Pollution Research, 28*(31), 41596–41611. https://doi.org/10.1007/s11356-021-14763-z

Rillig, M. C. (2018). Microplastic disguising as soil carbon storage. *Environmental Science & Technology, 52*(11), 6079–6080. https://doi.org/10.1021/acs.est.8b02338

Sahu, A., Mallick, K., & Das, A. P. (2024). *Environmental impact of single-use synthetic face mask and its recycling: A sustainable approach* (pp. 197–212). https://doi.org/10.1007/978-3-031-50840-0_11

Satapathy, S., Kar, D., Priyadarsini, S., Das, A. P., & Kuanar, A. (2024). Application of microorganisms in emerging waste management strategies and remediation of synthetic and cellulosic microfibers. In *Sustainable microbial technology for synthetic and cellulosic microfiber bioremediation* (pp. 183–206). https://doi.org/10.1007/978-3-031-62600-5_10

Shen, M., Li, Y., Song, B., Zhou, C., Gong, J., & Zeng, G. (2021). Smoked cigarette butts: Unignorable source for environmental microplastic fibers. *Science of the Total Environment, 791*, Article 148384. https://doi.org/10.1016/j.scitotenv.2021.148384

Shruti, V., Pérez-Guevara, F., & Kutralam-Muniasamy, G. (2021). Wet wipes contribution to microfiber contamination under COVID-19 era: An important but overlooked problem. *Environmental Challenges, 5*, Article 100267. https://doi.org/10.1016/j.envc.2021.100267

Singh, N., Tang, Y., & Ogunseitan, O. A. (2020a). Environmentally sustainable management of used personal protective equipment. *Environmental Science & Technology, 54*(14), 8500–8502. https://doi.org/10.1021/acs.est.0c03022

Singh, R. P., Mishra, S., & Das, A. P. (2020b). Synthetic microfibers: Pollution toxicity and remediation. *Chemosphere, 257*, Article 127199. https://doi.org/10.1016/j.chemosphere.2020.127199

Tchounwou, P. B., Yedjou, C. G., Patlolla, A. K., & Sutton, D. J. (2012). Heavy metal toxicity and the environment. In *EXS* (pp. 133–164). https://doi.org/10.1007/978-3-7643-8340-4_6

Textile Exchange. (2023). *Materials market report*. Textile Exchange. Accessed December 10, 2023. https://textileexchange.org/app/uploads/2023/11/Materials-Market-Report-2023.pdf

Thiel, M., Gallardo, C., Hinojosa, I. A., Luna, N., Morales, N., Ory, N., Pacheco, A. S., & Zavalaga, C. (2018). Impacts of marine plastic pollution from continental coasts to subtropical gyres—Fish, seabirds, and other vertebrates in the SE Pacific. *Frontiers in Marine Science, 5*, Article 359654. https://doi.org/10.3389/fmars.2018.00238

Tiwari, M. P., Rathod, T., Ajmal, P., Bhangare, R., & Sahu, S. K. (2019). Distribution and characterization of microplastics in beach sand from three different Indian coastal environments. *Marine Pollution Bulletin, 140*, 262–273. https://doi.org/10.1016/j.marpolbul.2019.01.055

Unice, K. M., Weeber, M., Abramson, M., Reid, R., Van Gils, J., Markus, A., Vethaak, A., & Panko, J. M. (2019). Characterizing export of land-based microplastics to the estuary - Part I: Application of integrated geospatial microplastic transport models to assess tire and road wear particles in the Seine watershed. *Science of the Total Environment, 646*, 1639–1649. https://doi.org/10.1016/j.scitotenv.2018.07.368

Vassilenko, E., Watkins, M., Chastain, S., Mertens, J., Posacka, A. M., Patankar, S., & Ross, P. S. (2021). Domestic laundry and microfiber pollution: Exploring fiber shedding from consumer apparel textiles. *PLoS ONE, 16*(7). https://doi.org/10.1371/journal.pone.0250346

Wang, Z., An, C., Chen, X., Lee, K., Zhang, H., & Feng, Q. (2021). Disposable masks release microplastics to the aqueous environment with exacerbation by natural weathering. *Journal of Hazardous Materials, 417*, Article 126036. https://doi.org/10.1016/j.jhazmat.2021.126036

Wu, P., Li, J., Xiao, L., Tang, Y., & Cai, Z. (2022). Release of tens of thousands of microfibers from discarded face masks under simulated environmental conditions. *Science of the Total Environment, 806*, Article 150458. https://doi.org/10.1016/j.scitotenv.2021.150458

Yang, L., Qiao, F., Lei, K., Li, H., Kang, Y., Cui, S., & An, L. (2019). Microfiber release from different fabrics during washing. *Environmental Pollution, 249*, 136–143. https://doi.org/10.1016/j.envpol.2019.03.011

Zhang, K. (2022). Microfibers released into the air from a household tumble dryer. *Environmental Science and Technology Letters, 9*(2), 120–126. https://doi.org/10.1021/acs.estlett.1c00911

Zhang, Y., Lykaki, M., Alrajoula, M. T., Markiewicz, M., Kraas, C., Kolbe, S., Klinkhammer, K., Rabe, M., Klauer, R., Bendt, E., & Stolte, S. (2021). Microplastics from textile origin—Emission and reduction measures. *Green Chemistry, 23*(15), 5247–5271. https://doi.org/10.1039/d1gc01589c

Zhao, M., Huang, L., Arulmani, S., Jia, Y., Wu, L., Wu, T., Zhang, H., & Xiao, T. (2022). Adsorption of different pollutants by using microplastic with different influencing factors and mechanisms in wastewater: A review. *Nanomaterials, 12*(13), 2256. https://doi.org/10.3390/nano12132256

Bioremediation as a Greener Technology for Municipal Waste-to-Energy Conversion

Suprada Swain, Sailaja Priyadarsini, Amel Gacem, Sunil Kumar Behera, Maha Awjan Alreshidi, Krishna Kumar Yadav, and Alok Prasad Das

1 Introduction

Our consumption of other resources has increased due to the concerning global population growth and rising energy demand. Given this, the two most important factors influencing a country's development and the state of its economy are industrialization and urbanization (Das et al., 2025; Reddy Koteswara et al., 2025). They do, however, come with several detrimental environmental repercussions. Due to rising consumption patterns that pollute the environment, the pattern of solid waste output worldwide has increased from 1.4 billion tonnes in 2012 to 2.3 billion tonnes in 2025. Soil pollution is a hidden concern since it is often impossible to evaluate directly. Among other things, the chemical industries contaminate the air, water, and soil and interfere with natural processes (Ghosh & Das, 2020; Mishra et al., 2024b). Industry-derived heavy metals eventually taint food and interfere with photosynthesis

S. Swain · S. Priyadarsini · A. P. Das (✉)
Department of Life Sciences, Rama Devi Women's University, Bhubaneswar, Odisha, India
e-mail: alok1053@gmail.com

A. Gacem
Department of Physics, Faculty of Sciences, University 20 Août 1955, Skikda, Algeria

S. K. Behera
Department of Biotechnology, Berhampur University, Berhampur, Odisha, India

M. A. Alreshidi
Department of Chemistry, College of Science, University of Ha'il, Ha'il, Saudi Arabia
e-mail: m.alrashedi@uoh.edu.sa

K. K. Yadav
Department of Environmental Science, Parul Institute of Applied Sciences, Parul University, Vadodara, Gujarat, India

Environmental and Atmospheric Sciences Research Group, Scientific Research Center, Al-Ayen University, Nasiriyah, Thi-Qar, Iraq

A. P. Das et al. (eds.), *Sustainable Municipal Waste Management with a Zero Waste Approach*, Sustainable Environmental Waste Management Strategies,
https://doi.org/10.1007/978-3-032-05950-5_11

by piercing membranes (Mishra et al., 2022a). Under the current circumstances, biological variety is declining, and land resources are becoming scarcer. To combat the rapid deterioration of the environment and ensure high agricultural productivity, improved soil health, and the genetic and biological capacities of microbes and crop plants, sustainable agriculture is being promoted by gradually reducing the use of synthetic agrochemicals and placing more emphasis on the use of substances derived from biowaste (Das & Ghosh, 2022; Das et al., 2024c; Dey et al., 2023; Mishra & Das, 2023; Mohanty et al., 2018). All forms of soil contamination require the creation of a novel, scientific remediation approach, such as the recently popular bioremediation method (Kumar et al., 2023a). Effective MSW management requires a thorough system for reducing, collecting, recycling, treating, and disposing of garbage, according to the US Environmental Protection Agency (USEPA). Integrated solid waste management (ISWM) is the term used to describe this approach. Finding the sources of waste generation, characterizing the waste, estimating its physical and chemical characteristics to select the best treatment technology, figuring out the potential materials recovery, assisting in the design of waste processing equipment, and maintaining compliance with environmental regulations are all crucial steps in a successful waste management program (Bala et al., 2024a; Das & Ghosh, 2018; Mishra et al., 2024a; Tripathy et al., 2024). For waste-to-energy applications, which are essential to MSW management systems, to have a significant impact on the low-carbon development society, accurate and exact MSW characterization data is required. Therefore, the current study is to evaluate the fluctuation in MSW's composition and qualities (physical and chemical) as well as characterize it on a seasonal basis. Additionally, it aims to offer a new approach to waste characterization that makes use of accurate data in a waste-to-energy plant strategy.

Polycyclic aromatic hydrocarbons (PAHs) arc a class of chemical compounds made up of two or more fused aromatic rings (Behera et al., 2021, 2022; Kanungo et al., 2023; Mohanta et al., 2024). They are heteroatom PAHs because they may have atoms such as oxygen, nitrogen, and sulphur in their ring structures. The characteristics of PAHs vary according to their molecular weight, but they are typically nonpolar, hydrophobic compounds. There are two types: high molecular weight (HMW) PAHs that have four- to six-ring structures, and low molecular weight (LMW) PAHs with two- and three- ring structures. The presence of several alkyl substituents of methyl and ethyl groups is also present in some PAHs. Their inherent properties, including hydrophobicity and heterocyclic aromatic ring structures, have indeed made them very durable and resistant to degradation. In general, soil contamination has been found to range from 1 μg/kg to 300 g/kg of PAH. Because of both natural and man-made activities, such as petrogenic and pyrogenic activities, PAH is found to be widely distributed in the environment. Additionally, incomplete fossil fuel combustion, coal and wood burning, and other environmental factors cause it. Furthermore, the extensive buildup of PAH in the environment may also be caused by natural events like forest fires and volcanic eruptions.

The bielectrons on either side of the benzene ring cannot be attacked by nucleophiles. This means that these products are the most persistent in the environment. They have high melting and boiling temperatures, poor water solubility, low vapor pressure, and are lipophilic. Additionally, the molecular weight of the molecule determines its microbial breakdown and persistence in soil. The type of contamination, its concentration, and the interaction of the polluted site's chemical, geological, physical, and biological features all affect how long a pollutant remains in the environment. Compared to recently contaminated soil, older PAH-contaminated soil has a higher persistence of PAH (Gundlapalli et al., 2024). Significant developments and creative strategies have been made in the field of bioremediation in recent years, especially regarding LMW PAHs. These advancements seek to address the particular difficulties presented by LMW PAHs by improving the efficacy and efficiency of biodegradation processes. There are several innovative methods included in bioremediation which show significant results in elimination of PAHs.

2 Bioremediation

Bioremediation has a high public acceptability rate and is a very economical procedure (Das et al., 2015; Paul et al., 2023; Pradhan et al., 2023). Although there are no technical requirements for this procedure, it requires experience and skill to be implemented after taking the bioremediation site's characteristics into account (Das et al., 2024b; Dey et al., 2024b; Kanungo et al., 2024; Nag et al., 2023). Both native species and microbial species isolated from a place other than the contaminated area can be used in bioremediation (Bhattacharjee et al., 2021; Das et al., 2023; Mohanty et al., 2024). The technique of biodegradation employing microbial isolates to treat the contamination is known as bioaugmentation. Hazardous materials like pesticides, heavy metals, xenobiotics, organic halogens, industrial effluents, and chlorinated compounds can be eliminated, detoxified, and their toxicity reduced via the processes involved in bioremediation. By investigating the utilization of microbial consortia and genetically modified microorganisms to target a larger variety of toxins and increase degradation efficiency, innovative research has increased the use of bioremediation. The promise of synthetic biology to engineer microbial strains with improved pathways for pollution degradation has been emphasized in recent research, which would notably advance the field. In addition, the application of phytoremediation—utilizing vegetation for the support of microbial processes—along with bioremediation has shown promise in the cleanup of difficult contamination scenarios and presents an additional technique for environmental restoration. Additional work has been carried out on the application of machine learning and bioinformatics to predict and optimize bioremediation processes. Researchers can determine key factors affecting the effectiveness of bioremediation and better fine-tune interventions by peering through vast databases on environmental and microbial genetic states. This data-driven strategy is a noteworthy improvement in increasing the bioremediation process's efficiency and economy. With advancing technology, bioremediation also continues to evolve, becoming more practical and efficient in decontaminating polluted areas. Leveraging microorganisms' ability to shield and purify our ecosystems remains a key strategy for handling the numerous issues brought on by environmental pollution.

Complexity of Contaminants: Priya et al., (2024) state that most of the pollutants found in contaminated areas have intricate chemical compositions that are hard for bioremediation methods to fully break down. Some pollutants, like persistent organic pollutants and heavy metals, could require genetically modified microorganisms or microbial consortia of specialists for efficient degradation.

Heterogeneity of Sites: Heterogeneity of Sites: In affected locations, there may be considerable differences in the pH, temperature, water content, and nutritional status of the environment, establishing means of bioremediation. Identifying strategies that can evolve in response to such changes and still yield reliable results may be problematic (Ceballos-Escalera et al., 2024).

Slow Degradation Rates: When using poisons that are difficult to remove, bioremediation methods can take time (Geed & Verma, 2024). Bacteria can take some time to acclimate to having contaminants present and build up populations that can break them down effectively. Also, conditions such as contaminated environments or a low nutrient supply can make the degradation slower.

Limited Use with Certain Pollutants: Though bioremediation proves useful for most organic pollutants, it is less efficient for some contaminants, including radioactive and heavy metal substances. It may be required to combine bioremediation with additional remediation methods for an overall cleanup, and these impurities may require distinct treatment processes.

Risk of Secondary Contamination: Half-degraded intermediates or transformation byproducts, potentially more harmful or transportable compared to the initial pollutants, may occasionally be released because of bioremediation processes. For bioremediation processes not to unintentionally exacerbate environmental circumstances, it is critical to comprehend and minimize the risk of secondary contamination (Liu et al., 2024).

Regulatory Challenges: Regional differences in the laws that regulate the application of bioremediation methods may complicate implementation, ensuring adherence to regulations. Regulations, obtaining necessary permits, and managing liability issues could complicate bioremediation initiatives to be more expensive (Prasad, 2024).

Monitoring and Verification: Strict monitoring and verification processes are required to assess the efficiency of bioremediation procedures and guarantee the accomplishment of the cleanup objective. Yet, the precision, sensitivity, and real-time data-capture abilities of the existing methods for in-situ monitoring of pollutant degradation and microbial activity can be constrained.

3 Important Goals for Bioremediation

Waste recycling: Bioremediation is used to clean contaminated soil, recycle wastewater, and treat industrial effluents.

Pollutant Detoxification: Bioremediation uses microorganisms and plants to change toxic pollutants into molecules that are less toxic or nontoxic. The hazards to the environment and human health connected to contaminated places are lessened by this detoxification procedure.

Mineralization of Organic Pollutants: The goal of bioremediation is the full mineralization of organic contaminants. To completely remove contaminants from the environment, complex organic materials must dissolve into their component constituents, including carbon, hydrogen, and several other elements, as well as carbon dioxide, water, and microbial biomass.

Ecosystem Restoration: By eliminating pollutants that upset ecological equilibrium, bioremediation aids in the restoration of natural ecosystems. This process improves ecosystem health and biodiversity by restoring the conditions necessary for different species to live and thrive.

Sustainable and Eco-Friendly Approach: Bioremediation is naturally sustainable and environmentally benign since it depends on biological processes. It causes much less environmental disturbance and requires fewer hazardous agents in contrast to conventional physical and chemical repair techniques.

Regulatory Standard Compliance: By ensuring that pollutant levels are lowered to acceptable levels, bioremediation protects both the environment and human health. For remediated areas to be safely used for residential, agricultural, or industrial purposes, regulatory compliance must be attained.

Cost-Effectiveness: Compared to traditional cleanup techniques, bioremediation frequently turns out to be more economical. It usually uses less energy and resources, which makes it a cost-effective choice for extensive environmental cleanup efforts.

4 Recent Advances in Bioremediation

New approaches to enhancing the efficiency of bioremediation have been suggested by new studies:

Genetically engineered microorganisms: Through genetic engineering, researchers have created microbial strains with improved capacities to break down particular contaminants. A researcher claims that the improved strains increase the breadth and efficacy of bioremediation procedures.

Bioaugmentation in Groundwater Remediation: To improve bioremediation processes, the process of adding specific microbial strains to contaminated locations is used. The formation of specialized microbial consortia targeted at pollutant breakdown has been made easier by recent developments.

Molecular biology and genetic engineering (Borchert et al., 2021) highlighted the critical role that bioaugmentation plays in purifying groundwater that has been contaminated by hydrocarbons, other industrial contaminants, such as chlorinated solvents, in their investigation of a number of microorganisms and plant-based bioremediation technologies. These studies demonstrate the efficacy of bioaugmentation techniques in improving bioremediation results and promoting environmental cleaning initiatives.

Nanotech Boost: Nanotech Boost: By employing cutting-edge methods to accelerate the breakdown of contaminants, nanotechnology has become an essential tool for improving bioremediation processes. The groundbreaking nature of nanoparticles in this sector is that they have many applications in bioremediation. They can serve as carriers, making it easier for microbial communities to receive nutrients or electron acceptors. Their metabolic activity would rise as a result, optimizing the breakdown of pollutants. The inherent characteristics of nanoparticles also increase their bioavailability and susceptibility to microbial enzymatic breakdown by allowing them to engage directly with contaminants. An advanced method of addressing the problems of environmental contamination is to combine nanotechnology and bioremediation techniques. Researchers can create customized solutions that increase cleanup performance in a variety of situations by utilizing the special properties of nanoparticles, such as their high surface area-to-volume ratio and adjustable surface chemistry. Cleaner ecosystems and more environmentally friendly management practices could be encouraged by this advancement.

This development has the potential to promote cleaner ecosystems and more sustainable environmental management techniques.

6 Bioremediation Types

Based on the location of remediation, bioremediation can be roughly divided into two types.

In-Situ Bioremediation

One crucial method for eliminating hazardous materials from the environment is in-situ bioremediation. Remediation in-situ is the term applied to describe cleanup that happens right at the source of the pollution. If both plants and animals are used in the cleanup process, it is called bioremediation. Contaminated soil or groundwater can be cleaned up utilizing in-situ bio-processing, sometimes referred to as on-site bio-processing, instead of costly and time-consuming cleanup techniques. This technique effectively gets rid of chemical pollutants by utilizing harmless plants, animals, and microbes. Using natural processes, reducing shipping costs, and requiring less physical effort are some advantages of local bioremediation (Bala et al., 2024b; Dey et al., 2024a, 2024c). On-site bioremediation techniques include natural attenuation, composting, sucking, bioventing, sparking, and microorganisms that aid in phytoremediation. There are three fundamental methods of in-situ bioremediation: bioaugmentation, biostimulation, or bioattenuation, which are all supported by these techniques. Despite its benefits, several issues could limit the efficacy of bioremediation in-situ soil tends to be situ. Soil is often contaminated by numerous organic contaminants, especially where there is industrialization. Additionally, pollutants, particularly in areas that are industrialized. Furthermore, a range of conditions such as, low temperatures, anaerobic environments, shortages of co-substrates and nutrients, the absence, of degradability, and bioavailability of organic molecules, may hamper the degradation of organic molecules. Environmental factors have a major impact on the effectiveness of in-situ bioremediation operations. Important aspects to take into account include seasonal temperature variations, soil composition, neutral pH level, nutritional supplements, oxygen tension levels, and gene expression (Sharma et al., 2024). In-situ bioremediation has the potential to become a potent instrument for reducing environmental contamination and promoting ecosystem renewal if these conditions are properly met.

In-Situ Bioremediation Methods

Through the assistance of microorganisms, which aid in their removal and biotransformation, poisons are extracted during the in-situ bioremediation process (Paul et al., 2024; Satapathy et al., 2024). Many soil remediation techniques have been developed in recent decades, including traditional physical and chemical techniques that are well-known for their effectiveness and affordability. However, because of their affordability, environmental friendliness, and comparatively faster turnaround times, biological treatments including phytoremediation, bioremediation, and term remediation have become more popular. Even though these natural processes and their results have received a lot of attention, little is known about how different soil purification methods interact.

Intrinsic In-Situ Bioremediation

Without the need for outside assistance, intrinsic bioremediation uses the innate ability of local microorganisms to break down pollutants. Using the innate ability of the current microbial populations, this mechanism transforms environmental contaminants into innocuous forms. Intrinsic bioremediation is used in subterranean areas like underground fuel tanks and is especially useful in soil conditions that are continuously contaminated by chemicals and contaminants. However, it is essential to confirm and evaluate the microbial community's innate capacity to break down contaminants by in-vitro testing before putting intrinsic bioremediation into practice.

Intrinsic/Natural Bioremediation Techniques

Without the need for outside scientific intervention, bioremediation is a natural process that uses the innate capacities of plants and animals to reduce the levels of toxins found in contaminated lithosphere, hydrosphere, and atmosphere. Because some species naturally exhibit bioremediation, this technique is incredibly dependable, economical, and environmentally benign. Constant observation, evaluation, and the important components of this process include the addition of nutrients, either native or exogenous microorganisms, and consideration of the physical and geochemical characteristics of the soil, several forms of intrinsic or natural bioremediation. To forecast successful site mitigation, it is essential to isolate native bacteria and assess physical attributes. Aerobic processes are an efficient method for in-situ bioremediation, and strategies like bioventing can speed up the bioremediation process. These procedures use a variety of techniques, such as composite sparking, to supply oxygen to the soil.

7 Types of Natural and Intrinsic Bioremediation Methods

Using Plants to Clean up (Phytoremediation)

Phytoremediation technologies decontaminate pollutants from polluted soil, air, and water using living vegetation. Efficient means of decontaminating heavy metal-polluted soil involve phytoextraction, which employs plants to remove and extract heavy metals, and phytostabilization, in which plants are utilized to reclaim heavy metals from the earth. Among the many common phytoremediation methods are phytodegradation, phytovolatilization, phytoextraction, and phytotransformation.

Phytodegradation

By converting organic pollutants into a less toxic form, phytodegradation utilizes the ability of plants to purify soil that has been contaminated with heavy metals and toxins. Phytodegradation degrades organic breakdown materials enzymatically through plant roots, upon which the plant uptakes it. This leads to a process known as phytotransformation, where polluted and worthless soil is transformed into productive, arable soil. By permitting sap to move upward, transpiration, an essential physiological process in plants, facilitates easier contamination absorption. The transpiration rate is influenced by a number of ecological factors, including temperature, pH, nutrition availability, soil moisture content, and so on. This illustrates the complex interrelationship between plants and environmental factors and physiology in phytodegradation processes.

Through the transformation of organic contaminants into less harmful forms, phytodegradation makes use of plants' capacity to cleanse soil that has been contaminated by heavy metals and toxins. This process breaks down organic materials enzymatically by plant roots, after which the plant absorbs it. Consequently, a process called phytotransformation takes place, converting contaminated and useless soil into productive, cultivable soil. By allowing sap to flow upward, transpiration, a vital physiological activity in plants, makes it easier for contaminants to be absorbed. Numerous ecological factors, including soil moisture, temperature, pH, and nutrient availability, affect the rate of transpiration. This illustrates the complex interplay between plant physiology and environmental factors during phytodegradation processes.

Phytostabilization

By converting organic contaminants into less toxic forms, phytodegradation utilizes the ability of plants to purify soil that has been tainted by pollutants and heavy metals. The process degrades organic breakdown materials enzymatically through plant roots, which are then absorbed by the plant. Thus, a phenomenon known as phytotransformation occurs, transforming polluted and barren soil into fertile, arable soil. By permitting sap to move towards the top, transpiration, an essential plant physiological process, facilitates easier contamination absorption. Transpiration is influenced by many factors in the ecological environment, such as temperature, pH, nutrition, moisture of the soil, and more, which influence transpiration speed. This emphasizes how intricate the process is between physiology of plants and environmental conditions in phytodegradation processes.

Phytovolatilization

By transforming soil contaminants into volatile compounds released into the air, phytovolatilization is a process that can detoxify the soil. Because of their excellent performance in this process, Brassicaceae plants are useful in the volatilization of soil contaminants. It is crucial to remember that contaminants are efficiently transferred from the ground to the atmosphere by phytovolatilization, although it does not completely eliminate poisons from the ground (Sharma et al., 2024). As plant roots absorb heavy metals through transpiration, phytovolatilization begins. The metals change into ionic forms inside the plant after absorption. Following their transformation, these chemicals are discharged into the environment as volatile gases. Through slow elimination of poisons from the ground, this process reduces their concentration and any harm to the ecosystem. By converting soil pollutants into gaseous compounds that are discharged into the atmosphere, a process known as phytovolatilization can detoxify the soil. Due to their remarkable effectiveness in this process, Brassicaceae plants are helpful in the volatilization of soil pollutants. It is crucial to remember that phytovolatilization efficiently moves contaminants from the soil to the atmosphere, even while it does not totally remove toxins from the soil (Sharma et al., 2024). When heavy metals are absorbed by plant roots through transpiration, phytovolatilization starts. Following absorption, these metals change into ionic forms inside the plant. These chemicals are discharged into the environment as volatile gases after undergoing transformation. By removing toxins from the soil gradually, this method lowers their concentration and any negative effects on the ecosystem.

Phytoextraction

One of the most popular bioremediation techniques is called phytoextraction, which attempts to eliminate pollutants from the soil, especially those that are at or above ground level or above. Phytosequestration, alternatively referred to also known as phytoaccumulation, phytoextraction, and metal compounds that increase the availability-increasing compounds, may occur of metals in the soil, can happen naturally. The species of the plant determines how much metal is absorbed by the plant and accumulates in it; other factors that impact the rate of phytoextraction include plant performance, soil metal availability, and plant selection.

Phytostimulation

Phytostimulation is the process of improving the soil's microbial flora to increase their capacity to effectively break down contaminants. This process usually takes place in the rhizosphere, the area of soil that is impacted by the plant's roots. According to Ratajczak et al., (2023), plants produce biomolecules into the rhizosphere, including sugars and acids, which promote the growth of bacteria that break down pollutants. These macromolecules stimulate the metabolism and growth of soil microbes by acting as strong phytostimulants. The breakdown of contaminants in the soil is accelerated by increased microbial activity. In contaminated situations, phytostimulation has proven to be very helpful in boosting microbial populations' ability to break down hydrocarbon pollutants. The overall approach, phytostimulation, is a realistic means to enhance the spontaneous processes for remediation of soil ecosystems.

Rhizofiltration

Using a dense thick network of plant roots, rhizofiltration is a phytoremediation process that filters for filtering water and eliminates contaminants and nutrients. Plants are initially grown hydroponically impure water to provide them with a strong root system before prior to producing rhizofiltration. The plants are then transferred to polluted water for acclimation when their roots are once they have developed fully formed roots, and then they are placed back in clean unpolluted water. Following acclimation, the plants are sent to contaminated regions transported to polluted areas where they remove contaminants such as extract pollutants like heavy metals from the soil or water. These plants uptake heavy metals in their roots, which are subsequently collected and disposed of appropriately then harvested and discarded in a suitable way to ensure the toxins are managed properly disposed. Rhizofiltration is particularly effective at cleaning contaminated water in two ways: either by direct planting at the site or by transferring the contaminated water to a controlled rhizofiltration system for groundwater.

Bioventing

Bioventing became a prominent soil remediation technology in the 1990s and has since been widely used in industry. By providing oxygen to native microorganisms, bioventing speeds up aerobic degradation by allowing them to decompose organic contaminants taken in soil when oxygen levels are low. Using vertical and horizontal wells, air is injected directly into the contaminated zone with the goal of using just enough air to promote deterioration while reducing the amount of contaminants that volatilize and leak into the environment. By drawing air in or forcing it through wells drilled into polluted soil, through bioventing, oxygen can enter the subsurface above the water table. Additionally, by converting toxins into less dangerous forms, nitrogen and phosphorus can be added to increase microbial growth rates and create an environment that is conducive to bioremediation. In unsaturated soil zones, bioventing is frequently employed to remove aerobically degradable organic substances, such as fuel residues. Although its efficacy decreases in a variety of soil types where pollutants may remain in less permeable layers, it is correlated with soil permeability. Its efficacy may be further reduced in soils with high iron and organic matter concentrations as a result of iron oxidation or competition with organic matter. Typically, bioventing requires a lengthier remediation time than Soil Vapor Extraction (SVE), sometimes lasting decades, especially in sandy soils that are homogeneous.

Biosparging

Similar to bioventing, this method involves injecting air into the soil to try to boost microbial activity and facilitate the removal of pollutants from contaminated areas. Unlike bioventing, biosparging entails pumping air into the soil's saturated zone. This process makes it possible for operation to allow hydrocarbon vapors to ascend into the unsaturated zone, which promotes an aspect that stimulates biodegradation even more. Soil porosity, which affects porosity of the soil, which influences the bioavailability and biodegradability of the microbiological contaminants, is one of two significant criteria that determine how effective biosparging is. Biosparging accelerates the natural biodegradation of aerobic chemicals by supplying oxygen to the soil's microorganisms. In the polluted saturated soil zone, air injection improves biodegradation and raises the oxygen content. Biosparging maintains sufficient oxygen level and provides a good supply of oxygen for microbial activity by using at low airflow rates. It is an established method for eliminating organic contaminants that decompose aerobically, such as waste from fuel on saturated soils. In a variety of soil types, biosparging is less effective when pollutants are present in less permeable layers. However, the permeability of the soil promotes its success. Because organic oxidation competes with contaminant oxidation, biosparging is less successful in soils that include substantial amounts of organic matter and iron. The ability of microorganisms to decompose contaminants is determined by the permeability of the soil; however, the addition of air to the soil enhances that capacity.

Composting

Composting is a successful bioremediation technique that uses microorganisms to transform organic waste and soil contaminants into less dangerous byproducts.

When composting, high temperatures promote biodegradation and hasten the breakdown of hazardous compounds. This technology is primarily used to control polluted soils or sediments by increasing the porosity of the compost mixture and composting contaminated soils for efficient breakdown wood, vegetable waste, animal manure, and certain organic matter are added to soil while thermophilic temperatures above 50 °C are maintained in order to promote the breakdown of organic materials. The most economical method for clearing individual "batches" of silt or contaminated soil is windrow composting.

8 Engineered Bioremediation and Its Types

Bio-slurping by supplying oxygen to promote biodegradation, this technique indirectly treats soil and groundwater by combining vacuum-assisted pumping, soil vapor extraction, and bioventing. Its primary objective is to recover free-phase pollutants from capillary, unsaturated, and saturated zones of liquids (LNAPLs), such as light non-aqueous phase. Furthermore, soils containing volatile and semi-volatile organic contaminants can be remedied with this method. Similar to pulling liquid through a straw from a tank, the method uses a "slurp" action to extract fluids (free products and soil gases) and extend the free product layer. LNAPLs are raised to the surface by the pumping mechanism in order to separate from water and air. Similar to conventional bioventing systems, this method enables complete removal of contaminants that are easily accessible. However, excessive soil moisture might hinder microbial activity, which is essential for levels of oxygen and the effectiveness of transpiration.

Biostimulation

The practice of adding electron acceptors, electron donors, or nutrients to soil to increase the biodegradation activity of native microbial communities is referred to as "biostimulation." The extra nutrients encourage bacterial populations, which improves their capacity to transform macromolecules into vital traces and microelements. This method of cleanup is economical, ecologically responsible, and efficient. Biostimulation involves supplying extremely concentrated nutrients, such as oxygen, phosphorus, nitrogen, and electron donors, to heavily polluted areas to encourage the breakdown of dangerous and deadly microorganisms.

Bioaugmentation

Bioaugmentation is the process of introducing naturally existing or genetically modified microbial strains into contaminated areas to enhance the breakdown or eradication of undesirable bacteria. To assist in the cleanup of pollution, certain microflora cultures are added and grown. Bioaugmentation employs additional uses extra microorganisms, such as bacteria or archaea, in order to speed up accelerate the breakdown of pollutants, whereas degradation of pollutants, while biostimulation employs nutrients to support indigenous nourish native bacteria. Because mixed bacterial cultures are more adaptable and degradable, and flexible than pure bacterial strains, they are frequently used. Thus, in settings affected by oil spills, bioaugmentation is commonly used widely utilized as a bioremediation alternative option.

Ex-Situ Bioremediation

Excavating and removing pollutants from contaminated regions for additional treatment is a component of ex-situ bioremediation techniques. The contaminants, the site, and the topography of the contaminated area all affect a number of factors, including the cost of treatment, the degree of pollution, the kind and concentration of ex-situ bioremediation techniques, and more. To choose the best ex-situ bioremediation strategy, performance factors are also essential. These techniques cover a variety of strategies developed for particular kinds of pollutants. Soil biopiling, composting, and landfilling are examples of solid-phase soil management practices. However, for soil slurry-phase treatment, the slurry phase needs to be processed in a bioreactor. Numerous hydrocarbon pollutants, such as aliphatic chains, lubricating oils, carbon compounds, ketones, phenols, cresols, polycyclic aromatic hydrocarbons, and semi-volatile organic compounds (SVOCs), may be successfully eliminated by bioremediation in-situ. It is a useful instrument for cleaning up polluted areas where these pollutants are present.

Techniques for Ex-Situ Bioremediation

Ex-situ biodegradation techniques are used to remove the pollutants from the contaminated areas and transport them to a different location for treatment. According to Atlas and Philp (2005), biodegradation methodologies and performance measures serve as the foundation for ex-situ selection. Numerous criteria, such as the kind and density of contaminants, treatment costs, the location and geography of the polluted site, and others, affect the choice of ex-situ bioremediation approaches. Ex-situ biodegradation techniques are used to transport pollutants from contaminated areas to another site for treatment. Atlas and Philp (2005) state that ex-situ selection is based on performance metrics biodegradation techniques. The kind and concentration of contaminants, treatment costs, the location and geography of the polluted site, and other factors all influence the choice of ex-situ bioremediation techniques.

Slurry-Phase Bioremediation

The method begins with an analysis of the pollutant components, the level of biodegradation, as well as physical properties of the soil. On this knowledge, water is added to the poured on the dug dirt that has been dug to create a slurr, in order to make a slurr. The slurry usually contains 5–40% solids by weight, depending on the specific requirements of the bioreactor. The wash water is processed in a slurry bioreactor after the soil has been washed to get rid of fine impurities. Depending on the treatment circumstances, the bioreactor is then filled with the required nutrients, oxygen, microorganisms, and pH adjusters (either acid or alkali). The slurry is dewatered and the resultant liquids are filtered and clarified once biodegradation is finished. Pesticide-damaged soils can be effectively bioremediated using slurry-phase bioreactors.

Solid-Phase Bioremediation

As an ex-situ technique, solid-phase bioremediation entails digging up polluted soil and accumulating it into piles. Organic waste, including leaves, animal manure, agricultural residues, and waste from homes, businesses, and municipalities, is added to these piles. Pipes are carefully placed throughout the heaps to encourage air circulation, which is necessary for ventilation, microbial respiration, and growth. In order to accomplish full cleanup, solid-phase systems need a lot more room and time than slurry-phase methods. Solid-phase treatment methods include land farming, composting, windrows, and biopiles (Kulshreshtha et al., 2014). These methods break down pollutants using natural microbial processes, albeit their effectiveness can vary depending on the type of toxin, aeration, and heap size.

Biopile

Biopiling is a state-of-the-art method that builds up excavated contaminated soil and then adds nutrients and, sometimes, aeration to increase microbial activity and improve bioremediation. Fertilizer, aeration, and irrigation. Among the key components of this method are treatment beds, leachate collection systems, and addition. Because it may biodegrade efficiently and affordably with the right management of nutrients, temperature, and aeration, its popularity is rising (Whelan et al., 2015). Improving the biopile's makeup can increase microbial activity and inhibit resiliency, which will hasten biodegradation. According to Sanscartier (2009), a biopile with enough humidity has a lower ultimate concentration of total petroleum hydrocarbons (TPH) because less biodegradable pollutants volatilize and leach. Furthermore, biopiles can manage substantial amounts of polluted soil in a small space (Chemlal et al., 2013). The contaminated soil needs to be sieved and aerated before processing in order for the biopile to work correctly. Wood fragments, hay, chips, bark, and other chemical-free materials are frequently utilized as bulking agents. Even though biopile systems are superior to conventional ex-situ bioremediation techniques, they do face unique difficulties like intricate engineering, upkeep, and operating expenses.

Windrows

As a bioremediation approach, windrows entail turning heaped soil on a regular basis to promote the microbial breakdown of hydrocarbons found in contaminated soils. Turning the trash piles on a regular basis enhances aeration, which speeds up decomposition and encourages microbial growth. Bioremediation can go more swiftly if aeration is improved, water is given, nutrients and pollutants are supplied consistently, and periodic rotations are used to adapt, convert, and mineralize the toxins. A high rate of hydrocarbon removal has been demonstrated by this technique.

Land Farming

One simple bioremediation technique that is frequently used in agricultural contexts is land farming. Although it is usually carried out as ex-situ bioremediation, in-situ methods may occasionally be used. Because land farming is generally inexpensive and requires little equipment, its success primarily depends on the level of contamination. During ex-situ land cropping, native microorganisms help the dangerous soil contaminants on the soil surface undergo aerobic biodegradation. Areas contaminated with hydrocarbons, especially polyaromatic hydrocarbons, are frequently cleaned with this technique. According to Da Silva et al. (2012) and Cerqueira et al. (2014), the two main mechanisms involved in the degradation of contaminants are biodegradation and volatilization. Ex-situ bioremediation methods, such as land farming, have the benefit of not requiring thorough pre-cleaning inspections of the polluted region, which reduces the time, labor, and cost.

Biofilters

Biofilters are useful instruments in environmental bioremediation because of their proven ability to purify wastewater and lower air pollution (Kumar et al., 2023b; Mishra & Das, 2021, 2022; Mishra et al., 2022b). These filters, which employ a bed of medium where microorganisms degrade pollutants, are among the first technologies created for this purpose. Some materials that can be used as bedding include gravel, stones, tree twigs, manure, organic waste, and synthetic fibers. A particular kind of biofilter called a trickling filter is frequently used to treat wastewater and other liquid effluents. This filter usually consists of a vertical tank with a spherical shape that is composed of aggregate, ceramic, or plastic.

A central vertical tubing that rotatably connects to spray nozzles at the top of the tank allows wastewater to be dispersed throughout it. Wastewater pollutants are broken down by microorganisms in the biofilms that grow on the filter media's surface.

Bioreactors

According to a researcher, a bioreactor is a container where organic materials undergo a variety of biological interactions that result in specific byproducts (Das & Ghosh, 2023c; Priyadarsini & Das, 2025). These reactors can operate in a variety of ways, including batch, fed-batch, sequencing batch, continuous, and multi-stage processes. The perfect circumstances for promoting the growth of bioremediation are provided by the regulated environment found in a bioreactor. Bioreactors have several benefits when it comes to cleaning up polluted soil. These consist of accurate pH level control, reduced turbulence, enhanced aeration, controlled temperature, and appropriate management of substrate and inoculum concentrations. According to another research scholar, the flexible design of bioreactors maximizes biological degradation while minimizing non-biological losses.

9 Merits of Bioremediation

In-Situ Bioremediation

For environmental cleanup projects, in-situ bioremediation is increasingly the preferred technique because of its many potent advantages. Primarily for in-situ cleanup activities, in-situ bioremediation is becoming the favored method due to its numerous strong benefits. Most importantly, it is more cost-effective and less expensive than ex-situ methods since approaches as it removes expenses related to eliminates construction of infrastructure and material transportation of materials. Additionally, by controlling large amounts of pollutants at the source, it streamlines the logistical difficulties associated with material transfer. Furthermore, by managing significant quantities of contaminants at the point of origin, it simplifies the logistical challenges of material transfer. Moreover, by producing byproducts which can be utilized on-site, in-situ bioremediation promotes fosters resource conservation and sustainable practices while yielding major recycling benefits. However, there are certain drawbacks to this tactic. Compared to development and achieves significant recycling advantages. But there are some disadvantages to this strategy. In comparison to other techniques, its reliance on natural processes results in a longer repair period. In addition, seasonal fluctuations in microbial activity, which are made worse by erratic weather patterns and particular pollutant properties, may present difficulties for characteristics can pose challenges to in-situ bioremediation. As a result, the decontamination process may take years, particularly if less easily biodegradable substances are included. Despite involved. Despite these challenges, in-situ bioremediation is still a crucial component and a vital part of environmental restoration initiatives since it provides a sustainable and affordable solution for damaged places.

Ex-Situ Bioremediation

Ex-situ bioremediation is a flexible and very successful strategy for environmental cleaning projects because it offers several advantages over in-situ methods. Its capacity to handle a greater range of contaminants and soil types is a key benefit. In contrast to in-situ methods, this flexibility enables ex-situ technologies to manage a range of contaminated situations more successfully. Ex-situ bioremediation also facilitates regulatory compliance and resource management. Unhindered access to resources is ensured by the availability of infrastructure created especially for remediation, allowing for thorough and efficient cleanup processes. On-site evaluation data can frequently be used to quickly evaluate the viability of ex-situ bioremediation, which facilitates quick decision-making and the timely execution of corrective actions. Additionally, ex-situ bioremediation is adaptable enough to handle different concentrations of polluted material. The procedure can be modified to effectively manage varying amounts of contamination, regardless of how big or small the problem is.

Finally, the capacity to ex-situ homogenizes polluted material.

Ex-situ bioremediation enjoys some advantages over in-situ methods, thus emerging as a flexible and very effective method for environmental cleaning efforts. One such major advantage is its capacity to treat a larger range of contaminants and types of soil. As compared to in-situ technologies, this adaptability allows ex-situ technologies to better deal with a variety of contaminated conditions. Free access to resources is guaranteed by the presence of infrastructure developed specifically for remediation, which enables efficient and effective cleanup operations. On-site data can often be utilized to rapidly assess the feasibility of ex-situ bioremediation, enabling rapid decision-making and the prompt implementation of corrective measures. Moreover, ex-situ bioremediation is versatile to deal with various levels of contaminated material. The process can be adjusted so that it can effectively treat various amounts of contamination, no matter how large or small the issue is. Lastly, the ability to ex-situ homogenize contaminated material. Figure 2 describes the different types of bioremediation techniques.

Waste-to-Energy Conversion

By using microorganisms to decompose organic wastes and produce biofuels, bioremediation techniques are essential for turning trash into energy (Priyadarsini & Das, 2024).

Anaerobic digestion is one of the most used techniques, where in an oxygen-free environment, microorganisms break down organic waste, including food scraps, agricultural wastes, and sewage sludge (Das et al.,). Biogas, which is mostly made up of methane, is produced by this process and can be utilized for cooking, heating, or power generation. Digestate, the byproduct, also functions as a fertilizer that is high in nutrients.

Microbial fuel cells (MFCs), which use microorganisms to break down organic materials and produce power, are another cutting-edge method. By moving electrons from microbial metabolic activities to an electrode, these fuel cells provide an electrical current that has the twin advantages of treating wastewater and producing electricity (Das & Ghosh, 2023a; Ghosh et al., 2023; Nag et al., 2023). MFCs have

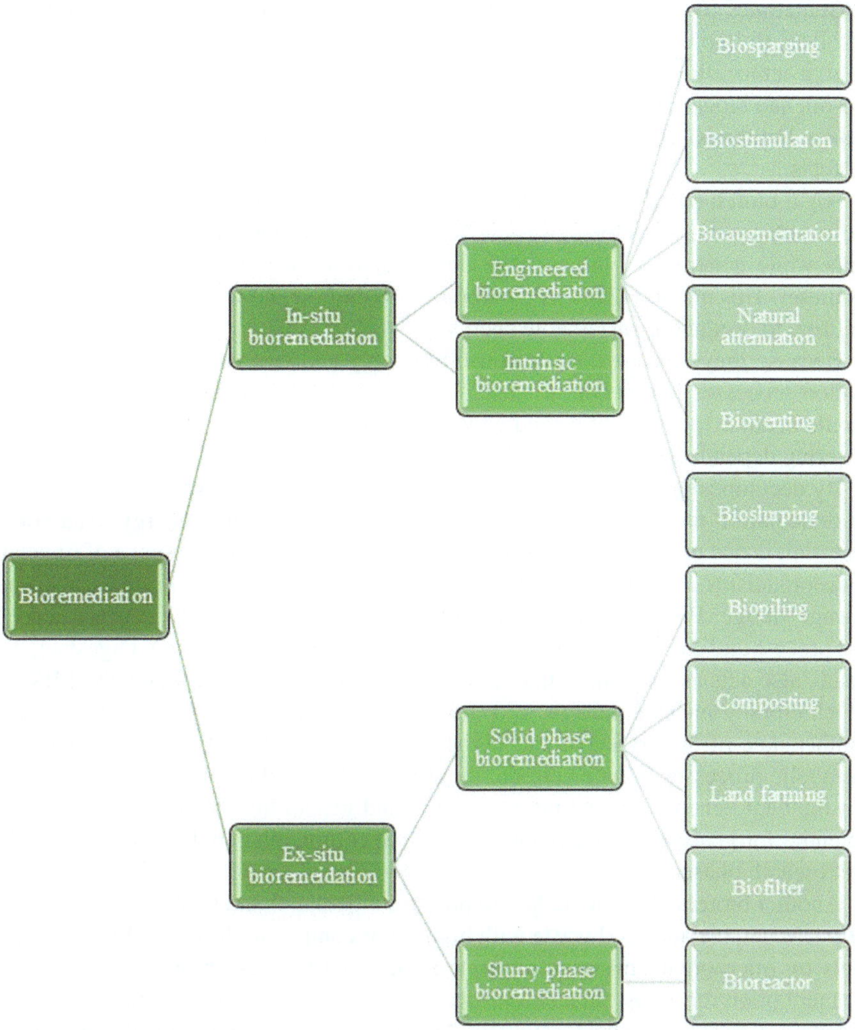

Fig. 2 Types of bioremediation techniques

promise for producing sustainable energy, but large-scale applications are still at the research stage.

Another bioremediation technique for creating bioethanol and biobutanol is fermentation. Food scraps and crop wastes that are high in sugar are fermented into alcohol-based fuels by microorganisms like bacteria and yeast. These biofuels are a cleaner alternative for industrial and transportation applications since they may be blended with or used in place of conventional fossil fuels.

Sometimes microbial pretreatment is used in conjunction with gasification and pyrolysis to improve the breakdown of complex organic waste prior to conversion at high temperatures. Syngas, a combination of hydrogen and carbon monoxide, bio-oil, and biochar is produced by these thermochemical processes and can be used to heat buildings, generate power, or be further processed into more sophisticated biofuels.

Algal bioremediation is also becoming a more successful method for producing biofuel. Algae may develop in wastewater, absorbing carbon dioxide and other pollutants while generating biomass that can be converted into biogas, bioethanol, or biodiesel. This method helps with carbon sequestration and wastewater treatment in addition to producing sustainable energy.

Because they convert trash into useful energy sources, each of these bioremediation processes is essential to sustainable waste management. These techniques should become more effective and widely used as research and technology develop, lowering dependency on fossil fuels and tackling environmental pollution.

By decomposing organic wastes using microorganisms and generating biofuels, bioremediation methods are crucial in converting garbage into energy. Anaerobic digestion is one of the most employed methods, whereby, in the absence of oxygen, microorganisms degrade organic waste, such as food waste, agricultural wastes, and sewage sludge. Biogas, which is primarily composed of methane, is generated by this process and can be used for cooking, heating, or power generation. Digestate, the waste, also acts as a fertilizer that is rich in nutrients. Microbial fuel cells (MFCs), which employ microorganisms to degrade organic matter and generate power, are another state-of-the-art technique. Through the transfer of electrons from microbial metabolic processes to an electrode, these fuel cells yield an electrical current that has the dual benefits of treating wastewater and generating electricity. MFCs hold potential for generating sustainable energy, but commercial applications are still in the research phase.

Another bioremediation method to produce bioethanol and biobutanol is fermentation. Vegetable and food waste with high sugar content are fermented into alcohol fuels by microorganisms such as bacteria and yeast. These biofuels constitute a cleaner option for industrial and transportation uses because they can be mixed with or substituted for traditional fossil fuels. At times, microbial pretreatment is combined with gasification and pyrolysis to enhance the degradation of complex organic waste before conversion at high temperatures. Syngas, a mixture of hydrogen and carbon monoxide, bio-oil, and biochar are generated by these thermochemical processes and can be utilized for warming buildings, producing energy, or being processed further into more complex biofuels. Algal bioremediation is increasingly becoming an efficient way of manufacturing biofuel. Algae can grow in wastewater, taking up carbon dioxide and other pollutants as they produce biomass that can be used to manufacture biogas, bioethanol, or biodiesel. It assists in the sequestration of carbon as well as treating wastewater while also generating renewable energy. Since they transform garbage into valuable sources of energy, each of these bioremediation processes is critical to sustainable waste management. These methods should improve and become more prevalent as research and technology advance, reducing reliance on fossil fuels and addressing environmental pollution.

10 Conclusion

The ecosystem and the quality of life are significantly impacted by environmental degradation. Many strategies have been used over time to lessen pollution, and one of the more promising ones created is bioremediation. This is the use of microorganisms to detoxify toxic substances found in contaminated wastewater as well as heavy metals. In addition to bacteria, alternative bioremediation techniques include vermiculture, which uses duckweed, mushrooms, earthworms, and even genetically modified organisms like endophytic fungi, which have demonstrated the ability to remove contaminants, including plastic waste. These approaches are a component of an expanding collection of strategies that address environmental contamination by combining physical, chemical, and biological processes. Bioremediation is particularly prized for its capacity to detoxify solid waste, lower environmental impact, and affordability. Bioremediation provides an effective and environmentally friendly waste management solution when paired with additional technologies such as chemical treatment, pyrolysis, or plasma incineration. Although there are still a number of issues, such as site-specific conditions and unclear byproduct results, there is still interest in tackling environmental contamination.

Environmental degradation has a significant effect on the ecosystem and quality of life. Various strategies have been applied over time to reduce pollution, and one of the more hopeful ones developed is bioremediation. This refers to the utilization of microorganisms to detoxify harmful contaminants present in contaminated wastewater as well as heavy metals. Besides bacteria, other bioremediation methods involve vermiculture, which employs duckweed, mushrooms earthworms, and even genetically engineered organisms such as endophytic fungi, which have shown capability to eliminate pollutants, including plastic waste. These methods are part of an increasing array of strategies that solve environmental pollution through the integration of physical, chemical, and biological processes. Bioremediation is especially valued for its potential to detoxify solid waste, reduce environmental load, and cost-effectiveness. Bioremediation is a suitable and green waste management solution when combined with other technologies like chemical treatment, pyrolysis, or plasma incineration. Though there are still several issues, like site-specific conditions and uncertain byproduct outcomes, there is still interest in addressing environmental pollution.

References

Bala, K., Anshul, Verma, S., Dogra, A., Das, A. P., Kumar, R., & Nandini. (2024a). Sustainable management for urban plastic waste generation. In *Sustainable management of urban plastic waste through circular economic approaches* (pp. 211–236). https://doi.org/10.1201/9781003465447-13

Bala, K., Dogra, A., Lata, P., & Das, A. P. (2024b). Discussing sources and biological-chemical remediation approaches for microplastic pollution. In *Sustainable microbial technology for synthetic and cellulosic microfiber bioremediation* (pp. 37–59). https://doi.org/10.1007/978-3-031-62600-5_3

Behera, I. D., Nayak, M., Biswas, S., Meikap, B. C., & Sen, R. (2021). Enhanced biodegradation of total petroleum hydrocarbons by implementing a novel two-step bioaugmentation strategy using indigenous bacterial consortium. *Journal of Environmental Management, 292*, Article 112746. https://doi.org/10.1016/J.JENVMAN.2021.112746

Behera, I. D., Nayak, M., Mishra, A., Meikap, B. C., & Sen, R. (2022). Strategic implementation of integrated bioaugmentation and biostimulation for efficient mitigation of petroleum hydrocarbon pollutants from terrestrial and aquatic environment. *Marine Pollution Bulletin, 177*, Article 113492. https://doi.org/10.1016/J.MARPOLBUL.2022.113492

Bhattacharjee, J., Mishra, S., & Das, A. P. (2021). Recent advances in sensor-based detection of toxic dyes for bioremediation application: A review. *Applied Biochemistry and Biotechnology, 2021*, 1–20. https://doi.org/10.1007/S12010-021-03767-7

Borchert, E., Hammerschmidt, K., Hentschel, U., & Deines, P. (2021). Enhancing microbial pollutant degradation by integrating eco-evolutionary principles with environmental biotechnology. *Trends in Microbiology, 29*(10), 908–918. https://doi.org/10.1016/J.TIM.2021.03.002

Ceballos-Escalera, A., Pous, N., Bañeras, L., Balaguer, M. D., & Puig, S. (2024). Advancing towards electro-bioremediation scaling-up: On-site pilot plant for successful nitrate-contaminated groundwater treatment. *Water Research, 256*, Article 121618. https://doi.org/10.1016/J.WATRES.2024.121618

Chemlal, R., Abdi, N., Lounici, H., Drouiche, N., Pauss, A., & Mameri, N. (2013). Modeling and qualitative study of diesel biodegradation using biopile process in sandy soil. *International Biodeterioration & Biodegradation, 78*, 43–48. https://doi.org/10.1016/J.IBIOD.2012.12.014

Das, A. P., & Ghosh, S. (2018). Bioleaching of Manganese from mining waste materials. *Materials Today: Proceedings, 5*(1), 2381–2390. https://doi.org/10.1016/J.MATPR.2017.11.459

Das, A. P., & Ghosh, S. (2023a). Evolution of biomining technology. In *Manganese mining microorganisms* (pp. 5–19). https://doi.org/10.1016/B978-0-12-822147-1.00012-1

Das, A. P., & Ghosh, S. (2023b). Microorganisms in manganese biomining processes. In *Manganese mining microorganisms* (pp. 55–61). https://doi.org/10.1016/B978-0-12-822147-1.00013-3

Das, A. P., & Ghosh, S. (2023c). Mixed culture bioleaching: an insight into manganese biomining process and efficacy. In *Manganese mining microorganisms* (pp. 129–141). https://doi.org/10.1016/B978-0-12-822147-1.00001-7

Das, A. P., Ghosh, S., Mohanty, S., & Sukla, L. B. (2015). Advances in manganese pollution and its bioremediation. In L. B. Sukla, N. Pradhan, S. Panda, & B. K. Mishra (Eds.), *Environmental microbe biotechnology* (1st ed., pp. 313–328). Springer. https://doi.org/10.1007/978-3-319-19018-1_16

Das, A. P., Mohanty, A. K., van Hullebusch, E. D., Figueiredo, G., Das, A. P., Mohanty, A. K., van Hullebusch, E. D., & Figueiredo, G. (2023). Bioremediation as an emerging technology for the removal of synthetic microplastic pollutants from marine ecosystem. *MarPB, 194*, Article 115297. https://doi.org/10.1016/J.MARPOLBUL.2023.115297

Das, A. P., Nayak, B., & Behera, I. D. (2024a). Applications of microorganisms in advanced geomicrobiological and biotechnological exploration. *Geomicrobiology Journal, 41*(4), 309–311. https://doi.org/10.1080/01490451.2024.2327059/ASSET//CMS/ASSET/DE2A1F67-08C1-475A-ABC7-2B3F4C9E8BCF/01490451.2024.2327059.FP.PNG

Das, A., Dey, S., Bala, K., Gnanasekaran, R., & Das, A. P. (2024b). Microbial bioremediation technology for sustainable treatment and management of synthetic microfiber waste. In *Sustainable microbial technology for synthetic and cellulosic microfiber bioremediation* (pp. 221–238). https://doi.org/10.1007/978-3-031-62600-5_12

Das, A., Dey, S., Gacem, A., Yadav, K. K., Nath, H., Ram, D. K., & Das, A. P. (2025). Revolutionizing solid hydrocarbon sourcing: Plastic's journey from waste to upcycled treasure. *Environmental Science and Engineering, Part, F218*, 331–351. https://doi.org/10.1007/978-3-031-81615-4_14

Das, A., & Ghosh, S. (2022). Role of microorganisms in extenuation of mining and industrial wastes. *Geomicrobiology Journal, 39*, 173–175. https://doi.org/10.1080/01490451.2022.2038953

Das, A., Ray, S. K., Mohanty, M., Mohanty, J., Dey, S., & Das, A. P. (2024c). Ecotoxicity of microplastic wastes and their sustainable management: A review. *Environmental Chemistry and Ecotoxicology, 6*, 144–152. https://doi.org/10.1016/J.ENCECO.2024.05.003

Dey, S., Bissoyi, A., Koteswara Reddy, G., & Das, A. P. (2024a). Synthetic microfibers: An urban emerging pollutant demanding for sustainable environmental solutions. In *Sustainable microbial technology for synthetic and cellulosic microfiber bioremediation* (pp. 107–121). https://doi.org/10.1007/978-3-031-62600-5_7

Dey, S., Das, A., Mallick, K., Sahu, A., & Das, A. P. (2023). *Environmental petroleum waste: Pollution, toxicity, sustainable remediation* (pp. 159–175). https://doi.org/10.1007/978-3-031-48220-5_7

Dey, S., Dash, A., & Das, A. P. (2024b). Bioremediation as a sustainable and circular solution for mining waste management and bioenergy generation. In *Sustainable management of mining waste and tailings: A circular economy approach* (pp. 107–119). https://doi.org/10.1201/978 1003442455-6/BIOREMEDIATION-SUSTAINABLE-CIRCULAR-SOLUTION-MINING-WASTE-MANAGEMENT-BIOENERGY-GENERATION-SUDESHNA-DEY-AKANSHYA-DASH-ALOK-PRASAD-DAS

Dey, S., Dash, N., Ray, S. K., Priyadarsini, S., Chhotaray, C., & Das, A. P. (2024c). Occurrence and quantification of tire wear plastics pollutants from highway city dust of Bhubaneswar city of Odisha. In *Sustainable microbial technology for synthetic and cellulosic microfiber bioremediation* (pp. 77–89). https://doi.org/10.1007/978-3-031-62600-5_5

Geed, S. R., & Verma, J. S. (2024). Integrated approach for efficient crude oil bioremediation: Bacterial consortium development, mathematical modelling and scalable bioprocess design. *Fuel, 358*, Article 130260. https://doi.org/10.1016/J.FUEL.2023.130260

Ghosh, S., & Das, A. P. (2020). Microbial metagenomics: current advances in investigating microbial ecology and population dynamics. In S. K. Nayak & B. B. Mishra (Eds.), *Frontiers in soil and environmental microbiology* (1st ed., pp. 247–254). CRC Press.

Ghosh, S., Tripathy, B., Dey, S., & Das, A. P. (2023). Comparative investigation of fungal and bacterial manganese biomining mechanisms. In *Manganese mining microorganisms* (pp. 99–115). https://doi.org/10.1016/b978-0-12-822147-1.00016-9

Gundlapalli, M., Sivagami, K., Gopalakrishnan, M., Harshini, P., Janjaroen, D., & Ganesan, S. (2024). Biodegradation of low molecular weight polycyclic aromatic hydrocarbons in soil: Insights into bacterial activities and bioremediation techniques. In *Sustainable chemistry for the environment* (Vol. 7). Elsevier B.V. https://doi.org/10.1016/j.scenv.2024.100146

Guo, W., Ren, H., Jin, Y., Chai, Z., & Liu, B. (2024). The bioremediation of the typical persistent organic pollutants (POPs) by microalgae-bacteria consortia: A systematic review. *Chemosphere, 355*, Article 141852. https://doi.org/10.1016/J.CHEMOSPHERE.2024.141852

Jiang, C., Wang, Y., Chen, Y., Wang, S., Mu, C., & Shi, X. (2024). The phytoremediation potential of 14 salix clones grown in Pb/Zn and Cu mine tailings. *Forests, 15*(2). https://doi.org/10.3390/f15020257

Kanungo, J., Sahoo, T., Bal, M., & Behera, I. D. (2024). Performance of bioremediation strategy in waste lubricating oil pollutants: A review. *Geomicrobiology Journal, 41*(4), 360–373. https://doi.org/10.1080/01490451.2023.2245395

Yang, L., Wang, J., Yang, Y., Li, S., Wang, T., Oleksak, P., Chrienova, Z., Wu, Q., Nepovimova, E., Zhang, X., & Kuca, K. (2022). Phytoremediation of heavy metal pollution: Hotspots and future prospects. *Ecotoxicology and Environmental Safety, 234*, Article 113403. https://doi.org/10.1016/J.ECOENV.2022.113403

Zapata-Peñasco, I., Avelino-Jiménez, I. A., Mendoza-Pérez, J., Vázquez Guevara, M., Gutiérrez-Ladrón de Guevara, M., Valadez- Martínez, M., Hernández-Maya, L., Garibay-Febles, V., Fregoso-Aguilar, T., & Fonseca-Campos, J. (2024). Environmental stressor assessment of hydrocarbonoclastic bacteria biofilms from a marine oil spill. *Biotechnology Reports, 42*, e00834. https://doi.org/10.1016/J.BTRE.2024.E00834

Innovative Strategies for Sustainable Municipal Waste Management

Dolly Bhue, Sailaja Priyadarsini, Amel Gacem, Aditya Kishor Das, Kiran Bala, Maha Awjan Alreshidi, Krishna Kumar Yadav, and Alok Prasad Das

1 Introduction

Municipal waste, also referred to as urban waste, is defined as "waste from households and waste from other sources, such as retail, administration, education, health services, accommodation and food services, and other services and activities, which are similar in nature and composition to waste from households." 2018/851, an EU directive (Dey et al., 2023; Mishra et al., 2023). It represents about 27% of the total garbage generated in the EU (Fabio Fiorillo, 2024). Urbanization and population growth are the main causes of the significant increase in trash generation throughout time (Priyadarsini & Das, 2024, 2025). There are serious threats to the environment and human health from improper waste management techniques (2023a; Ghosh et al.,

D. Bhue · S. Priyadarsini · A. P. Das (✉)
Department of Life Sciences, Rama Devi Women's University, Bhubaneswar, Odisha, India
e-mail: alok1053@gmail.com

A. Gacem
Department of Physics, Faculty of Sciences, University 20 Août 1955, Skikda, Algeria

A. K. Das
Chemistry Department, ITER, Siksha 'O' Anusandhan University, Bhubaneswar, India

K. Bala
Department of Zoology, Deshbandhu College, University of Delhi, Delhi, India

M. A. Alreshidi
Department of Chemistry, College of Science, University of Ha'il, Ha'il, Saudi Arabia
e-mail: m.alrashedi@uoh.edu.sa

K. K. Yadav
Department of Environmental Science, Parul Institute of Applied Sciences, Parul University, Vadodara, Gujarat, India

Environmental and Atmospheric Sciences Research Group, Scientific Research Center, Al-Ayen University, Nasiriyah, Thi-Qar, Iraq

A. P. Das et al. (eds.), *Sustainable Municipal Waste Management with a Zero Waste Approach*, Sustainable Environmental Waste Management Strategies,
https://doi.org/10.1007/978-3-032-05950-5_12

2018; Rapeepan Pitakaso, 2024). By 2050, trash production is expected to expand by 70% worldwide, with developing countries accounting for a large share of this increase, according to World Bank research (Asmae El jaouhari, 2025). In 2016, it was estimated that the global annual generation of waste was around 2.01 billion metric tons. The majority of this trash was produced in Asia, with North America, Europe, and Africa coming in second and third. By 2050, the previously estimated amount is anticipated to rise to 3.4 billion metric tons. A large amount of the country's solid waste is either burned or dumped in unprotected regions due to inadequate infrastructure for collection and disposal, damaging the soil, water, and atmosphere (Das & Ghosh, 2022; Mishra & Das, 2023; Mohanty et al., 2018). Disease-carrying insects and rodents have increased as a result of the accumulation of trash in public areas, which helps spread illnesses like cholera and dengue fever. The tourism business has been adversely affected by improper waste management, as seen by the contamination of many well-known tourist destinations by trash and litter (Das et al., 2024d; Mishra et al., 2019b). The state of garbage management in big cities is particularly worrisome. Landfills are overflowing, rubbish is accumulating on sidewalks and streets, and lakes and streams are becoming poisoned as a result of the large amount of waste produced in metropolitan areas and the absence of infrastructure for collection and disposal. The management of solid waste is an essential part of sustainable development since it has a big influence on people's everyday lives, the environment, and the accomplishment of international goals (Mishra & Das, 2021; Mishra et al., 2024a; Tripathy et al., 2024). Addressing urgent global issues and advancing sustainable practices are made easier with the help of the United Nations' Sustainable Development Goals (SDGs). Water security, climate change mitigation, responsible consumption and production, ecosystem preservation, sustainable urban and community development, and health promotion are just a few of the Sustainable Development Goals (SDGs) that highlight how important efficient solid waste management is to achieving these goals (Mujtaba,).Waste management is the process of collecting, moving, monitoring, disposing of, and recycling waste produced by both individuals and corporations. The environment and public health are protected by this procedure, which guarantees that trash is handled securely from the point of origin until it is properly disposed of or repurposed. Due to its significant effects on the environment, public health, and public investments, waste management is a primary duty of local governments and occupies a prominent place on the urban policy agenda (Bala et al., 2024a; Kumar et al., 2023; Mishra & Das, 2022). Additionally, it poses a significant obstacle for EU institutions. Due to the lofty goals set by recent EU directives for municipal garbage collection and recycling, significant investments in infrastructure, technologies, capacities, and procedures will be required soon. Efficient waste management is a complex issue that is impacted by a wide range of laws, rules, sociocultural settings, environmental circumstances, financial considerations, and available resources. Waste management involves many stakeholders, each of whom has particular interests and worries. Urban waste management and the intricate relationships between various parties and processes must be coordinated by local governments to protect the environment and preserve resources (Das & Ghosh, 2018; Ray et al., 2024; Sahu et al., 2024). Achieving aggressive recycling goals entails a careful balancing act between expenses and earnings, compliance with EU legislation, public education, and smart investments in technology and infrastructure. They employ several laws and initiatives to minimize landfills, boost recycling, and decrease waste to guarantee effective waste management (The Waste Framework Directive 2008/98/EC). Stakeholders in the waste management sector include private businesses that oversee the collection, transportation, and disposal of trash. Their services either compete with or supplement garbage management programs sponsored by the government or municipalities. While they may participate in public–private partnerships (PPPs) or bid against municipalities for service contracts, these businesses typically function independently of government waste management organizations. Households produce waste and are worried about convenient and prompt rubbish collection services. To cut waste and lessen their influence on the environment, they might be interested in recycling initiatives and actively participate in coproduction activities. Businesses generate waste, and they may have an interest in following waste management laws and economically disposing of waste. Additionally, some companies might want to decrease trash production and increase sustainability. Waste management laws are established and enforced by government organizations at all levels, including local, regional, and federal. By making sure that trash legislation (The trash Framework Directive 2008/98/EC) are followed, they hope to safeguard the environment and public health (Fabio Fiorillo, 2024).For waste-to-energy applications, which are essential to MSW management systems, to significantly influence the development of a low-carbon civilization, precise, and reliable MSW characterization data is required. Therefore, the current study is to evaluate the fluctuation in MSW's composition and qualities (physical and chemical) as well as characterize it on a seasonal basis. It also seeks to provide a novel waste characterization method that uses precise data in an integrated waste-to-energy facility manner. The approach taken in this study offers a customized, workable solution that improves the precision and applicability of waste characterization for better waste-to-energy conversion processes, in contrast to current methods that do not sufficiently account for the variability in waste composition typical of developing countries. The selected methodology's novel feature is the way it combines improved sample precision with a regionally specific ASTM D5231-92 standard modification. The development of more efficient and sustainable waste management techniques is aided by this adaptation, which more properly captures seasonal variations in waste composition. By doing this, our study closes a large gap in the literature and provides a model that may be applied to comparable settings around the world, supporting larger initiatives in sustainable waste management (Karim Emara, 2024). Life Cycle Assessment (LCA) studies focuses at the most significant issues that come up while choosing different waste treatment solutions, and it is widely used to identify and minimize environmental impacts. When used to waste management, Life Cycle Assessment (LCA) is a useful method for reducing environmental impacts since it highlights the serious consequences of ineffective waste management systems. By collecting and analyzing inputs and outputs, life cycle assessment (LCA) assesses a product system's environmental impact over the course of its entire life cycle. However, life cycle assessment (LCA), which can be computer-based with the appropriate software, can also be an important environmental organizational tool within a defined system boundary. It helps identify environmental issues and their potential effects over the course of waste lives. The Life Cycle Assessment (LCA) tool is used to assess the life cycle of products and waste management systems for recycling and waste disposal. An LCA assessment is essential in this case before deciding on any technology, rules, or implementation tactics. Software tools that can be used to calculate LCA performances include SimaPro, GaBi, IWM-2 (integrated waste management II), WRATE (waste resources assessment tool for the environment), EASE WASTE (environmental assessment of solid waste systems and technologies), ORWARE (organic waste research), and WISARD (waste-integrated systems for assessment of recovery and disposal). Developing countries like India primarily employ the life cycle assessment (LCA) technique for the management of environmental health and safety, regulatory impacts, and the effective use of resource allocation throughout their existence. LCA helps identify connected systems and characteristics that can lessen their effects and is used to evaluate environmental impact assessments pertaining to climate change, toxicity potential, and carcinogenic behavior. However, in low- to middle-income nations, where implementing suitable solid waste management practices is typically necessary to lessen global environmental consequences, this life cycle analysis technique should be promoted. To meet the goals of sustainable waste management, it is essential to strategically integrate various waste management technologies (Anchal Sharma, 2023). The management of municipal waste has emerged as a crucial concern for cities worldwide as urbanization and population growth persist (Ghosh & Das, 2017; Pradhan et al., 2023). Conventional waste management techniques are no longer adequate to handle the scope of the issue; instead, creative solutions must be investigated and put into practice. This chapter looks at several innovative and sustainable approaches to municipal garbage management, with an emphasis on new laws, regulations, and tactics that can lower waste production, increase recycling rates, and lessen environmental effects. Through an analysis of existing practices and an investigation of novel ideas, this study will offer valuable insights into how communities might transition to a more sustainable future.

2 Types of Municipal Waste

On a very conservative estimate, at least 33% of the 2.01 billion tons of municipal solid garbage produced worldwide each year is not managed in an environmentally responsible way. Waste production per person per day varies greatly, ranging from 0.11 to 4.54 kilos, with an average of 0.74 kg worldwide. Despite making up only 16% of the world's population, high-income nations produce over 34% of the world's garbage, or 683 million tons. Municipal waste encompasses the discarded materials generated by households, commercial establishments, institutions and small scale industries within urban and semi urban areas.

Organic Waste

Organic garbage, which originates from a number of sources, including food, animal and human waste, garden and wood products, and metals, papers, plastics, and glasses, makes up around half of the waste produced globally. The increase in organic solid waste (OSW) generation from municipal, industrial, medical, and agricultural sources has led to a global waste management crisis, which is ascribed to inadequate treatment techniques. Waste is often disposed of in landfills, which creates sustainability issues such as pollution of the environment, greenhouse gas emissions, and degradation of land quality (Aliakbar Dehghan, 2025).

Plastic Waste

Certainly, plastics play a crucial role in the goods utilized in contemporary society. Plastics are utilized in practically every part of daily living, such as apparel, accessories, food, and medical equipment (Dey et al., 2024b; Priyadarshini et al., 2024a; Singh Jadaun et al., 2022). The packaging of everyday items accounts for well over one-third of plastic usage, with over one-third coming from building supplies like vinyl cladding and plastic pipes (Jagatee et al., 2024; Mishra et al., 2020; Singh et al., 2020). Over 40 million tons of polymers are converted into textile fiber, including polyester, acrylic, and nylon, which are subsequently utilized in the global apparel sector. Poly-cotton blend clothing uses a lot of PET plastic, as do many other high-performance textiles that are made solely of plastics, such as nylons, polyethers, and fluoropolymers (Dey et al., 2024a; Mishra et al., 2024b; Paul et al., 2024). Because they are made to be used only once, these polymers are known as single use plastics (SUPs). The consumption trends of the five most widely used plastics across a range of industries appear to be steady in the industrialized world. The patterns of plastic use, however, may differ slightly in poorer countries (Mishra et al., 2021, 2022a, 2022b; Satapathy et al., 2024). For instance, in India, 42% of resin use is attributable to the packaging industry. Plastics are utilized extensively in the production of toys, furniture, and industrial uses. Due to lower unit costs and better performance criteria, plastics are increasingly being utilized in developed nations to replace materials including paper, metals, wood, and glass (Das et al., 2024b, 2024c; Mishra et al., 2019a; Priyadarshini et al., 2024b). Plastics have gained a lot of popularity as a commodity and have shown themselves to be adaptable in a wide range of designs and shapes (2024a; Das et al., 2023; Mallick et al., 2024). Among the many unique properties of plastics are their versatility in terms of temperature, resistance to light and chemicals, and capacity to be manipulated as a hot melt (Bala et al., 2024b; Mishra et al., 2022a; Mohanty et al., 2024). The earth would utilize as least 1800 million tons of plastics annually by 2050 if this pattern continues, albeit at a moderate average expansion rate of 5%. This anticipated development is mostly the result of rising consumer demand for plastics (Nalini Singh Chauhan, 2022).

Electronic Waste (E-waste)

Technological progress has led to a major improvement in living conditions with the quick proliferation of electronic equipment such as TVs, laptops, mobile phones, and washing machines (Prabhakar et al., 2019; Raj & Das, 2023). But worries about electronic garbage, or "e-waste," and its by-products have also increased consequently of this expansion. A growth in e-waste, or abandoned electronic devices like phones, TVs, and appliances, is a result of both the quick development of technology and consumers' thirst for the newest devices (Priyadarsini & Das, 2024). The amount of e-waste produced worldwide increased by 21% between 2014 and 2019, reaching 53.6 million tons, and is expected to rise to 74 million tons by 2030. The corporate and public sectors are responsible for almost 70% of the e-waste, according to Kumar's analysis. Household garbage is freely abolished by local rag pickers, where it is managed incorrectly and causes multiple negative health impacts on the recyclers, even if large institutions and government agencies are somewhat required to dispose of their rubbish to the appropriate dealer. A variety of commodities, including glass, metals, and polymers, can be recovered in a systematic manner from e-waste, making it a raw material resource. Hazardous, non-hazardous, and precious metals are all present in e-waste. Common non-hazardous metals are iron (Fe), manganese (Mn), tin (Sn), copper (Cu), and aluminum (Al). However, e-waste can also contain toxic materials like sulfur (S), cadmium (Cd), beryllium oxide (BeO), brominated flame retardants, lead (Pb), lithium (Li), nickel (Ni), and mercury (Hg); chlorofluorocarbon; organic materials; rare earth elements like yttrium (Y), lanthanum (La), cerium (Ce), neodymium (Nd), gallium arsenide (GaAs), and zinc sulfide (ZnS); halogenated compounds; and radioactive materials like americium; and toner dust (Das & Ghosh, 2023; Das et al., 2011; Ghosh & Das, 2018; Mohanty et al., 2016). Additionally, the non-metallic portion of e-waste includes materials like wood, glass, ceramics, and plastic. More than 25% of the metallic fraction is composed of metals like iron and nickel, with the remainder being made up of glass, ceramics, polymers, and other metals. The demand for the base metals Zn, Pb, Cu, and Ni, as well as the rare earth minerals Yt, La, Ce, and Nb, as well as the precious metals Pt, Au, Ag, and Pd, has grown dramatically in recent years due to electronic applications. Over 60 distinct metals, including basic, precious, rare earth, and valuable metals, are commonly found in e-waste, which is valued for its overall metal composition. The value share of precious metals (Pt, Pd, Au, and Ag) in London was reported to range from 85% (PCBs) to 93% (mobile phones), while less than 10% of base metals such as Fe, Al, and Cu were used. E-waste contributes to toxic heavy metals and organic pollutants, both of which are extremely harmful to human health (Das et al., 2015a, 2015b; Das & Ghosh, 2018; Ghosh et al., 2023b). Humans are greatly exposed to organic contaminants like flame retardants, PCBs (polychlorinated biphenyls), PAHs (polycyclic aromatic hydrocarbons), and dioxin-related substances through outdated processing techniques like manual disassembly, open burning, and plastic recycling. The lipid content and DNA of humans are oxidatively damaged by PAHs (Kansal, 2002) as well as being carcinogenic (Zhechen Zhang, 2024). Breast milk can contain PCBs, exposing nursing new-borns to these dangerous substances at a crucial developmental stage. Potential negative consequences on new-borns, such as compromised immunological function, hormone imbalances, and poor neurodevelopment, have been connected to this exposure. While long-term exposure to high dioxin levels is associated with impairments to the immunological, neurological, endocrine, and reproductive systems, short-term exposure can result in skin lesions and decreased liver function. Dioxins are considered a recognized human carcinogen and are most dangerous to developing fetuses and new-borns. Because of their high toxicity, precautions must be taken to reduce exposure (Kaniz Fatema, 2025).

Biomedical Waste

India produces a huge volume of biomedical waste (BMW) every day. Hospital visits by people of different ages, sexes, ethnicities, and religions contribute to the growing diversity and abundance of BMW production. About 1.5–2 kg per bed per day are produced by BMW in India. Sharps, laboratory trash, anatomical waste, and other garbage are included in BMW, which can be lethal if not properly separated. Furthermore, improper separation of waste plastic, a recyclable and cytotoxic substance, could be detrimental to our environment (Khatun et al., 2023; Sadasivuni et al., 2019). BMW was not regarded as a danger to the environment or to people in the past. Fears of coming into contact with dangerous microbes such the hepatitis B virus (HBV) and the human immunodeficiency virus (HIV) led individuals to think about the possible risks of BMW in the 1980s and 1990s (Khatun et al., 2023). Because it contains possible viruses or other disease-causing microorganisms, BMW is dangerous. Blood bags, cotton swabs, needles, bedding, dressing supplies, and human samples are among the objects that contain it. This makes BMW's bad management a public health concern. Additionally, specific actions by the public are required to reduce the growing environmental harm brought on by BMW management's carelessness (Himani & Bansod, 2023).

Textile Waste

One of the biggest sources of garbage in the world is the textile sector, which accounts for a sizable amount of municipal solid waste globally. According to the EEA, the sector has more than doubled its output in recent decades, yet the amount of time that clothing is used before being thrown away has dropped by roughly 40%. Figure 1; Bar table showing Rate of Municipal Waste Generation.

Consuming textiles has a direct correlation with a number of harmful environmental effects during the production, use, and disposal processes. An estimated 1.5 trillion liters of water are used annually by the garment sector alone. A study by the European Economic Community (2023) found that the average amount of textiles

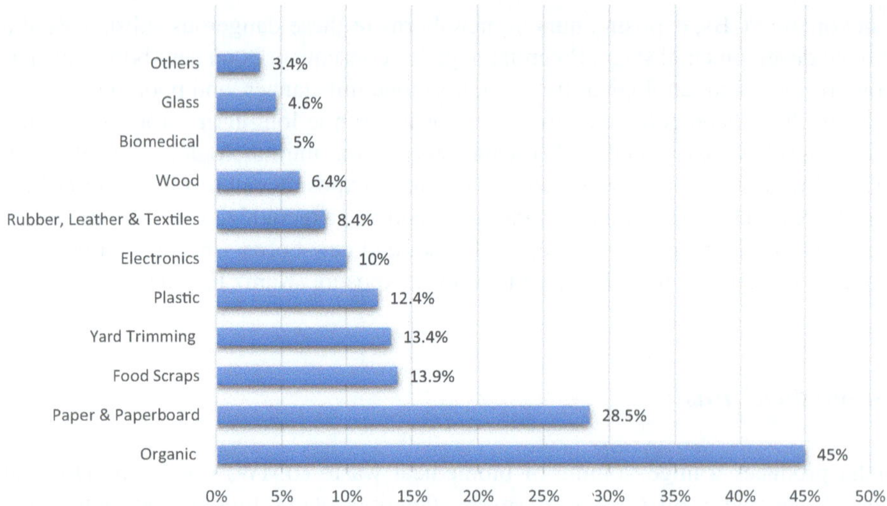

Fig. 1 Bar table showing rate of municipal waste generation

consumed in Europe uses 400 m^3 of land, 9 m^3 of water, and 391 kg of raw materials per person. With an approximate carbon footprint of 270 kg, the textile industry ranks fourth in terms of its detrimental life cycle impact on the environment and climate change, behind the transportation, housing, and food sectors. Concerns about the ecology are further aggravated by synthetic microfibers released into the atmosphere during use. Approximately 16–35% of micro plastics discharged worldwide, or 8% in Europe, are caused by washing synthetic textiles. Textile waste management is a challenge in addition to the effects on the stages of manufacturing and use. 12% of textile waste is recycled, one percent is repurposed, and seventy-three percent is landfilled or burned. Just 22% of post-consumer textile waste in Europe is collected separately for recycling or reuse. Cleaning cloths, filling for mattresses, and insulation materials are the most common uses for recycled textiles. Since technology to recycle clothing into virgin fibers are still in their infancy, just 1% of used clothing is turned into new clothing (Inna Pitak, 2025).

3 Strategies for Sustainable Municipal Waste Management

MCDM Techniques

The management of municipal waste is a significant and complex issue for cities and towns around the world. Managing, processing, and disposing of trash requires effective and sustainable ways as the amount of waste produced rises with urbanization.

Cost-effectiveness, environmental impact, technological feasibility, and social acceptance are just a few of the conflicting considerations that decision-makers—from private sector stakeholders to city planners and environmental agencies—frequently have to weigh. Techniques such as Multi-Criteria Decision-Making (MCDM) offer an organized and methodical way to assess and choose the best options from a range of options in order to handle this complexity. The term "MCDM" describes a group of decision-making techniques that assess several, frequently opposing criteria in order to find a solution that best balances different aspects. When it comes to municipal waste management, MCDM techniques give decision-makers the ability to evaluate different options for waste treatment, recycling, disposal, and facility location based on a variety of factors, such as economic cost, social acceptability, operational efficiency, and environmental impact. Through the integration of these many aspects, MCDM procedures provide a framework for decision-making that is more sustainable, transparent, and balanced. Several MCDM strategies are frequently used in municipal waste management to assist decision-makers in navigating the trade-offs between various options. Despite taking different approaches, all methods have one thing in common: they all call for a methodical assessment of the available options using predetermined standards in order to determine which one best satisfies the waste management process's overall goals. Three MCDM techniques were employed in this study: PROMETHEE II, the Technique for Order of Preference by Similarity to Ideal Solution (TOPSIS), and the Analytical Hierarchy Process (AHP) with Fuzzy. The Analytic Hierarchy Process (AHP) will be applied alongside fuzzy logic to give the criteria and scenarios weights according to their various relevance levels. By using fuzzy logic, decision-makers will find it easier to handle ambiguous and imprecise facts. In municipal waste management, one of the most used MCDM approaches is the Analytical Hierarchy Process (AHP). It assists in organizing complicated decision-making issues into a hierarchical framework, with the decision's objective at the top and the criteria and alternatives gradually broken down into smaller sub-criteria. By allocating relative relevance (weights) to each criterion through pairwise comparisons, AHP enables decision-makers to evaluate many alternatives against one another. In order to determine the best choice based on the combined assessments, the weighted sum of each alternative's performance scores is computed before the final decision is made. The scenarios are assessed using the PROMETHEE II and TOPSIS approaches. The optimal choice is the one that is most similar to the ideal response and most different from the anti-ideal one in the Technique for Order of Preference by Similarity to Ideal Solution (TOPSIS). TOPSIS evaluates options according to how close they are to the optimal solution, which is the one that meets all requirements the best. Every option is assessed according to how closely it resembles both this ideal solution and the "negative ideal solution," which stands for the worst-case scenario. TOPSIS creates a ranking that ranks alternatives according to how close they are to the ideal by calculating the distance between each option and the ideal and negative ideal solutions. Several waste disposal techniques (such as recycling, incineration, or landfilling) can be assessed using TOPSIS according to a number of factors, including price, environmental effect, and community acceptability. A straightforward and efficient method of ranking options is offered by TOPSIS, which employs a geometric distance-based methodology. It does not, however, always reflect the reality of waste management decisions, where some considerations (such as environmental impact) may be more essential than others. This is because it presupposes that all criteria are equally relevant. The assessment is predicated on seven factors. In MCDM, the relative relevance of the criteria is determined using the fundamental scale of comparison. Pairwise comparisons between several criteria are made easier with the help of this widely used tool. Each number on the scale has a distinct significance and spans from 1 to 9 (Mujtaba et al.,).

Solid Waste Management Legislation

Low rates of trash recycling and waste mismanagement are mostly caused by insufficient or non-existent regulation combined with lax enforcement. According to respondents in Anchorage, the absence of a state mandate and a local rule requiring curb side recycling are obstacles to raising recycling rates. Mandatory recycling laws, according to the interviewees, may have allowed waste efforts to continue even after local government leadership changed because individuals' involvement in recycling programs is essential to their efficacy. The implementation of national waste laws and regulations changed local waste management practices in Murmansk and Tromsø. The waste prevention and recycling legislation that Norway put into place beginning in the 1990s included a landfill charge (1999–2015), an incinerator tax (1999–2010), a landfill ban on biodegradable waste (2009), and an 80% recovery target for municipal rubbish (1991). These national rules prompted the Tromsø City Council to pass legislation on recycling glass (1992), textiles (1993), paper, and cardboard (1995). In 1997, Tromsø also began segregating organic trash and abandoned the practice of dumping of household waste in landfills. As per a Remiks interviewee, the implementation of organic waste separation was a pre-emptive step that local authorities took in anticipation of the nationwide landfill ban on biodegradable trash. In 2014, the Russian government introduced revisions to Federal Law No 89-FZ that prohibited the elimination of trash including recyclable elements in landfills. Similar to Norway, this was done to enhance MSWM (Government of the Russian Federation, 2014). After that, in 2017, the federal government passed Decree No 1589-r, which prohibited the disposal of a list of production and consumer waste types that had significant recycling components (Nadezhda Filimonova, 2024).

Life Cycle Sustainability Assessment LCSA Approach

The three primary LCT tools used in this sort of analysis (eLCA, LCC, and sLCA) differ in how they employ the effect categories, which is why the LCSA approach is used for MSWMSs. Generally speaking, the same midpoint impact categories should include all the indicators and factors that result in the same kinds of affects. The economic costs and revenues are frequently categorized based on when they occur during the life cycle of the management waste facility, even if this is standard procedure for the environmental categories. This approach is intended to educate municipalities on the long-term viability of different waste management techniques, maximizing resource use, mitigating adverse effects, and promoting favorable results for environmental health and human well-being. Assessing the ecological effects of various waste management techniques is the main focus of the environmental component of LCSA for municipal waste management. LCSA determines the direct and indirect environmental effects by monitoring the inputs and outputs at each step of the waste management process, including waste collection, sorting, recycling, energy recovery, and disposal. This entails examining pollution, water use, emissions (including greenhouse gases), energy consumption, and the depletion of natural resources. Recycling initiatives, for instance, may lessen the need to extract raw materials, whereas incineration may increase air pollution. LCSA evaluates these effects throughout their entire life cycle in order to determine each waste management option's overall environmental impact. The objective is to find methods that reduce the overall environmental load by minimizing harmful emissions, conserving resources, and encouraging material recovery. LCSA's economic dimension (e-LCA) assesses the long-term viability of various waste management techniques. The pressure on municipalities to lower waste management expenses while maintaining efficient and just systems is increasing. LCSA includes life cycle costing (LCC), which focuses on the direct and indirect expenses related to waste management procedures. Waste collection, transportation, sorting, and treatment are examples of operating costs; landfill monitoring and waste-to-energy facility upkeep are examples of long-term maintenance costs; and capital expenditures include things like infrastructure and technological investments. Additionally, LCSA takes into account externalities, such as the financial consequences of pollution-related health effects or the possible income from energy recovery or the resale of recovered materials. The social component of LCSA (s-LCA) investigates how waste management systems affect society as a whole, which is sometimes overlooked in conventional environmental or economic analyses. Municipal trash management impacts communities, public health, working conditions, and general social equality in addition to the environment and economics. When evaluating social outcomes, LCSA takes into account elements including the health and safety of waste management employees, the effect that waste treatment plants have on the areas in which they are located, and how advantages and disadvantages are distributed among various social groups. Additionally, it examines the ways in which waste management systems support or undermine social inclusion (e.g., access to recycling programs in low-income communities), public health (e.g., air and water quality, disease prevention), and job development (Annachiara Ceraso, 2024).

Managing Municipal Waste During Emergencies

A dynamic and changing research landscape is revealed by the investigation of AI in trash classification, especially in the difficult situation of disaster scenarios. The research in this field emphasizes the benefits and drawbacks of existing AI techniques, suggesting that more specific approaches are required for catastrophe waste management. The effectiveness of AI techniques, especially deep learning and convolutional neural networks (CNNs), has been demonstrated by recent waste classification research. These methods have been effectively used to classify a variety of waste categories with accuracy on datasets such as TrashNet, Waste Picture, and HPU_WASTE. Data scarcity has been handled by innovations including data augmentation techniques, such as DCGAN and albumentation, which have improved model performance. But there are also issues like deep learning's high processing requirements and the requirement for larger datasets that capture the complexity of the real world. All things considered, these studies demonstrate AI's promise for trash classification, but they also highlight the necessity of continued study to improve these models. The application of convolutional neural networks for garbage classification, such as AlexNet, DenseNet121, and SqueezeNet. Their results show the promise of deep learning methods in solid waste classification, especially the high accuracy attained by DenseNet121. Nevertheless, this study mostly concentrates on common trash kinds without exploring the intricacies of disaster garbage, which frequently consists of a mixture of mud and other materials. A major advancement in handling the complexities of waste classification is the use of a dual ensemble deep learning architecture that integrates multiple image-segmentation techniques with heterogeneous CNN outputs. This method has outperformed other models in the categorization of general solid waste, demonstrating how well our AI-enhanced model distinguishes and classifies garbage with previously unheard-of accuracy. A significant step toward more flexible and effective waste management techniques is the incorporation of these AI-enhanced models into municipal waste management regulations. By streamlining waste management processes to handle the unexpected surge of waste that follows disasters, this strategic inclusion can improve disaster planning and response. In addition to guaranteeing the continuation of waste management services during emergencies, this strategy also helps make municipal waste management systems more sustainable overall (Asmae El Jaouhari, 2025; Rapeepan Pitakaso, 2024).

Anaerobic Digestion (AD)

Waste management and energy insecurity are addressed by the widespread use of anaerobic digestion, or AD, to treat organic waste in residences, workplaces, and urban areas. AD reduces greenhouse gas emissions by the use of municipal and kitchen solid waste for biogas production, nutrient recycling, and energy recovery from digested biomass. Organic matter found in kitchen waste (KW) and municipal solid waste (MSW) is broken down by bacteria during anaerobic digestion to create biogas and the energy they require to survive. The effective disposal of a number of solid wastes, such as cooked rice waste, cabbage waste, and Irish potato peels, has drawn criticism. In the meantime, there is a lot of potential for biogas production from these organic wastes from MSW and KW. Substrates having at least 50% moisture content are suitable for bioconversion processes. One major problem with AD in fruit and vegetable waste is rapid acidification due to low pH and increased formation of volatile fatty acids, which limits methanogenation in the digester. Variables that impact process efficiency and biogas production include pH, the carbon–nitrogen (CN) ratio of the substrate, total volatile solids (TVS), moisture content (MC), total solids (TS), fluid retention time (HRT), and the presence of hazardous metals in the substrate. Moisture content facilitates methanogenic bacteria's movement and nutrient transport while also encouraging their growth. Total volatile solids and total solids can be used to gauge the efficiency of the anaerobic process and the amount of biogas that will be generated. High volatile solid content does not necessarily equate to high biogas generation because of the lack of lignin material's volatile solids. A high output of 91% moisture content and 9% total solids. As a successful waste stabilization technique that transforms bio-waste into digestate rich in nutrients and sustainable energy, anaerobic digestion is crucial to agricultural and environmental sustainability. The anaerobic breakdown of organic materials in landfills, sewage sludge, and bio-waste digesters produces biogas (Das & Ghosh, 2023). Biogas contains carbon dioxide (30–45%), hydrogen sulfide (1–2%), nitrogen (0–1%), siloxanes, water vapor, a trace quantity of oxygen, methane (55–77%), and carbon monoxide. The output may not always be feasible, and not all wastes are appropriate for the production of biogas. Thus, characterization is required in order to assess the substrate's viability for biogas generation (Register Mrosso, 2023).

The Concept of "Zero Waste" Systems

In 1973, Dr. Paul Palmer came up with the term "zero waste" to describe the repurposing of chemicals. The spirit of "zero waste" ideas cannot be sufficiently encapsulated in a single definition. The Zero Waste International Alliance, on the other hand, defines zero waste in an organized manner as the design and administration of goods and procedures to recover and conserve all resources while methodically preventing and eliminating material waste. Reducing, reusing, redesigning, regenerating, recycling, repairing, remanufacturing, and reselling are just a few of the concepts that have been developed for sustainable waste management systems. Examples include full life cycle of design systems, zero landfill and incineration of waste, and more. Therefore, the zero waste design strategy puts reducing waste and reusing things first, followed by recycling and composting the rest. Urban planners and architects are growing increasingly interested in zero waste ideas and how to implement them by modernizing recycling infrastructure and redesigning the urban system with "zero waste" in order to create "low-to-no carbon" city districts. In order to achieve a 100% recycling rate for systems that manage municipal solid waste, the zero waste concept calls for the recovery of all resources from waste products. A lack of community desire to pay, procurement against sustainability, inconsistent state legislation, short-term thinking by producers and consumers, and the notion that the lowest bids get commissioned are the main barriers to zero waste. This new holistic paradigm will need us to start the shift to a low-carbon world and give up the goal of increasing our consumption, such as purchasing more and more goods. This entails facilitating the availability of improved, more effective technologies as well as encouraging behavioral and attitude adjustments. Indeed, behavioral change will be required to account for 25% of the emissions decrease. The new ecological paradigm of urban development and commercial operations will focus on integrating systems and fostering innovation at all scales. To achieve zero waste city goals, we suggest five interrelated fundamental principles that must be implemented concurrently to turn a city into a zero waste city. The guiding ideas are: Changes in behavior and sustainable consumption, extended accountability for producers and consumers, all municipal solid garbage is recycled 100% of the time, Zero waste disposal and incineration laws, 100% resource recovery from waste. These five guiding concepts are essential for turning cities become waste-free environments. Furthermore, all five principles must be applied simultaneously for the transformation process to be successful. However, A vision for a long-term zero-waste metropolis would be necessary based on the collective application of each principle, as the goal of a zero-waste city is not only very challenging to achieve, but also requires sustained efforts to do so (Atiq Uz Zaman, 2011).

Municipal Waste Management in the Circular Economy (CE)

The EU's shift to the CE model includes municipal garbage management. The EC published its first CE document, which set the "zero waste program" target for Europe. For the EU to adopt the CE model, this message is essential because it outlines precisely what needs to be done to go from a linear economy to a circular economy. The EC's goal was to end the landfilling of municipal garbage. Improvements in municipal waste management have also been shown to boost the social, economic, and environmental benefits in EU member states. The second CE document, known as the CE Action Plan or CE Package, was released in 2015. Long-term initiatives to decrease waste disposal (including unlawful dump sites) and improve trash preparation for recycling and reuse were proposed in the CE Action Plan. Here, the EC stated that packaging waste and municipal garbage are the two main waste streams. A second component of the CE Package was a proposal to update the Waste Framework Directive (WFD) and other key EU waste laws. Waste Electrical and Electronic Equipment (WEEE) Directive, Landfill Directive, Packaging Directive, and Directives on Batteries, Accumulators, and End-of-Life Vehicles. Furthermore, the eco-design directive, which serves as the foundation for sustainable product design, is a significant document pertaining to energy-related objects that can be turned into municipal waste. 2018 saw the implementation of the new Waste Directive, which amended the previous one. The monitoring framework's requirements were met by proposing a set of pertinent indicators that were categorized into the CE's four primary areas: waste management, secondary raw materials, production and consumption, and competitiveness and innovation (Marzena Smol, 2020).

Pay-As-You-Throw (PAYT) Scheme

By addressing the pricing of garbage as a potential barrier to deploying weight-based pay-as-you-throw (PAYT) systems, this work carries on the legacy of relatively little PAYT research. The methodology outlined in this study helps ensure that garbage costs are distributed fairly among citizens and can be adjusted to take advantage of lower subsidy rates or to fully cover recyclables collection in residual waste prices. These are the study's two primary findings. First, in order to get a more comprehensive understanding of the effects of weight-based PAYT schemes in various circumstances, additional investigation and real-world testing are required. Waste pricing is a crucial component of effective weight-based PAYT systems, but there are a number of other crucial elements as well: Understanding regional variations in waste management systems, keeping an eye on the system's operation, and communicating clearly and fairly. Furthermore, there are certain logistical and administrative issues that must be resolved when integrating weighing and identifying technology into waste management systems. In order to assess the output of various collection methods and estimate the impact of the weight-based PAYT system, precise waste generation and composition statistics are required. It is challenging to assess the suitability of specific PAYT solutions due to the contextual variations in waste management among nations. For instance, in some locations, the legislative environment may impede the meaningful growth of PAYT programs. Second, expanding the use of weight-based PAYT may be a significant component of the recycling-friendly policy mix in various nations that seek to increase the national rate of MSW recycling or waste reduction. Due to the EU's aggressive recycling goals, many EU member states are reportedly required to enact a number of new laws in order to meet the standards. To encourage eco-friendly conduct, policy tools that are both educational and regulatory are required in addition to financial incentives. For MSW management to be more sustainable, weight-based PAYT pricing must be understood as a component of a broader systemic shift. Local waste management operators are vital to this transformation and need to be given the tools and resources they need to succeed (Aino Ukkonen, 2021). Figure 2 shows diagrammatic Representation of Innovative Strategies for Sustainable Municipal Waste Management.

Fig. 2 Diagrammatic representation of innovative strategies for sustainable municipal waste management

4 Conclusion

Sustainable municipal waste management is a pressing challenge that requires immediate and innovative solutions. As urbanization continues to rise globally, the volume of waste generated is increasing, which poses significant environmental, economic, and social risks. Through this study, we have explored various types of waste management strategies and highlighted several innovative approaches that can address these issues effectively. Innovative strategies such as waste-to-energy technologies, circular economy principles, and implementing innovative waste management systems stand out as viable solutions for establishing a more effective and sustainable waste management system process. Additionally, community engagement and public–private partnerships are essential to guaranteeing the effectiveness of these tactics. Municipalities must implement comprehensive and integrated waste management plans that encourage resource recovery, recycling, and reuse in addition to lowering waste production. The involvement of advanced technologies, policy development, and education of the public are pivotal components for achieving a sustainable future. The study emphasizes the need for continuous innovation, collaboration, and investment to ensure that municipalities can effectively manage their waste and reduce their environmental footprint. Ultimately, adopting innovative waste management strategies will lead to cleaner, greener cities and a more sustainable future for generations to come.

References

Aino Ukkonen, O. S. (2021). Weight-based pay-as-you-throw pricing model: Encouraging sorting in households through waste fees. *Waste Management, 135*, 372–380. https://doi.org/10.1016/j.wasman.2021.09.011

Aliakbar Dehghan, V. O. (2025, February). Machine learning-based prediction of the C/N ratio in municipal organic waste. *Environmental Technology & Innovation, 37*, 103977. https://doi.org/10.1016/j.eti.2024.103977

Anchal Sharma, R. G. (2023, November 3). Life cycle assessment of municipal solid waste generated from hilly cities in India—A case study. *Heliyon, 9*(11), e21575. https://doi.org/10.1016/j.heliyon.2023.e21575

Annachiara Ceraso, A. C. (2024, September). Life Cycle Sustainability Assessment of municipal solid waste management systems: A review. *Journal of Environmental Management, 368*, 1–18. https://doi.org/10.1016/j.jenvman.2024.122143

Asmae El jaouhari, A. S.-w. (2025, December 9). Turning trash into treasure: Exploring the potential of AI in municipal waste management—An in-depth review and future prospects. *Journal of Environmental Management, 373*, 123658. https://doi.org/10.1016/j.jenvman.2024.123658

Atiq Uz Zaman, S. L. (2011, November 02). Challenges and opportunities in transforming a city into a "Zero waste city". *Challenges, 2*, 73–93. https://doi.org/10.3390/challe2040073

Bala, K., Anshul, Verma, S., Dogra, A., Das, A. P., Kumar, R., & Nandini. (2024a). Sustainable management for urban plastic waste generation. In *Sustainable management of urban plastic waste through circular economic approaches* (pp. 211–236). https://doi.org/10.1201/9781003465447-13

Bala, K., Dogra, A., Lata, P., & Das, A. P. (2024b). Discussing sources and biological-chemical remediation approaches for microplastic pollution. In *Sustainable microbial technology for synthetic and cellulosic microfiber bioremediation* (pp. 37–59). https://doi.org/10.1007/978-3-031-62600-5_3

Das, A. P., Behera, I. Dipamitra., & Das, N. Prasad. (2024a). *Renewable energy generation and value addition from environmental microfiber pollution through advanced greener solution* (p 278).

Das, A. P., Dutta, K., Khatun, R., Behera, I. D., Singh, S., & Mishra, S. (2023). Microfiber pollution and its microbial mitigation: A review on current trends and future prospects. *Journal of the Taiwan Institute of Chemical Engineers*, 105104. https://doi.org/10.1016/J.JTICE.2023.105104

Das, A. P., & Ghosh, S. (2018). Bioleaching of Manganese from mining waste materials. *Materials Today: Proceedings, 5*(1), 2381–2390. https://doi.org/10.1016/J.MATPR.2017.11.459

Das, A. P., & Ghosh, S. (2023). Microorganisms in manganese biomining processes. In *Manganese mining microorganisms* (pp. 55–61). https://doi.org/10.1016/B978-0-12-822147-1.00013-3

Das, A. P., Ghosh, S., Mohanty, S., & Sukla, L. B. (2015a). Advances in manganese pollution and its bioremediation. In L. B. Sukla, N. Pradhan, S. Panda, & B. K. Mishra (Eds.), *Environmental microbe biotechnology* (1st ed., pp. 313–328). Springer. https://doi.org/10.1007/978-3-319-19018-1_16

Das, A. P., Ghosh, S., Mohanty, S., & Sukla, L. B. (2015b). Consequences of manganese compounds: A review. *Toxicological & Environmental Chemistry, 96*(7), 981–997. https://doi.org/10.1080/02772248.2015.1005428

Das, A. P., Sukla, L. B., Pradhan, N., & Nayak, S. (2011). Manganese biomining: A review. *Bioresource Technology, 102*(16), 7381–7387. https://doi.org/10.1016/J.BIORTECH.2011.05.018
Das, A., Dey, S., & Das, A. P. (2024b). Ecotoxicological impacts of synthetic microfiber pollutants and development of sustainable mitigation strategies. *Environmental Chemistry and Ecotoxicology*. https://doi.org/10.1016/J.ENCECO.2024.12.003
Das, A., Dey, S., Bala, K., Gnanasekaran, R., & Das, A. P. (2024c). Microbial bioremediation technology for sustainable treatment and management of synthetic microfiber waste. In *Sustainable microbial technology for synthetic and cellulosic microfiber bioremediation* (pp. 221–238). https://doi.org/10.1007/978-3-031-62600-5_12
Das, A., & Ghosh, S. (2022). Role of microorganisms in extenuation of mining and industrial wastes. *Geomicrobiology Journal, 39*, 173–175. https://doi.org/10.1080/01490451.2022.2038953
Das, A., Ray, S. K., Mohanty, M., Mohanty, J., Dey, S., & Das, A. P. (2024d). Ecotoxicity of microplastic wastes and their sustainable management: A review. *Environmental Chemistry and Ecotoxicology, 6*, 144–152. https://doi.org/10.1016/J.ENCECO.2024.05.003
Dey, S., Bissoyi, A., Koteswara Reddy, G., & Das, A. P. (2024a). Synthetic microfibers: An urban emerging pollutant demanding for sustainable environmental solutions. In *Sustainable microbial technology for synthetic and cellulosic microfiber bioremediation* (pp. 107–121). https://doi.org/10.1007/978-3-031-62600-5_7
Dey, S., Das, A., Mallick, K., Sahu, A., & Das, A. P. (2023). *Environmental petroleum waste: Pollution, toxicity, sustainable remediation* (pp. 159–175). https://doi.org/10.1007/978-3-031-48220-5_7
Dey, S., Dash, N., Ray, S. K., Priyadarsini, S., Chhotaray, C., & Das, A. P. (2024b). Occurrence and quantification of tire wear plastics pollutants from highway city dust of Bhubaneswar city of Odisha. In *Sustainable microbial technology for synthetic and cellulosic microfiber bioremediation* (pp. 77–89). https://doi.org/10.1007/978-3-031-62600-5_5
Fabio Fiorillo, E. M. (2024, April). Municipal strategies, fiscal incentives and co-production in urban waste management. *Socio-Economic Planning Sciences, 92*, 101817. https://doi.org/10.1016/j.seps.2024.101817
Ghosh, S., & Das, A. P. (2017). Bioleaching of manganese from mining waste residues using Acinetobacter sp. *Geology, Ecology, and Landscapes, 1*(2), 77–83. https://doi.org/10.1080/24749508.2017.1332847
Ghosh, S., & Das, A. P. (2018). Metagenomic insights into the microbial diversity in manganese-contaminated mine tailings and their role in biogeochemical cycling of manganese. *Scientific Reports, 8*(1), 1–12. https://doi.org/10.1038/s41598-018-26311-w
Ghosh, S., Kumar, M. S., Bal, B., & Das, A. P. (2018). Application of bioengineering in revamping human health. In *Synthetic biology: Omics tools and their applications* (pp. 21–37). https://doi.org/10.1007/978-981-10-8693-9_2
Ghosh, S., Sinha, J. K., Ghosh, S., Vashisth, K., Han, S., & Bhaskar, R. (2023a). Microplastics as an emerging threat to the global environment and human health. *Sustainability, 15*(14), 10821. https://doi.org/10.3390/SU151410821
Ghosh, S., Tripathy, B., Dey, S., & Das, A. P. (2023b). Comparative investigation of fungal and bacterial manganese biomining mechanisms. In *Manganese mining microorganisms* (pp. 99–115). https://doi.org/10.1016/b978-0-12-822147-1.00016-9
Jagatee, S., Priyadarshini, S., Rath, C. C., & Das, A. P. (2024). Synthetic microfiber: An enduring environmental problem linked to sustainable development. *Environmental Science and Engineering, Part, F2333*, 93–112. https://doi.org/10.1007/978-3-031-51792-1_6
Himani, S., Bansod, P. D. (2023, February 03). Biomedical waste management and its importance: A systematic review. *Cureus, 15*(2), e34589. https://doi.org/10.7759/cureus.34589
Inna Pitak, A. S. (2025, March). Transforming textile waste into alternative fuel by thermochemical methods: A mini-review. *Cleaner Waste Systems, 10*, 100199. https://doi.org/10.1016/j.clwas.2024.100199

Kaniz Fatema, M. N. (2025, January 15). E-waste recycling in an optimized way for copper recovery by leaching and a case study on E-waste generation and management in Dhaka city. *Heliyon, 11*(1), e41453. https://doi.org/10.1016/j.heliyon.2024.e41453

Kansal, A. (2002). Solid waste management strategies for India. *Indian Journal of Environmental Protection*, 22.

Karim Emara, A. O. (2024, December). An innovative approach to municipal solid waste characterization for waste-to-energy processes in Egypt. *Cleaner Waste Systems, 9*, 100161. https://doi.org/10.1016/j.clwas.2024.100161

Khatun, R., Singh, S., Dubey, N. K., & Prasad Das, A. (2023). A review on marine-based phytochemicals and their application in biomedical research. In *Recent frontiers of phytochemicals: Applications in food, pharmacy, cosmetics, and biotechnology* (pp. 383–395). Elsevier. https://doi.org/10.1016/B978-0-443-19143-5.00013-X

Kumar, M., Pandit, S., Patel, V., Khanna, N., Nag, M., Lahiri, D., Ray, R. R., Das, A., & Das, D. (2023). Maximization of energy recovery from starch processing wastewater by thermophilic dark fermentation coupled with microbial fuel cell technology. *Geomicrobiology Journal*, 1–12. https://doi.org/10.1080/01490451.2023.2209555

Mallick, K., Sahu, S., Sahu, A., Dey, S., & Das, A. P. (2024). A critical review of marine microfiber pollution routes, toxicity, and its sustainable remediation. *Environmental Science and Engineering, Part, F2333*, 189–211. https://doi.org/10.1007/978-3-031-51792-1_10

Marzena Smol, J. D.-K. (2020, June 03). Transformation towards circular economy (CE) in municipal waste management system: Model solutions for Poland. *Sustainability, 12*(11), 2–25. https://doi.org/10.3390/su12114561

Mishra, S., & Das, A. P. (2021). Current treatment technologies for removal of microplastic and microfiber pollutants from wastewater. In *Wastewater treatment: Cutting-edge molecular tools, techniques and applied aspects* (pp. 237–251). https://doi.org/10.1016/B978-0-12-821881-5.00011-8

Mishra, S., & Das, A. P. (2022). Treatment of the wastewater polluted with synthetic microfiber released from washing machine. In B. B. Das, H. Hettiarachchi, P. K. Sahu, & S. Nanda (Eds.), *Lecture notes in civil engineering* (Vol. 207, pp. 109–117). Springer Science and Business Media Deutschland GmbH. https://doi.org/10.1007/978-981-16-7509-6_9/COVER

Mishra, S., & Das, A. P. (2023). Microbial remediation of synthetic microfiber contaminated wastewater. In M. P. Shah (Ed.), *Microbial technologies in industrial wastewater treatment* (pp. 337–350). Springer. https://doi.org/10.1007/978-981-99-2435-6_17

Mishra, S., Dash, D., Al-Tawaha, A. R. M. S., & Das, A. P. (2022a). A review on heavy metal ion adsorption on synthetic microfiber surface in aquatic environments. *Applied Biochemistry and Biotechnology, 194*(10), 4639–4654. https://doi.org/10.1007/S12010-022-04029-W/METRICS

Mishra, S., Dash, D., & Das, A. P. (2022b). Detection, characterization and possible biofragmentation of synthetic microfibers released from domestic laundering wastewater as an emerging source of marine pollution. *Marine Pollution Bulletin, 185*, Article 114254. https://doi.org/10.1016/J.MARPOLBUL.2022.114254

Mishra, S., Dash, D., & Das, A. P. (2024a). Isolation and molecular characterization of polyester-degrading bacterial strains from municipal wastewater and their application in microplastic degradation in contaminated wastewater. *Geomicrobiology Journal*. https://doi.org/10.1080/01490451.2024.2376101

Mishra, S., Dash, D., & Das, A. P. (2024b). Microbial degradation of polyester microfibers using indigenously isolated bacterial strain Exiguobacterium Sp. *CLEAN—Soil, Air, Water, 52*(12), e202300343. https://doi.org/10.1002/CLEN.202300343

Mishra, S., Ghosh, S., van Hullebusch, E. D., Singh, S., & Das, A. P. (2023). A Critical review on the recovery of base and critical elements from electronic waste-contaminated streams using microbial biotechnology. *Applied Biochemistry and Biotechnology, 2023*, 1–30. https://doi.org/10.1007/S12010-023-04440-X

Mishra, S., Rath, C. C., & Das, A. P. (2019a). Marine microfiber pollution: A review on present status and future challenges. *Marine Pollution Bulletin, 140*, 188–197. https://doi.org/10.1016/J.MARPOLBUL.2019.01.039

Mishra, S., Rout, P. K., & Das, A. P. (2019b). Solar photovoltaic panels as next generation waste: A review. *Biointerface Research in Applied Chemistry, 9*(6).

Mishra, S., Rout, P. K., & Das, A. P. (2021). Emerging microfiber pollution and its remediation. In R. Prasad (Ed.), *Environmental and microbial biotechnology* (pp. 247–266). Springer. https://doi.org/10.1007/978-981-15-5499-5_9

Mishra, S., Singh, R. P., Rath, C. C., & Das, A. P. (2020). Synthetic microfibers: Source, transport and their remediation. *Journal of Water Process Engineering, 38*, Article 101612. https://doi.org/10.1016/J.JWPE.2020.101612

Mohanty, M., Tripathy, B., Koteswara Reddy, G., & Das, A. P. (2024). Greener technologies for establishing restraint over microplastic pollution. In *Sustainable microbial technology for synthetic and cellulosic microfiber bioremediation* (pp. 239–258). https://doi.org/10.1007/978-3-031-62600-5_13

Mohanty, S., Ghosh, S., Bal, B., & Das, A. P. (2018). A review of biotechnology processes applied for manganese recovery from wastes. *Reviews in Environmental Science and Bio/Technology, 17*(4), 791–811. https://doi.org/10.1007/S11157-018-9482-1

Mohanty, S., Ghosh, S., Nayak, S., & Das, A. P. (2016). Isolation, identification and screening of manganese solubilizing fungi from low-grade manganese ore deposits. *Geomicrobiological Journal, 34*(4), 309–316. https://doi.org/10.1080/01490451.2016.1189016

Mujtaba, M. A., et al. (2024a, February 10). Evaluating sustainable municipal solid waste management scenarios: A multicriteria decision making approach. *Heliyon, 10*(4), e25788. https://doi.org/10.1016/j.heliyon.2024.e25788

Mujtaba, M. A., et al. (2024b, February 29). Evaluating sustainable municipal solid waste management scenarios: A multicriteria decision making approach. *Heliyon, 10*(4), e25788. https://doi.org/10.1016/j.heliyon.2024.e25788

Nadezhda Filimonova, S. J. (2024, December). Sustainable municipal solid waste management: A comparative analysis of enablers and barriers to advance governance in the Arctic. *Journal of Environmental Management, 371*, 123111. https://doi.org/10.1016/j.jenvman.2024.123111

Nalini Singh Chauhan, A. P. (2022). In P. S. Arif Ahamad (ed.), *Role of education and society in dealing plastic pollution in the future* (First ed.). John Wiley & Sons. https://doi.org/10.1002/9781119800897

Paul, A., Priyadarsini, S., Karakuş, S., Behera, J. K., Gnanasekaran, R., & Das, A. P. (2024). Fate and accumulation of synthetic microfibers in the environment. In *Sustainable microbial technology for synthetic and cellulosic microfiber bioremediation* (pp. 61–76). https://doi.org/10.1007/978-3-031-62600-5_4

Prabhakar, A., Mishra, S., & Das, A. P. (2019). Isolation and identification of lead (Pb) solubilizing bacteria from automobile waste and its potential for recovery of lead from end of life waste batteries. *Geomicrobiological Journal, 36*(10), 894–903. https://doi.org/10.1080/01490451.2019.1654044

Pradhan, G., Tripathy, B., Ram, D. K., Digal, A. K., & Das, A. P. (2023). Bauxite mining waste pollution and its sustainable management through bioremediation. *Geomicrobiology Journal*, 1–10. https://doi.org/10.1080/01490451.2023.2235353

Priyadarshini, B., Santhosh Kumar, M., & Das, A. P. (2024a). Microbial degradation of synthetic plastics released from COVID-19 generated single use disposable face masks. *Environmental Science and Engineering, Part, F2827*, 1–16. https://doi.org/10.1007/978-3-031-50840-0_1

Priyadarshini, S., Jagatee, S., & Das, A. P. (2024b). Synthetic fabrics and microfiber pollution–An assessment of their global impact. *Environmental Science and Engineering, Part, F2333*, 137–157. https://doi.org/10.1007/978-3-031-51792-1_8

Priyadarsini, S., & Das, A. P. (2024). Analytical and structural characterization of waste lithium-ion batteries for their effective recycling strategy. *Environmental Chemistry and Ecotoxicology.* https://doi.org/10.1016/j.enceco.2024.12.004

Priyadarsini, S., & Das, A. P. (2025). Lithium bioleaching: A review on microbial-assisted sustainable technology for lithium bio-circularity. *Journal of Water Process Engineering, 69*, Article 106744. https://doi.org/10.1016/j.jwpe.2024.106744

Raj, K., & Das, A. P. (2023). Lead pollution: Impact on environment and human health and approach for a sustainable solution. *Environmental Chemistry and Ecotoxicology, 5*, 79–85. https://doi.org/10.1016/J.ENCECO.2023.02.001

Rapeepan Pitakaso, T. S.-D. (2024, May 30). Optimization-driven artificial intelligence-enhanced municipal waste classification system for disaster waste management. *Engineering Applications of Artificial Intelligence, 133*, 108614. https://doi.org/10.1016/j.engappai.2024.108614

Ray, S. K., Dey, S., Das, A. P., & John, A. (2024). Microplastic waste management through greener technologies: A review. In *Sustainable management of urban plastic waste through circular economic approaches* (pp. 237–261). https://doi.org/10.1201/9781003465447-14

Register Mrosso, A. C. (2023, May). Characterization of kitchen and municipal organic waste for biogas production: Effect of parameters. *Heliyon, 9*(5), e16360. https://doi.org/10.1016/j.heliyon.2023.e16360

Sadasivuni, K. K., Rattan, S., Waseem, S., Brahme, S. K., Kondawar, S. B., Ghosh, S., Das, A. P., Chakraborty, P. K., Adhikari, J., Saha, P., & Mazumdar, P. (2019). Silver nanoparticles and its polymer nanocomposites—Synthesis, optimization, biomedical usage, and its various applications. In *Lecture notes in bioengineering* (pp. 331–373). https://doi.org/10.1007/978-3-030-04741-2_11

Sahu, A., Dey, S., Mallick, K., Priyadarsini, S., Behera, I. D., & Das, A. P. (2024). A review on advanced technology for sustainable management of synthetic microplastic waste. In *Sustainable management of urban plastic waste through circular economic approaches* (pp. 262–279). https://doi.org/10.1201/9781003465447-15

Satapathy, S., Kar, D., Priyadarsini, S., Das, A. P., & Kuanar, A. (2024). Application of microorganisms in emerging waste management strategies and remediation of synthetic and cellulosic microfibers. In *Sustainable microbial technology for synthetic and cellulosic microfiber bioremediation* (pp. 183–206). https://doi.org/10.1007/978-3-031-62600-5_10

Singh Jadaun, J., Bansal, S., Sonthalia, A., Rai, A. K., & Singh, S. P. (2022). Biodegradation of plastics for sustainable environment. *Bioresource Technology, 347*, Article 126697. https://doi.org/10.1016/J.BIORTECH.2022.126697

Singh, R. P., Mishra, S., & Das, A. P. (2020). Synthetic microfibers: Pollution toxicity and remediation. *Chemosphere, 257*, Article 127199. https://doi.org/10.1016/j.chemosphere.2020.127199

Tripathy, B., Dey, S., Dubey, N. K., Lo, W.-C., & Das, A. P. (2024). *An omics-based approach for managing the risk of COVID-19-generated wastes* (pp. 51–66). https://doi.org/10.1007/978-3-031-50840-0_4

Zhechen Zhang, Z. C.-S. (2024, June 20). Municipal solid waste management challenges in developing regions: A comprehensive review and future perspectives for Asia and Africa. *Science of the Total Environment, 930*, 172794. https://doi.org/10.1016/j.scitotenv.2024.172794

Integrating Municipal Solid Wastes as Refuse-Derived Fuels in Cement Co-processing: A Circular Economy Approach for Sustainable Resource Management

Himansu Sekhar Patra, Aditya Kishore Dash, Kabir Mohan Sethy, and Alok Prasad Das

1 Introduction

Solid waste management due to on-going rapid population growth, urbanization, and industrialization has become an emerging global issue. With the world population surpassing 8 billion in 2023 (Worldometer, 2023), MSW generation has reached an estimated 2.01 billion tons annually, projected to increase 3.40 billion tons by the year 2050 (Ellis, 2018). This sharp rise in waste generation, combined with inadequate disposal methods, has led to severe environmental hurdles, including pollutions like air, water, and soil, and also the exacerbation of climate change through greenhouse gas emissions (Dey et al., 2023; Mishra et al., 2023). Improper disposal and uncontrolled burning of solid waste contribute to air pollution, along with other potential sources such as motor vehicles, industries, agriculture, and domestic activities, posing serious health risks (Priyadarsini & Das, 2024, 2025). Currently, a significant portion of MSW is disposed of in landfills, both regulated and unregulated or are illegally dumped (Sauve & Van Acker, 2020). These practices are unsustainable and contribute to CH_4 emissions, a potent manmade greenhouse gas with 34

H. S. Patra
Environment and Sustainability, MSP Steel & Power Ltd., Raigarh, India

A. K. Dash (✉)
Chemistry Department, ITER, Siksha 'O' Anusandhan University, Bhubaneswar, India
e-mail: adityadash@soa.ac.in

K. M. Sethy
Geography Department, Utkal University, Vani Vihar, Bhubaneswar, India

A. P. Das
Department of Life Sciences, Rama Devi Women's University, Bhubaneswar, India

A. P. Das et al. (eds.), *Sustainable Municipal Waste Management with a Zero Waste Approach*, Sustainable Environmental Waste Management Strategies,
https://doi.org/10.1007/978-3-032-05950-5_13

times potential for global warming of CO_2. Landfilling also demands vast land areas, resulting for land scarcity, leachate generation, and also soil contamination. Thus, improper disposal of solid waste into the soil causes land contamination. However certain wastes like fly ash (Dash et al., 2015; Pradhan et al., 2015) and rice mill waste water (Pradhan et al., 2016) can support the growth of plant species as they contain few essential nutrients required for plant growth. Soil fertility study (Nayak et al., 2018) and evaluation of enzyme activities near solid waste dump site (Panigrahi et al., 2024) play an important role in this context (Dash & Patra, 2016; Dash et al., 2016). Moreover, the environmental impact of improper waste disposal extends to marine ecosystems (Das et al., 2024). Out of the 50 world's largest dumpsites, 38 are indirect contributors to marine pollution, causing coastal erosion and severe damage to marine ecosystems (Maalouf et al., 2020). Besides marine pollution, improper disposal of solid waste can also cause contamination of water environment since it contains various toxic materials. This can have both short-term and long-term effects on the ecosystem including on human life (Das & Ghosh, 2022; Mishra et al., 2019). Physico-chemical methods for treatment of contaminant from water/waste water is cost intensive which attracts the scientist to analyze the water/waste water samples and to go for low-cost biological treatments (Amreen et al., 2024; Dash, 2012; Dash & Mishra, 1997, 1999a, 1999b; Jena et al., 2022; Mahapatra et al., 2019; Mishra et al., 2012; Parida et al., 2023; Pradhan et al., 2013, 2015a, 2015b, 2017; Shreya et al., 2022). These environmental challenges underscore the urgent need for innovative and sustainable SWM solutions. Important greenhouse gases (GHG) such as CH_4 and 45% CO_2 are emitted from the decomposition of organic wastes in landfill sites. People living near to solid waste dumping or composting sites often complain about foul odor and polluted air quality, thus leading to affect from various respiratory-related disease (Mishra & Das, 2023; Mohanty et al., 2018). Therefore it is important to know the air quality, determination of the air quality index and the dispersion of pollutants of these area for which it is important to monitor the ambient air and to find out suitable control measures (Dash & Dash, 2017, 2018a). Effective SWM systems emphasize waste recycling/reduction, and resource recovery to mitigate environmental impacts. Plastic is one of the major non-biodegradable components of the MSW. Besides the major hazardous elements, the plastic component of solid wastes is also a threat to the environment. A lot of literatures are available on the occurrence, fate, environmental impacts, and management of plastic/micro plastics wastes worldwide (Dash et al., 2024; Naik et al., 2024a, 2024b; Parida & Dash, 2024; Panigrahi & Dash, 2024). Technological innovations, including smart waste management solutions and sustainable waste-to-energy (WtE) systems, are driving advancements in this sector.

Rapid economic and population growth has significantly increased the demand for electricity and construction materials like cement (WBCSD Guidelines, 2014). The cement industry is a cornerstone of India's infrastructure and economic development, supporting critical sectors such as construction, real estate, and urbanization. India's cement sector plays a vital role in the ongoing smart city, housing and other construction projects. Cement making is resource-intensive processes, where a lot of electricity, water, and limestone is used, thus posing significant environmental problems. The cement production is especially an energy-consuming sector with temperatures exceeding 1450 °C and energy inputs between 3000 and 6500 MJ, thus increases the production costs (Karki & Dahal, 2020).

To overcome this challenge, both government and cement industry are prioritizing on sustainable use of MSW as alternative energy resources. Recovering recyclable and combustible materials from MSW is one of the most promising approaches to create refuse-derived fuel. Use of RDF offers several benefits such as carbon neutrality apart from lower sulfate and nitrate emissions with comparison to traditional fossil fuels like coal (Yi & Jang, 2018). This paper tries to highlights the role of Indian cement industry in economic growth apart from the challenge it is facing from resource efficiency point of view. It looks at continuous initiatives to implement various sustainable practices that are progressively changing this sector into a more ecologically friendly model, such as waste recycling, alternative fuels, and energy-efficient technologies.

A CE strategy shows promise as a means of reducing resource depletion apart from promoting economic growth. The CE seeks to manage waste by redefining it as the source of resources or assets rather than as simple trash. The 3R principles, i.e., reduction, recycling, and reuse serves as a practical framework for implementing circular consumption. Co-processing of waste in cement kilns is one of effective application of this principle. It not only reduces the industry's reliance on fossil fuel but also reduces the CO_2 emissions thus providing a safe and environmental responsible solution for management of municipal waste. Most existing research primarily focuses on broader themes, such as CE in relation to alternative fuels and RDF, the evaluation of pollutants emitted during the solid wastes co-processing in cement kilns, and technological solutions for the waste-derived utilization. One way to utilize MSW is by converting it into RDF that acts as a substitute fuel in different industries, particularly cement manufacturing. Toxic elements from solid wastes and other type of wastes cause environmental contamination (Patra & Dash, 2024). Detection (Das et al., 2014a; 2014b), biodegradation (Das & Mishra, 2010), health assessment (Das & Mishra, 2008; Das & Singh, 2011), and environmental effect (Das et al., 2011, 2014a, 2014b) of various elements in the environment and their impacts have been studied.

Calorific value, low moisture, low ash content, and the absence of hazardous substances are some of the criteria for using RDF in cement co-processing thus ensuring adherence to environmental and technical standards (Tihin et al., 2023). The need of the sustainable biomass fuel transition policies in cement industry is being promoted, which emphasizes the acceleration of the transition process, promotion of innovative ideas, and restructuring of systems by challenging existing regimes that impede systemic change. Supporting biomass fuel-based cement plant at various locations in policies has been promoted (Mateus et al., 2023). Jąderko-Skubis (2021) has identified key properties of waste-derived fuels, the suitable ratio of municipal as well as industrial waste streams, and analyzed a set of variables and interactions that influence the design process. The potential use of various wastes in the cement industry has been explored and highlighted by various authors. Shehata et al., 2022 have highlighted the role of RDF in supporting CE, thus achieving the united nations sustainable development goals apart from analyzing the key challenges that hinders its adoption. Future research directions and metrics for integrating solid waste into the CE and achieving SDGs are proposed by the scientist. Aranda Uson et al. (2013) have studied the technical, environmental, and economic assessment of the cement manufacturing industry's utilization of MSW, meat and bone meal, sewage sludge, biomass, and end-of-life tires. Cement industry has a vital role in the CE by waste utilization from other industries as alternative fuels and raw materials. Biomass, RDF, used tires, and hazardous waste that have been identified as promising options for alternative fuel (Kukreja et al., 2023).

Trend in Generation of Solid Waste and Its Management in Indian context: India, with a population of approximately 1.21 billion, has an urban population of over 377 million, with 31.16% of entire population as per, Census of India, 2011. Rapid urbanization has created substantial socio-economic and environmental challenges for municipalities, driven by increased migration and the depletion of natural resources. As reported by State Pollution Control Boards (SPCB) and Pollution Control Committees (PCCs), the country generated a total of 170,339 tons of solid waste per day (TPD) during 2021–22. Of this, 156,449 TPD was collected, reflecting an overall collection efficiency of 92%. Furthermore, 91,511 TPD waste was treated, accounting for 54% of the waste generated. Over the past 5 years, the average per person waste generation in India has risen from 98.79 g/capita/day—123.45 g/capita/day, as reported by (CPCB, 2022). This data underscores the growing hurdles for managing solid waste in the face of urbanization and increasing per capita waste production. As per the Solid Waste Management Rules, 2016, processing of waste involves scientifically managing segregated solid waste for reuse, recycling, or conversion into new items. This includes activities like separation, sorting, washing, and compaction. Waste treatment refers to processes aimed at altering the physical, chemical, or biological properties of waste to reduce its quantity and possible harm. Common treatment methods include incineration, composting, vermin-composting, biogas production, and anaerobic digestion. In India, various systems are employed for waste processing and treatment, including windrow composting, vermin-composting, pit composting, bio-methanation, organic waste converters, RDF/pelletization plants, facilities for material recovery, WtE plants, and incinerators. The trend over the past 5 years (2017–2018 to 2021–2022) shows a steady increase in the waste processing or treating, rising percentage from 36.46% during 2017–2018 and 54% during 2021–2022 (CPCB, 2022).

Despite these advancements, significant amounts of waste are still landfilled. As of 2021–2022, 41,455 TPD, or 24% of the generated waste, is disposed of in landfills. Thirteen WtE plants are currently operational in India. Additionally, 1244 sites are identified as sanitary landfills, with 669 constructed, 135 under construction, and 645 in operation (CPCB, 2022). However, 10 sanitary landfills have already been exhausted. Gaps in solid waste management remain, particularly concerning waste that is neither processed nor securely disposed of in sanitary landfills. Bridging these gaps requires immediate action and coordinated efforts from all levels of governance, including local authorities, district administrations, and state departments, to ensure effective waste management practices.

Characteristics and Compositions of Solid Wastes at Indian Context: Study of characteristics and composition of MSW is essential before selecting WtE options for producing RDF. Such factors play a significant role in optimizing energy recovery from waste. Typically, MSW is generated from a variety of sources, including residential areas, commercial establishments, industrial activities, construction sites, institutions, agriculture, and municipal services. Residential areas are the largest contributors to solid waste generation. In urban regions, MSW primarily consists of waste food -39.78%, paper-13.28%, plastics-16.67%, diapers-7.89%, garden waste-4.81%, glass-3.40%, metal-3.15%, rubber-1.93%, textiles-3.30%, tetra pak-1.96%, wood-1.44%, and leather-0.37% (CPCB, 2022). The high organic content of MSW makes it an excellent candidate for energy conversion. Wet organic fractions can be processed into biogas through anaerobic digestion, while other organic materials can be transformed into biofuels like bioethanol, biodiesel, and syngas using fermentation or pyrolysis methods.

In India, about 6% of collected MSW undergoes composting at mechanical biological treatment plants, the most widely used processing technology. A significant development in West to energy initiatives is the mass-burn combustion facility at the Okhla landfill site in New Delhi, established through a public–private partnership. These efforts aim to minimize reliance on landfills and advance sustainable waste management practices. Government schemes like Jawaharlal Nehru National Urban Renewal Mission actively promote West to Energy projects. The cement industry's minimal participation in PPPs highlights a missed opportunity for broader collaboration in waste management and resource recovery.

Cement Manufacturing Process: India which is the second-largest cement manufacturing globally, contributes over 8% of the world's installed capacity (Indian Cement Industry Report, 2024). With significant growth anticipated in infrastructure and construction, the cement industry is poised to benefit from rising demand. The robust recovery of the industrial sector, post COVID-19 has further boosted cement demand, signaling strong long-term market potential. Government initiatives, such as the development of 98 smart cities, are expected to further catalyze the sector's growth. The industry's expansion is also supported by the abundant availability of critical raw materials, including limestone and coal, essential for cement production. In 2023, the Indian cement industry achieved a market size of 3.96 billion tons, with projections estimating growth to 5.99 billion tons by 2032, driven by an annual compound growth of 4.7% from 2024 to 2032 (Indian Cement Industry Report, 2024). During FY23, India produced 374.55 million tons of cement, reflecting a year-on-year growth of 6.83%. The above trend shows the pivotal role of cement industry in propelling Indian economy and infrastructure advancement underscoring its importance in meeting the nation's growing construction and industrial needs.

Cement production is a multi steped process comprising various activities. The process begins with extraction of raw material such as dolomite from quarries. The process progresses through pyro-processing thus forming clinkers. Initially, raw materials undergo conditioning through crushing, grinding, drying, and blending. Limestone is combined with various additives to introduce essential components such as calcium Ca, SiO_2, Al, and Fe_2O_3. This blended mixture, referred to as raw meal, is heated in pre calciner and pre heater, a key element in present cement manufacturing to start the dissociation of $CaCO_3$ into CaO and CO_2. After pre-calcination, which processes approximately 20–40% of the raw material and enhances energy efficiency, the raw meal enters into the kiln. Within the kiln, temperatures reach up to 1450 °C, causing calcium oxide to react with other compounds to form Ca_2SiO_4 and aluminates. The output from this stage is clinker, a nodular material that is cooled and followed by ground with about 5% gypsum or alternative materials like slag, fly ash, pozzolans, and industrial by-products from steel production, ceramics, or construction and demolition waste. The addition of gypsum or other materials is essential for the final cement product, which is ground to the desired fineness and packaged for distribution. Ensuring high-quality cement production is particularly important when incorporating alternative raw materials. The clinker's typical composition must be achieved through the precise mixing of raw materials and ash. To maintain quality, the composition of ash must be carefully analyzed before it is added to the raw meal, ensuring that the manufacturing process meets stringent standards. The detailed steps in cement manufacturing and the chemical reactions involved are as follows:

2 Raw Material Preparation:

The raw materials after quarried from the rocks are crushed and shifted to pre-blending storage and materials like iron, slag, bauxite, sand, and fly ash are mixed and maintained the chemical composition.

Homogenization

To reduce the particle size, raw materials are grinded in the raw mill following homogenization for production of clinker.

Preheating

Raw material mixture is heated into oxides of ion pre-heated cyclones in the presence of natural gas, coal, or petroleum.

Kiln Phase

Clinker requires a multi stepped heat treatment and is done in kiln. There are five steps in the kiln phases:

1. *Free water evaporation*: Water from raw materials requires to evaporate at above the atmospheric pressure and in the process, evaporation stops gradually with a temperature above 120 °C.
2. *Decomposition phase*:

    (a) Decomposition of clay: The common clay material alkalis of the raw material is kaolinite [$Al_2Si_2O_5$ $(OH)_4$]. At high temperature, the kaolinite reacts with acid gases in the kiln.

$$Si_2Al_2O_5(OH)_2 \rightarrow 2SiO_2 + Al_2O_3 + 2H_2O_{(g)}$$

$$\underset{\text{(Orthoclase)}}{KAlSi_3O_8} + 0.5SO_2 + 0.25O_2 \rightarrow 3SiO_2 + 0.5Al_2O_3 + 0.5K_2SO_4$$

(b) Decomposition of dolomite: Dolomite, silicate, carbonate are the magnesia of the raw-mill. In the process, dolomite reacts like;

$$CaMg(CO_3)_2 \rightarrow CaCO_3 + MgO + CO_2$$

(c) In the process, non-carbonate magnesium compounds react as follows:

$$\underset{\text{(Phlogopite)}}{KMg_3AlSi_3O_{10}(OH)_2} + 0.5SO_2 + 0.25O_2 \rightarrow 0.5K_2SO_4 + 3MgO$$
$$+ 0.5Al_2O_3 + 3SiO_2 + H_2O_{(g)}$$

(d) Besides these steps, pure calcite present in the kiln is decomposed at 650 °C.

3. *Combination of decomposition products*:

$$2CaCO_3 + SiO_2 \rightarrow Ca_2SiO_4 + 2CO_2$$

(a) The reactive products obtained from clay and dolomite decomposition combine to liberate CO_2 as shown in the equation. In addition, magnesia reacts with silica to form forsterite.

$$2MgO + SiO_2 \rightarrow \underset{\text{(forsterite)}}{Mg_2SiO_4}$$

(b) Phosphorus reacts with free silica to produce whitlockite and this stage completes as all the silica is used up.

$$\underset{\text{(whitlockite)}}{Ca_5(PO_4)3OH} + 0.25SiO_2 \rightarrow 1.5Ca_3(PO_4)_2 + 0.25Ca_2SiO_4 + 0.5H_2O_{(g)}$$

(c) Reaction between alumina and iron oxide: in this process, $CaCO_3$ reacts with other oxides without forming the free lime. $Ca_{12}A_{14}O_{33}$ is formed, though tricalcium aluminate is stable.

$$12CaCO_3 + 7Al_2O_3 \rightarrow Ca_{12}Al_{14}O_{33} + 12CO_2$$
$$4CaCO_3 + Al_2O_3 + Fe_2O_3 \rightarrow Ca_4Al_2Fe_2O_{10} + 4CO_2$$
$$4CaCO_3 + Al_2O_3 + Mn_2O_3 \rightarrow Ca_4Al_2Mn_2O_{10} + 4CO_2$$

4. *Decomposition of left calcite*: The remaining calcite forms free lime.

$$CaCO_3 \rightarrow CaO + CO_2$$

5. *Sintering*: In this process, when the kiln temperature reaches 1300 °C, clinker flux is formed. When liquid is formed due to variation in the rate of melting, alite begins to form as per the following reaction:

$$Ca_2SiO_4 + CaO \rightarrow Ca_3SiO_5$$

The alite formed above acts as an ion exchange among the solid phase. It drags the reactants by surface tension to make them together. As a result, the mixtures of solids are converted into spherical nodules which are present in clinker.

3 Cooling Phase and Storing

The mixture kiln is quickly cooled to 100–200 °C from 2000 °C producing clinker which is then stored for cement production.

Table 1 shows the composition of cement highlighting the important ingredients. Important ingredients include, lime, silica, sulfur trioxide, alumina, iron oxide, calcium sulfate, magnesia etc.

Co-processing in Cement Kilns as an Option for Management of Municipality Solid Wastes: The cement industry is categorized as a resource intensive industry consuming substantial amounts of natural resources for raw materials and fuel. Globally, the sector accounts for approximately 3.84% of total primary energy consumption and nearly 5% of industrial energy consumption (CII & Shakti Sustainable Energy Foundation report, 2011). In India, coal remains the primary energy source for cement production; however, the rapid depletion of fossil fuels necessitates the exploration of alternative energy sources. Additionally, the industry faces increasing pressure to meet its social responsibility by adopting sustainable practices, reducing fossil fuel dependency, and minimizing greenhouse gas emissions. The diminishing availability of natural raw materials further complicates the situation, making it increasingly difficult to secure land for mineral deposits critical to cement production. As the demand for cement in India continues to rise, it is imperative for the industry to adopt alternative fuels and raw materials to ensure long-term sustainability. Simultaneously, waste management has become a pressing challenge in India due to rapid urbanization and industrial expansion. Lack of suitable infrastructure is a major hurdle in efficient waste management in India hence leading to both environmental and social concerns. Recognizing the critical importance of waste management, the Prime Minister's National Action Plan on Climate Change has recognized urban waste management as a key pillar of sustainable economic development. Addressing this issue is essential not only for supporting the ecological balance but also for ensuring a sustainable future for industries and communities alike.

Co-processing is a sustainable and eco-friendly waste disposal method that offers significant advantages over conventional practices like landfilling and incineration. This approach minimizes emissions and eliminates residual waste post-treatment.

Table 1 Cement composition highlighting the important ingredients

Sl. No.	Ingredients	Importance in cement manufacturing
1	Lime–Calcium oxide–CaO/ Calcium hydroxide–$Ca(OH)_2$	Form aluminates and silicates of calcium and the insufficient reduces the cement strength and can set the cement quickly and excess of it makes unsound and can cause the cement disintegrate and expand
2	Silica–Silicon dioxide–SiO_2	It provides strength to the cement for binding, which should be present about 30%
3	Sulfur trioxide–SO_3	Should not be higher than 2% since it causes unsound to cement
4	Alumina–Aluminum oxide–Al_2O_3	It provides fast setting property to the cement and excess can reduce the cement strength. Appropriate quantity can reduce the clinkering temperature
5	Iron oxide–Fe_2O_3	It gives cement color and behaves as a flux. Under higher temperature, reacts with calcium and aluminum forming tricalcium alumino-ferrite which gives more strength to the cement
6	Calcium sulfate–$CaSO_4$	Added in the form of gypsum which retards the setting property of the cement
7	Magnesia–Magnesium oxide–MgO	Should not be present more than 2%, otherwise it will reduce the cement strength
8	Alkaline	It should not exceed more than 1%, because excess can cause efflorescence

Lime, silica, sulfur trioxide, alumina, iron oxide, calcium sulfate, magnesia are the important ingredients

It involves using waste materials as different fuels and raw materials in industrial processes, particularly in cement production, to recover both energy and materials. Figure 1 shows the various stages for cement manufacturing and the processing and integration of RDF in the process (Hajinezhad et al., 2016).

Cement kilns, operating at extremely high temperatures, enable the efficient disposal of various waste types without generating harmful emissions. Recognized under the Basel Convention, co-processing is acknowledged as a safe and sustainable solution for managing a wide array of wastes, including hazardous and municipal solid waste. Consequently, numerous countries have adopted co-processing as a standard practice for environmentally responsible waste disposal.

Fig. 1 Different stages of cement manufacturing and the processing and integration of RDF in the process

4 Suitability of Municipality Solid Waste Use as Alternative Fuel Resource in Cement Plant

Various waste materials are increasingly being used as alternative fuels, but their specifications must be carefully studied to ensure compliance with the standards for health, safety, and environmental before being introduced into the cement units. These requirements include a proper chemical composition, adequate organic content, and physical properties that are sufficiently uniform and manageable for shifting, such as size and density (Madlool et al., 2011). Additionally, the fuels must possess a high value of calorie, capable of releasing a significant energy into the combustion chamber during the process.

Cement manufacturing is both carbon and energy intensive techniques which contributes to CO_2 emissions (Wang et al., 2023). More than 50% of the natural CO_2 emissions arise from the calcination process, where $CaCO_3$ is heated to produce CaO. An additional 40% of CO_2 emissions result from the combustion of fossil-based solid fuels used to heat the clinker to high temperatures and power pre-heaters and other operational systems (Lehne & Preston, 2018). The remaining 10% of CO_2 emissions stem from the transportation of raw materials. Approximately 0.9 tons of

CO_2 are emitted for each ton of cement production. Additionally, significant energy resources are required to operate the pre-calciner tower and the kiln's final stages.

Globally, solid fuels that are fossil-based like coal are predominantly used to generate thermal energy for the calcination process in the preheater tower and clinkerization process in kiln. These energy requirements constitute 30–40% of the industry's operating costs (Rahman et al., 2013). Rising operating costs driven by fluctuations in natural energy prices have prompted cement production globally to explore alternative feed stocks as potential replacements for fossil-based fuels (Nakomcic-Smaragdakis et al., 2016). Table 2 shows the various components of municipal solid wastes and techniques applied to prepare refuse-derived fuel.

RDF such as tires, plastics, sewage sludge, petroleum pellets, paper waste, household waste, agricultural waste, and wood, offer a viable choice as alternative biofuels for cement production. One key advantage of RDF is its cost-effectiveness, as it is more economical to accommodate a cement kiln for RDF usage than to construct a dedicated waste incineration facility (Horsley et al., 2016). The resources are renewable compared to coal. MSW is also considered as a re-burning fuel and re-generable biofuel in the process of co-firing since about 10–20% of total input heat-reduced NOx emissions by about 70% (Nanda & Berruti, 2021).

In the European Union, the breakdown of alternative fuels used are like plastics (37%), mixed industrial wastes (18%), tires (15%), other fossil-based wastes (7%), biomass (7%), animal bone meal and fats (6%), solvents (5%), impregnated sawdust

Table 2 Components of municipal solid wastes and the techniques applied to prepare refuse-derived fuel

Sl. No.	Concept	Technology applied	Components of MSW
1	Drum screen, disc screen, vibrating screen	Size classification	Glass, plastic, fines, organics, paper
2	Magnetic separation	Magnetic properties of the components in a mixture	Ferrous metals from nonferrous metals
3	Physical separation	Personal observation	Large-sized materials, PVA
4	Air classification	Particle separation as per their size represented by particle equivalent diameter	Plastic, metals, paper from mixed waste
5	Eddy current separation	Conductivity	Nonferrous metals which are caught and other materials falls off because of gravity
6	Ballistic separation	Mechanical motion	Paper and light-weight packing from construction and demolition mixed wastes
7	Shredding	Cutting and tearing	Films, hard plastic, wood, paper
8	Pulverization	Crushing and grinding	Ore, glass, ceramics, grains, medicines

(3%), and waste oil (2%) (European recycling policy and the circular economy–2015). The co-processing of various waste types in cement kilns as alternative raw materials and fuels is a proven global solution that significantly addresses waste management challenges while helping the cement industry mitigates resource depletion concerns. This approach represents a win–win solution to both the cement industry and society as a whole.

5 Conversion Process of Municipality Solid Waste to Refuse-Derived Fuel

Refuse-derived fuel is the product of processing MSW and is a widely used another fuel in many European Nations. Nations such as Italy, Belgium, and Netherlands have minimum one cement kiln utilizing RDF (Tihin et al., 2023). Before being fed into cement kilns, MSW must be kept to remove recyclable, inert, and sometimes moist fractions. Remaining material, which constitutes approximately 20–50% of the total MSW weight, can either be incinerated or pelletized for use. RDF sized < 25 mm (two dimensions) is baled and shifted to cement plant. In the plant, it is de packed and fired in the burner either pneumatically or by mechanical carrying.

The first method involves source separation, where waste is sorted into different bags or bins by type, using color-coded systems. The second method relies on mechanical treatment at specialized facilities, which involves processes such as crushing, shredding, screening, separation through magnetic method, eddy current technology for separation, air current separation, and the removal of biodegradable components, metals, and glass (Climate Technology Centre & Network, 2015). Pre-treatment techniques are essential for maximizing the recovery of recyclable materials, reducing waste moisture content, and homogenizing and stabilizing its combustibility. These steps simplify handling, shifting, and storage processes.

The composition, typology like fluff, dry fluff, and pellets, and technological features of processing facilities are key factors influencing the life-cycle performance of RDF and the direct emissions from WtE facilities. These elements are critical for the effective utilization of RDF in cement industry (Longo et al., 2020). Therefore, it is essential to implement strict internal control protocols to analyze the specifications and characteristics of RDF, ensuring compliance with global standards and meeting the minimum quality criteria for fuel standards.

RDF primarily comprises 9.3–42.3 wt% of plastics, 5.3–25.6 wt% of paper or biogenic waste, and 2.1–18.8 wt% of textiles (Tihin et al., 2023). Apart from that, RDF contains inorganic components (non-combustible), as ash content, which ranges from 5.7 to 30.6 wt% (Aldrian et al., 2020). RDF is classified by its energy content, usually between 10 and 25 MJ/kg, and contains a considerable biogenic fraction of approximately 50–65% on an energy basis. The MSW are pretreated either through mechanical or mechanical–biological treatment thus enhancing its energy recovery potential and optimization of wastes type for cycling in to energy.

Mechanical treatment plants typically include steps such as mechanical sorting, size reduction, separation and screening, sieving, blending, drying, and pelletizing. These methods collectively improve the RDF quality and composition by integrating advanced technologies and systematic approaches for waste sorting, recycling, and recovery.

The inherent heterogeneity of MSW presents challenges for energy recovery, underscoring the importance of thorough fractionation and sorting during the pre-treatment phase. Proper segregation ensures the efficient use of energy conversion technologies, such as thermo-chemical processes. Collaboration between public, government, and private sector is required for successful sorting and segregation of wastes. This collaboration should give priority to separation of waste at source level and sorted it as per types and composition. Development of National standards and guidelines has been made in countries like China and India for pre-treatment of domestic waste to promote RDF production (Martignon, 2020). This effort highlights the role of RDF in energy recovery and sustainable waste management. This initiatives aims to managing huge volume of MSW apart from meeting the energy demands of cement industry. The guidelines are designed to address the market barriers caused by absence of uniform procedures and criteria for ensuring RDF quality and quantity.

6 Potential Benefits of Co-processing RDF

Cement furnaces continuously operate, sustaining decomposition conditions and stable material temperatures over time. Their large heat capacity ensures that even during an emergency shutdown, the temperature fluctuates gradually for several minutes thus allowing complete incineration of fuels. The combustion process in rotary cement furnaces is totally waste-free, as the ash generated during fuel incineration is incorporated into the resulting clinker. Due to alkaline environment of cement furnaces, the acid gases emitted during use of alternative fuels combustion are captured and neutralized before its release in to atmosphere. This process efficiently cleans the emission apart from removing heavy metal emissions as the metals condense on to dust particles and are captured by dust removal system in later stage. They are again re integrated into the process ultimately becoming a part of the clinker.

Addressing Local Environmental Issues: Co-processing RDF from MSW significantly decreases the amount of waste sent to landfills. This method helps address local waste management challenges, preserves land resources, and lowers the costs of landfill construction and maintenance.

CO_2 Reduction Potential: Co-processing RDF in cement kiln drastically reduces greenhouse gas such as methane emissions from landfill area. Similarly due to replacement of fossil fuels in cement kilns results in reduction of CO_2 emissions with the potential to save approximately 1.7 million tons of CO_2 annually (Mitra et al., 2024).

Coal Saving Potential of RDF: Annually million tons of coal can be replaced by RDF in the cement industry. Co-processing 1.37 million tons of RDF annually results in a Thermal Substitution Rate of 3.2% which indicates the proportion of fossil fuel replaced (Mitra et al., 2024). This approach offers a sustainable recovery of energy value of waste thus achieving solid waste management. Co- processing has been recognized as an alternative to the traditional landfilling method in accordance with various international environmental agreements such as Stockholm and Basel Conventions. Formalizing co-processing through launching a national-level mission will give it recognition as a priority recovery method. This initiative would play a crucial role in advancing the goals of the National Mission on Sustainable Habitats, Nationally Appropriate Mitigation Actions & National Action Plan on Climate Change thus contributing to India's broader sustainability and climate objectives.

7 Co-processing of MSW in Cement Plant: An Approach in Circular Economy

In present time, CE concept, has garnered widespread recognition as a key solution to waste management issues thus achieving environmental sustainability. The 3R approach: reduce, reuse, and recycle tries to emphasize on minimizing waste and pollution thus extending the lifespan of products apart from regenerating the natural systems (Bala et al., 2024; Mishra & Das, 2021). CE principles are gaining recognition in various sectors including solid and liquid waste management at international level by integrating environmental well-being with economic activity. Unlike traditional "take-make-dispose" approach of linear economy, CE focuses on transforming waste into its original form without compromising its quality through true recycling-transforming principle (Ray et al., 2024; Sahu et al., 2024).

This approach not only supports sustainable business models but also reduces emissions and enhances the efficient use of natural resources. The CE framework for managing municipal solid and liquid waste received a significant boost with the launch of the Swachh Bharat Mission-Urban in 2014. Figure 2 depicts the concept of circular economic in the context of concrete/cement industry.

In response to waste generation challenges, NITI Aayog identified 11 focus areas that require a comprehensive CE approach. The cement industry has emerged as a key player in sustainable waste management through co-processing, which combines the recycling of mineral materials with energy recovery in the cement manufacturing process. Life cycle assessment studies consistently show that co-processing is a more effective waste management solution compared to landfill and incineration. This process involves using waste materials as fuel substitute or raw materials in industrial operations, such as cement plants and thermal power stations, to recover both energy and materials. MSW, non-recyclable hazardous waste, plastic waste, tire waste, and biomass represent key waste streams with significant potential for use in

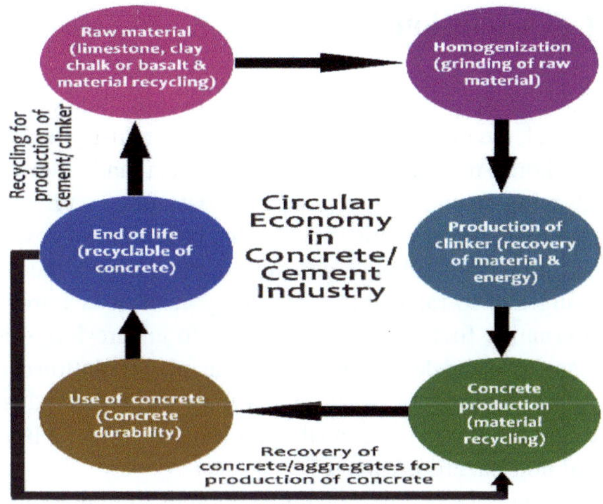

Fig. 2 Circular economy in concrete/cement industry. It involves raw material, homogenization, clinker production, concrete production, concrete use, and end of life along with their integration

cement plants, with MSW expected to play a key role in alternative fuel utilization within the cement industry.

8 Key Challenges

Refuse-derived fuel is commonly used in cement industry due to its compatibility with co-processing in kilns and the industry's significant thermal energy requirements. However, several challenges hinder the effective utilization of MSW as RDF in the Indian cement sector. These challenges include waste quality, infrastructure limitations, environmental concerns, techno-economic feasibility, and fuel feeding issues. One primary concern is the inconsistent quality of waste supplied to cement plants. High moisture and ash content in non-homogeneous waste may cause in equipment failures thus interfering with cement production processes. Certain physio-chemical requirements, such as having a high heating value, little ash, and low chloride value of RDF need to be maintained to make it compatible for use in cement kiln. Many RDF suppliers fail to adhere the above mentioned standards apart from development of adequate infrastructure for processing of RDF.

Though ongoing advancements in co-processing of RDF offer hope for future improvements, ensuring its usage are environmentally sustainable is another challenge. Lack of comprehensive data and studies on the techno-economic analysis of RDF is another key challenge for its adoption. The higher mass flow rates requirement of co-processing RDF can lead to operational issues more particularly in fuel feeding systems. Similarly excessive chlorine content in RDF poses risks of corrosion during clinker production. The above challenges need to be addressed for optimizing use of RDF in cement industry thus achieving sustainable waste management practices.

9 Conclusion

There is an urgent need for exploration of alternative resources in cement industry as coal, the primary source of energy in industry is depleting day by day. The corporate house must also prioritize their social and environmental responsibility through adopting sustainability practices apart from addressing the scarcity of traditional source of energy. This includes the reduction of fossil fuel consumption thus minimizing the greenhouse gas emissions. Though the demand for cement in India is rising due to rapid infrastructure growth, it is imperative for the industry to seek alternative fuels and raw materials to ensure long-term viability as both are non-renewable and declining fast. Increasing population and urbanization is putting stress on management of waste. Apart from this rapid industrial growth which is vital for economic growth is contributing to generation of large volumes of both hazardous and non-hazardous waste.

India is facing significant challenges in managing the generation of voluminous waste due to inadequate technology, space, and infrastructure. While economic growth is a frequent topic of debate, less attention has been put on management of this waste. Effective waste management is critical in achieving countries sustainable development apart from reducing the global greenhouse gas emissions. Co-processing of MSW in cement kilns gives a suitable and eco-friendly solution for management of waste. Apart from partially substituting the fossil fuels demand of a plant, this approach also has the potential for efficient management of waste. The cement industry can contribute significantly to sustainability and waste management by adoption of co-processing. The MoEFCC recently replace the Municipal Solid Wastes (Management and Handling) Rules, 2000 with Solid Waste Management Rules, 2016 which highlights the promotion of WtE in cement plants adopting co-processing systems. Mandatory use of solid waste-based RDF and replacing at least 5 percent of their future requirement by RDF is the key highlights of this rule which signify the intention of government to address the solid waste management issue. The CPCB has also formulated a comprehensive set of pre-processing and co-processing guidelines of hazardous and other wastes for cement plant. The whole objective behind this to ensure management of tones of MSW either properly recycled or managed sensibly to minimize India's overall environmental footprint. This increasing demand of fuel and raw material for the cement industry can be partially satisfied through co-processing of waste. Most of the cement plants in the country have already adopted and few are in pipeline of setting up processes to use AFR in their kiln.

References

Aldrian, A., Viczek, S. A., Pomberger, R., & Sarc, R. (2020). Methods for identifying the material-recyclable share of SRF during co-processing in the cement industry. *MethodsX*. https://doi.org/10.1016/j.mex.2020.100837

Amreen, S., Dash, S. K., Naik, B., Dash, A. K., Verma, A. K., & Pradhan, A. (2024). Marine macro algae-derived activated carbon an efficient bio-sorbent for Malachite green removal from wastewater. *Environmental Quality Management*. https://doi.org/10.1002/tqem.22213

Aranda Uson, A., Lopez-Sabir, A. M., Ferreira, G., & Llera Sastresa, E. (2013). Uses of alternative fuels and raw materials in the cement industry as sustainable waste management options. *Renewable and Sustainable Energy Reviews, 23*, 242–260. https://doi.org/10.1016/j.rser.2013.02.024

Bala, K., Anshul, Verma, S., Dogra, A., Das, A. P., Kumar, R., & Nandini. (2024). Sustainable management for urban plastic waste generation. In *Sustainable management of urban plastic waste through circular economic approaches* (pp. 211–236). https://doi.org/10.1201/9781003465447-13

Cast study manual on alternative fuels & raw material use in Indian cement Industry. Confederation of Indian Industry & Shakti Sustainable Energy Foundation, 2011.

Central Pollution Control Board. (2022). Annual report on solid waste management, 2021–2022. CPCB.

Climate Technology Centre & Network. (2015). Feasibility study to use waste as fuel for cement factories, 2015. https://www.ctc-n.org/system/files/dossier/3b/output_1_technical_study_reviewed.pdf. Accessed July 11, 2023.

Das, A. P., Kumar, P. S., & Swain, S. (2014a). Recent advances in biosensor based endotoxin detection. *Biosensors & Bioelectronics, 51*, 62–75.

Das, A. P., & Mishra, S. (2010). Biodegradation of the metallic carcinogen hexavalent chromium Cr (VI) by an indigenously isolated bacterial strain. *Journal of Carcinogenesis, 9*, 6. https://doi.org/10.4103/1477-3163.63584

Das, A. P., Ghosh, S., Mohanty, S., & Sukla, L. B. (2014b). Consequences of manganese compounds: A review. *Toxicological and Environmental Chemistry, 96*(7), 981–997.

Das, A. P., & Singh, S. (2011). Occupational health assessment of chromite toxicity among Indian miners. *Indian Journal of Occupational and Environmental Medicine, 15*(1), 6–17.

Das, A. P., Sukla, L. B., Pradhan, N., & Nayak, S. (2011). Manganese biomining: A review. *Bioresearch Technology, 102*(16), 7381–7387.

Das, A., & Ghosh, S. (2022). Role of microorganisms in extenuation of mining and industrial wastes. *Geomicrobiology Journal, 39*, 173–175. https://doi.org/10.1080/01490451.2022.2038953

Das, A., & Mishra, S. (2008). Hexavalent Chromium (VI): Health hazards and environmental pollutant. *Journal of Environment Research and Development, 2*, 386–392.

Das, A., Ray, S. K., Mohanty, M., Mohanty, J., Dey, S., & Das, A. P. (2024). Ecotoxicity of microplastic wastes and their sustainable management: A review. *Environmental Chemistry and Ecotoxicology, 6*, 144–152. https://doi.org/10.1016/J.ENCECO.2024.05.003

Dash, A. K. (2012). Impact of domestic wastewater on seed germination and physiological parameters of wheat and rice. *International Journal of Recent Research and Applied Sciences, 12*(2), 280–286.

Dash, A. K., & Dash, S. K. (2017). Atmospheric pollution load assessment through air quality index: A case study. *Indian Journal of Environmental Protection, 37*(9), 736–741.

Dash, A. K., & Mishra, P. C. (1997). Blue–green alga in sewage—amended paper mills waste water. *International Journal of Environmental Studies, 53*, 9–10.

Dash, A. K., & Mishra, P. C. (1999a). Role of the blue–green alga, *Westiellopsis prolifica* in reducing pollution load from paper mill wastewater. *Indian Journal of Environmental Proctection, 19*(1), 1–15.

Dash, A. K., & Mishra, P. C. (1999b). Growth response of the blue-green alga, *Westiellopsis prolifica* in Sewage enriched paper mill waste water. *Revista Internacional De Contaminación Ambiental, 15*(2), 79–83.

Dash, A. K., & Patra, H. S. (2016). Floristic composition and biological spectrum study near the mining cluster at Keonjhar district, Odisha, India. *International Journal of Engineering Sciences and Research Technology, 5*(2), 441–447.

Dash, A. K., Pradhan, A., Das, S., & Mohanty, S. S. (2015). Fly ash as a potential source of soil amendment in agriculture and a component of integrated plant nutrient supply system. *Journal of Industrial Pollution Control, 31*(2), 251–259.

Dash, A. K., Pradhan, A., & Patra, H. S. (2016). Soil physico- chemical characteristics near the mining areas of Keonjhar District, Odisha, India. *International Journal of Current Research, 8*(5), 30306–30309.

Dash, A. K., Pradhan, A., & Sukla, L. B. (2024). Environmental occurrence and contemporary health issues of micro plastics. In A. P. Das, I. D. Behera, & N. P. Das (Eds.), *Renewable energy generation and value addition from environmental microfiber pollution through advanced greener solution. Environmental science and engineering* (pp. 113–136). Springer. ISBN: 978-3-031-51792-1. https://doi.org/10.1007/978-3-031-51792-1_7

Dash, S. K., & Dash A. K. (2018a). Air pollution tolerance index to assess the pollution tolerance level of plant species in industrial areas. *Asian Journal of Chemistry, 30*(1), 219–222. https://doi.org/10.14233/ajchem.2018.20991

Dash, S. K., & Dash, A. K. (2018b). Seasonal variation of air quality index and assessment. *Global Journal of Environmental Science and Management, 4*(4), 483–492. https://doi.org/10.22034/gjesm.2018.04.008

Dash, S. K., Pradhan, A., Panda, B. P., & Dash, A. K. (2023). Ambient air quality monitoring and assessment of an industrial area as a tool for sustainable environmental management. *Material Today Proceedings*. https://doi.org/10.1016/j.matpr.2023.10.024

Dash, S. K., Sahu, H. B., & Dash, A. K. (2019). Atmospheric dispersion modeling using AERMOD to predict the impact of PM_{10} near Bileipada, Odisha. *Indian Journal of Environmental Protection, 39*(4), 299–306.

Dey, S., Das, A., Mallick, K., Sahu, A., & Das, A. P. (2023). *Environmental petroleum waste: Pollution, toxicity, sustainable remediation* (pp. 159–175). https://doi.org/10.1007/978-3-031-48220-5_7

Ellis, C. (2018). World bank: Global waste generation could increase 70% by 2050. *WasteDive*. https://www.wastedive.com/news/world-bank-globalwaste-generation-2050/533031/

Hajinezhad, A., Halimehjani, E. Z., & Tahani, M. (2016). Utilization of refuse derived fuel (RDF) from urban waste as an alternative fuel for cement factory: A case study. *International Journal of Renewable Energy Research, 6*(2), 114205057. https://doi.org/10.20508/ijrer.v6i2.3170.g6837

Horsley, C., Emmert, M. H., & Sakulich, A. (2016): Influence of alternative fuels on trace element content of ordinary Portland cement. *Fuel, 184*: 481–489. https://doi.org/10.1016/j.fuel.2016.07.038

Indian cement industry report. By: India Brand Equity Foundation, November, 2024.

Jąderko-Skubis, K. (2021). Production of alternative fuels from waste: assumptions for the design of new fuel recipes. *International Journal of Sustainable Engineering, 14*, 1157–1169. https://doi.org/10.1080/19397038.2020.1836063

Jena, P. S., Pradhan, A., Nanda, S. P., Dash, A. K., & Naik, B. (2022). Biosorption of heavy metals from wastewater using Saccharomyces cerevisiae as a biosorbent: A mini review. *Materials Today Proceedings, 67*(8), 1140–1146. https://doi.org/10.1016/j.matpr.2022.07.306

Karki, K. K., & Dahal, K. R. (2020). Reduction of coal consumption by using municipal combustible waste as a fuel in cement Kiln with special focus on Hetauda cement industry in Nepal. *Journal of Civil, Construction and Environmental Engineering, 5*, 35. https://doi.org/10.11648/j.jccee.20200503.11

Kukreja, K., Soni, M. K., Mohapatra, B., & Rao, M. V. R. (2023). De-carbonising of the Indian Cement Industry through alternative fuels—Challenge of transfer chute jamming. *Asian Journal of Water, Environment and Pollution, 5*, 71–77. https://doi.org/10.3233/AJW230067

Kumar, D. S., Bhushan, S. H., & Kishore, D. A. (2018). Atmospheric dispersion model to predict the impact of gaseous pollutant in an industrial and mining cluster. *Global Journal of Environmental Science and Management, 4*(3), 351–358. https://doi.org/10.22034/gjesm.2018.03.008

Lehne, J., & Preston, F. (2018). Making concrete change: Innovation in low-carbon cement and concrete. The Royal Institute of International Affairs. Chatham House Report Series, www.chathamhouse.org/sites/default/files/publications/research/2018-06-13-makingconcrete-c, 138. www.chathamhouse.org

Longo, S., Cellura, M., & Girardi, P. (2020). Life Cycle Assessment of electricity production from refuse derived fuel: A case study in Italy. *Science of the Total Environment, 738*, Article 139719. https://doi.org/10.1016/j.scitotenv.2020.139719

Maalouf, A., Mavropoulos, A., & El-Fadel, M. (2020). Global municipal solid waste infrastructure: Delivery and forecast of uncontrolled disposal. *Waste Management and Research, 38*(9), 1028–1036. https://doi.org/10.1177/0734242X20935170

Madlool, N. A., Saidur, R., Hossain, M. S., & Rahim, N. A. (2011). A critical review on energy use and savings in the cement industries. *Renewable and Sustainable Energy Reviews, 15*, 2042–2060. https://doi.org/10.1016/j.rser.2011.01.005

Mahapatra, B., Dhal, N. K., Dash, A. K., Panda, B. P., Panigrahi, K. C. S., & Pradhan, A. (2019). Perspective of mitigating atmospheric heavy metal pollution: Using mosses as biomonitoring and indicator organism. *Environmental Science and Pollution Research, 26*(29), 29620–29638. https://doi.org/10.1007/s11356-019-06270-z

Martignon, G. P. (2020). Trends in the use of solid recovered fuels, 2020. https://www.ieabioenergy.com/wp-content/uploads/2020/05/Trends-in-use-of-solidrecovered-fuels-Main-Report-Task36.pdf. Accessed July 11, 2023.

Mateus, M. M., Neuparth, T., & Cecflio, D. M. (2023). Modern Kiln burner technology in the current energy climate: Pushing the limits of alternative fuel substitution. *Fire.* https://doi.org/10.3390/fire6020074

Mishra, P. C., Dash, A. K., & Pradhan, K. (2012). Metals in environmental segments at Hirakud of Odisha, India. *ISCA Journal of Biological Sciences, 1*(1), 7–27.

Mishra, S., & Das, A. P. (2021). Current treatment technologies for removal of microplastic and microfiber pollutants from wastewater. In *Wastewater treatment: Cutting-edge molecular tools, techniques and applied aspects* (pp. 237–251). https://doi.org/10.1016/B978-0-12-821881-5.00011-8

Mishra, S., & Das, A. P. (2023). Microbial remediation of synthetic microfiber contaminated wastewater. In M. P. Shah (Ed.), *Microbial technologies in industrial wastewater treatment* (pp. 337–350). Springer. https://doi.org/10.1007/978-981-99-2435-6_17

Mishra, S., Ghosh, S., van Hullebusch, E. D., Singh, S., & Das, A. P. (2023). A critical review on the recovery of base and critical elements from electronic waste-contaminated streams using microbial biotechnology. *Applied Biochemistry and Biotechnology, 2023*, 1–30. https://doi.org/10.1007/S12010 023 04440 X

Mishra, S., Rout, P. K., & Das, A. P. (2019). Solar photovoltaic panels as next generation waste: a review. *Biointerface Research in Applied Chemistry, 9*(6).

Mitra, A., Sharma, S., Bhattacharya, S., Garg, A., Devotta, S., & Sen, K. (2024). *Climate change and india uncertainty reduction in greenhouse gas inventory estimates.* University Press. ISBN (Print) 81-7371-517-3.

Mohanty, S., Ghosh, S., Bal, B., & Das, A. P. (2018). A review of biotechnology processes applied for manganese recovery from wastes. *Reviews in Environmental Science and Bio/Technology, 17*(4), 791–811. https://doi.org/10.1007/S11157-018-9482-1

Naik, B., Sukla, L. B., & Dash, A. K. (2024a). Sources of synthetic microfiber pollutants in the environment and their fate and toxicity. In A. P. Das, I. D. Behera, & D. Bhanja (eds.), *Sustainable microbial technology for synthetic and cellulosic microfiber bioremediation.* Springer. https://doi.org/10.1007/978-3-031-62600-5_1

Naik, B., Sukla, L. B., & Dash, A. K. (2024b). Innovative methods and greener technology for remediation of microfiber pollutants. In *Prospects of science, technology and applications* (pp. 295–315). https://doi.org/10.1201/9781003489443-32

Nakomcic-Smaragdakis, B., Cepic, Z., Senk, N., Doric, J., & Radovanović, L. Z. (2016). Use of scrap tires in cement production and their impact on nitrogen and sulfur oxides emissions. *Energy Sources Part A, 38*, 485–493. https://doi.org/10.1080/15567036.2013.787473

Nanda, S., & Berruti, F. (2021). A technical review of bioenergy and resource recovery from municipal solid waste. *Journal of Hazardous Materials, 403*, Article 123970. https://doi.org/10.1016/j.jhazmat.2020.123970

Nayak, K. P., Mishra, A., Pradhan, A., Dash, A. K., & Saren, S. (2018). Soil fertility study as a tool for sustainable agriculture. *"NPI" International Research Journal of Industrial and Environmental Biotechnology (IRJIEB), 1*(1), 57–74.

Panigrahi, B., & Dash, A. K. (2024). Sustainable management for reduction of plastic waste through circular economy. In A. P. Das, & A. John (eds.), *Sustainable management of urban plastic waste through circular economic approaches.* CRC Press. https://doi.org/10.1201/9781003465447

Panigrahi, B. R., Pradhan, S. C., Dash, A. K., & Mishra, P. C. (2024). Evaluation of protease activity in the soil of municipal solid waste dumping sites in Balasore district of Odisha, India. *Pollution Research, 43*(3–4), 267–269.

Parida, B., & Dash, A. K. (2024). Sources, characteristics, environmental impact, and sustainable remediation of synthetic microfiber. In A. P. Das, I. D. Behera, & D. Bhanja, (eds.), *Sustainable microbial technology for synthetic and cellulosic microfiber bioremediation.* Springer. https://doi.org/10.1007/978-3-031-62600-5_2

Parida, B., Dash, A. K., & Verma, A. (2023). Technologies for treatment of landfill leachate: A brief review. *Recent Innovations in Chemical Engineering, 16*(4), 260–284. https://doi.org/10.2174/0124055204276847231013040437

Parida, S., Das, M., and & Dash, A. K. (2015). Ambient air study near the industrial complex 'Nisha' at Angul, Odisha, India. In *Advances in environmental sciences and engineering* (pp. 514–537). Daya Publishing House, a division of Astral Int. Pvt., Ltd. ISBN: 978-93-5124-336-6.

Patra, H. S., & Dash, A. K. (2024). Management and recycling of mining wastes with zero waste adaptation technology as a tool for sustainable environmental management. In A. P. Das, E. D. van Hullebusch, & A. Akçil (eds.), *Sustainable management of mining waste and tailings* (1st ed., 358 pages). CRC Press. https://doi.org/10.1201/9781003442455. eBook ISBN9781003442455.

Pradhan, A., Dash, A. K., Mohanty, S. S., & Das, S. (2015s). Potential use of fly ash in floriculture: A case study on the photosynthetic pigments content and vegetative growth of *Tagetes erecta* (MARIGOLD). *Ecology Environment and Conservation, 21*(August Supplement), 369–376.

Pradhan, A., Kumari, S., Dash, S., Biswal, D. P., Dash, A. K., & Panigrahi, K. C. S. (2017). Heavy metal absorption efficiency of two species of mosses (*Physcomitrella patens* and *Funaria hygrometrica*) studied in mercury treated culture under laboratory condition. *IOP Conference Series: Materials Science and Engineering, 225*, 012225. https://doi.org/10.1088/1757-899X/225/1/012225

Pradhan, A., Sahu, S. K., & Dash, A. K. (2013). Changes in pigment content (chlorophyll and carotenoid), enzyme activities (catalase and peroxidase), biomass and yield of rice plant (*Oriza sativa* L.) following irrigation of rice mill wastewater under pot culture conditions. *International Journal of Scientific Engineering and Research, 4*(6), 2706–2717.

Pradhan, A., Sahu, S. K., & Dash, A. K. (2015b). Changes in physico-chemical characteristics of soil following application of rice mill wastewater under field irrigation conditions. In *Advances in environmental sciences and engineering* (pp. 480–492). Daya Publishing House, a division of Astral Int. Pvt., Ltd. ISBN: 978-93-5124-336-6.

Pradhan, A., Sahu, S. K., & Dash, A. K. (2016). Changes in physico-chemical characteristics of soil following application of rice mill wastewater under pot irrigation conditions. *Advanced Science Letters, 22*(2), 504–510. https://doi.org/10.1166/asl.2016.6838

Priyadarsini, S., & Das, A. P. (2024). Analytical and structural characterization of waste lithium-ion batteries for their effective recycling strategy. *Environmental Chemistry and Ecotoxicology.* https://doi.org/10.1016/j.enceco.2024.12.004

Priyadarsini, S., & Das, A. P. (2025). Lithium bioleaching: A review on microbial-assisted sustainable technology for lithium bio-circularity. *Journal of Water Process Engineering, 69*, Article 106744. https://doi.org/10.1016/j.jwpe.2024.106744

Rahman, A., Rasul, M. G., Khan, M. M. K., & Sharma, S. (2013). Impact of alternative fuels on the cement manufacturing plant performance: An overview. *Procedia Engineering, 56*, 393–400. https://doi.org/10.1016/j.proeng.2013.03.138

Ray, S. K., Dey, S., Das, A. P., & John, A. (2024). Microplastic waste management through greener technologies: A review. In *Sustainable management of urban plastic waste through circular economic approaches* (pp. 237–261). https://doi.org/10.1201/9781003465447-14

Sahoo, D., Dash, A. K., & Sahu, S. K. (2017). Ambient air quality monitoring and health impact study of air pollution near Joda of Keonjhar, Odisha, India. *International Journal of Engineering Sciences and Research Technology, 6*(1), 429–434. https://doi.org/10.5281/zenodo.260069

Sahu, A., Dey, S., Mallick, K., Priyadarsini, S., Behera, I. D., & Das, A. P. (2024). A review on advanced technology for sustainable management of synthetic microplastic waste. In *Sustainable management of urban plastic waste through circular economic approaches* (pp. 262–279). https://doi.org/10.1201/9781003465447-15

Sahu, S. K., Mishra, A., & Dash, A. K. (2015). Ambient air quality around OCL India Limited at Rajgangpur, Odisha, India. In *Advances in environmental sciences and engineering* (pp. 459–479). Daya Publishing House, a division of Astral Int. Pvt., Ltd. ISBN: 978-93–5130-300-8.

Sauve, G., & Van Acker, K. (2020). The environmental impacts of municipal solid waste landfills in Europe: A life cycle assessment of proper reference cases to support decision making. *Journal of Environmental Management, 261*, Article 110216. https://doi.org/10.1016/j.jenvman.2020.110216

Shehata, N., Obaideen, K., Sayed, E. T., Abdelkareem, M. A., Mahmoud, M. S., El-Salamony, A. L. H. R., Mahmoud, H. M., & Olabi, A. G. (2022). Role of refuse-derived fuel in circular economy and sustainable development goals. *Process Safety and Environment Protection, 163*, 558–573. https://doi.org/10.1016/j.psep.2022.05.052

Shreya, Verma, A. K., Gilron, J., Oren,Y., Ronen, Z., Dash, A. K., Bhunia, P., & Dash, R. R. (2022). Effect of biological treatment on perfluoroalkyl and polyfluoroalkyl (PFASs) substances degradation. In *Current developments in biotechnology and bioengineering: Sustainable treatment technologies for pre and poly-flourakyl substances*. https://doi.org/10.1016/B978-0-323-99906-9.00004-8

Tihin, G. L., Mo, K. H., Onn, C. C., Ong, H. C., Taufiq-Yap, Y. H., & Lee, H. V. (2023). Overview of municipal solid waste- derived refuse-derived fuels for cement co-processing. *Alexandria Engineering Journal, 84*, 153–174. https://doi.org/10.1016/j.aej.2023.10.043

Wang, Y., Yi, H., Tang, X., Wang, Y., An, H., & Liu, J. (2023). Historical trend and de-carbonization pathway of China's cement industry: A literature review. *Science of the Total Environment, 891*, Article 164580. https://doi.org/10.1016/J.SCITOTENV.2023.164580

World Business Council for Sustainable Development (WBCSD). (2014). Cement sustainability initiative (CSI), guidelines for co-processing fuels and raw materials in cement manufacturing, version 2.0. ISBN: 978-2-940521-19-7.

Worldometer. (2023). World population clock: 8 billion people (LIVE, 2023). Worldometer. https://www.worldometers.info/world-population/. Accessed July 11, 2023.

Yi, S., Jang, Y. C. (2018). Life cycle assessment of solid refuse fuel production from MSW in Korea. *Journal of Material Cycles and Waste Management, 20*, 19–42. https://doi.org/10.1007/s10163-016-0541-9

Life Cycle Assessment (LCA) Studies to Evaluate the Environmental and Economic Impacts of Different Waste Management Strategies and Technologies

A. R. Pilapitiya and M. A. P. C. Piyathilaka

1 Introduction

Introduction to Life Cycle Assessment (LCA)

As defined in the ISO 14040 standards, Life Cycle Assessment (LCA) is a comprehensive analytical method used to evaluate the environmental impacts of a product, both manufactured and consumed, throughout its entire life cycle. Here, the term "product" includes services as well. This encompasses all life cycle stages from raw material extraction, production, use, end-of-life treatment, recycling, and final disposal. The primary aim of LCA is to provide a holistic assessment of environmental burdens, enabling stakeholders to make informed decisions to minimize negative environmental impacts. By systematically analyzing inputs and outputs at each stage, LCA identifies opportunities for improvement, promoting sustainable practices across various industries (ISO 14040, 2006).

An LCA can assist in,

- "Identifying opportunities to improve the environmental performance of products at various points in their life cycle,
- Informing decision-makers in industry, government, or non-government organizations,
- The selection of relevant indicators of environmental performance, including measurement techniques, and
- Marketing" (ISO 14000, 2006).

A. R. Pilapitiya · M. A. P. C. Piyathilaka (✉)
Department of Environmental Technology, Faculty of Technology, University of Colombo, Technocity, Homagama, Sri Lanka
e-mail: poorna@et.cmb.ac.lk

A. P. Das et al. (eds.), *Sustainable Municipal Waste Management with a Zero Waste Approach*, Sustainable Environmental Waste Management Strategies,
https://doi.org/10.1007/978-3-032-05950-5_14

Life cycle assessment (LCA) has a set of unique characteristics that have the potential to address some problems that no other assessment tool can address (Guinée et al., 2002). Four major characteristics identified in such a way are:

(a) **Takes a life cycle perspective**

The life cycle concept is borrowed from the field of biology into LCA. The main reason why a life cycle is considered in LCAs is that it allows to identify and prevent the threat that is shifted between lifecycle stages (Hauschild et al., 2018).

(b) **Covers a broad range of environmental issues**

Most of the tools used for the quantification of environmental impacts widely focus on climate change and global warming where LCA addressed more than 15 environmental concerns including climate change, land occupation and transformation, eco-toxicity on ecosystems and humans, eutrophication, etc. (ISO 14044, 2006). The key reason to address multiple issues through this approach is to avoid the burden on the environment when shifting from one life cycle stage to another because trying to lower one type of environmental issue can unintentionally increase many other types of environmental impacts (Hauschild et al., 2018).

(c) **Quantitative**

Results gained through an LCA always answer the question "How much does a product system potentially impact the environment?". What is meant by the quantitative nature of an LCA is that the results can be used to compare the environmental impacts of different processes and product systems (Bandara, 2020).

(d) **Based on science**

The quantification of the environmental impact of a product or system through an LCA is rooted in natural science. Flows are generally based on measurement whereas the models of relationships between emission and impact are based on proven causalities such as chemical reaction schemes using nitrogen oxides and volatile organic compounds in the formation of atmospheric ground-level ozone (Finnveden et al., 2009). Other than the scientific base, LCA requires value judgment, which is reflected by assigning weights to different types of environmental problems to evaluate the overall impact associated with a product system which is discussed below further (Hauschild et al., 2018).

The following table shows the strengths and limitations of an LCA. As the table describes, LCA takes a holistic approach by considering all stages of a product's life cycle, from raw material extraction, manufacturing, transportation, use, and disposal. This ensures that environmental impacts are assessed across the entire supply chain rather than focusing on a single phase. Unlike the methods that focus on specific aspects, LCA evaluates multiple environmental impacts such as global warming potential, resource depletion, water pollution, land use, toxicity, biodiversity, climate change impact, etc. This broad perspective helps in identifying trade-offs and unintended consequences. Furthermore, an LCA is designed to provide an objective assessment based on scientific data and standardized methodologies. It aims to minimize biases by using the best available data and assumptions, ensuring fair comparisons between different products or systems. While LCA provides comparative insights, it does not define whether a product is truly sustainable on its own. Sustainability depends on broader factors such as policy frameworks, economic viability, and social acceptability, which LCA does not directly measure. Thus, LCA is best suited for relative comparisons rather than determining absolute sustainability. Therefore, LCA is a comprehensive and objective tool for environmental assessment, but it has inherent limitations when it comes to defining absolute sustainability. This methodology is instrumental in understanding the cumulative environmental effects associated with products and services, thereby facilitating the development of strategies for environmental impact reduction. LCA is one of the different environmental management systems such as risk assessment, environmental performance evaluation, environmental auditing, and environmental impact assessment (EIA). Therefore, an LCA might not be applicable in all situations as it does not address the economic or social aspects of a product (ISO 14044, 2006).

History of LCA and Development of LCA Standards

The concept of LCA first emerged in the 1960s when society's concern was drawn towards environmental degradation and limited access to resources. LCA was rooted in the packaging industry first where it has mainly focused on energy utilization and a few emissions, inducing a massive un-coordinated method development in the US and Northern Europe (Hellweg & Milà i Canals, 2014). Primary studies were carried out in companies that had adopted the above methods internally and made few communications to stakeholders as well. After a period of quietness in the 1970s, 80s, and 90s, an increase in methodological development and international collaboration and coordination in the scientific community was observed (Simonen, 2014). Universities initiated the increased development of methods for LCA. With the strengthening of the methodologies, LCAs were applied to a wide range of products and systems by both industry and governments and the results were communicated through academic papers and industry and government reports (Guinee et al., 2011). Still, the methodological development is continued and more attention is given to the central parts of the LCA methodology and standardization-related approaches (Hauschild et al., 2018).

Importance of LCA in Evaluating Environmental Impacts

LCA is often considered a sustainability assessment tool. When the history of the concept "sustainability" is considered, it has many definitions but it gained momentum with the Brundtland Commission's report "Our Common Future" in 1987 (Hauschild et al., 2018). It comprised four dimensions in its interpretation of the term sustainability: (1) measures of welfare, (2) inter-generational equity, (3) intra-generational equity, and (4) interspecies equity. Through this, the relevance of environmental protection for dimensions 2 and 4 are then demonstrated, along with the strategy of LCA to achieve environmental protection through the reduction of environmental impacts per delivery of a function (Hauschild et al., 2018).

When environmental sustainability is considered, it has to address how we can conserve the environment? and what are the overall drivers that lead to environmental degradation? which was first addressed by Holdren and Ehrlich (1974) through the IPAT equation.

$$I = P.A.T$$

where

(I) is the environmental impact,
(P) is the population,
(A) is the per capita affluence, and
(T) is the technology factor.

Here, LCA depicts how a specific functionality can be utilized in the most eco-friendly way, or how to improve the eco-efficiency of a product. In this manner, LCA can be used as a methodology that can guide to make decisions on improving one of the three dimensions in the IPAT equation; namely the technology (T) dimension (Hauschild et al., 2018).

ISO 14044 standards on "Environmental Management—Life Cycle Assessment—Requirements and guidelines," are used by the practitioners to carry out LCAs. An LCA focuses on the environmental aspects and potential environmental impacts throughout a product's life cycle starting from raw material extraction through production, use, end-of-life treatment, recycling, and final disposal which is also known as the cradle-to-grave approach (ISO 14044, 2006).

There are four phases in an LCA study (ISO 14044, 2006):

(a) Goal and scope definition phase:

The scope of an LCA includes the system boundary and level of detail depending on the subject and the intended use of study, whereas the goal defines the depth and breadth of a particular LCA.

(b) Life Cycle Inventory (LCI) analysis phase:

LCI is an inventory consisting of input/output data such as raw materials, utilities used, wastes generated, effluents to water, air, land, etc. with reference to the system that is being studied. The data that is being collected for the inventory should satisfy the needs of the goals defined for the study.

(c) Life Cycle Impact Assessment (LCIA) phase:

The purpose of carrying out an LCIA is to present additional information in assessing the LCI results of a product system to better understand the environmental significance of that particular system.

(d) Interpretation phase:

This is the final phase of an LCA procedure, where the results of an LCI or LCIA, or both are being summarized and discussed to derive at conclusions, recommendations, and decision-making aligned with the defined goal and scope (ISO 14044, 2006).

LCA's Role in Waste Management

Environmental sustainability is considered as one of the major concerns today (UNEP, 2011). Companies are encouraged to adopt green technologies, cleaner production systems, and technologies to reduce their environmental footprints (Mazzi, 2020). As discussed above, many environmental management tools are available today but LCA is one of the tools among them that is capable of addressing a broad range of environmental impacts. With a life cycle perspective, the whole system is analyzed including the evaluation of the product's entire life cycle along with a long-term time horizon and a multidimensional view (Mazzi, 2020). Impact minimization can be achieved through five levers from a life cycle perspective namely lifetime extension, dematerialization, manufacturing efficiency, substitution, and recovery (Olivetti & Cullen, 2018). This is why Life Cycle Thinking (LCT) is important. Decisions made based on a full life cycle perspective and broader implications on the environment, economic, and social pillars will address unintended trade-offs between these pillars and focus on the key drivers of change resulting in the progress of faster and more efficient sustainable development (LCI, 2017).

In the realm of waste management, LCA is invaluable for assessing and comparing the environmental performance of various strategies and technologies. Waste management encompasses a range of activities, including collection, treatment, recycling, and disposal. Each of these activities has distinct environmental implications (Giesen et al., 2020).

By applying LCA, stakeholders can:

- **Compare Treatment Options**

Evaluate the environmental trade-offs between landfilling, incineration, recycling, composting, and other waste treatment methods.

- **Optimize Resource Recovery**

Determine the most efficient ways to reclaim materials and energy from waste streams, thereby reducing the need for virgin resources.

- **Informed Policy and Planning**

Aid in the development of waste management policies and infrastructure planning that align with environmental sustainability goals.

- **Enhance System Efficiency**

Identify inefficiencies and areas for improvement within existing waste management systems to minimize environmental impacts.

For instance, an LCA might reveal that recycling a particular material consumes less energy and produces fewer emissions than producing the material from virgin resources, thereby supporting recycling initiatives. Conversely, it might show that, for certain waste types, energy recovery through incineration could be more environmentally beneficial than landfilling. Therefore, an LCA provides a comprehensive framework for evaluating the environmental and economic impacts of different waste management strategies and technologies, guiding stakeholders toward more sustainable practices (Denison, 1996).

2 Overview of Waste Management Strategies and Technologies

Traditional Waste Management Techniques

Traditional waste management techniques have been widely used for decades to handle municipal solid waste, industrial waste, and hazardous waste. These methods primarily focus on waste disposal rather than resource recovery (Kosoe et al., 2019; Dey et al., 2023; Mishra et al., 2019, 2023). They often involve landfilling, open dumping, incineration, composting, and mechanical recycling. While effective in managing large volumes of waste, these methods come with environmental concerns such as pollution, greenhouse gas emissions, and resource depletion (Das et al., 2024; Das & Ghosh, 2022; Mohanty et al., 2018).

- **Landfilling**: The most common method of waste disposal, where waste is buried in designated sites. Modern landfills are designed with liners, leachate collection systems, and gas recovery mechanisms to mitigate environmental impacts. Over the past few decades, landfills have significantly improved due to stricter regulations and advancements in environmental science (Nanda & Berruti, 2021). Modern landfills are highly engineered and managed facilities designed to comply with federal regulations while minimizing environmental harm (O'Leary & Tchobanoglous, 2002). These landfills are strategically located to avoid environmentally sensitive areas and are equipped with monitoring systems that detect groundwater contamination and landfill gas emissions

(Vaverková, 2019). Additionally, landfills are subject to stringent design, operation, and closure requirements under the Resource Conservation and Recovery Act (RCRA) (RCRA, 2011).

The general procedure at a landfilling site is as, waste is first collected from households, industries, and commercial areas and transported to landfill sites. Upon arrival, the waste is weighed, inspected, and categorized (if not have done before collection) to ensure hazardous materials are properly handled (Date, 1998). The waste is then spread in layers and compacted using heavy machinery to reduce volume. Each layer is covered with soil or alternative cover material to minimize odor, pests, and wind-blown debris. Advanced landfills incorporate systems for leachate collection, gas recovery, and environmental monitoring to prevent contamination of surrounding soil and groundwater. This type of waste management strategy is used for municipal solid waste, industrial waste, and hazardous waste, particularly in urban areas where alternative disposal methods are limited (Josimović & Marić, 2012).

There are different types of landfills. Landfills are regulated under RCRA Subtitle D (solid waste) and Subtitle C (hazardous waste) or under the Toxic Substances Control Act (TSCA). Subtitle D focuses on state and local governments as the primary planning, regulating, and implementing entities for the management of nonhazardous solid waste, such as household garbage and nonhazardous industrial solid waste. The Subtitle D landfills include the following (RCRA, 2011):

- **Municipal Solid Waste Landfills (MSWLFs)**—Specifically designed to receive household waste, as well as other types of nonhazardous wastes.
  - Bioreactor Landfills—A type of MSWLF that operates to rapidly transform and degrade organic waste.
- **Industrial Waste Landfill**—Designed to collect commercial and institutional waste (i.e. industrial waste), which is often a significant portion of solid waste, even in small cities and suburbs.
  - Construction and Demolition (C&D) Debris Landfill—A type of industrial waste landfill designed exclusively for construction and demolition materials, which consists of the debris generated during the construction, renovation, and demolition of buildings, roads, and bridges. C&D materials often contain bulky, heavy materials, such as concrete, wood, metals, glass, and salvaged building components.
  - Coal Combustion Residual (CCR) landfills—An industrial waste landfill used to manage and dispose of coal combustion residuals (CCRs or coal ash). EPA established requirements for the disposal of CCR in landfills and published them in the Federal Register published on April 17, 2015 (Kron & Nelson, 2019).

The Subtitle C establishes a federal program to manage hazardous wastes from cradle to grave. The objective of the Subtitle C program is to ensure that hazardous waste is handled in a manner that protects human health and the environment. To this end, there are Subtitle C regulations for the generation, transportation and treatment, storage or disposal of hazardous wastes. Subtitle C landfills including the following (RCRA, 2011):

- **Hazardous Waste Landfills**—Facilities used specifically for the disposal of hazardous waste. These landfills are not used for the disposal of solid waste.
  - Polychlorinated Biphenyl (PCB) landfills—PCBs are regulated by the Toxic Substances Control Act. While many PCB decontamination processes do not require EPA approval, some do require approval.
- **Open Dumping**: Open dumping remains a dominant municipal solid waste (MSW) management practice, with approximately 2.24 billion tons of solid waste generated globally each year. Around 40% of the world's waste is disposed of in open dumpsites, particularly in cities within middle- and lower-income countries that lack proper waste collection systems. For instance, in many African cities, up to 90% of waste is openly dumped, while in Latin America and the Caribbean, the figure stands at 45%. Although open dumping is a low-cost waste management option, it poses severe environmental and public health risks. Emissions from such sites release hazardous substances like dioxins, furans, and mercury, contributing to air, water, and soil pollution (Clemente et al., 2024). Individuals working in these sites and nearby communities face high exposure to toxic pollutants, increasing health risks such as respiratory issues and diseases spread by vectors attracted to waste accumulation. While open dumping varies regionally, improper waste management remains a pressing global concern, exacerbating environmental crises (Rasel Kabir, 2015).

Generally, waste is collected and transported to open sites where it is dumped without segregation (Pattnaik & Reddy, 2010). Since there are no engineered barriers or leachate collection systems, waste decomposes naturally, often leading to groundwater contamination, air pollution from spontaneous fires, and disease outbreaks due to pest infestations. Informal waste pickers often scavenge for recyclable materials, further exacerbating health risks. Normally, this type of dumping is often found in regions lacking waste management infrastructure, particularly in low-income urban and rural areas (Remigios, 2010).

- **Incineration**: The European Environment Agency defines waste incineration as the process of burning solid waste under controlled conditions to reduce its weight and volume, and often to produce energy. Communities are tasked with formulating waste management strategies that encompass options such as waste reduction, landfilling, recycling, reuse, and composting (Knox, 2005). While the communities are tasked to focus on alternatives to incineration, it's important to note that integrating methods like recycling with incineration can modify the composition of waste streams directed to incineration facilities. However, evaluating the comparative effectiveness of these waste management options is beyond the scope of this committee's mandate (Sabbas et al., 2003).

Typically, an incineration facility encompasses several key processes: waste storage and handling, waste preparation, combustion, air pollution control, and management of residual ash. The types of waste incineration facilities examined in this report include incinerators, industrial boilers, furnaces, and kilns. These technologies vary widely, ranging from large stationary facilities designed to process millions of tons of waste annually from extensive regions to mobile incinerators intended for the remediation of hazardous waste at specific contaminated sites. Three types of waste to which incineration is applied extensively are municipal solid waste, hazardous waste, and medical waste (Makarichi et al., 2018).

Waste is sorted to remove hazardous and recyclable materials before being fed into a high-temperature furnace, where it undergoes combustion at temperatures exceeding 850 °C. The process results in the production of ash, flue gases, and heat energy. Flue gases pass through air pollution control systems, such as scrubbers and filters, to remove harmful emissions before being released into the atmosphere. Some incineration plants integrate heat recovery systems to produce electricity or district heating (Makarichi et al., 2018).

- **Composting**: Composting is the natural process of recycling organic matter, such as leaves and food scraps, into a valuable fertilizer that can enrich soil and plants. Anything that grows decomposes eventually; composting simply speeds up the process by providing an ideal environment for bacteria, fungi, and other decomposing organisms (such as worms, sowbugs, and nematodes) to do their work. The resulting decomposed matter, which often ends up looking like fertile garden soil, is called compost. Fondly referred to by farmers as "black gold," compost is rich in nutrients and can be used for gardening, horticulture, and agriculture (Sayara et al., 2020).

Organic discards can be processed in industrial-scale composting facilities, in smaller-scale community composting systems, and in anaerobic digesters, among other options. Home composting, is also a great way to keep your organic discards out of the waste stream and produce a valuable soil amendment for your own use (Ayilara et al., 2020).

The organic waste, such as food scraps and yard waste, is first shredded and mixed to achieve an optimal carbon-to-nitrogen ratio. The material is then placed in composting bins, aerated piles, or windrows where microorganisms break down organic matter in the presence of oxygen. Regular turning, moisture control, and temperature monitoring are essential to facilitate efficient decomposition. The process takes several weeks to months, depending on the method used (Ayilara et al., 2020; Sayara et al., 2020).

Advanced Waste Management Technologies

As the global population grows and urbanization accelerates, the demand for effective and sustainable waste management technologies has increased dramatically

(Priyadarsini & Das, 2024, 2025). Traditional waste management methods, such as landfilling and open burning, have proven inadequate in addressing the environmental challenges of modern waste generation (Mishra & Das, 2023; Mishra et al., 2024; Tripathy et al., 2024). Consequently, innovative waste management technologies have emerged, each offering unique solutions to minimize environmental impacts, optimize resource recovery, and promote sustainability. These advanced technologies vary in approach, ranging from mechanical processes to biological and chemical treatments, and they are being adopted across the world to address the diverse waste management needs of different regions (Mishra & Das, 2021, 2022; Mishra et al., 2022b).

One of the most promising advancements in waste management is Waste-to-Energy (WTE) technology, which has gained significant attention in countries like Japan, Sweden, and Germany. WTE involves converting non-recyclable waste materials into energy through processes such as incineration, gasification, and pyrolysis (Tun et al., 2020). The primary goal of WTE is to reduce the volume of waste sent to landfills while simultaneously generating electricity or heat. Incineration is the most common WTE method and is highly efficient in reducing waste volume, often up to 90%. In some cases, the energy generated from incineration can be used for district heating or fed into the national grid (Awasthi et al., 2019). However, the environmental impacts of WTE, such as air emissions and the production of harmful by-products like dioxins and heavy metals, have raised concerns. As a result, stricter regulations and advanced filtration technologies are being employed to minimize these risks, making WTE an increasingly viable solution when managed appropriately (Awasthi et al., 2019; Cui et al., 2024).

Another notable technology is mechanical biological treatment (MBT), which combines mechanical sorting processes with biological treatments, such as composting or anaerobic digestion, to manage waste. MBT is primarily used for the treatment of municipal solid waste (MSW), to extract recyclable materials and divert organic waste from landfills. The mechanical phase of MBT typically involves sorting waste into categories like metals, plastics, and paper, while the biological phase can either compost biodegradable materials or use anaerobic digestion to break down organic waste and produce biogas (Lornage et al., 2007). The biogas generated during anaerobic digestion can be utilized for energy production, contributing to a more sustainable and circular waste management system. Countries like Germany and the UK have adopted MBT technologies as part of their waste diversion strategies, contributing to reduced landfill use and the generation of renewable energy (Montejo et al., 2013).

In addition to mechanical and biological technologies, pyrolysis and gasification are increasingly recognized as advanced thermal treatment methods. Both technologies involve the heating of waste materials in an oxygen-limited environment, resulting in the conversion of waste into valuable by-products, such as synthetic fuels, chemicals, or gases. Pyrolysis is particularly useful for treating plastic waste, which is difficult to recycle using traditional methods. It can produce oil and gas that can be used as energy sources, as well as carbon char, which can have industrial applications (Akhtar et al., 2018). Gasification, on the other hand, involves the partial oxidation of waste at high temperatures to produce syngas, a mixture of carbon monoxide, hydrogen, and methane, which can be used to generate electricity or converted into chemicals. These advanced thermal treatment technologies have gained attention for their ability to handle various waste streams, including plastics, biomass, and hazardous waste, offering a promising alternative to traditional incineration (Digman et al., 2009).

Anaerobic digestion (AD) is another advanced waste management technology that has gained global recognition, especially for managing organic waste such as food scraps, agricultural residues, and sewage sludge. Anaerobic digestion involves the breakdown of organic matter by microorganisms in the absence of oxygen, producing biogas as a by-product (Vyas et al., 2022). The biogas is primarily composed of methane, which can be captured and used for energy production. AD is particularly effective in diverting organic waste from landfills and reducing greenhouse gas emissions, as it captures methane that would otherwise be released during the decomposition of organic waste in landfills. Countries like Sweden, Canada, and the United States have implemented large-scale anaerobic digestion facilities to handle organic waste and produce renewable energy. In addition to biogas production, the process generates a nutrient-rich digestate, which can be used as a soil conditioner, contributing to the circular economy by returning valuable nutrients to the land (Paranjpe et al., 2023).

Composting is an advanced biological treatment technology, often used in combination with other waste management methods. While traditional composting has been practiced for centuries, modern advancements in composting technology, such as controlled aerobic composting, have made it a more efficient and sustainable method of managing organic waste (Ayilara et al., 2020). Composting is an environmentally friendly way to recycle food scraps, yard waste, and other biodegradable materials, turning them into nutrient-rich compost that can improve soil health and reduce the need for synthetic fertilizers. In countries like the Netherlands, composting is widely practiced, and advanced composting systems are used in urban and agricultural settings to manage organic waste sustainably. The adoption of large-scale composting operations has helped reduce landfill dependency and promote the recovery of valuable organic resources (Zhou et al., 2022).

The rise of the circular economy has also influenced the development and adoption of waste management technologies. There, the waste is treated as a resource, and efforts are focused on reducing, reusing, and recycling materials to extend their life cycle (Bala et al., 2024; Das & Ghosh, 2018; Mishra et al., 2022a). Technologies such as advanced recycling methods, including chemical recycling, are gaining traction as a way to address the growing plastic waste crisis (Ferronato et al., 2019). Chemical recycling technologies, such as depolymerization and solvolysis, break down plastics into their chemical components, allowing for the creation of new plastic products without the loss of material quality. This innovation holds promise for solving the problem of mixed and contaminated plastics, which are difficult to recycle using traditional mechanical methods (Lahane & Kant, 2022; Mondal et al., 2023).

Finally, the integration of smart waste management systems is another key trend shaping the future of waste management. Smart systems use sensors, data analytics, and the Internet of Things (IoT) to monitor and optimize waste collection, sorting, and processing. These systems can track the waste generation patterns of cities, helping authorities to better allocate resources, improve recycling rates, and minimize operational costs. Smart waste management solutions are already being implemented in cities like Singapore and Barcelona, where they help manage waste more efficiently and reduce the environmental footprint of waste operations (Lakhouit, 2025).

Advanced waste management technologies are transforming the way we handle waste worldwide. These technologies not only offer solutions to reduce waste volume and minimize environmental impacts but also promote resource recovery and energy production, aligning with the principles of sustainability and the circular economy. As technological advancements continue, it is expected that waste management systems will become even more efficient, integrated, and adaptable to the growing challenges of global waste management. By adopting these cutting-edge technologies, countries can make significant strides toward achieving sustainable waste management and reducing their environmental footprint.

Circular Economy Approaches

Circular economy (CE) approaches are gaining traction worldwide as countries seek to reduce waste, conserve resources, and mitigate environmental degradation. Unlike the traditional linear economy, which follows a "take-make-dispose" model, the circular economy promotes a system where resources are kept in use for as long as possible, minimizing waste and encouraging the continuous use of materials (Elisha, 2020). By embracing circular economy principles, countries and industries are developing more sustainable business models, enhancing product life cycles, and reducing the environmental impacts associated with production and consumption. These approaches are being adopted at various levels, from national policies to corporate strategies, across diverse sectors including manufacturing, agriculture, and waste management (Rashid & Malik, 2023).

One of the most widely recognized circular economy practices is resource recovery from waste. Countries like the Netherlands, Sweden, and Japan have pioneered the development of advanced waste sorting and recycling systems that aim to recover valuable materials from waste (Singh & Ordoñez, 2016). These systems focus on diverting waste from landfills and incinerators, promoting the recycling and reuse of materials such as metals, plastics, and paper. For example, in the Netherlands, the concept of "circular cities" is being implemented, where urban waste is treated as a resource, and a significant proportion of waste is recovered and reused in the production of new goods. Similarly, Japan has developed an advanced waste sorting system, which allows for high levels of recycling, especially in cities like Kamikatsu, Katsuura District, Tokushima Prefecture, Japan, which has a goal of zero waste to landfill (Savini, 2019; Williams, 2019).

In addition to waste recovery, product design is a fundamental pillar of the circular economy. The circular design encourages manufacturers to create products with longer lifespans, ease of disassembly, and recyclability in mind. In countries such as Germany, circular design is integral to industry standards, with companies focusing on producing durable goods that can be repaired, refurbished, or repurposed at the end of their life cycle (Ávila-Gutiérrez et al., 2019). The European Union has introduced several directives, such as the eco-design directive, to promote circularity in product development. These policies require companies to design products with a focus on minimizing resource use, enhancing product longevity, and ensuring products can be recycled or reused after their useful life. As a result, industries in the European Union are increasingly adopting circular design principles to reduce environmental impact and boost resource efficiency (Barkhausen et al., 2022).

Another important circular economy strategy is extended producer responsibility (EPR), which shifts the responsibility for the entire life cycle of products to the producers. This includes not only the production and sale of goods but also their collection, recycling, and disposal after use (Maitre-Ekern, 2021). Several countries have implemented EPR schemes, particularly for products like electronics, packaging, and vehicles. For instance, in countries such as France and Germany, producers are required to take responsibility for the collection and recycling of packaging materials through take-back schemes or deposit return systems. EPR is becoming a key tool for encouraging companies to design products with end-of-life management in mind, ensuring that products are returned to the economy rather than ending up as waste in landfills (Dawson, 2019).

In food waste management, the circular economy approach is also making significant strides. With the global food system being one of the largest contributors to waste and greenhouse gas emissions, many countries are adopting circular principles to reduce food waste at all stages of the supply chain (Parfitt et al., 2010). In places like the United Kingdom, food waste is increasingly seen as a resource for creating energy and compost. Through anaerobic digestion (AD) technology, organic food waste is converted into biogas, which can be used for electricity generation or as a renewable energy source. The by-product of this process, digestate, can be used as a soil conditioner, returning nutrients to agricultural land. Additionally, initiatives to redistribute surplus food to those in need, rather than sending it to landfills, have been widely implemented in regions like the United States and Europe (Gonçalves & Maximo, 2023; Tamasiga et al., 2022).

Sharing and leasing models are gaining traction in the context of the circular economy, particularly in industries such as fashion, electronics, and transportation. In many European countries, companies are adopting business models that focus on renting, leasing, or sharing products rather than selling them outright. For example, in the fashion industry, companies like H&M and Patagonia have introduced clothing rental services, allowing customers to rent clothes for specific occasions instead of purchasing new items. This reduces the demand for new raw materials and promotes the reuse of garments, extending their life cycle. Similarly, car-sharing and bike-sharing schemes in cities around the world, such as those in Amsterdam, Berlin, and New York, are reducing the need for personal vehicle ownership, thereby lowering the environmental impact associated with car manufacturing and disposal (Arora et al., 2023).

Another crucial circular economy practice is industrial symbiosis, which involves the exchange of materials, energy, and by-products between different industries. The aim is to create a closed-loop system where the waste or by-products from one industry become inputs for another, reducing waste and optimizing resource use. One of the most famous examples of industrial symbiosis is found in Kalundborg, Denmark, where various industries have collaborated to share resources. For instance, the power plant in Kalundborg supplies excess heat to a nearby fish farm, while the fish farm's waste is used as fertilizer by a nearby agricultural business. This system of shared resources has led to significant reductions in waste, energy consumption, and material costs, providing economic benefits for all involved parties. Industrial symbiosis has been adopted in several other regions worldwide, including in the UK and China, where circular supply chains are being developed to reduce industrial waste and increase overall resource efficiency (Branca et al., 2020; Castellet-Viciano et al., 2022; Martin, 2020).

Biological cycles also play a crucial role in the circular economy, particularly in agriculture and food production. Circular agriculture aims to minimize the use of synthetic inputs such as fertilizers and pesticides by leveraging natural systems and processes to maintain soil fertility, increase biodiversity, and reduce waste (Hamam et al., 2021). Techniques like composting and crop rotation are part of a broader strategy to create closed-loop systems where organic waste is recycled back into the soil, improving soil health and reducing the need for chemical inputs. Countries such as France, the Netherlands, and Australia are pioneers in integrating circular agriculture principles into their farming systems, helping to reduce environmental impact while maintaining food production levels (Agus et al., 2021; Muscat et al., 2021).

Circular supply chains are another integral part of the global transition to a circular economy. Companies worldwide are beginning to rethink traditional supply chains, which typically operate linearly, to adopt more sustainable, circular models. This involves using recycled materials in production, designing products for easier reuse or refurbishment, and adopting more efficient logistics systems that reduce waste. In places like the United States and Japan, companies in the electronics and automotive industries are incorporating recycled components and raw materials into their production processes, driving innovation in the use of secondary materials (Genovese et al., 2017).

Circular economy approaches are rapidly transforming global waste management, manufacturing, and consumption patterns. These practices not only reduce waste and pollution but also conserve resources, reduce dependency on raw materials, and create new economic opportunities. From resource recovery and product design to extended producer responsibility and industrial symbiosis, circular economy principles are being implemented across various industries and sectors worldwide. As these approaches gain momentum, they offer a path towards a more sustainable future, fostering economic resilience while addressing critical environmental challenges. With continued innovation and adoption, the circular economy has the potential to reshape how societies produce, consume, and manage waste, ultimately creating a more sustainable and resource-efficient world.

Integrated Waste Management Systems

Integrated Waste Management (IWM) refers to the comprehensive and coordinated approach to managing solid waste, which encompasses a combination of strategies aimed at reducing, recycling, treating, and sustainably disposing of waste (Ikhlayel & Nguyen, 2017). This approach emphasizes the importance of managing waste streams from the point of generation to final disposal, while simultaneously considering environmental, economic, and social impacts (Seadon, 2006). Integrated systems ensure that waste is handled in a way that reduces its negative effects on human health, minimizes environmental pollution, and maximizes resource recovery. The IWM approach takes into account various waste types, such as municipal solid waste, industrial waste, hazardous waste, and organic waste, and aims to manage each waste stream appropriately through a combination of techniques and technologies (Seadon, 2006).

A key principle of Integrated Waste Management is the **hierarchical approach** to waste management, which prioritizes waste minimization, followed by reuse, recycling, and recovery, with disposal as a last resort (Ikhlayel & Nguyen, 2017). This hierarchy encourages the reduction of waste generation at the source, promoting practices like eco-design, product stewardship, and waste prevention. The next priority is to reuse products or materials as much as possible, followed by recycling or recovering materials from waste streams. In cases where recycling is not feasible, recovery methods such as composting, anaerobic digestion, or energy recovery are used to extract valuable resources or energy from waste. Finally, disposal methods like landfilling and incineration are considered only when all other options have been exhausted (AlManssoor, 2020).

One of the most significant aspects of an Integrated Waste Management System is the **combination of technologies** used to handle different waste streams (Seadon, 2006). For example, organic waste can be diverted from landfills through processes such as composting or anaerobic digestion, which turn food scraps and yard waste into useful products like compost and biogas. In contrast, recyclables such as plastics, metals, and paper are typically separated from the waste stream using sorting technologies such as mechanical sorting systems or manual sorting in recycling facilities. Non-recyclable and residual waste can be treated using thermal methods such as **Waste-to-Energy (WTE)** technologies, including incineration, pyrolysis, or gasification, which reduce waste volume and recover energy (AlManssoor, 2020).

Landfills, although considered a last resort in the IWM hierarchy, are still a critical component of waste management systems, especially in regions where waste diversion strategies are not fully implemented. Modern landfills, however, are designed with advanced techniques such as leachate management and landfill gas recovery systems to minimize environmental impacts. Landfill gas, primarily methane, is captured and used as a source of energy, and leachate is treated to prevent contamination of groundwater and soil (Ikhlayel & Nguyen, 2017).

In **developed countries**, Integrated Waste Management Systems are highly advanced, with comprehensive infrastructure for waste collection, sorting, recycling, and treatment. Countries such as Germany, Sweden, and the Netherlands have established well-integrated systems that rely on strict regulations, efficient waste segregation at the source, and robust recycling programs. For instance, Germany has a nationwide **Green Dot system** where producers are responsible for the recycling and disposal of their packaging (Rousso & Shah, 1994). The country also has an extensive system for collecting and recycling e-waste and other materials, which has led to one of the highest recycling rates in the world. In Sweden, around 99% of waste is recycled, converted into energy, or composted, with only a small fraction sent to landfills. This success has been driven by a combination of policies, public awareness, and innovative waste treatment technologies, including WTE plants and district heating systems (Rousso & Shah, 1994).

In **developing countries**, IWM systems face significant challenges due to limited infrastructure, lack of resources, and informal waste management practices. However, some countries have made progress by adopting low-cost, decentralized approaches to waste management. For example, **India** has implemented community-level composting initiatives and established waste-to-energy plants to address the growing volume of waste in urban areas (Somani et al., 2021). In **Africa**, informal waste pickers play a crucial role in waste collection and recycling, and there is increasing interest in formalizing this sector to improve recycling rates and reduce the environmental impact of unmanaged waste. In these regions, IWM often focuses on cost-effective solutions such as waste segregation, promoting community-based recycling programs, and building local capacity for waste management (Karani & Jewasikiewitz, 2007).

The success of Integrated Waste Management Systems depends not only on technology and infrastructure but also on effective **policy and regulation**. Government support plays a vital role in ensuring that waste management strategies are well coordinated and that appropriate regulations are enforced (Sakai et al., 2011). Policies such as Extended Producer Responsibility (EPR), which requires producers to take responsibility for the entire life cycle of their products, including collection and recycling, have been successful in promoting waste reduction and recycling in many countries. The European Union's Circular Economy Action Plan is an example of a comprehensive policy that integrates waste management with sustainability goals, encouraging the transition to a circular economy by reducing waste generation and promoting resource recovery (Memon, 2010; Sakai et al., 2011).

Public **education and awareness** are also crucial components of IWM systems. In many successful cases, such as in Japan and South Korea, the public has been educated on the importance of waste segregation and recycling, and citizens are actively engaged in waste management efforts. The implementation of public awareness campaigns and waste separation at the source has been key to increasing recycling rates and reducing contamination in recyclable materials. In countries like Switzerland and the UK, local governments have incentivized waste sorting through schemes that reward households and businesses for proper waste separation, further improving participation rates in recycling and composting programs (Debrah et al., 2021).

Economic considerations also play a significant role in shaping IWM systems. The costs of waste management can vary depending on the type of waste, the technologies used, and the scale of the system. While advanced waste-to-energy plants and recycling facilities require significant upfront investments, they can provide long-term economic benefits by recovering valuable materials, generating energy, and reducing landfill disposal costs. Similarly, the costs associated with waste collection, transportation, and sorting are influenced by the level of infrastructure development, geographic location, and population density. In many cases, integrated waste management systems are seen as a more cost-effective solution in the long term due to their emphasis on resource recovery and the reduction of waste sent to landfills (Cobo et al., 2018).

Therefore, Integrated Waste Management Systems offer a comprehensive approach to addressing the global waste crisis by combining various waste management strategies and technologies to minimize environmental impacts and maximize resource recovery. By prioritizing waste prevention, recycling, energy recovery, and responsible disposal, IWM systems create a more sustainable model for managing waste in both developed and developing countries. As waste management technologies continue to advance, and as policies and public engagement improve, the future of IWM will focus on achieving greater efficiency, resource conservation, and environmental protection, helping move towards a more circular and sustainable economy worldwide.

3 LCA Phases in Evaluating the Impacts of Waste Management

When performing an LCA, the requirements of ISO 14044 include goal and scope definition, life cycle inventory analysis, life cycle impact assessment, and life cycle interpretation. These phases are comprehensively described below (ISO 14044, 2006).

Goal and Scope Definition

The goal of an LCA states the intended application, the reasons for carrying out the study, the intended audience, i.e. to whom the results of the study are intended to be communicated, and whether the results are intended to be used in comparative assertions intended to be disclosed to the public.

The scope should be sufficiently well defined to ensure that the breadth, depth and detail of the study are compatible and sufficient to address the stated goal. The scope includes the product system to be studied; the functions of the product system or, in the case of comparative studies, the systems; the functional unit; the system boundary; allocation procedures; impact categories selected and methodology of impact assessment, and subsequent interpretation to be used; data requirements; assumptions; limitations; initial data quality requirements; type of critical review, if any; type and format of the report required for the study. LCA is an iterative technique, and as data and information are collected, various aspects of the scope may require modification in order to meet the original goal of the study.

Life Cycle Inventory Analysis (LCI)

Data for each unit process within the boundary of the system can be classified under major headings, including energy inputs, raw material inputs, ancillary inputs, other physical inputs, products, co-products and waste, emissions to air, discharges to water and soil, and other environmental aspects. Data collection can be a resource-intensive process. Practical constraints on data collection should be considered in the scope and documented in the study report.

Following the data collection, calculation procedures, including validation of data collected, the relating of data to unit processes, and the relating of data to the reference flow of the functional unit, are needed to generate the results of the inventory of the defined system for each unit process and for the defined functional unit of the product system that is to be modeled.

The calculation of energy flows should take into account the different fuels and electricity sources used, the efficiency of conversion and distribution of energy flow, as well as the inputs and outputs associated with the generation and use of that energy flow.

Few industrial processes yield a single output or are based on a linearity of raw material inputs and outputs. In fact, most processes yield more than one product, and they recycle intermediate or discarded products as raw materials. Consideration should be given to the need for allocation procedures when dealing with systems involving multiple products and recycling systems.

Life Cycle Impact Assessment (LCIA)

The impact assessment phase of LCA is aimed at evaluating the significance of potential environmental impacts using the LCI results. In general, this process involves associating inventory data with specific environmental impact categories and category indicators, thereby attempting to understand these impacts. The LCIA phase also provides information for the life cycle interpretation phase. The impact assessment may include the iterative process of reviewing the goal and scope of the LCA study to determine if the objectives of the study have been met, or to modify the goal and scope if the assessment indicates that they cannot be achieved.

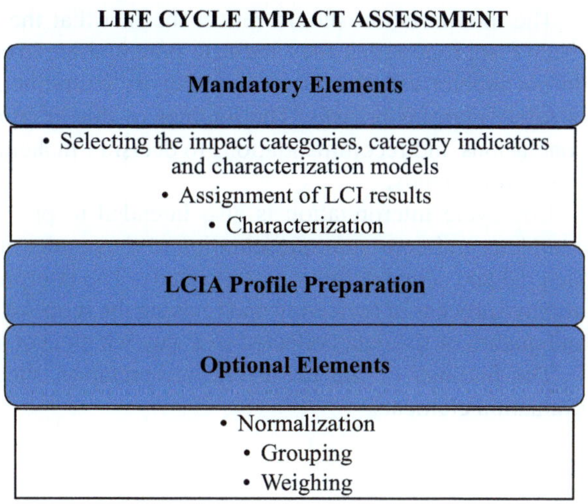

Fig. 1 Elements of LCIA. Adapted from: ISO 14040 Standard (https://www.iso.org/standard/37456.html)

Issues such as choice, modeling, and evaluation of impact categories can introduce subjectivity into the LCIA phase. Therefore, transparency is critical to the impact assessment to ensure that assumptions are clearly described and reported.

The elements of the LCIA phase is given in Fig. 1. Separation of the LCIA phase into different elements is helpful and necessary for several reasons, as follows:

(a) each LCIA element is distinct and can be clearly defined;
(b) the goal and scope definition phase of an LCA can consider each LCIA element separately;
(c) a quality assessment of the LCIA methods, assumptions and other decisions can be conducted for each LCIA element;
(d) LCIA procedures, assumptions and other operations within each element can be made transparent for critical review and reporting;
(e) the use of values and subjectivity (hereafter referred to as value-choices), within each element, can be made transparent for critical review and reporting.

The level of detail, choice of impacts evaluated, and methodologies used depend on the goal and scope of the study.

Life Cycle Interpretation

Interpretation is the phase of LCA in which the findings from the inventory analysis and the impact assessment are considered together or, in the case of LCI studies, the findings of the inventory analysis only. The interpretation phase should deliver results that are consistent with the defined goal and scope and which reach conclusions, explain limitations and provide recommendations.

The interpretation should reflect the fact that the LCIA results are based on a relative approach, that they indicate potential environmental effects, and that they do not predict actual impacts on category endpoints, the exceeding of thresholds or safety margins or risks. The findings of this interpretation may take the form of conclusions and recommendations to decision-makers, consistent with the goal and scope of the study.

Life cycle interpretation is also intended to provide a readily understandable, complete, and consistent presentation of the results of an LCA, in accordance with the goal and scope definition of the study. The interpretation phase may involve the iterative process of reviewing and revising the scope of the LCA, as well as the nature and quality of the data collected in a way which is consistent with the defined goal.

The findings of the life cycle interpretation should reflect the results of the evaluation element.

4 Environmental Impacts of Different Waste Management Strategies and Technologies

Ecosystem Impacts

Waste management strategies and technologies have varying impacts on ecosystems, influencing biodiversity, soil and water quality, and atmospheric conditions. Traditional landfilling, one of the most widely used methods, can lead to habitat destruction, groundwater contamination through leachate, and methane emissions that contribute to climate change. Incineration reduces waste volume and generates energy but releases pollutants such as dioxins, heavy metals, and particulate matter, which can harm air quality and terrestrial and aquatic ecosystems (Budjav, 2022). Open burning, a widely practiced but highly unsustainable method, emits toxic pollutants, including polycyclic aromatic hydrocarbons (PAHs) and fine particulate matter (PM2.5), which degrade air quality and pose severe risks to both ecosystems and human health. Composting and anaerobic digestion, considered more sustainable alternatives, minimize landfill waste and enrich soil health, though they may produce localized emissions of greenhouse gases like methane if not properly managed (Shmelev & Powell, 2006). Recycling reduces raw material extraction and associated environmental degradation but requires energy-intensive processing, sometimes leading to secondary pollution. Emerging technologies, such as pyrolysis and gasification, aim to mitigate traditional waste management impacts by converting waste into energy with lower emissions. However, their long-term ecological effects are still under study (Patil et al., 2024). An integrated approach, combining multiple waste management techniques with proper regulations and technological advancements, is crucial to minimizing ecosystem degradation while promoting resource efficiency and sustainability.

Global Warming Potential (GWP)

Naturally, the solar radiation that reaches the earth's atmosphere leaves the earth through reflection and infrared radiation resulting in an equilibrium in the atmosphere. If the fragmentation of the solar radiation received by the Earth's surface is considered, 21% of the sunlight reaching the Earth's atmosphere is reflected back to space as air molecules, clouds, and Earth's surface (oceans and icy regions), whereas 28% is absorbed by the greenhouse gases in the atmosphere and 50% by the Earth's atmosphere of the remaining portion (Bandara, 2020). The latter heats the Earth's surface and releases into the atmosphere as infrared radiation which is partially absorbed by the greenhouse gases (GHGs) and persistent in the atmosphere causing an increase in both atmospheric temperatures together with GHGs content in the atmosphere (Hauschild et al., 2018). As per the research done by scientists, it was discovered that the beginning of this warming trend has started back in the eighteenth century when the use of coal was first started. And, the trend has increased gradually with the further use of fossil fuels including gasoline, diesel, kerosene, natural gas, and petrochemicals with industrialization (Shafer, 2017). The mainly considered greenhouse gases are water vapor, carbon dioxide (CO_2), methane (CH_4), nitrous oxides (NO_x), and ozone (O_3). The effects caused by excessive GHG emissions can be stated as below.

1. Transport, transformation, and distribution of GHGs in the atmosphere,
2. Disturbance of the radiation balance-radiative forcing,
3. Increase in global temperature of atmosphere and surface,
4. Increase in sea level due to heat expansion and melting of land-based ice,
5. Increased water vapor content of the atmosphere causing more extreme weather,
6. Negative effects on ecosystems and human health (Hauschild et al., 2018).

In the general context, the issue of GHG emission is regarded at different levels such as global, national, corporate, or organizational. When it comes to the Global Warming Potential (GWP), it is quantified by the unit "kgCO_2eq" (kilogram carbon dioxide equivalent) in calculations. The increase of GHG emissions at the above-mentioned different levels is a consequence of emissions from supply chains, within/between businesses or nations. They reflect the GHG emissions of goods and services emitted at different stages of processes, materials used, and decisions taken at different phases of particular product or service life cycles that also potentially affect climate change (PAS2050, 2011).

The global warming potential (GWP) of different waste management strategies and technologies is primarily influenced by greenhouse gas (GHG) emissions, particularly carbon dioxide (CO_2), methane (CH_4), and nitrous oxide (N_2O). **Landfilling**, one of the most common methods, has a high GWP due to anaerobic decomposition of organic waste, which generates methane—a greenhouse gas approximately 28 times more potent than CO_2 over 100 years. **Open burning**, often used in regions with limited waste management infrastructure, releases large amounts of CO_2, black carbon, and other short-lived climate pollutants, significantly contributing to global warming. **Incineration**, while reducing waste volume and generating energy, emits CO_2 from the combustion of organic and plastic waste (Cherubini et al., 2009). However, modern waste-to-energy plants with energy recovery can offset fossil fuel-based emissions, lowering net GWP. **Composting and anaerobic digestion** provide more climate-friendly alternatives, as composting mainly emits CO_2 (which is biogenic and considered carbon–neutral), while anaerobic digestion captures methane for energy production, reducing its overall climate impact. **Recycling** significantly lowers GWP by reducing the demand for virgin material extraction and energy-intensive manufacturing processes, thereby avoiding emissions from fossil fuel combustion. **Advanced thermal treatment technologies**, such as pyrolysis and gasification, produce syngas and biochar, which can be used as alternative energy sources or carbon sequestration materials, potentially reducing GHG emissions. The overall GWP impact of waste management depends on the efficiency of these technologies, waste composition, and integration of circular economy principles to minimize emissions across the waste life cycle (Nanda & Berruti, 2021).

Waste-to-energy (WTE) plants play a complex role in global warming potential (GWP) by reducing landfill waste while simultaneously generating emissions. These facilities incinerate municipal solid waste (MSW) to produce electricity or heat, thereby offsetting fossil fuel use. However, their climate impact depends on multiple factors, including the type of waste burned, energy recovery efficiency, and emission control technologies. On one hand, WTE plants emit **carbon dioxide (CO_2)** from the combustion of organic materials and plastics. While CO_2 from biogenic sources (e.g., food and paper waste) is considered carbon–neutral, emissions from plastic waste (derived from fossil fuels) contribute to global warming. Additionally, incineration can release **nitrous oxide (N_2O)**, a potent greenhouse gas with a GWP nearly 300 times that of CO_2. On the other hand, WTE plants reduce **methane (CH_4) emissions** by diverting organic waste from landfills, where anaerobic decomposition would generate methane—a greenhouse gas 28 times more potent than CO_2 over 100 years. Modern WTE facilities with **advanced air pollution control systems** also minimize emissions of particulates, dioxins, and heavy metals (Mayer et al., 2019).

The net GWP impact of WTE depends on **energy efficiency and grid integration**. If WTE electricity replaces coal-fired power, it can lead to net emissions reductions. However, if it replaces cleaner energy sources like wind or solar, its relative climate benefit diminishes. Additionally, **carbon capture and storage (CCS) technologies** in WTE plants could further lower emissions, making them a more climate-friendly option in the future. Ultimately, while WTE can reduce landfill methane emissions and provide energy recovery, it is not a zero-emission solution. Its role in climate mitigation is most effective when integrated with waste reduction, recycling, and composting in a circular economy framework (Chen, 2018).

Air, Water, and Soil Pollution

Waste management strategies and technologies have significant environmental impacts, particularly on soil, water, and air quality. The extent of pollution varies depending on the method used, waste composition, and management efficiency (Budjav, 2022).

Landfills can degrade soil quality through leachate seepage, which contains heavy metals, organic pollutants, and hazardous chemicals. Over time, this contamination reduces soil fertility and may alter microbial communities. If improperly lined, landfills generate leachate that percolates into groundwater and nearby water bodies, introducing toxic substances and pathogens. This can pose risks to aquatic ecosystems and drinking water supplies. Decomposing organic waste in landfills emits methane (CH_4), a potent greenhouse gas, along with volatile organic compounds (VOCs) and hydrogen sulfide (H_2S), contributing to air pollution and unpleasant odors (Khanam et al., 2023).

Residual ash from open burning contains heavy metals, dioxins, and polycyclic aromatic hydrocarbons (PAHs), which can accumulate in soil and affect plant growth. Pollutants from burnt waste can be transported by rainwater runoff, contaminating surface and groundwater with hazardous chemicals. Open burning releases high levels of CO_2, CO, black carbon, dioxins, furans, and particulate matter (PM2.5), severely degrading air quality and posing health risks (Budjav, 2022).

Properly managed incineration produces ash residues that require safe disposal. Bottom ash can sometimes be reused in construction, but fly ash contains toxic metals and dioxins, requiring hazardous waste treatment. Modern incineration facilities use scrubbers to remove pollutants, but improper wastewater management can lead to contamination of water bodies. Incineration emits CO_2, NO_x, SO_2, and trace amounts of heavy metals and dioxins. However, advanced filtration systems can significantly reduce these emissions (Rădoi, 2021).

Composting enriches soil with organic matter and nutrients, improving soil health and microbial diversity. However, if contaminated waste (e.g., plastic or heavy metals) is used, it can introduce pollutants into the soil. Poorly managed composting sites may generate leachate with excess nutrients, which can cause eutrophication in nearby water bodies. Composting produces CO_2 and, in some cases, small amounts of methane and ammonia. Anaerobic digestion, however, captures methane for energy use, reducing its atmospheric impact (Skenderovic et al., 2015).

Recycling minimizes waste disposal, reducing the risk of soil contamination. However, some recycling processes (e.g., e-waste recycling) can generate hazardous waste residues. Industrial recycling processes may produce wastewater containing chemicals and heavy metals, which require treatment before discharge. Recycling reduces the need for raw material extraction, lowering emissions from mining and manufacturing. However, energy-intensive recycling processes can still contribute to air pollution if powered by fossil fuels (Ludwig et al., 2012).

By converting waste into energy and biochar, these technologies reduce landfill dependence, minimizing soil contamination risks. Biochar can improve soil quality if used properly. Wastewater from gasification processes requires treatment to remove contaminants before disposal. Compared to incineration, these processes emit fewer pollutants, as they operate at lower oxygen levels, reducing dioxin and furan formation. However, CO_2 and syngas emissions still contribute to air pollution if not managed properly.

Each waste management strategy affects soil, water, and air quality differently. Traditional methods like landfilling and open burning pose severe environmental risks, while modern technologies such as WTE, anaerobic digestion, and recycling offer more sustainable solutions. An integrated waste management approach, prioritizing waste reduction, resource recovery, and pollution control technologies, is essential for minimizing environmental impacts and promoting sustainability.

Resource Depletion

Waste management strategies and technologies play a crucial role in resource conservation or depletion, depending on their efficiency in recovering valuable materials and reducing the need for raw material extraction. The extent of resource depletion varies based on how effectively a method promotes recycling, energy recovery, and sustainable material use (Sakai et al., 2017).

Landfilling results in the permanent loss of valuable materials, including metals, plastics, and organic matter, which could otherwise be recycled or repurposed. Once waste is buried, recovering materials or energy from it is nearly impossible, leading to increased demand for virgin resources (Singh et al., 2014).

Open burning destroys materials that could be recycled or processed into new products, increasing pressure on natural resources. The uncontrolled combustion of plastics and organic matter does not generate usable energy, making it an unsustainable waste disposal method (Singh et al., 2014).

Waste-to-Energy (WTE) plants generate energy from waste, reducing the need for fossil fuels. However, burning waste eliminates the possibility of material recovery. While bottom ash can be reused in construction, a significant portion of the material becomes non-recoverable, contributing to resource depletion (Song et al., 2015).

Composting converts organic waste into nutrient-rich compost, improving soil quality and reducing the need for synthetic fertilizers. Anaerobic digestion captures methane for energy use, reducing reliance on fossil fuels while recovering organic matter.

Recycling reduces the demand for virgin raw materials, preserving finite resources such as metals, paper, and plastics that helps in significant resource conservation. This process often requires less energy compared to producing materials from raw sources, reducing fossil fuel use and associated environmental impacts (Chen et al., 2024).

Advanced waste treatment technologies such as pyrolysis and gasification technologies convert waste into fuels (syngas, biochar, or oil), reducing dependence on fossil fuels and raw materials emphasizing energy and resource recovery. If efficiently managed, these processes can support circular economy principles by converting waste into valuable secondary resources (Chen et al., 2024).

Waste management technologies significantly influence resource depletion. While traditional methods like landfilling and open burning exacerbate resource loss, modern strategies such as recycling, composting, and energy recovery help conserve materials and reduce reliance on virgin resources. Implementing an integrated waste management system that prioritizes resource recovery is essential for long-term sustainability.

Energy Use

Energy use in waste management varies depending on the strategy or technology applied. Some methods consume large amounts of energy for processing, while others generate energy that can offset fossil fuel dependence. The efficiency of energy use is critical in determining the sustainability of each approach.

Landfilling itself requires minimal direct energy for waste transportation, compaction, and site maintenance. Traditional landfills do not recover energy, but modern landfill gas (LFG) recovery systems capture methane emissions to generate electricity and heat, improving energy efficiency.

Open burning destroys waste without harnessing its energy potential. The uncontrolled combustion process releases heat into the atmosphere without any means of capturing or utilizing it, making it highly inefficient.

WTE plants require energy for waste processing, combustion, and pollution control systems. Incineration with energy recovery converts waste into electricity or heat, partially offsetting fossil fuel consumption. However, energy efficiency varies, typically ranging from 15 to 30%. Modern WTE plants incorporate combined heat and power (CHP) systems, improving energy efficiency by utilizing both electricity and thermal energy (Tsui & Wong, 2019).

Composting primarily relies on microbial decomposition, requiring minimal mechanical energy for aeration and turning. Organic waste breakdown in anaerobic digesters produces biogas (methane-rich gas), which can be used for electricity, heat, or as a renewable fuel. This offsets fossil fuel energy use.

Recycling processes consume energy for material collection, sorting, and reprocessing, with energy intensity depending on the material type. **Paper and plastic recycling** require moderate energy but save more compared to virgin material production whereas **metal recycling** (e.g., aluminum) is highly energy-efficient, using up to 95% less energy than primary metal extraction. Recycling significantly reduces the need for energy-intensive raw material extraction and refining, contributing to net energy conservation (Tozlu et al., 2016).

Pyrolysis and Gasification technologies require controlled high-temperature environments to break down waste into useful byproducts. **Pyrolysis** produces bio-oil and syngas, which can be used as alternative fuels. **Gasification** generates syngas for electricity or industrial applications, reducing dependence on fossil fuels. Energy recovery in these setups depends on system design, feedstock quality, and integration with other energy systems (Hasan et al., 2021).

Different waste management strategies have varying impacts on energy use. While traditional methods like landfilling and open burning offer little to no energy recovery, modern technologies such as WTE, anaerobic digestion, and advanced thermal treatments improve energy efficiency. Recycling minimizes overall energy demand by reducing reliance on virgin material production. A sustainable waste management system should prioritize energy-efficient practices while integrating energy recovery technologies to minimize environmental impact.

5 Economic Impacts of Different Waste Management Strategies and Technologies

Cost–Benefit Analysis of Different Waste Management Strategies and Technologies

Cost–Benefit Analysis (CBA) is a systematic approach used to evaluate the economic feasibility and efficiency of different projects, policies, or technologies by comparing their costs and benefits. It helps decision-makers determine whether an investment or strategy is worth pursuing by assessing financial, environmental, and social impacts. CBA is conducted to ensure that resources are allocated efficiently and that waste management strategies provide maximum economic and environmental benefits while minimizing costs. It helps policymakers, municipalities, and private organizations decide which waste management approach is the most cost-effective and sustainable in the long run (Soesanto et al., 2021).

The steps of carrying out a CBA is as follows (Soesanto et al., 2021);

1. Identifying Costs and Benefits: All direct and indirect costs (e.g., capital investment, operational costs, maintenance, environmental remediation) and benefits (e.g., resource recovery, energy savings, and environmental protection) are quantified.
2. Monetizing Costs and Benefits: Assigning monetary values to all identified factors, including intangible environmental and social benefits.
3. Discounting Future Costs and Benefits: Since benefits and costs occur over time, discount rates are applied to compare present and future values.
4. Comparing Net Present Value (NPV): If benefits outweigh costs (i.e., a positive NPV), the project is considered economically viable.
5. Sensitivity Analysis: Evaluating how changes in key variables (e.g., energy prices, waste generation rates) affect the results to ensure robustness.

Waste management decisions involve significant financial investment and long-term environmental implications. Conducting a CBA ensures that, public funds and private investments are allocated effectively, the most economically and environmentally sustainable waste management methods are prioritized, trade-offs between short-term costs and long-term benefits are well understood, and externalities, such as pollution and climate impacts, are accounted for in decision-making.

Given below (Table 1) is a CBA done for different waste management strategies and technologies that are being widely used worldwide. According to the table, different waste management technologies and strategies vary in costs and benefits. **Landfilling** has low initial costs but requires long-term environmental management, with potential revenue from methane recovery. **Open burning** has minimal operational costs but causes severe environmental and health issues. **Incineration and waste-to-energy** requires high capital investment but reduces landfill dependency and generates energy. **Composting and anaerobic digestion** incur infrastructure and operational costs but produce compost or biogas, benefiting agriculture and reducing methane emissions. **Recycling** involves collection and processing expenses but saves energy, reduces raw material demand, and creates jobs. **Pyrolysis and gasification** require high-tech investments but convert waste into valuable fuels with lower emissions than incineration.

While landfilling has a lower upfront cost, long-term environmental costs (e.g., pollution, land use, methane emissions) make it less economically viable in the long run. Despite the low direct cost associated with open burning, it has substantial hidden costs due to air pollution, health risks, and ecosystem degradation, making it an unsustainable option. Although expensive upfront, WTE plants can be financially viable in regions with high energy demand, stringent landfill regulations, and strong pollution control policies. Composting and anaerobic digestion provide long-term economic and environmental benefits, making them highly cost-effective for managing organic waste. Recycling is economically viable when markets for recovered materials are stable and when integrated with efficient collection and sorting systems. While costly, pyrolysis and gasification technologies can be profitable with proper policy incentives, research, and energy recovery integration (Jamasb & Nepal, 2010).

A cost–benefit analysis of waste management strategies highlights the trade-offs between economic investment, environmental sustainability, and long-term benefits. Traditional methods like landfilling and open burning have lower immediate costs but impose significant long-term environmental and social burdens. On the other hand, modern waste management strategies such as WTE, composting, anaerobic digestion, and recycling provide higher long-term returns by recovering resources and reducing pollution. Governments and industries should prioritize integrated waste management systems that balance cost-effectiveness with sustainability for a more circular economy.

Table 1 Cost benefit analysis of different waste management strategies and technologies

Waste management technology/strategy	Costs	Benefits
Landfilling	• Land acquisition and infrastructure development • Leachate and gas management systems • Long-term environmental monitoring and remediation costs	• Low initial capital costs compared to other waste treatment technologies • Potential revenue from landfill gas (methane) recovery for energy generation
Open burning	• Severe environmental and health damage leading to increased public healthcare costs • High air pollution and climate change impacts	• Minimal infrastructure or operational costs
Incineration and waste-to-energy	• High capital investment for plant construction and maintenance • Air pollution control and regulatory compliance costs • Potential public opposition due to environmental concerns	• Reduces landfill dependency and waste volume • Generates electricity and heat, providing economic returns • Potential for carbon credits if emissions are well managed
Composting and anaerobic digestion	• Infrastructure for composting sites or anaerobic digesters • Labor and operational costs	• Produces valuable compost or biogas, reducing reliance on chemical fertilizers and fossil fuels • Reduces landfill methane emissions • Supports soil health and sustainable agriculture
Recycling	• Collection, sorting, and processing infrastructure • Market volatility for recycled materials	• Reduces demand for raw material extraction, lowering production costs • Saves energy compared to manufacturing new materials • Creates employment opportunities in the recycling industry
Pyrolysis and gasification	• High initial capital investment for advanced processing plants • Technological complexity and operational costs	• Converts waste into valuable fuels and chemicals, reducing reliance on fossil fuels • Lower emissions compared to incineration • Potential revenue from syngas and biochar production

Sources Jamasb and Nepal (2010), Soesanto et al. (2021)

Life Cycle Costing (LCC)

Life Cycle Costing (LCC) is a comprehensive financial assessment method that evaluates the total costs associated with a product, service, or infrastructure over its entire lifespan. In waste management, LCC is used to assess the long-term economic viability of different strategies by considering not only initial investment costs but also operational, maintenance, and end-of-life costs. LCC helps decision-makers understand the true financial implications of waste management strategies over time. It ensures that short-term cost savings do not lead to higher long-term expenses due to environmental damage, inefficient resource use, or expensive remediation efforts (Martinez-Sanchez et al., 2015).

Listed below are the basic steps of carrying out an LCC (Sharma & Chandel, 2021).

1. **Identifying Cost Components**: Costs are categorized into capital investment, operational expenses, maintenance costs, environmental mitigation, and end-of-life disposal costs.
2. **Estimating Costs Over Time**: Forecasting future expenses, including waste treatment, labor, energy consumption, and environmental compliance.
3. **Applying Discounting Methods**: Since costs occur at different points in time, discount rates are used to bring all costs to present values for accurate comparison.
4. **Comparing Alternatives**: Different waste management options are evaluated to determine the most cost-effective choice in the long term

Life cycle costing should be needed in waste management strategies and technologies based on different perspectives including, budgeting and long-term financial planning, encouragement for the adoption of sustainable waste management practices by identifying hidden costs, ensuring that waste management decisions align with economic and environmental goals, and providing a basis for comparing traditional and advanced waste treatment technologies (Sharma & Chandel, 2021).

Life Cycle Costing provides a holistic economic evaluation of waste management strategies, considering both short-term expenses and long-term financial sustainability. While traditional methods like landfilling and open burning appear low-cost initially, they impose significant future liabilities. On the other hand, modern solutions such as WTE, composting, anaerobic digestion, and recycling offer better long-term value by reducing environmental impact and generating economic returns. Governments and waste management authorities should integrate LCC into decision-making to ensure sustainable and cost-effective waste management solutions.

Influence of Market Factors

Market factors play a critical role in determining the feasibility, efficiency, and sustainability of different waste management strategies and technologies. Economic conditions, policy incentives, resource demand, technological advancements, and consumer behavior directly influence the viability of waste management systems. A well-functioning market can drive sustainable waste management by creating financial incentives for recycling, waste-to-energy projects, and circular economy initiatives (Zheng et al., 2016).

Key Market Factors Affecting Waste Management (Markina et al., 2024):

1. **Material and Resource Prices**—The demand and price of raw materials and secondary resources affect the profitability of recycling and waste recovery.
2. **Energy Market Dynamics**—The cost of fossil fuels and renewable energy alternatives impact waste-to-energy (WTE) projects.
3. **Government Policies and Regulations**—Subsidies, carbon pricing, landfill taxes, and extended producer responsibility (EPR) programs shape waste management investments.
4. **Technological Advancements**—Innovations in waste sorting, treatment, and recycling impact the cost-effectiveness of different waste management strategies.
5. **Consumer and Industry Behavior**—Public participation in waste segregation, corporate sustainability initiatives, and industrial waste generation patterns influence waste management approaches.
6. **Global and Local Economic Conditions**—Economic downturns and financial crises affect investment in waste infrastructure and sustainability programs.

Market factors significantly shape the economic and operational feasibility of different waste management strategies. The evolving regulatory landscape, energy market fluctuations, and consumer preferences are pushing the industry toward more sustainable solutions such as recycling, composting, and WTE. Governments, businesses, and communities must align their strategies with market trends to ensure an economically viable and environmentally responsible waste management system.

Overview of the Economical Sustainability of Different Waste Management Strategies

Economic sustainability in waste management refers to the ability of waste management strategies and technologies to deliver long-term financial benefits without compromising environmental integrity or social equity. It encompasses both the direct costs of waste management (e.g., collection, transportation, processing) and indirect benefits, such as resource recovery, energy generation, and reduced environmental damage. Economic sustainability requires an approach that balances immediate financial constraints with the long-term financial, environmental, and social outcomes of waste management systems (Morrissey & Browne, 2004).

Economic sustainability ensures that waste management practices are viable in the long run without exhausting financial resources or creating economic burdens for future generations. It allows governments and industries to allocate resources efficiently, reduce reliance on expensive landfill sites, and foster the recovery of valuable materials from waste. A sustainable system reduces long-term environmental and public health costs, supports the creation of green jobs, and fosters circular economy principles (Shekdar, 2009).

Economic Sustainability can be assessed in several steps as given below (Morrissey & Browne, 2004).

1. **Cost Analysis**: Analyzing capital, operational, and maintenance costs to determine the immediate financial viability of waste management systems.
2. **Revenue Generation**: Identifying potential revenue streams from resource recovery (e.g., recycled materials, energy generation, compost), and assessing profitability over time.
3. **Long-Term Financial Impacts**: Assessing the financial implications of environmental liabilities (e.g., pollution control, remediation, carbon offset) and future waste disposal needs.
4. **Policy Incentives and Subsidies**: Evaluating the impact of government incentives, carbon taxes, and other policy frameworks on the financial sustainability of waste management systems.
5. **Market Trends**: Considering market prices for recycled materials, energy, and byproducts in assessing long-term financial viability.

Economic Sustainability of Different Waste Management Strategies

1. **Landfilling**

    - **Economic Sustainability**:

        – Landfilling is typically the least expensive in terms of capital investment compared to more advanced waste treatment options.
        – However, over time, landfilling incurs high operational and environmental costs due to land acquisition, leachate management, methane control, and the long-term environmental monitoring required.
        – Landfills also face increasing regulatory costs due to stricter environmental laws, contributing to their diminishing economic sustainability.

    - **Revenue Opportunities**: Limited, primarily through methane gas recovery for energy generation.
    - **Long-Term Sustainability**: Landfilling is economically unsustainable in the long run due to growing environmental liabilities and the limited availability of suitable land (Nanda & Berutti, 2021).

2. **Open Burning**

    - **Economic Sustainability**:

        – Open burning has very low operational costs, as it requires minimal infrastructure and labor. However, it leads to significant long-term social and environmental costs due to its harmful effects on air quality and human health.

  - The lack of a waste-to-energy or resource recovery component further diminishes the economic sustainability of open burning.

- **Revenue Opportunities**: None, as open burning does not recover energy or materials.
- **Long-Term Sustainability**: Open burning is economically unsustainable due to high hidden costs, such as healthcare expenses, environmental damage, and lost productivity from pollution (Tan et al., 2015).

3. **Incineration and Waste-To-Energy (WTE) Plants**

   - **Economic Sustainability**:

     - WTE plants require significant initial capital investment and ongoing operational costs. These facilities are economically sustainable in areas with high energy demand, as they can generate electricity or heat.
     - Energy recovery reduces the dependence on landfilling, and in some cases, can be financially profitable by selling the generated energy. However, WTE plants are only economically sustainable if they are efficiently integrated with local waste streams and energy markets.

   - **Revenue Opportunities**: Energy generation (electricity, heat), carbon credits, and potential revenue from waste materials recovery.
   - **Long-Term Sustainability**: The high initial investment and operational costs make WTE less financially attractive unless supported by favorable energy prices, government incentives, and high waste volumes (Tan et al., 2015).

4. **Composting and Anaerobic Digestion**

   - **Economic Sustainability**:

     - Both composting and anaerobic digestion have lower capital and operational costs compared to other waste treatment options.
     - They can generate valuable byproducts, such as compost (used for soil enrichment) or biogas (used for energy production), which offset operational costs.
     - These technologies are economically sustainable when integrated into local agricultural and energy markets, providing valuable resources that enhance soil health and reduce the need for chemical fertilizers.

   - **Revenue Opportunities**: Sale of compost, biogas, and carbon credits.
   - **Long-Term Sustainability**: Composting and anaerobic digestion are financially sustainable in the long run, especially with rising demand for organic farming, renewable energy, and sustainable waste management solutions (Abubakar et al., 2022).

5. **Recycling**

   - **Economic Sustainability**:
     - Recycling can be a highly sustainable waste management strategy if the necessary infrastructure and markets for recovered materials are in place.
     - The costs of collection, sorting, and processing are offset by the revenue generated from selling recycled materials, reducing the need for raw material extraction and lowering manufacturing costs.
     - Market fluctuations in the price of raw materials and recycled goods, however, can impact the profitability of recycling systems.
   - **Revenue Opportunities**: Sale of recycled materials (e.g., metals, plastics, paper), reduced production costs for manufacturers.
   - **Long-Term Sustainability**: Recycling is one of the most economically sustainable waste management practices when supported by robust market demand for recycled materials and long-term policy incentives (Abubakar et al., 2022).

6. **Advanced Waste Treatment Technologies (Pyrolysis & Gasification)**

   - **Economic Sustainability**:
     - Pyrolysis and gasification technologies offer potential long-term economic sustainability by converting waste into valuable fuels or chemicals, thereby reducing reliance on fossil fuels.
     - However, the high initial investment and operational costs make these technologies economically viable only in areas with high waste generation and energy demand.
     - These technologies can become economically sustainable through market demand for the products they generate and government support for renewable energy projects.
   - **Revenue Opportunities**: Sale of biofuels, chemicals, and energy recovery from waste.
   - **Long-Term Sustainability**: Pyrolysis and gasification have the potential for long-term economic sustainability if properly integrated with energy markets, carbon pricing systems, and circular economy models (Abubakar et al., 2022).

The economic sustainability of different waste management strategies varies widely depending on initial capital investment, operational costs, revenue generation opportunities, and long-term environmental liabilities. Landfilling and open burning, while economically attractive in the short term, present long-term financial challenges due to environmental costs and limited resource recovery. In contrast, strategies like recycling, composting, anaerobic digestion, and WTE plants offer more sustainable financial outcomes, particularly when integrated with markets for recycled materials and renewable energy. Advanced waste treatment technologies, although expensive upfront, hold promise for economic sustainability in regions with high energy demands and government support. A sustainable waste management system should be designed with long-term financial, environmental, and social goals in mind, ensuring that both immediate costs and future benefits are balanced effectively.

6 Comparative LCA Studies of Different Waste Management Technologies

Landfilling Versus Incineration

The environmental impacts of **landfilling** and **incineration** can be assessed using **Life Cycle Assessment (LCA)** through a detailed and systematic analysis of the processes involved in each waste management strategy.

Table 2 shows, how assessing key environmental impacts of waste management strategies is crucial for sustainable decision-making. **Landfilling** and **incineration** are evaluated across impact categories such as **greenhouse gas emissions, air pollution, resource recovery, land use, water contamination, and long-term costs**. Landfilling contributes significantly to climate change due to methane emissions from anaerobic decomposition, while incineration emits CO_2 and air pollutants like dioxins and NO_x. Resource recovery is minimal in landfilling but more efficient in incineration through energy generation. Land use impacts are higher for landfills due to site expansion and post-closure management, whereas incineration requires less space but produces toxic ash requiring proper disposal. Water pollution risks arise from landfill leachate and incineration ash. The LCA framework systematically quantifies these impacts, integrating energy offsets, environmental burdens, and long-term sustainability considerations to identify the most effective waste management strategy.

Here is a breakdown of how the key impacts such as greenhouse gas emissions, air pollution, resource recovery, land use, leachate, and long-term sustainability are assessed in an LCA framework (Morris, 2005).

Table 3 captures the key differences in environmental impacts and sustainability factors of landfilling and incineration, providing an overview of their comparative performance in an LCA context. As stated in Table 3, the LCA framework evaluates the environmental trade-offs between landfilling and incineration, considering key impact categories such as greenhouse gas emissions, air pollution, resource recovery, land use, water contamination, and cost implications. Landfilling is associated with high methane emissions, significant land use requirements, and long-term environmental risks from leachate contamination, despite its lower initial costs. In contrast, incineration emits CO_2 but reduces methane emissions, offers higher energy recovery, and requires less land. However, it generates toxic air pollutants and residual ash, requiring effective pollution control measures. While landfilling has lower upfront costs, its long-term management expenses are high, whereas incineration demands

Table 2 Assessment of key impacts in LCA framework

Environmental impact	Landfilling	Incineration
Greenhouse gas emissions	Methane (CH_4) emissions from anaerobic decomposition. Assessed using **emission factors** based on waste composition, landfill design, and methane recovery systems	CO_2 emissions from combustion. Energy recovery is assessed for **net CO_2emissions** considering energy produced and fossil fuel use
Air pollution and human health	Air pollutants like **VOC, H_2S**, and **ammonia**. Emission factors used, and **mitigation measures** (e.g., gas recovery) are considered for air pollution reduction	**Dioxins, NOx, SO_2**, particulate matter, and heavy metals. Emissions are assessed with respect to **air pollution control technologies** (scrubbers, filters) and their effectiveness
Resource recovery and energy use	Limited energy recovery from landfill gas (LFG). Energy recovery efficiency and **methane utilization** are assessed	Energy recovery (electricity, heat) from waste combustion. Energy output, **carbon offset**, and recovery efficiency are considered
Land and space requirements	Large land area required for disposal. Land degradation, **site expansion**, and **long-term post-closure** management (monitoring, leachate treatment) are included	Less land required for incineration plants. Assessment includes land use for plant construction, **waste storage**, and **ash disposal**
Leachate/water pollution	Leachate production from decomposing waste, potentially contaminating groundwater. **Leachate management systems** are assessed for their effectiveness	Ash disposal generates leachate risks if not properly managed. **Ash treatment and disposal** systems are considered for potential water pollution
Cost implications	Lower initial costs, but high long-term environmental management costs (landfill closure, post-closure monitoring)	High initial costs for construction and operation, but revenue from **energy generation** can offset costs
Long-term environmental impact	Long-term risks of **methane emissions, land degradation**, and **groundwater contamination**. Long-term costs of post-closure care are included	Long-term **air pollution** risks, **toxic ash** disposal, and **resource recovery** potential. Environmental impacts of residual waste and ash handling are considered

Source Morris (2005)

Table 3 Overview of the performance of landfilling and incineration in an LCA context

Environmental impact	Landfilling	Incineration
Greenhouse gas emissions	High methane (CH_4) emissions from anaerobic decomposition	CO_2 emissions from combustion, lower methane
Air pollution and human health	Low to moderate air pollution (VOC, H_2S, ammonia)	Higher air pollution (dioxins, NO_x, SO_2, particulate matter, heavy metals)
Resource recovery and energy use	Limited energy recovery (methane for energy)	High energy recovery (electricity, heat) from waste combustion
Land and space requirements	Requires large land areas for waste disposal and expansion	Requires less land area compared to landfilling, but significant infrastructure needs
Leachate/water pollution	High risk of groundwater contamination from leachate	Potential for ash contamination; lower risk of water pollution compared to landfills
Cost implications	Lower initial costs, high long-term environmental management costs	High initial costs, but potential for revenue from energy generation
Long-term environmental impact	High risk of long-term pollution (methane, leachate)	Environmental impact reduced with energy recovery and emission control, but still significant pollution risks

Source Morris (2005)

higher investment but can offset costs through energy generation. By integrating these factors, the LCA approach helps identify the most environmentally and economically sustainable waste management strategy based on impact mitigation and resource efficiency.

Ultimately, the choice between landfilling and incineration depends on specific regional factors, including waste composition, environmental regulations, technological capabilities, and societal goals. While incineration can offer greater energy recovery, landfilling remains a simpler but less sustainable option with serious long-term environmental consequences.

Recycling Versus Composting

Both recycling and composting offer significant environmental benefits when compared to more traditional waste management strategies like landfilling and incineration (Barlaz et al., 2003). From a Life Cycle Assessment (LCA) perspective:

- Recycling excels in reducing greenhouse gas emissions, conserving natural resources, and saving energy. By recovering valuable materials like metals, paper, and plastics, recycling significantly reduces the need for virgin material extraction and minimizes energy-intensive production processes. However, it does require energy for processing and transportation, and air pollution can be a concern, though it is generally lower than that of incineration. Land use for recycling facilities is more efficient than landfills but still significant depending on the scale and material types. Over the long term, recycling helps mitigate environmental degradation by reducing waste and supporting resource conservation.

- Composting, on the other hand, is an excellent strategy for managing organic waste by turning it into nutrient-rich compost that benefits soil health and enhances carbon sequestration. It has relatively low energy demands and minimal greenhouse gas emissions, especially when managed properly under aerobic conditions. Air pollution from composting is minimal, though ammonia and methane may be emitted if conditions are not carefully controlled. The land space required for composting is more substantial than for recycling but remains far less than landfilling. Composting also provides long-term environmental benefits by reducing the need for chemical fertilizers, improving soil structure, and promoting healthier ecosystems (Di Maria & Sisani, 2017).

Tables 4 and 5 show how the LCA framework evaluates the environmental performance of recycling and composting, highlighting their role in resource efficiency and waste reduction. As stated in the table, recycling is energy-intensive but significantly reduces CO_2 emissions by lowering the demand for virgin materials. It also has high energy recovery potential, saving resources in material production, though it requires large processing facilities and careful leachate management. Composting, on the other hand, has lower greenhouse gas emissions, with minimal methane release and potential carbon sequestration in soil. While it lacks energy recovery, it supports nutrient cycling, improving soil health and reducing chemical fertilizer use. Composting requires more land but has lower initial costs compared to recycling facilities. Over the long term, both strategies contribute to sustainability, with recycling focusing on resource conservation and composting enhancing soil quality and ecosystem health, making them key components in a circular economy approach within the LCA framework.

These tables compare recycling and composting from an LCA perspective, showing the key differences in greenhouse gas emissions, air pollution, energy use, land requirements, water pollution, costs, and long-term environmental impacts. Each method has its strengths, and the choice between them depends on specific waste management goals and local conditions.

While recycling is highly effective for material recovery, energy savings, and resource conservation, composting shines in managing organic waste and improving soil health. Both methods contribute to reducing the overall environmental footprint, with recycling better suited for materials recovery and composting being a sustainable solution for organic waste and soil enhancement. The decision between the two depends on local waste composition, infrastructure, and environmental goals. Ideally, both strategies should be integrated into a holistic waste management system to maximize environmental sustainability.

Table 4 Assessment of key impacts in LCA framework for recycling and composting

Environmental impact	Recycling	Composting
Greenhouse gas emissions	Energy-intensive processes (e.g., transportation, processing), but **reduction** in CO_2 emissions from recycling materials instead of producing virgin materials	Lower CO_2 emissions compared to landfill or incineration. **Methane (CH_4)** emissions from anaerobic conditions are low, with **carbon sequestration** in soil
Air pollution and human health	Air emissions from the processing plants (e.g., VOCs, NO_x). **Emissions** reduced compared to virgin material production	Minimal air pollution; however, the **composting process** may release small amounts of ammonia and VOCs if not managed
Resource recovery and energy use	High **energy recovery** potential from recycling materials (metal, glass, plastics). **Energy savings** from producing recycled materials instead of new ones	Limited energy recovery. Composting mainly focuses on organic waste decomposition, returning nutrients to the soil, with minimal energy use compared to other waste treatments
Land and space requirements	**Space-efficient** compared to landfilling, but facilities for **sorting and processing** can be large	Requires significant land for composting facilities, especially for large-scale operations or if **aerobic conditions** need to be controlled
Leachate/water pollution	Potential for leachate generation if improperly managed at recycling facilities. Risk of contaminating **groundwater** from sorting and washing processes	Minimal leachate, but improper composting practices (e.g., inadequate moisture management) can cause **runoff contamination**. Proper management of the composting site is key
Cost implications	High initial setup costs for recycling facilities, but **cost savings** from reduced need for virgin materials, waste diversion, and energy recovery	Low initial costs for setting up **composting facilities**; however, efficiency increases with scale. **Revenue** can be generated from compost sales
Long-term environmental impact	**Resource conservation** over the long term, reducing the demand for raw materials. Reduces the environmental footprint of producing new products	**Nutrient cycling** into soil, improving soil health and carbon sequestration. Reduced need for chemical fertilizers and lower long-term environmental impact on ecosystems

Source Barlaz et al. (2003), Di Maria and Sisani (2017)

Waste-to-Energy Versus Material Recovery

When evaluating the environmental performance of Waste-to-Energy (WTE) technologies and Material Recovery from a Life Cycle Assessment (LCA) perspective, the analysis considers multiple stages of the waste management process, from waste generation and collection to disposal or recovery. Each approach has its advantages

Table 5 Overview of performance of recycling and composting in an LCA context

Environmental impact	Recycling	Composting
Greenhouse gas emissions	**Recycling processes** (e.g., aluminum, paper, plastics) save significant **GHG emissions** compared to producing materials from raw resources, but still contribute some GHGs due to the energy-intensive processing	**Minimal emissions** during aerobic composting, but **greenhouse gases** (mainly **methane**) are produced if **anaerobic conditions** are present. Carbon sequestration occurs when the compost is used to enhance soil
Air pollution and human health	Air pollution from materials like **plastics** or **paper** during processing. The impact can be reduced with better technologies (e.g., enclosed systems)	Air pollutants can be released in small quantities, such as **ammonia** and **methane**, but are far less toxic compared to incineration. Effective **compost management** can mitigate these issues
Resource recovery and energy use	**Energy recovery** through recycling reduces the demand for raw material extraction and lowers energy consumption. Recycled aluminum, for instance, saves significant energy compared to new aluminum production	**Energy use** in composting is minimal; however, the environmental benefit is **resource recovery** through nutrient-rich compost that improves soil health and reduces chemical fertilizer needs
Land and space requirements	**Efficient** in land use compared to landfills but requires significant land for sorting and processing materials. **Space needs** vary by material type	Requires **land for composting** facilities and can have a significant spatial footprint for large-scale operations. **Space efficiency** is lower than recycling but still much better than landfilling
Leachate/water pollution	Potential **leachate risks** from contaminants like plastics or metals, if not properly handled. Proper **separation and cleaning** processes mitigate this	Minimal leachate production. If improperly managed, **runoff** from wet composting processes can cause contamination of nearby water sources
Cost implications	**High upfront costs** for setting up recycling infrastructure, but **long-term savings** from less reliance on virgin materials and energy savings	Lower initial investment. Operational costs depend on the **scale of the facility**. Can generate income from selling the compost, and it is more cost-effective than landfilling or incineration
Long-term environmental impact	Long-term **resource conservation** reduces mining impacts and energy consumption. Recycling lessens the environmental **burden** compared to producing new materials	Composting contributes to **soil health**, enhances **carbon sequestration**, and decreases the need for chemical fertilizers. The long-term impact includes ecosystem health improvements

Source Di Maria and Sisani (2017)

and disadvantages in terms of environmental impacts, resource use, and long-term sustainability.

Waste-to-Energy (WTE) refers to the process of generating energy, typically in the form of electricity or heat, through the combustion or gasification of waste materials. From an LCA perspective, WTE has both positive and negative environmental impacts depending on the technology used and the types of waste being processed (Bueno et al., 2015).

Greenhouse Gas Emissions: WTE systems emit greenhouse gases (GHGs), primarily in the form of CO_2, during the combustion of organic waste. However, WTE also generates energy that can offset the use of fossil fuels, thereby reducing the overall carbon footprint. The net GHG emissions of WTE depend on the efficiency of the plant and the type of waste being burned. Waste materials such as plastics and non-recyclable materials may contribute higher emissions compared to organic waste (Giugliano et al., 2011).

Resource Recovery and Energy Use: The key advantage of WTE is its ability to recover energy from waste. This can be used to generate electricity or provide district heating, thus reducing the need for additional energy sources. However, WTE plants are generally less energy-efficient than recycling in terms of resource recovery, as only a fraction of the waste's material value can be recovered through energy production. Non-combustible materials such as glass and metals are not utilized for energy recovery and are instead left behind as bottom ash (Bueno et al., 2015).

Air Pollution and Human Health: The combustion process in WTE facilities produces air pollutants such as particulate matter, dioxins, nitrogen oxides (NO_x), and sulfur dioxide (SO_2), which can have adverse effects on both human health and air quality. Advanced emission control technologies like scrubbers, filters, and electrostatic precipitators are used to mitigate these impacts, but risks remain, especially when facilities are not optimally managed. These emissions are a significant concern in WTE plants, requiring regular monitoring to ensure compliance with environmental standards (Bueno et al., 2015).

Land Use and Space Requirements: WTE plants require significant land for the facility itself, as well as for waste storage and ash disposal. The land requirement can be substantial, especially in urban areas. However, compared to landfills, WTE is more space-efficient in that it can process waste and generate energy within a relatively smaller footprint (Bueno et al., 2015).

Long-Term Environmental Impact: In the long term, WTE's main environmental challenge is the generation of toxic ash, which must be carefully disposed of. WTE plants also contribute to resource depletion because they burn materials that could otherwise be recycled and reused. Furthermore, the overall sustainability of WTE is questioned in the context of a circular economy, as it prioritizes energy recovery over material recovery, potentially leading to more consumption of non-renewable resources (Bueno et al., 2015).

Material recovery, or recycling, involves the separation and processing of waste materials, such as metals, plastics, paper, and glass, to convert them back into raw materials that can be used for manufacturing. From an LCA perspective, material recovery is typically seen as a more environmentally sustainable option compared to WTE, especially in terms of resource conservation, energy savings, and long-term environmental impacts (Holmgren & Henning, 2004).

Greenhouse Gas Emissions: One of the most significant environmental advantages of material recovery is the reduction in greenhouse gas emissions. Recycling materials like aluminum, paper, and plastics typically requires less energy than producing new materials from raw resources, resulting in a net reduction in GHG emissions. For example, recycling aluminum saves up to 95% of the energy required to produce new aluminum. However, the emissions associated with sorting, transporting, and processing materials in recycling facilities must still be considered (Holmgren & Henning, 2004).

Resource Recovery and Energy Use: Material recovery plays a crucial role in the circular economy, allowing valuable materials to be recovered, recycled, and reused. This helps to conserve natural resources by reducing the demand for raw materials (Holmgren & Henning, 2004). Recycling also results in significant energy savings, particularly for materials like metals, plastics, and paper, as less energy is required for their production compared to virgin materials. Energy use in material recovery is typically lower than in WTE, as it does not require combustion or the high-temperature processes used in energy recovery.

Air Pollution and Human Health: Material recovery generally produces fewer air pollutants compared to WTE, as there is no combustion involved. However, certain stages of the recycling process, particularly mechanical sorting and transportation, can generate particulate matter and dust, which can pose health risks to workers if not managed properly. Recycling facilities also have to ensure that toxic substances like lead, mercury, and other contaminants are separated to prevent hazardous materials from re-entering the material stream (Holmgren & Henning, 2004).

Land Use and Space Requirements: Material recovery requires substantial space for sorting, processing, and storing materials, though generally, recycling facilities take up less land than WTE plants (Holmgren & Henning, 2004). Sorting facilities need to accommodate specialized equipment for different materials, and large-scale operations may require significant space for storage and handling of recyclable materials. Land efficiency is higher in recycling compared to waste incineration or landfilling, but land availability can still be a challenge in urban areas with limited space.

Long-Term Environmental Impact: The long-term environmental benefits of material recovery are significant. By recycling materials, the need for virgin resource extraction is reduced, which in turn decreases the environmental impact associated with mining, logging, and drilling (Holmgren & Henning, 2004). Additionally, recycling helps reduce the amount of waste sent to landfills or incinerators, contributing to waste reduction and resource conservation. Over time, material recovery contributes to the establishment of a circular economy, where materials are reused, and environmental degradation is minimized.

From an LCA perspective, Material Recovery generally offers lower environmental impacts in terms of greenhouse gas emissions, resource conservation, and energy savings compared to WTE technologies. Material recovery focuses on recycling valuable materials, which reduces the need for virgin resources and saves energy, thus mitigating environmental degradation. It is also a crucial element in promoting a circular economy, supporting long-term sustainability by reusing materials and decreasing waste generation (Holmgren & Henning, 2004).

On the other hand, WTE serves as an important waste management solution when material recovery is not feasible, particularly for non-recyclable waste. WTE provides energy recovery and can significantly reduce the volume of waste sent to landfills. However, it produces greenhouse gas emissions, air pollutants, and generates toxic ash, making it less environmentally sustainable than material recovery in the long term (Holmgren & Henning, 2004).

The choice between WTE and material recovery depends on factors such as the composition of waste, the efficiency of recycling programs, and the availability of infrastructure. In many cases, a combined approach that integrates both WTE and material recovery could help optimize waste management strategies, enhancing sustainability, energy recovery, and resource conservation while minimizing environmental impacts.

Traditional Versus Circular Economy Approaches

Traditional waste management technologies and circular economy approaches represent two fundamentally different approaches to handling waste and managing resources. Traditional waste management is primarily focused on disposal and treatment of waste, often resulting in landfilling or incineration. In contrast, the circular economy aims to reduce, reuse, and recycle resources to keep them in use for as long as possible, thus minimizing waste and environmental impacts. Evaluating these approaches from a Life Cycle Assessment (LCA) perspective highlights their environmental sustainability, resource efficiency, and long-term benefits (Brändström & Saidani, 2022).

1. **Waste Generation and Resource Conservation**

In traditional waste management, waste is often treated as a byproduct of consumption and is disposed of through methods such as landfilling or incineration. These methods typically result in resource loss, as valuable materials are discarded, contributing to resource depletion. Landfills are a major repository for discarded goods, and incineration also leads to the loss of materials that could be recycled or reused. Traditional waste management does not prioritize resource recovery, leading to higher resource extraction rates to produce new goods.

The circular economy flips this model by emphasizing resource efficiency and waste reduction. In a circular economy, products and materials are designed for longer life, repair, recycling, or remanufacturing, allowing them to be reused in subsequent production cycles. This model helps conserve natural resources by reducing the demand for raw materials and promoting the recovery of valuable materials from used products. LCA in the circular economy focuses on keeping materials in use for as long as possible, thus reducing the need for virgin resources and preventing waste generation (Spreafico, 2022).

2. **Waste Treatment Technologies and Environmental Impacts**

Traditional waste treatment technologies, such as landfilling and incineration, have significant environmental drawbacks. Landfills, for example, produce methane emissions—a potent greenhouse gas—as organic waste decomposes anaerobically. Additionally, landfills can lead to soil contamination and leachate, which can seep into surrounding water sources. Incineration, on the other hand, generates air pollution (e.g., particulate matter, dioxins, and NO_x) and CO_2 emissions. Although modern incineration plants have advanced filtration systems, emissions can still pose risks to human health and air quality (Brändström & Saidani, 2022).

Circular economy approaches significantly reduce waste by prioritizing material recovery and recycling. By emphasizing closed-loop recycling, waste is diverted from landfills and incinerators, helping mitigate environmental impacts such as air pollution, soil contamination, and greenhouse gas emissions. For example, materials like plastics, metals, and paper can be recycled into new products, reducing the need for waste disposal and mitigating the environmental impact of both landfilling and incineration. Additionally, the use of WTE in circular economy models, when carefully managed, can offer energy recovery with reduced emissions compared to traditional incineration (Iqbal et al., 2020).

3. **Greenhouse Gas Emissions and Climate Change Impact**

Both landfills and incineration contribute to greenhouse gas emissions. In landfills, methane is emitted as organic waste decomposes anaerobically, with landfills being one of the largest sources of methane emissions globally. Methane is a potent greenhouse gas that significantly contributes to global warming. Incineration, while producing energy, still emits CO_2, along with other pollutants (Iqbal et al., 2020). The energy generated is often lower than the potential energy recovery from recycling processes, and emission control technologies in incinerators may not always capture all the harmful substances.

The circular economy aims to minimize greenhouse gas emissions by reducing the need for raw material extraction and lowering the emissions associated with production. Through recycling, materials are kept in circulation, preventing the need to produce new materials that typically require significant energy and resource consumption. Recycling certain materials, like aluminum, can reduce emissions by up to 95% compared to producing new aluminum from ore. Moreover, designing for durability and reuse minimizes the environmental footprint of products across their entire life cycle. The energy recovery aspect of circular economies, particularly through WTE systems, can offset some of the emissions produced by traditional methods, although this remains secondary to recycling in overall environmental impact (Iqbal et al., 2020).

4. **Economic Considerations and Resource Efficiency**

Traditional waste management technologies tend to be economically inefficient because they primarily focus on disposal rather than resource recovery. Landfills require significant land area, which has both direct and indirect costs. They also require long-term maintenance and monitoring due to the risk of leachate and methane emissions. Incineration, while providing energy, typically operates at lower energy efficiency compared to other forms of energy generation, and the cost of disposal remains high. Moreover, waste materials that could have been recycled are lost, reducing the potential economic benefits from resource recovery.

In the circular economy, waste is viewed as a resource rather than a byproduct, and economic efficiency is greatly improved by maximizing resource recovery and reducing the need for virgin material extraction. The focus on recycling, remanufacturing, and upcycling allows for the creation of new value from existing materials. This leads to job creation, economic resilience, and more sustainable economic growth. The economic cost of waste disposal is significantly lower in the circular economy, as less waste is generated, and more materials are reused, creating long-term cost savings. Additionally, circular systems like Waste-to-Energy (WTE) can help recover energy and reduce dependence on fossil fuels (Soto-Paz et al., 2023).

5. **Long-Term Environmental Sustainability**

The long-term sustainability of traditional waste management is questionable due to the linear nature of the approach. Landfills continue to expand, contributing to land degradation and pollution, while incineration remains a source of toxic emissions. Both methods are limited in their ability to reduce the demand for raw materials and to recover energy or resources from waste. As resource scarcity becomes more apparent, reliance on the traditional approach will strain both natural resources and the environment (Soto-Paz et al., 2023).

The circular economy is inherently more sustainable in the long term. By keeping materials in use, reducing waste, and promoting sustainable product design, it aims to close the resource loop and create a regenerative system. This approach not only conserves natural resources but also reduces pollution, energy consumption, and greenhouse gas emissions. Over time, as more industries adopt circular economy principles, the overall environmental impact of consumption and production is minimized, leading to a more resilient and sustainable future.

From an LCA perspective, circular economy approaches are far more environmentally sustainable than traditional waste management technologies. The circular economy's emphasis on resource recovery, recycling, and reduction of waste generation provides long-term environmental benefits, including lower emissions, energy savings, and resource conservation. Traditional waste management technologies, such as landfilling and incineration, focus on disposal, leading to resource depletion, greenhouse gas emissions, and pollution.

While traditional waste management may still be necessary in certain situations, particularly for non-recyclable or hazardous waste, a shift toward a circular economy is essential to address the growing challenges of resource scarcity and environmental degradation. By adopting circular economy principles, we can create a more sustainable and resource-efficient system that not only reduces the environmental impact of waste but also contributes to economic resilience and long-term sustainability.

7 Challenges and Advancements in LCA Tools and Models for Waste Management

Challenges in LCA Tools and Models for Waste Management

Life Cycle Assessment (LCA) has become an essential tool for evaluating the environmental impacts of various waste management strategies. However, despite its widespread application, there are several challenges associated with using LCA tools and models for waste management, particularly in the context of complex, dynamic systems like waste treatment and disposal. These challenges arise due to the complexity of waste systems, the variability in data quality, and the diverse environmental impacts involved in the waste management process (Christensen et al., 2007).

1. Complexity of Waste Systems

One of the primary challenges in applying LCA to waste management is the complexity of waste systems. Waste management is a multi-stage process that involves various technologies, such as collection, transportation, sorting, recycling, incineration, and landfilling. Each stage can have unique environmental impacts, which makes it difficult to assess the overall impact accurately. Moreover, waste systems often have interconnected variables, including waste composition, recycling rates, and local infrastructure, which can vary widely from one region to another.

LCA models must account for regional differences in waste management practices and infrastructure, which can significantly affect the results. The variability of waste types (e.g., organic, plastic, metal, or electronic waste) further complicates the modeling process, as each material may have a different environmental impact throughout its life cycle. To address this, LCA models must be flexible and capable of incorporating the complex interplay of variables in waste management systems (Christensen et al., 2020).

2. Data Availability and Quality

The accuracy of an LCA is heavily dependent on the availability and quality of the data used in the assessment. In the context of waste management, data on waste composition, emission factors, energy use, and waste treatment technologies are often limited or inconsistent, especially for small-scale operations or informal sectors in developing regions. The lack of standardized data across different waste management systems adds another layer of complexity to LCA studies.

Moreover, data can be difficult to obtain for specific waste management technologies, such as Waste-to-Energy (WTE) plants, where variations in technology, efficiency, and emissions may occur. Similarly, emission factors for landfill operations can vary depending on the type of waste, geographical location, and the management practices in place. Inadequate or outdated data can lead to inaccurate LCA results, thus affecting decision-making and the overall reliability of the assessment (Clavreul et al., 2012).

3. Uncertainty in Impact Assessment

LCA models are also subject to uncertainty due to the numerous assumptions that need to be made, particularly when predicting the long-term impacts of waste management practices. Uncertainty can arise from various sources, such as the variability in waste composition, the lifetime of products and materials, and the uncertainty in predicting future changes in waste management technologies or policies. The use of generic models and assumptions about technologies and processes can introduce systematic errors into the analysis.

Additionally, assessing long-term impacts, such as the potential for pollution accumulation or resource depletion, can be challenging in waste management. For example, the environmental impact of landfills may not be fully realized until many years after waste is disposed of. Similarly, emissions from incineration or Waste-to-Energy technologies may have long-term health and environmental impacts that are difficult to quantify accurately (Clavreul et al., 2012).

4. Impact Categories and Trade-offs

Another challenge in LCA for waste management is the selection and interpretation of appropriate impact categories. Waste management systems have a wide range of potential impacts, including climate change, human health, resource depletion, and ecosystem quality. These impacts can be interconnected, and changes in one area can lead to trade-offs in others. For example, while recycling may reduce the need for raw material extraction and conserve energy, it may also lead to other impacts, such as chemical pollution from the use of toxic substances in certain recycling processes.

Similarly, incineration may reduce the volume of waste but may lead to air pollution or health risks associated with the release of harmful substances. Understanding these trade-offs is essential, but can be difficult to quantify in a single LCA model. Moreover, the prioritization of certain impact categories (e.g., carbon emissions) over others (e.g., human health) may result in a skewed understanding of the true environmental impact of waste management systems (Clavreul et al., 2012).

5. Temporal and Spatial Considerations

LCA studies often face challenges when considering the temporal and spatial aspects of waste management. The environmental impacts of waste management practices can change over time, especially when considering long-term effects such as landfill gas emissions or the potential for the recovery of secondary materials through recycling. Additionally, spatial variability in environmental conditions (e.g., local climate, soil composition, and population density) can affect the performance of waste management systems. For instance, landfill methane emissions may vary depending on the climate, which can influence the carbon footprint of landfilling in different regions.

Some LCA models may not account for these temporal changes adequately, resulting in an incomplete picture of the overall environmental impact. Geographic differences in infrastructure, labor costs, and technology adoption can also lead to varying impacts from the same waste management technology, which makes it challenging to create a universal model that applies to all locations (Clavreul et al., 2012).

6. Socio-Economic Factors

While LCA focuses primarily on environmental impacts, it does not always capture the socio-economic factors that influence waste management decisions. Social acceptance, public behavior, political will, and economic factors can play a crucial role in the success or failure of a waste management strategy. For instance, high upfront costs for advanced recycling technologies or Waste-to-Energy (WTE) systems may be a significant barrier to adoption in developing countries, where the economic capacity for such investments is limited.

Moreover, social factors such as the availability of skilled labor, local regulations, and public awareness of the environmental benefits of recycling or waste reduction programs can influence the effectiveness of a particular waste management strategy. LCA models are often limited in their ability to incorporate these socio-economic dimensions, which are critical for understanding the real-world performance of waste management systems (Clavreul et al., 2012).

7. Integration with Policy and Decision-Making

LCA is a valuable tool for evaluating environmental impacts, but integrating its findings into policy development and decision-making processes can be challenging. Policy decisions regarding waste management often involve multiple stakeholders with competing interests, including environmental groups, industry representatives, and local governments. LCA can provide valuable insights into the environmental trade-offs of various waste management options, but translating these insights into actionable policies requires clear communication of complex findings (Clavreul et al., 2012).

Moreover, LCA models may not always reflect the economic feasibility or the social acceptability of different waste management options. For example, while a particular strategy may appear environmentally superior in an LCA, it may not be the most cost-effective or practical solution in a given context. This requires policymakers to consider LCA findings alongside other factors, such as economic costs, technological readiness, and social acceptability.

While Life Cycle Assessment (LCA) is a powerful tool for evaluating the environmental impacts of waste management systems, several challenges hinder its application in real-world scenarios. These challenges include the complexity of waste systems, data limitations, uncertainty in impact assessment, and trade-offs between different impact categories. Additionally, temporal and spatial considerations, socio-economic factors, and the integration of LCA with policy add to the complexity of using LCA effectively in waste management decision-making. Overcoming these challenges requires continuous improvement in LCA models, better data collection, and a more comprehensive approach to decision-making that considers environmental, economic, and social factors.

Advancements in LCA Tools and Models for Waste Management

In recent years, there have been significant advancements in Life Cycle Assessment (LCA) tools and models for waste management, driven by the increasing need for environmentally sustainable waste management and the growing complexity of waste systems. These advancements focus on improving the accuracy, flexibility, and reliability of LCA models to better reflect the environmental, economic, and social impacts of different waste management strategies. With these advancements, LCA has become a more robust tool for guiding policy decisions, improving waste management practices, and fostering circular economy initiatives.

1. Integration of Waste Management Technologies into LCA

One of the major advancements in LCA tools for waste management is the integration of a wide range of waste management technologies into the models. Modern LCA tools can now evaluate complex systems that include not only landfilling and incineration but also advanced recycling, Waste-to-Energy (WTE) technologies, and emerging solutions like biological treatment and composting. These tools are designed to capture the interactions between different waste treatment stages, considering not just the direct environmental impacts of each technology but also the synergies and trade-offs that may arise when technologies are combined in integrated waste management systems.

For instance, new LCA models allow for the evaluation of multi-stage processes such as waste sorting, pre-treatment, energy recovery, and resource recovery together, giving a more holistic view of the total environmental footprint. Additionally, these tools can assess how material recycling or energy recovery through WTE impacts the need for new resources and the energy demands of waste systems (Nurzhan et al., 2025).

2. Incorporation of Regional and Temporal Variability

Traditionally, LCA models for waste management were often limited in their ability to consider regional variability in waste generation, infrastructure, and local conditions. However, recent advancements have enabled models to integrate regional factors into their assessments. These factors include geographic differences, such as local infrastructure for recycling, waste composition, landfilling regulations, and climate variations, all of which can significantly affect the performance of different waste management technologies (Nurzhan et al., 2025).

For example, landfill methane emissions can vary depending on local climatic conditions (e.g., temperature and precipitation), which can influence the rate of waste decomposition and methane production. Similarly, LCA tools can now account for the energy grid mix in a particular region when assessing the impacts of Waste-to-Energy plants, considering the electricity generated from such plants and its substitution effects on local power generation.

Advancements in LCA models also allow for the temporal assessment of waste management practices, considering both short-term and long-term environmental impacts. This includes evaluating the effects of waste management technologies over decades or even centuries, which is particularly important for landfills that continue to generate environmental impacts such as leachate and methane emissions long after closure (Nurzhan et al., 2025).

3. Enhanced Data Collection and Modeling Capabilities

The quality and availability of data are critical to the success of any LCA. Over the past few years, there has been a significant improvement in data collection methods and modeling capabilities. The development of more robust databases and life cycle inventory (LCI) data sets for specific waste materials, waste management technologies, and environmental impacts has significantly enhanced the accuracy of LCA models. Organizations and research institutions are now compiling more detailed and region-specific data on waste streams, treatment technologies, and environmental impact factors.

Moreover, the rise of big data and IoT (Internet of Things) has enabled the collection of real-time data from waste management systems, improving the accuracy of models. Smart waste management systems equipped with sensors can now track waste composition, collection rates, energy usage, and emissions in real time, feeding this data directly into LCA models. This continuous flow of data allows for more dynamic and accurate assessments of waste management strategies and can support decision-making at both the local and global scales (Nurzhan et al., 2025).

4. Life Cycle Costing (LCC) and Integration with Environmental Impact

Another significant advancement in LCA for waste management is the integration of Life Cycle Costing (LCC) with environmental impact assessments. LCC allows for the evaluation of not only the environmental but also the economic impacts of different waste management technologies over their life cycles. This helps stakeholders make more informed decisions by providing a clearer picture of the cost-effectiveness of various technologies, such as recycling, composting, WTE, and incineration (Nurzhan et al., 2025).

By incorporating LCC into LCA models, it becomes possible to compare the costs and benefits of different strategies, considering both direct costs (e.g., capital and operational costs) and indirect costs (e.g., environmental damage, resource depletion). This integration ensures that the best environmental and economic outcomes are achieved while providing a more comprehensive understanding of trade-offs in waste management systems.

5. Modeling of Circular Economy Systems

LCA models have also advanced to better assess the impact of circular economy strategies. Circular economy approaches focus on extending the lifecycle of materials, reducing waste, and promoting resource efficiency. Recent LCA tools can model closed-loop systems, where materials are continuously reused, repaired, or recycled, with minimal waste sent to landfills. These models evaluate the environmental savings from material recovery and energy recovery and consider the benefits of product design that facilitates repairability and recyclability.

The modeling of circular supply chains is another advancement, where LCA tools are used to assess the entire lifecycle of a product, from design and production to disposal and recycling. This allows businesses and policymakers to understand the full impact of adopting a circular approach, including the reduction in raw material consumption, energy use, and emissions (Laurent et al., 2014).

6. Advanced Modeling of Multi-Criteria Decision Analysis (MCDA)

Another recent advancement is the integration of Multi-Criteria Decision Analysis (MCDA) with LCA. While LCA provides a valuable environmental perspective, it may not fully capture the social, economic, and technical aspects of waste management systems. MCDA helps decision-makers evaluate waste management strategies by considering multiple criteria beyond just environmental impacts, such as social acceptability, cost-effectiveness, and technological feasibility (Laurent et al., 2014).

For example, when comparing strategies like WTE and recycling, LCA can provide insights into their environmental impacts, while MCDA can help incorporate stakeholder preferences, policy goals, and cost–benefit analyses into the decision-making process. This integrated approach allows for a more holistic and inclusive evaluation of waste management strategies.

7. User-Friendly and Accessible LCA Tools

Advancements in LCA tools have also focused on user-friendliness and accessibility. Traditional LCA software can be complex and difficult to use, often requiring specialized knowledge of life cycle analysis techniques. However, newer LCA tools are being developed with intuitive user interfaces and simplified workflows, enabling a wider audience, including waste management professionals, local governments, and businesses, to use them effectively.

These user-friendly LCA tools are often web-based, allowing users to easily input data, run simulations, and interpret results without requiring in-depth technical expertise. This makes it easier for stakeholders to assess the sustainability of different waste management strategies and supports informed decision-making across various sectors (Laurent et al., 2014).

Advancements in LCA tools and models for waste management have made it possible to assess waste treatment systems with a much higher degree of accuracy, flexibility, and comprehensiveness. These advancements include the integration of a variety of waste management technologies, the incorporation of regional and temporal variability, the improved availability of data, the integration of Life Cycle Costing, and the modeling of circular economy systems. Furthermore, the inclusion of Multi-Criteria Decision Analysis (MCDA) and the development of more user-friendly tools make LCA more accessible and practical for stakeholders at all levels. These innovations enhance our ability to make more informed, sustainable decisions that support both environmental protection and resource conservation.

8 Use of LCAs in Policy Decision-Making and Future Directions of Waste Management

Importance of LCAs in Policy Decision Making

Life Cycle Assessment (LCA) has become an essential tool in the policy decision-making process for waste management. By providing a comprehensive analysis of the environmental, and economic impacts of different waste management strategies over their entire life cycle, LCA enables policymakers to make well-informed decisions that align with sustainability goals. In the context of waste management, LCA helps to identify not only the direct environmental consequences but also the indirect impacts that can occur at various stages of the waste management process, from collection to disposal. LCA's importance lies in its ability to guide policies that aim to optimize resource use, reduce waste, and minimize environmental damage (Tillman, 2000).

LCA offers a systematic approach to assessing the environmental impacts of waste management practices across multiple impact categories, including climate change, human health, eutrophication, and resource depletion. This allows policymakers to understand the full scope of potential environmental consequences of different strategies. For example, when deciding between recycling, landfilling, or incineration, LCA provides critical insights into which option has the lowest overall environmental footprint. Policymakers can use this data to prioritize waste management strategies that minimize greenhouse gas emissions, toxic pollutants, and other harmful environmental effects, thus aligning with broader sustainability targets (Tillman, 2000). For instance, WTE plants may appear to reduce waste volume and generate energy, but LCA can reveal the emissions and pollutants produced in the incineration process, helping to balance the benefits of energy recovery against environmental damage. Similarly, landfilling might reduce immediate waste volume but could have long-term consequences such as methane emissions. LCA provides a more accurate, holistic view of the environmental impact of these different approaches, supporting evidence-based policy decisions (Dong et al., 2018).

In the transition towards a circular economy, LCA plays a crucial role by enabling policymakers to evaluate the environmental benefits and challenges of circular economy strategies compared to traditional disposal options (Dong et al., 2018). Circular economy approaches focus on extending the lifecycle of materials, reducing waste, and promoting resource efficiency. For example, the carbon footprint of recycling processes, including the energy and emissions associated with transporting materials, can be compared to the impact of sending waste to a landfill. By assessing these factors, LCA helps identify strategies that optimize resource recovery, reduce material consumption, and minimize environmental degradation. Policymakers can then craft regulations and incentives that encourage closed-loop systems and reduce dependency on landfills and incineration, thereby supporting the broader goals of a circular economy (Dong et al., 2018).

Beyond environmental impacts, LCA also supports economic decision-making by helping policymakers identify the most cost-effective waste management strategies that achieve environmental goals while minimizing public spending. When waste management options are assessed through LCA, the life cycle costs, including capital investment, operational expenses, and potential environmental penalties, are considered alongside the environmental benefits (Wardenaar et al., 2012). This enables policymakers to evaluate which strategies provide the best value for money. In many cases, advanced recycling and Waste-to-Energy technologies may have higher initial costs but can provide long-term benefits through resource recovery and energy production. Alternatively, composting may be less costly in terms of implementation but may not generate as many economic returns. By incorporating Life Cycle Costing (LCC) into LCA, policymakers are equipped with a more comprehensive view of the economic and environmental trade-offs, helping them design policies that balance cost with sustainability goals (Dong et al., 2018).

Effective waste management policies require long-term planning that considers future population growth, consumption patterns, and technological advancements. LCA tools help policymakers account for the long-term environmental consequences of waste management systems, such as emissions from landfills or the depreciation of waste management infrastructure. With this foresight, policymakers can design adaptive policies that are resilient to changes over time, ensuring that waste management systems remain effective as environmental standards evolve (Wardenaar et al., 2012). For example, the development of Waste-to-Energy plants involves high upfront capital costs but can provide long-term benefits by reducing waste and generating energy. LCA helps policymakers determine whether such investments are financially and environmentally viable in the long run or if alternative strategies like improved material recovery or composting might be more effective in achieving waste management goals.

LCA also plays a vital role in addressing public and social concerns related to waste management. Policies that implement waste management systems can be met with resistance from communities due to fears about health risks, pollution, or aesthetic impacts, such as the placement of incinerators or landfills near residential areas. LCA can provide scientific evidence to address these concerns by assessing not just the environmental but also the health and social impacts of waste management technologies. For instance, incineration can generate concerns about air quality and health risks due to the release of harmful substances like dioxins and particulate matter. Through LCA, policymakers can assess the magnitude of these risks in comparison to other options such as recycling or composting, and use this information to improve public engagement and address social concerns. Furthermore, LCA helps policymakers understand the trade-offs between health and environmental goals, enabling more transparent and inclusive decision-making processes (Wardenaar et al., 2012).

Moreover, LCA supports the alignment of waste management policies with broader national and international regulations and agreements. As countries commit to achieving climate goals, such as reducing carbon emissions, LCA allows policymakers to evaluate which waste management strategies are most effective in meeting these goals. It provides a scientific basis for crafting environmental regulations that help meet sustainable development targets, such as those outlined in the Paris Agreement or the United Nations Sustainable Development Goals (SDGs). By using LCA to assess the climate impacts of waste management strategies, policymakers can prioritize approaches that contribute to carbon neutrality, such as those that promote recycling and energy recovery. This alignment ensures that waste management policies support larger environmental goals while contributing to the reduction of carbon footprints and fostering sustainable development (Wardenaar et al., 2012).

Finally, LCA improves policy transparency and accountability by providing a rigorous, data-driven methodology for evaluating waste management strategies. When governments and organizations adopt LCA, they can ensure that the policies and technologies they promote are based on solid scientific evidence, rather than assumptions or political interests. This transparency helps build public trust in waste management policies and ensures that environmental goals are not compromised by short-term political considerations. By publishing the results of LCA studies and involving the public in the decision-making process, governments can foster greater community involvement and support for waste management policies. This creates an environment where policy decisions are made based on evidence and transparency, leading to better outcomes for both the environment and the population.

Therefore, Life Cycle Assessment (LCA) is a vital tool for effective policy decision-making in waste management. It provides a comprehensive, scientific basis for evaluating the environmental, economic, and social impacts of different waste management strategies, enabling policymakers to make informed choices that align with sustainability goals. LCA helps to identify the most effective strategies for reducing waste, conserving resources, and minimizing environmental harm. By using LCA, policymakers can ensure that waste management systems are cost-effective, socially acceptable, and capable of achieving long-term environmental sustainability, ultimately contributing to a cleaner, more sustainable future.

Future Directions of LCA-Based Waste Management Impact Evaluation

The future directions of Life Cycle Assessment (LCA)-based waste management impact evaluation are pivotal in refining and improving waste management systems globally. As the world faces increasing pressure from growing populations, urbanization, and environmental degradation, the need for more sustainable waste management practices has become more urgent. LCA, with its ability to evaluate the full environmental, economic, and social impacts of waste management strategies, will continue to evolve to meet these challenges. One of the most significant future directions is the incorporation of dynamic life cycle modeling, which will allow for more accurate and real-time evaluation of waste management systems (Anand et al., 2017). This approach would consider fluctuating variables, such as changes in waste composition, technological advancements, or policy shifts, providing a more flexible and adaptive tool for decision-making.

Moreover, future LCA-based evaluations will likely place increased emphasis on holistic sustainability assessments, integrating not only environmental impacts but also social and economic dimensions in a more interconnected manner. Currently, LCA has often been focused primarily on environmental impacts, with some attention to cost and resource use. However, in the future, there is a growing recognition that waste management systems must be evaluated from a triple bottom-line perspective, which includes social equity, economic viability, and environmental sustainability. This integrated approach would allow policymakers to design waste management strategies that not only minimize environmental harm but also promote economic growth and social well-being, especially in underserved communities (Anand & Amor, 2017).

Another significant future direction lies in the development of regional-specific LCA methodologies. While LCA has typically been applied in broader or generic contexts, future studies will likely be more region-specific, accounting for local waste generation patterns, environmental characteristics, and societal needs. Different regions have different waste profiles, environmental conditions, and socio-economic contexts, all of which influence the effectiveness of waste management strategies. By incorporating regional nuances, future LCA tools could provide more precise and actionable recommendations for local policymakers, allowing for a more tailored approach to waste management. This would also enable more efficient resource allocation, ensuring that the most appropriate technologies and strategies are adopted for each context (Mirabella et al., 2019).

The integration of advanced technologies into LCA-based waste management evaluations is another promising area for future development. With the rapid advancement of artificial intelligence (AI), machine learning, and big data, LCA methodologies are likely to become more powerful, accurate, and scalable. These technologies can help automate the data collection and analysis process, reducing the time and cost associated with conducting LCAs. AI and machine learning, in particular, can be used to model and predict future waste generation patterns, optimize waste treatment processes, and identify the most effective waste management strategies in real-time (Anand et al., 2017). These tools could also improve the modeling of complex systems and interdependencies within waste management, leading to more robust and reliable conclusions.

Furthermore, future LCA frameworks will increasingly incorporate circular economy principles. As the world moves away from linear production and consumption models, the need to evaluate the environmental and economic impacts of circular approaches will intensify. LCA can play a key role in assessing the effectiveness of resource recovery, product life extension, and material reuse in circular systems. In the future, LCA will likely evolve to better account for the environmental benefits of these approaches, such as reductions in raw material extraction, energy use, and waste generation (Mirabella et al., 2019). Additionally, incorporating circular economy principles into LCA-based evaluations will help shift the focus from waste disposal and treatment to waste prevention, designing waste management systems that prioritize resource conservation and sustainability.

In line with the evolving understanding of sustainability, future LCA models are expected to adopt a more comprehensive approach by considering interdisciplinary factors. For instance, LCA will increasingly integrate knowledge from fields such as ecotoxicology, social science, and economics to offer a more nuanced perspective on the broader impacts of waste management strategies. This interdisciplinary approach will allow for a deeper understanding of the trade-offs and synergies between environmental protection, human health, and economic growth. Additionally, this will enable LCA to account for complex social and behavioral dynamics that influence waste generation and management, offering a more holistic assessment of policy effectiveness (Mirabella et al., 2019).

A major challenge moving forward will be the standardization and harmonization of LCA methodologies. While LCA has gained widespread use, the inconsistency in LCA methodologies across studies remains an issue, particularly when comparing results from different geographic regions, technologies, or waste types. Standardizing LCA frameworks will allow for more reliable and comparable assessments of different waste management strategies across multiple contexts (Mirabella et al., 2019). In this regard, international collaborations and agreements to create unified LCA standards will be essential. Such standardization will also facilitate the sharing of data and best practices across countries, supporting global efforts toward more sustainable waste management systems.

Lastly, the integration of economic and financial aspects into LCA-based waste management impact evaluations will become increasingly important. In the future, LCA is expected to adopt a more comprehensive view of cost–benefit analyses, where the true economic value of waste management strategies is assessed not just in terms of direct costs but also in terms of the long-term economic benefits and externalities, including environmental and health costs (Mirabella et al., 2019). This could lead to a better understanding of the true cost of waste management systems, allowing for the development of strategies that are not only environmentally sustainable but also economically feasible and socially equitable.

9 Conclusion

In conclusion, Life Cycle Assessment (LCA) serves as a crucial tool for evaluating the environmental and economic impacts of various waste management strategies and technologies as discussed in this chapter. By providing a comprehensive, multi-dimensional analysis of waste management practices across their entire life cycle, LCA offers invaluable insights that help policymakers, researchers, and industry leaders make informed decisions that balance sustainability goals with practical realities. Through its ability to assess not only environmental impacts such as resource depletion, emissions, and pollution but also economic factors like costs and benefits, LCA provides a holistic view of the trade-offs involved in waste management.

The growing emphasis on LCA highlights the need to shift towards more sustainable, resource-efficient practices in the waste management sector. By comparing strategies such as recycling, composting, incineration, landfilling, and emerging technologies like Waste-to-Energy (WTE) and material recovery, LCA helps identify the most effective solutions for minimizing environmental harm, conserving resources, and optimizing economic outcomes. As the global community faces increasing challenges related to waste generation, resource scarcity, and climate change, the role of LCA in shaping policies and technological innovations becomes ever more critical.

Looking ahead, LCA's future direction will likely involve integrating advanced technologies, circular economy principles, and interdisciplinary perspectives to provide more dynamic, region-specific, and comprehensive assessments. The incorporation of social, economic, and environmental factors in a unified approach will be essential for designing waste management systems that not only minimize adverse impacts but also enhance social equity and long-term economic viability. With continued advancements in LCA methodologies and the growing recognition of its value in policy decision-making, LCA will remain a cornerstone in the global pursuit of sustainable waste management solutions.

Ultimately, LCA offers an essential framework for understanding the full range of impacts associated with waste management strategies and technologies, enabling decision-makers to adopt the most sustainable and economically viable approaches.

By fostering a deeper understanding of environmental and economic trade-offs, LCA can guide the transition to more circular, efficient, and resilient waste management systems that contribute to a more sustainable future for all.

References

Abubakar, I. R., Maniruzzaman, K. M., Dano, U. L., AlShihri, F. S., AlShammari, M. S., Ahmed, S. M. S., Al-Gehlani, W. A. G., & Alrawaf, T. I. (2022). Environmental sustainability impacts of solid waste management practices in the global South. *International Journal of Environmental Research and Public Health, 19*(19), 12717.

Agus, C., Nugraheni, M., Pertiwiningrum, A., Wuri, M. A., Hasanah, N. A. I., Sugiyanto, C., & Primananda, E. (2021). Tropical biological natural resource management through integrated biocycles farming system. *Sustainable bioeconomy: Pathways to sustainable development goals* (pp. 209–238).

Akhtar, A., Krepl, V., & Ivanova, T. (2018). A combined overview of combustion, pyrolysis, and gasification of biomass. *Energy & Fuels, 32*(7), 7294–7318.

AlManssoor, A., (2020). Obstacles and opportunities in planning and implementing integrated sustainable municipal solid waste management (ISMSWM) in Kuwait–a Gulf Cooperation Council (GCC) Country. Available at: http://hdl.handle.net/10012/16202. Accessed: February 03, 2025.

Anand, C. K., & Amor, B. (2017). Recent developments, future challenges and new research directions in LCA of buildings: A critical review. *Renewable and Sustainable Energy Reviews, 67*, 408–416.

Arora, R., Mutz, D., & Mohanraj, P. (Eds.). (2023). *Innovating for the circular economy: Driving sustainable transformation.* CRC Press. https://doi.org/10.1201/9781003201816

Ávila-Gutiérrez, M. J., Martín-Gómez, A., Aguayo-González, F., & Córdoba-Roldán, A. (2019). Standardization framework for sustainability from circular economy 4.0. *Sustainability, 11*(22), 6490.

Awasthi, M. K., Sarsaiya, S., Chen, H., Wang, Q., Wang, M., Awasthi, S. K., Li, J., Liu, T., Pandey, A., & Zhang, Z. (2019). Global status of waste-to-energy technology. In *Current developments in biotechnology and bioengineering* (pp. 31–52). Elsevier.

Ayilara, M. S., Olanrewaju, O. S., Babalola, O. O., & Odeyemi, O. (2020). Waste management through composting: Challenges and potentials. *Sustainability, 12*(11), 4456.

Bala, K., Anshul, Verma, S., Dogra, A., Das, A. P., Kumar, R., & Nandini. (2024). Sustainable management for urban plastic waste generation. *Sustainable management of urban plastic waste through circular economic approaches*, 211–236. https://doi.org/10.1201/9781003465447-13

Bandara, W. A. L. M. (2020). *Quantification of lifecycle impact of secondary packaging: An empirical study with special reference to milk powder*.

Barkhausen, R., Durand, A., & Fick, K. (2022). Review and analysis of ecodesign directive implementing measures: Product regulations shifting from energy efficiency towards a circular economy. *Sustainability, 14*(16), 10318.

Barlaz, M. A., Kaplan, P. O., Ranjithan, S. R., & Rynk, R. (2003). Comparing recycling, composting and landfills. *Biocycle, 44*(9), 60–66.

Branca, T. A., Colla, V., Algermissen, D., Granbom, H., Martini, U., Morillon, A., Pietruck, R., & Rosendahl, S. (2020). Reuse and recycling of by-products in the steel sector: Recent achievements paving the way to circular economy and industrial symbiosis in Europe. *Metals, 10*(3), 345.

Brändström, J., & Saidani, M. (2022). Comparison between circularity metrics and LCA: A case study on circular economy strategies. *Journal of Cleaner Production, 371*, Article 133537.

Budjav, B. (2022). Evaluation of environmental pollution and waste management strategies on the ecosystem. *Journal of Enterprise and Business Intelligence, 2*(4), 223–234.
Bueno, G., Latasa, I., & Lozano, P. J. (2015). Comparative LCA of two approaches with different emphasis on energy or material recovery for a municipal solid waste management system in Gipuzkoa. *Renewable and Sustainable Energy Reviews, 51*, 449–459.
Castellet-Viciano, L., Hernández-Chover, V., Bellver-Domingo, Á., & Hernández-Sancho, F. (2022). Industrial symbiosis: A mechanism to guarantee the implementation of circular economy practices. *Sustainability, 14*(23), 15872.
Chen, F., She, W., Zeng, F., Gao, M., Wen, C., Liu, G., Wu, L., Wu, Y., & Zhang, C. (2024). Resource depletion and conservation: Balancing demands on resources for green recovery and sustainability. *Resources Policy, 89*, Article 104544.
Chen, Y. C. (2018). Evaluating greenhouse gas emissions and energy recovery from municipal and industrial solid waste using waste-to-energy technology. *Journal of Cleaner Production, 192*, 262–269.
Cherubini, F., Bargigli, S., & Ulgiati, S. (2009). Life cycle assessment (LCA) of waste management strategies: Landfilling, sorting plant and incineration. *Energy, 34*(12), 2116–2123.
Christensen, T. H., Bhander, G., Lindvall, H., Larsen, A. W., Fruergaard, T., Damgaard, A., Manfredi, S., Boldrin, A., Riber, C., & Hauschild, M. (2007). Experience with the use of LCA-modelling (EASEWASTE) in waste management. *Waste Management & Research, 25*(3), 257–262.
Christensen, T. H., Damgaard, A., Levis, J., Zhao, Y., Björklund, A., Arena, U., Barlaz, M. A., Starostina, V., Boldrin, A., Astrup, T. F., & Bisinella, V. (2020). Application of LCA modelling in integrated waste management. *Waste Management, 118*, 313–322.
Clavreul, J., Guyonnet, D., & Christensen, T. H. (2012). Quantifying uncertainty in LCA-modelling of waste management systems. *Waste Management, 32*(12), 2482–2495.
Clemente, E., Domingues, E., Quinta-Ferreira, R. M., Leitão, A., & Martins, R. C. (2024). European and African landfilling practices: An overview on MSW management, leachate characterization and treatment technologies. *Journal of Water Process Engineering, 66*, Article 105931.
Cobo, S., Dominguez-Ramos, A., & Irabien, A. (2018). From linear to circular integrated waste management systems: A review of methodological approaches. *Resources, Conservation and Recycling, 135*, 279–295.
Cui, W., Wei, Y., & Ji, N. (2024). Global trends of waste-to-energy (WtE) technologies in carbon neutral perspective: Bibliometric analysis. *Ecotoxicology and Environmental Safety, 270*, Article 115913.
Das, A., & Ghosh, S. (2022). Role of microorganisms in extenuation of mining and industrial wastes. *Geomicrobiology Journal, 39*, 173–175. https://doi.org/10.1080/01490451.2022.2038953
Das, A., Ray, S. K., Mohanty, M., Mohanty, J., Dey, S., & Das, A. P. (2024). Ecotoxicity of microplastic wastes and their sustainable management: A review. *Environmental Chemistry and Ecotoxicology, 6*, 144–152. https://doi.org/10.1016/J.ENCECO.2024.05.003
Das, A. P., & Ghosh, S. (2018). Bioleaching of Manganese from mining waste materials. *Materials Today: Proceedings, 5*(1), 2381–2390. https://doi.org/10.1016/J.MATPR.2017.11.459
Date, L. R. (1998). *Landfill standards: A guideline on the regulatory and approval requirements for new or expanding landfilling sites*. Available at: http://docs.files.ontario.ca/documents/1110/66-landfill-standards-en.pdf. Accessed: January 20, 2025.
Dawson, L. (2019). 'Our waste, our resources; A strategy for England'–Switching to a circular economy through the use of extended producer responsibility. *Environmental Law Review, 21*(3), 210–218.
Debrah, J. K., Vidal, D. G., & Dinis, M. A. P. (2021). Raising awareness on solid waste management through formal education for sustainability: A developing countries evidence review. *Recycling, 6*(1), 6.
Denison, R. A. (1996). Environmental life-cycle comparisons of recycling, landfilling, and incineration: A review of recent studies. *Annual Review of Energy and the Environment, 21*(1), 191–237.

Dey, S., Das, A., Mallick, K., Sahu, A., & Das, A. P. (2023). Environmental petroleum waste: Pollution. *Toxicity, Sustainable Remediation, 159–175*. https://doi.org/10.1007/978-3-031-48220-5_7

Di Maria, F., & Sisani, F. (2017). Greenhouse gas emissions and environmental impact from recycling the organic fraction of solid waste: Comparison of different treatment schemes from a life cycle perspective. *Recycling, 2*(3), 13.

Digman, B., Joo, H. S., & Kim, D. S. (2009). Recent progress in gasification/pyrolysis technologies for biomass conversion to energy. *Environmental Progress & Sustainable Energy: An Official Publication of the American Institute of Chemical Engineers, 28*(1), 47–51.

Dong, Y., Miraglia, S., Manzo, S., Georgiadis, S., Sørup, H. J. D., Boriani, E., Hald, T., Thöns, S., & Hauschild, M. Z. (2018). Environmental sustainable decision making–The need and obstacles for integration of LCA into decision analysis. *Environmental Science & Policy, 87*, 33–44.

Elisha, O. D. (2020). Moving beyond take-make-dispose to take-make-use for sustainable economy. *International Journal of Scientific Research in Education, 13*(3), 497–516.

Ferronato, N., Rada, E. C., Portillo, M. A. G., Cioca, L. I., Ragazzi, M., & Torretta, V. (2019). Introduction of the circular economy within developing regions: A comparative analysis of advantages and opportunities for waste valorization. *Journal of Environmental Management, 230*, 366–378.

Finnveden, G., Hauschild, M. Z., Ekvall, T., Guinée, J., Heijungs, R., Hellweg, S., Koehler, A., Pennington, D., & Suh, S. (2009). Recent developments in life cycle assessment. *Journal of Environmental Management, 91*(1), 1–21.

Genovese, A., Acquaye, A. A., Figueroa, A., & Koh, S. L. (2017). Sustainable supply chain management and the transition towards a circular economy: Evidence and some applications. *Omega, 66*, 344–357.

Giugliano, M., Cernuschi, S., Grosso, M., & Rigamonti, L. (2011). Material and energy recovery in integrated waste management systems. An evaluation based on life cycle assessment. *Waste Management, 31*(9–10), 2092–2101.

Gonçalves, M. L. M., & Maximo, G. J. (2023). Circular economy in the food chain: Production, processing and waste management. *Circular Economy and Sustainability, 3*(3), 1405–1423.

Guinée, J. B., Gorree, M., Heijungs, R., Huppes, G., Kleijn, R., Udo de Haes, H. A., Van der Voet, E., & Wrisberg, M. N. (2002). *Life cycle assessment. An operational guide to ISO standards*, Volume 1, 2, 3. Centre of Environmental Science, Leiden University (CML), The Netherlands.

Guinee, J. B., Heijungs, R., Huppes, G., Zamagni, A., Masoni, P., Buonamici, R., Ekvall, T., & Rydberg, T. (2011). Life cycle assessment: Past, present, and future. *Environmental Science & Technology, 45*(1), 90–96. https://doi.org/10.1021/es101316v

Hamam, M., Chinnici, G., Di Vita, G., Pappalardo, G., Pecorino, B., Maesano, G., & D'Amico, M. (2021). Circular economy models in agro-food systems: A review. *Sustainability, 13*(6), 3453.

Hasan, M. M., Rasul, M. G., Khan, M. M. K., Ashwath, N., & Jahirul, M. I. (2021). Energy recovery from municipal solid waste using pyrolysis technology: A review on current status and developments. *Renewable and Sustainable Energy Reviews, 145*, Article 111073.

Hauschild, M. Z., Rosenbaum, R. K., & Olsen, S. I. (2018). *Life cycle assessment*. Springer International Publishing, Cham. https://doi.org/10.1007/978-3-319-56475-3 Book.

Hellweg, S., & Milà i Canals, L. (2014). Emerging approaches, challenges and opportunities in life cycle assessment. *Science, 344*(6188), 1109–1113.

Holdren, J. P., & Ehrlich, P. R. (1974). Human population and the global environment: population growth, rising per capita material consumption, and disruptive technologies have made civilization a global ecological force. *American Scientist, 62*, 282–292.

Holmgren, K., & Henning, D. (2004). Comparison between material and energy recovery of municipal waste from an energy perspective: A study of two Swedish municipalities. *Resources, Conservation and Recycling, 43*(1), 51–73.

Ikhlayel, M., & Nguyen, L. H. (2017). Integrated approaches to water resource and solid waste management for sustainable development. *Sustainable Development, 25*(6), 467–481.

Iqbal, A., Liu, X., & Chen, G. H. (2020). Municipal solid waste: Review of best practices in application of life cycle assessment and sustainable management techniques. *Science of the Total Environment, 729*, Article 138622.
ISO 14000. (2006). *Environmental management-life cycle assessment-requirements and guidelines.* Available at: https://www.iso.org/standard/37456.html. Accessed: January 25, 2025.
ISO 14040. (2006). *Environmental management-life cycle assessment-requirements and guidelines.* Available at: https://www.iso.org/standard/37456.html. Accessed: January 25, 2025.
ISO 14044. (2006). *Environmental management-life cycle assessment-requirements and guidelines.* Available at: https://www.iso.org/standard/37456.html. Accessed: January 25, 2025.
Jamasb, T., & Nepal, R. (2010). Issues and options in waste management: A social cost–benefit analysis of waste-to-energy in the UK. *Resources, Conservation and Recycling, 54*(12), 1341–1352.
Josimović, B., & Marić, I. (2012). Methodology for the regional landfill site selection. *Sustainable Development-Authoritative and Leading Edge Content for Environmental Management*. https://doi.org/10.5772/45926
Karani, P., & Jewasikiewitz, S. M. (2007). Waste management and sustainable development in South Africa. *Environment, Development and Sustainability, 9*, 163–185.
Khanam, Z., Sultana, F. M., & Mushtaq, F. (2023). Environmental pollution control measures and strategies: An overview of recent developments. *Geospatial analytics for environmental pollution modeling: Analysis, control and management* (pp. 385–414).
Knox, A. (2005). An overview of incineration and EFW technology as applied to the management of municipal solid waste (MSW). *ONEIA Energy Subcommitte*. Available at: http://www.durhamenvironmentwatch.org/Incinerator%20Files%20IIOverviewOfIncinerationAndEFWKnox.pdf. Accessed: January 27, 2025.
Kosoe, E. A., Diawuo, F., & Osumanu, I. K. (2019). Looking into the past: Rethinking traditional ways of solid waste management in the Jaman South Municipality, Ghana. *Ghana Journal of Geography, 11*(1), 228–244.
Kron, N., & Nelson, E. (2019). *State vs Federal CCR rule regulations: Comparisons and impacts.* Available at: www.worldofcoalash.org. Accessed: January 12, 2025.
Lahane, S., & Kant, R. (2022). Investigating the sustainable development goals derived due to adoption of circular economy practices. *Waste Management, 143*, 1–14.
Lakhouit, A., (2025). Revolutionizing urban solid waste management with AI and IoT: A review of smart solutions for waste collection, sorting, and recycling. *Results in Engineering*, 104018.
Laurent, A., Clavreul, J., Bernstad, A., Bakas, I., Niero, M., Gentil, E., Christensen, T. H., & Hauschild, M. Z. (2014). Review of LCA studies of solid waste management systems–Part II: Methodological guidance for a better practice. *Waste Management, 34*(3), 589–606.
LCI. (2017). Life Cycle Initiative 2017–2022 Strategy Document. *Life cycle initiative*. Available at: https://www.lifecycleinitiative.org/wp-content/uploads/2017/11/Strategy-document-of-the-Life-Cycle-Initiative-Oct.-2017-ANNEXES.pdf. Accessed: January 01, 2025.
Lornage, R., Redon, E., Lagier, T., Hébé, I., & Carré, J. (2007). Performance of a low cost MBT prior to landfilling: Study of the biological treatment of size reduced MSW without mechanical sorting. *Waste Management, 27*(12), 1755–1764.
Ludwig, C., Hellweg, S., & Stucki, S. (Eds.). (2012). *Municipal solid waste management: Strategies and technologies for sustainable solutions.* Springer Science & Business Media.
Maitre-Ekern, E. (2021). Re-thinking producer responsibility for a sustainable circular economy from extended producer responsibility to pre-market producer responsibility. *Journal of Cleaner Production, 286*, Article 125454.
Makarichi, L., Jutidamrongphan, W., & Techato, K. A. (2018). The evolution of waste-to-energy incineration: A review. *Renewable and Sustainable Energy Reviews, 91*, 812–821.
Markina, L., Kovach, V., & Vlasenko, O. (2024). Analysis of the world market of waste management. *Technology Audit and Production Reserves, 3*(77), 36–43.
Martin, M. (2020). Industrial symbiosis networks: Application of the circular economy for resource efficiency. *Handbook of the circular economy* (pp. 50–60).

Martinez-Sanchez, V., Kromann, M. A., & Astrup, T. F. (2015). Life cycle costing of waste management systems: Overview, calculation principles and case studies. *Waste Management, 36*, 343–355.

Mayer, F., Bhandari, R., & Gäth, S. (2019). Critical review on life cycle assessment of conventional and innovative waste-to-energy technologies. *Science of the Total Environment, 672*, 708–721.

Mazzi, A., (2020). Introduction. Life cycle thinking. In *Life cycle sustainability assessment for decision-making* (pp. 1–19). Elsevier.

Memon, M. A. (2010). Integrated solid waste management based on the 3R approach. *Journal of Material Cycles and Waste Management, 12*, 30–40.

Mirabella, N., Allacker, K., & Sala, S. (2019). Current trends and limitations of life cycle assessment applied to the urban scale: Critical analysis and review of selected literature. *The International Journal of Life Cycle Assessment, 24*, 1174–1193.

Mishra, S., & Das, A. P. (2021). Current treatment technologies for removal of microplastic and microfiber pollutants from wastewater. *Wastewater Treatment: Cutting-Edge Molecular Tools, Techniques and Applied Aspects, 237–251*. https://doi.org/10.1016/B978-0-12-821881-5.00011-8

Mishra, S., & Das, A. P. (2022). Treatment of the wastewater polluted with synthetic microfiber released from washing machine. In B. B. Das, H. Hettiarachchi, P. K. Sahu & S. Nanda (Eds.), *Lecture notes in civil engineering* (Vol. 207, pp. 109–117). Springer Science and Business Media Deutschland GmbH. https://doi.org/10.1007/978-981-16-7509-6_9/COVER

Mishra, S., & Das, A. P. (2023). Microbial remediation of synthetic microfiber contaminated wastewater. In M. P. Shah (Ed.), *Microbial technologies in industrial wastewater treatment* (pp. 337–350). Springer, Singapore. https://doi.org/10.1007/978-981-99-2435-6_17

Mishra, S., Dash, D., & Das, A. P. (2022). Detection, characterization and possible biofragmentation of synthetic microfibers released from domestic laundering wastewater as an emerging source of marine pollution. *Marine Pollution Bulletin, 185*, Article 114254. https://doi.org/10.1016/J.MARPOLBUL.2022.114254

Mishra, S., Dash, D., & Das, A. P. (2024). Isolation and molecular characterization of polyester-degrading bacterial strains from municipal wastewater and their application in microplastic degradation in contaminated wastewater. *Geomicrobiology Journal*. https://doi.org/10.1080/01490451.2024.2376101

Mishra, S., Ghosh, S., van Hullebusch, E. D., Singh, S., & Das, A. P. (2023). A critical review on the recovery of base and critical elements from electronic waste-contaminated streams using microbial biotechnology. *Applied Biochemistry and Biotechnology, 2023*, 1–30. https://doi.org/10.1007/S12010-023-04440-X

Mishra, S., Rout, P. K., & Das, A. P. (2019). Solar photovoltaic panels as next generation waste: a review. *Biointerface Research in Applied Chemistry, 9*(6).

Mishra, S., Singh, R. P., Rout, P. K., & Das, A. P. (2022). Membrane bioreactor (MBR) as an advanced wastewater treatment technology for removal of synthetic microplastics. *Development in Wastewater Treatment Research and Processes, 45–60*. https://doi.org/10.1016/B978-0-323-85583-9.00022-3

Mohanty, S., Ghosh, S., Bal, B., & Das, A. P. (2018). A review of biotechnology processes applied for manganese recovery from wastes. *Reviews in Environmental Science and Bio/Technology, 17*(4), 791–811. https://doi.org/10.1007/S11157-018-9482-1

Mondal, S., Singh, S., & Gupta, H. (2023). Green entrepreneurship and digitalization enabling the circular economy through sustainable waste management-An exploratory study of emerging economy. *Journal of Cleaner Production, 422*, Article 138433.

Montejo, C., Tonini, D., del Carmen Márquez, M., & Astrup, T. F. (2013). Mechanical–biological treatment: Performance and potentials. An LCA of 8 MBT plants including waste characterization. *Journal of Environmental Management, 128*, 661–673.

Morris, J. (2005). Comparative LCAs for curbside recycling versus either landfilling or incineration with energy recovery (12 pp). *The International Journal of Life Cycle Assessment, 10*, 273–284.

Morrissey, A. J., & Browne, J. (2004). Waste management models and their application to sustainable waste management. *Waste Management, 24*(3), 297–308.

Muscat, A., de Olde, E. M., Ripoll-Bosch, R., Van Zanten, H. H., Metze, T. A., Termeer, C. J., van Ittersum, M. K., & de Boer, I. J. (2021). Principles, drivers and opportunities of a circular bioeconomy. *Nature Food, 2*(8), 561–566.

Nanda, S., & Berruti, F. (2021). Municipal solid waste management and landfilling technologies: A review. *Environmental Chemistry Letters, 19*(2), 1433–1456.

Nurzhan, A., Ruan, X., & Chen, D. (2025). A review of life cycle assessment application in municipal waste management: Recent advances, limitations, and solutions. *Sustainability, 17*(1), 302.

O'Leary, P. R., & Tchobanoglous, G. (2002). Landfilling. *Handbook of solid waste management* (pp. 14–1).

Olivetti, E. A., & Cullen, J. M. (2018). Toward a sustainable materials system. *Science, 360*(6396), 1396–1398. https://doi.org/10.1126/science.aat6821

Paranjpe, A., Saxena, S., & Jain, P. (2023). A review on performance improvement of anaerobic digestion using co-digestion of food waste and sewage sludge. *Journal of Environmental Management, 338*, Article 117733.

Parfitt, J., Barthel, M., & Macnaughton, S. (2010). Food waste within food supply chains: Quantification and potential for change to 2050. *Philosophical Transactions of the Royal Society b: Biological Sciences, 365*(1554), 3065–3081.

PAS2050:2011. (2011). Specification for the assessment of the life cycle greenhouse gas emissions of products and services. Available at: https://biolatina.com/wp-content/uploads/2018/08/PAS2050.pdf. Accessed: January 03, 2025.

Patil, K. D., De, J., Patil, V. K., & Kulkarni, M. M. (2024). Environmental effects and threats of waste: Understanding threats and challenges to ecosystem, health, and sustainability and mitigation strategies. In *From waste to wealth* (pp. 37–69). Springer Nature Singapore.

Pattnaik, S., & Reddy, M. V. (2010). Assessment of municipal solid waste management in Puducherry (Pondicherry), India. *Resources, Conservation and Recycling, 54*(8), 512–520.

Priyadarsini, S., & Das, A. P. (2024). Analytical and structural characterization of waste lithium-ion batteries for their effective recycling strategy. *Environmental Chemistry and Ecotoxicology.* https://doi.org/10.1016/j.enceco.2024.12.004

Priyadarsini, S., & Das, A. P. (2025). Lithium bioleaching: A review on microbial-assisted sustainable technology for lithium bio-circularity. *Journal of Water Process Engineering, 69*, Article 106744. https://doi.org/10.1016/j.jwpe.2024.106744

Rădoi, M. I. (2021). Environmental protection: Soil pollution and waste management. Revista de Ştiinţe Politice. *Revue des Sciences Politiques, 69*, 167–176.

Rasel Kabir, M., (2015). *Conversion of municipal solid waste (MSW) into safe landfilling material.*

Rashid, S., & Malik, S. H. (2023). Transition from a linear to a circular economy. *Renewable energy in circular economy* (pp. 1–20). Springer International Publishing.

RCRA Orientation Manual. (2011). *The U.S. Environmental Protection Agency.* Available at: https://www.anthc.org/wp-content/uploads/2015/12/IEH_RCRA_Orientation_Manual_2014.pdf. Accessed: February 25, 2025.

Remigios, M. V. (2010). An overview of the management practices at solid waste disposal sites in African cities and towns. *Journal of Sustainable Development in Africa, 12*(7), 233–239.

Rousso, A. S., & Shah, S. P. (1994). Packaging taxes and recycling incentives: The German green dot program. *National Tax Journal, 47*(3), 689–701.

Sabbas, T., Polettini, A., Pomi, R., Astrup, T., Hjelmar, O., Mostbauer, P., Cappai, G., Magel, G., Salhofer, S., Speiser, C., & Heuss-Assbichler, S. (2003). Management of municipal solid waste incineration residues. *Waste Management, 23*(1), 61–88.

Sakai, S. I., Yano, J., Hirai, Y., Asari, M., Yanagawa, R., Matsuda, T., Yoshida, H., Yamada, T., Kajiwara, N., Suzuki, G., & Kunisue, T. (2017). Waste prevention for sustainable resource and waste management. *Journal of Material Cycles and Waste Management, 19*, 1295–1313.

Sakai, S. I., Yoshida, H., Hirai, Y., Asari, M., Takigami, H., Takahashi, S., Tomoda, K., Peeler, M. V., Wejchert, J., Schmid-Unterseh, T., & Douvan, A. R. (2011). International comparative

study of 3R and waste management policy developments. *Journal of Material Cycles and Waste Management, 13*, 86–102.
Savini, F. (2019). The economy that runs on waste: Accumulation in the circular city. *Journal of Environmental Policy & Planning, 21*(6), 675–691.
Sayara, T., Basheer-Salimia, R., Hawamde, F., & Sánchez, A. (2020). Recycling of organic wastes through composting: Process performance and compost application in agriculture. *Agronomy, 10*(11), 1838.
Seadon, J. K. (2006). Integrated waste management–Looking beyond the solid waste horizon. *Waste Management, 26*(12), 1327–1336.
Shafer, D. M. (2017). Climate change primer. Available at: https://www.warmheartworldwide.org/uploads/1/1/7/3/11737361/climate-change-primer-2017.pdf. Accessed: February 03, 2025.
Sharma, B. K., & Chandel, M. K. (2021). Life cycle cost analysis of municipal solid waste management scenarios for Mumbai, India. *Waste Management, 124*, 293–302.
Shekdar, A. V. (2009). Sustainable solid waste management: An integrated approach for Asian countries. *Waste Management, 29*(4), 1438–1448.
Shmelev, S. E., & Powell, J. R. (2006). Ecological–economic modelling for strategic regional waste management systems. *Ecological Economics, 59*(1), 115–130.
Simonen, K. (2014). *Life cycle assessment*. Routledge. Available at: https://www.routledge.com/LifeCycleAssessment/Simonen/p/book/9780415702423?srsltid=AfmBOopgIDCArTtXoI3Wde4NQXAc6aTu9dTpgQzUdYnNc1IdKDpkUgfd. Accessed: February 03, 2025.
Singh, J., Laurenti, R., Sinha, R., & Frostell, B. (2014). Progress and challenges to the global waste management system. *Waste Management & Research, 32*(9), 800–812.
Singh, J., & Ordoñez, I. (2016). Resource recovery from post-consumer waste: Important lessons for the upcoming circular economy. *Journal of Cleaner Production, 134*, 342–353.
Skenderovic, I., Kalac, B., & Becirovic, S. (2015). Environmental pollution and waste management. *Balkan Journal of Health Science, 3*(1), 2–10.
Soesanto, Q. M. B., Utomo, L. R., Rachman, I., & Matsumoto, T. (2021). Life cycle assessment and cost benefit analysis of municipal waste management strategies. *LIFE, 7*, 30–2021.
Somani, P., Navaneethan, R. D., & Thangaiyan, S. (2021, May). Integrated solid waste management in urban India: A mini review. In *Journal of physics: Conference series* (Vol. 1913, No. 1, p. 012084). IOP Publishing.
Song, Q., Li, J., & Zeng, X. (2015). Minimizing the increasing solid waste through zero waste strategy. *Journal of Cleaner Production, 104*, 199–210.
Soto-Paz, J., Arroyo, O., Torres-Guevara, L. E., Parra-Orobio, B. A., & Casallas-Ojeda, M. (2023). The circular economy in the construction and demolition waste management: A comparative analysis in emerging and developed countries. *Journal of Building Engineering, 78*, Article 107724.
Spreafico, C. (2022). An analysis of design strategies for circular economy through life cycle assessment. *Environmental Monitoring and Assessment, 194*(3), 180.
Tamasiga, P., Miri, T., Onyeaka, H., & Hart, A. (2022). Food waste and circular economy: Challenges and opportunities. *Sustainability, 14*(16), 9896.
Tan, S. T., Ho, W. S., Hashim, H., Lee, C. T., Taib, M. R., & Ho, C. S. (2015). Energy, economic and environmental (3E) analysis of waste-to-energy (WTE) strategies for municipal solid waste (MSW) management in Malaysia. *Energy Conversion and Management, 102*, 111–120.
Tillman, A. M. (2000). Significance of decision-making for LCA methodology. *Environmental Impact Assessment Review, 20*(1), 113–123.
Tozlu, A., Özahi, E., & Abuşoğlu, A. (2016). Waste to energy technologies for municipal solid waste management in Gaziantep. *Renewable and Sustainable Energy Reviews, 54*, 809–815.
Tripathy, B., Dey, S., Dubey, N. K., Lo, W.-C., & Das, A. P. (2024). *An omics-based approach for managing the risk of COVID-19-generated wastes*, 51–66. https://doi.org/10.1007/978-3-031-50840-0_4

Tsui, T. H., & Wong, J. W. (2019). A critical review: Emerging bioeconomy and waste-to-energy technologies for sustainable municipal solid waste management. *Waste Disposal & Sustainable Energy, 1*(3), 151–167.

Tun, M. M., Palacky, P., Juchelkova, D., & Síťař, V. (2020). Renewable waste-to-energy in Southeast Asia: Status, challenges, opportunities, and selection of waste-to-energy technologies. *Applied Sciences, 10*(20), 7312.

UNEP. (2011). Towards a green economy: Pathways to sustainable development and poverty eradication. Nairobi, Kenya, United Nations Environment Program, ISBN: 978-92-807-3143-9.

Van Der Giesen, C., Cucurachi, S., Guinée, J., Kramer, G. J., & Tukker, A. (2020). A critical view on the current application of LCA for new technologies and recommendations for improved practice. *Journal of Cleaner Production, 259*, Article 120904.

Vaverková, M. D. (2019). Landfill impacts on the environment. *Geosciences, 9*(10), 431.

Vyas, S., Prajapati, P., Shah, A. V., Srivastava, V. K., & Varjani, S. (2022). Opportunities and knowledge gaps in biochemical interventions for mining of resources from solid waste: A special focus on anaerobic digestion. *Fuel, 311*, Article 122625.

Wardenaar, T., Van Ruijven, T., Beltran, A. M., Vad, K., Guinée, J., & Heijungs, R. (2012). Differences between LCA for analysis and LCA for policy: A case study on the consequences of allocation choices in bio-energy policies. *The International Journal of Life Cycle Assessment, 17*, 1059–1067.

Williams, J. (2019). Circular cities: Challenges to implementing looping actions. *Sustainability, 11*(2), 423.

Zheng, S., Liu, W., & Zhi, Q. (2016). Cleaner waste management: A review based on the aspects of technology, market and policy. *Energy Procedia, 104*, 492–497.

Zhou, Y., Xiao, R., Klammsteiner, T., Kong, X., Yan, B., Mihai, F. C., Liu, T., Zhang, Z., & Awasthi, M. K. (2022). Recent trends and advances in composting and vermicomposting technologies: A review. *Bioresource Technology, 360*, Article 127591.

Sustainable Waste Management Practices and Their Profound Impact on Environmental Conservation

Subhadra Rajpoot and Alok Prasad Das

1 Introduction

Any material or substance that is discarded after its use and no longer kept is referred to as waste and in other sense material that we don't know how to reuse is referred to as waste (Arena & Gregorio, 2014). Waste is an unavoidable byproduct of most human cultural and developmental activities like industrial, commercial, and residential infrastructure (Wan et al., 2019). Waste material can exist in all physical states of matter as solid, liquid, semi-solid, or containerized gas material. Solid waste refers to discarded material or abdicated material that is neither liquid nor gaseous (Abdel-Shafy & Mansour, 2018; Dey et al., 2023; Mishra et al., 2019, 2023). It includes hazardous materials, industrial waste, chemicals, agricultural waste, commercial waste, biomedical waste, sanitary waste, institutional waste, demolition waste, construction waste, sewage waste, market and catering waste, packaging materials, and moreover solid and semi-solid residential waste (Das et al., 2024b; Das & Ghosh, 2022; Mohanty et al., 2018).

Thrown waste material is simply a raw material which is a valuable resource that is kept in the wrong places (Wu et al., 2022). A survey reported that the production of disposable material from that area is proportional to the residential average income (Qu et al., 2009). Moreover, higher solid waste generation is majorly found in high-income countries. The rate of higher income and the rate of urbanization are correlated with each other which leads to generating more waste material. So, the production of waste and income level are absolutely linked to each other. Additionally, the quantity of waste production is directly equivalent to the usage of resources (Mishra et al., 2024; Mishra & Das, 2021, 2023; Tripathy et al., 2024).

S. Rajpoot (✉) · A. P. Das
Department of Applied Sciences, Amity University Greater Noida, Greater Noida, India
e-mail: srajpoot@gn.amity.edu

A. P. Das et al. (eds.), *Sustainable Municipal Waste Management with a Zero Waste Approach*, Sustainable Environmental Waste Management Strategies,
https://doi.org/10.1007/978-3-032-05950-5_15

Historically, solid waste consisted predominantly of biodegradable; the solid natural wastes were non-hazardous and contained more organic material. Technological innovations have led to the emergence of new types of synthetic materials. However, technological advancements, such as the Industrial and Green Revolutions, have significantly altered the composition of solid waste. Over the last 50 years, there has been a sharp rise in the amount of paper and cardboard waste, including books, office papers, newspapers, magazines, commercial printing, tissue paper, and packaging materials.

At the same time, the percentage of plastic waste has also surged, with materials like polyvinyl chloride (PVC), polypropylene (PP), polystyrene (PS), low-density polyethylene (LDPE), and high-density polyethylene (HDPE), and other multilayered plastic polymers are common components of municipal solid waste (MSW) (Bhardwaj, 2022). Although plastic consumption will keep expanding, its growth is anticipated to be more gradual than in the last 25 years. Modern waste also includes bulky items such as furniture, electronics, lamps, bookcases, file cabinets, stereos, radios, computers, television sets, stoves, refrigerators, dryers, tires, and appliances. Traditional societies produce minimal and simple waste, while modern technological societies generate large amounts of complex waste.

Countries like China, India, and the USA are the largest producers of MSW (Nanda & Berruti, 2021). Globally, India stands as the third largest producer of waste (Mahajan, 2019). India, in particular, faces unique challenges due to its large volume of waste, a high proportion of organic content, and an informal, and unregulated waste collection sector. Central Pollution Control Board (CPCB) reported India produced an average of 143,449 metric tons of MSW per day (CPCB, 2013). In 2021, India generated approximately 62 million tons of MSW annually, of which only 43 million tons were collected and a fraction of approximately 12 million tons were treated, and the rest with most ending up in landfills (Mohanty et al., 2022). India produces about 150,000 tons of solid waste every day, and this amount is projected to grow by 5% each year (David et al., 2019).

Worldwide, MSW production reached 2.01 billion tons annually (Lino et al., 2023). Approximately 0.74 kg of waste are produced by a person per day globally (Kaza et al., 2018). High-income nations, despite comprising only 16% of the world's population, contribute 34% of the world's total waste and may increase by 19% by 2050 (Tayeh et al., 2020). Based on recent statistics, the global production of municipal solid waste (MSW) amounted to approximately 2.01 billion metric tons, and it is projected to increase to 3.4 billion metric tons by 2050 (Iqbal et al., 2021). This figure is expected to rise significantly more than three times by 2050, with low-income nations experiencing the most drastic increases due to inadequate waste management systems. The present study examines the composition of solid waste, its types, sources, management, challenges, and implications for environmental sustainability and human health. Solid waste management and its components are well presented in Fig. 1.

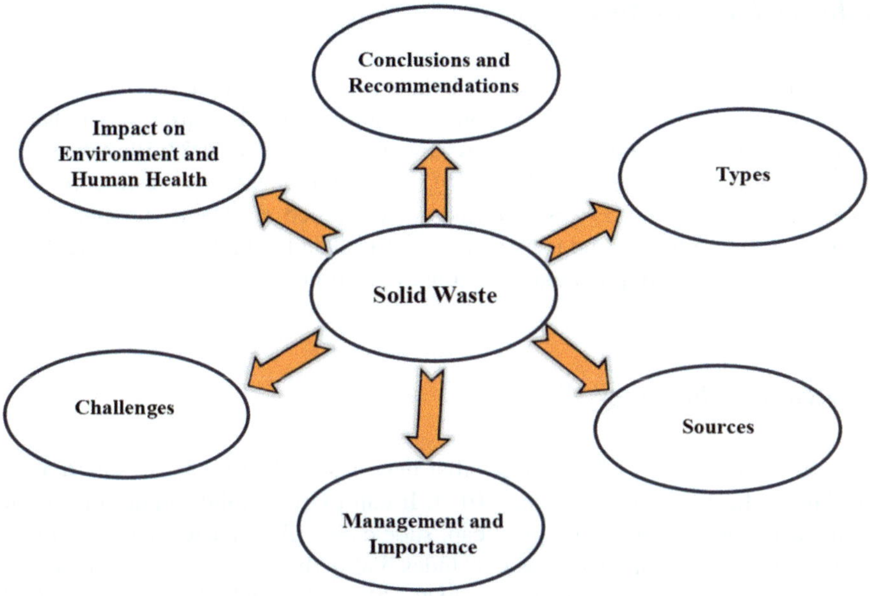

Fig. 1 Solid waste management and its components

2 Types of Solid Waste

Solid waste can be classified into organic and inorganic waste and the classification is based on its characteristics and origin (Arafat et al., 2015). Plastics, paper, wood, textiles, and food waste are classified as organic waste while metals and glass materials are inorganic wastes.

Municipal Solid Waste (MSW) and Biodegradable Waste

MSW denotes the regular waste produced by households, schools, colleges, offices, and public spaces (Ogwueleka, 2009). This waste includes organic and inorganic waste items like food scraps, paper, plastic, clothes, heavy metals, etc. Biodegradable waste consists only of organic waste such as wood, paper, and food waste (Katinas et al., 2019). This waste can be decomposed by microbes over time and minimizes their environmental impact. These types of waste are usually managed by local government authorities or private waste management companies.

Industrial and E-Waste

Industrial waste is generated by factories, industries, mills, and mines (Krishna et al., 2020). They may be hazardous or non-hazardous materials. Hazardous industrial waste includes toxic chemicals and heavy metals (Alam et al., 2023; Bhardwaj et al., 2023a; Priyadarsini & Das, 2024, 2025), while non-hazardous waste consists of paper, wood, and glass. E-waste consists of discarded electronic devices such as computers, smartphones, and televisions. Industrial and e-waste are growing day by day due to the increasing population, industrialization, and urbanization.

Medical and Sanitary Waste

Medical waste consists of substances that may pose risks to both the environment and human health (Hossain et al., 2011). It can exist in solid, liquid, or gaseous forms. It includes a microbiological cap, shoe cover, gloves, used syringes, plastic bottles, and lab coat. Sanitary waste includes bodily fluids or waste such as sanitary napkins, diapers, and bandages (Khatun et al., 2023a, 2023b; Sadasivuni et al., 2019). It is considered hazardous because it can contain pathogens and harmful bacteria (Padmanabhan & Barik, 2019). Improper disposal of these wastes can lead to the spread of disease and contamination of water sources. Proper disposal methods for sanitary waste include using specialized containers or collection systems, treatment facilities, and incineration.

Construction and Radioactive Waste

Construction waste consists of materials that either do not decompose or decompose very slowly over time (Taurino et al., 2023). It includes construction debris such as bricks, rubble, glass, plastic containers, and concrete. Radioactive waste is generated by nuclear power plants, medical treatments, scientific research, and certain industrial processes (Deng et al., 2020). It is very dangerous and can cause skin cancer and mutation. So, it requires careful and specialized handling.

Legacy Waste

It has been stored or collected improperly for years in landfills, abandoned mines, or industrial sites. It can pose several environmental and health risks because it is often disposed of in an uncontrolled and unscientific manner. It can be categorized into four types: contained and/or stored waste, buried waste, contaminated soil and groundwater, and contaminated building materials and structures (Chandrappa & Das, 2024). Some challenges associated with this waste include radiological characterization, safety issues, leachate management, and fire control. Some processing methods for legacy waste include Bioremediation, Biomining, Stabilization, and Screening. In India, there are around 3,084 legacy waste dumpsites that cover almost 10,000 ha (Ojha et al., 2024). The Swachh Bharat Mission 2.0 was launched in 2021 to achieve a "Garbage-Free Status" for all cities by 2026 (Gupta et al., 2022). However, only 470 out of 2424 dumpsites have been completely remediated, and only 16% of land has been reclaimed.

3 Source of Solid Waste

In developing solid waste management programs, it is crucial to identify the sources of solid waste. Solid waste originates from various places, including garbage from restaurants, households, and hotels, trash from offices, hospitals, and clinics, waste from industries such as rubber, foam, textiles, automobiles, leather, plastics, and metals, and discarded items from slaughterhouses, as well as fruit and vegetable markets (Bala et al., 2024; Kumar et al., 2023; Mishra et al., 2024; Mishra & Das, 2022). The sources of solid waste within a community are clearly outlined in Table 1.

4 Solid Waste Management (SWM) and Its Importance

Waste management is also key to advancing a circular economy, where materials are continuously reused (Das & Ghosh, 2018; Ray et al., 2024). The systematic handling of waste materials from production to disposal is known as solid waste management (SWM) or municipal solid waste management (MSWM). SWM is a complex process that incorporates various disciplines and technologies. The management of solid waste involves several steps, including on-site handling, processing, transport, collection, monitoring, recycling, and disposal. SWM is therefore defined as the discipline focused on controlling the generation, storage, transport, processing, collection, and disposal of society's waste. It is a vital aspect of urban environmental management, impacting both daily life and the overall quality of the environment. Effective waste management is essential for preserving a clean and sustainable environment (Akram et al., 2021).

SWM system includes several steps: waste production, handling, storage, transportation, treatment, and disposal. Solid waste is produced from residential, commercial, and industrial sources. At the point of origin, waste is separated and stored in containers. Waste handling encompasses sorting, segregation, collection, loading, secondary storage, material recovery, baling, shredding, crushing, unloading, processing, transportation, and disposal of solid waste. Waste is collected from storage bins and transported to waste processing facilities. The collected waste

Table 1 Sources of solid waste generation within a community

S.No.	Source	Typical waste generators	Types of solid waste
1	Occupied waste	Single-family and multifamily residences	Food, paper, cardboard, plastics, wood, leather, yard waste, textiles, glass, tin cans, metals, ashes, sanitation waste, special wastes (such as consumer electronics, sanitary pads and diapers, household hazardous wastes, bulky items, batteries, white goods, tires, and oil)
2	Industrial waste	Heavy and light manufacturing, construction sites, fabrication, chemical plants, and power generation plants	Packaging food wastes, housekeeping waste, demolition and construction materials, ashes, and chemical wastes
3	Commercial waste	Office buildings, hotels, restaurants, markets, stores, print shops, service stations, auto repair shops, etc.	Wood, plastics, metals, cardboard, paper, food wastes, and hazardous chemical wastes
4	Institutional waste	Schools, universities, and government research centers	Paper, wood, food waste, chemicals, broken glasses, microbiological hazardous waste
5	Road, building, and construction sites	Road repair, destruction of buildings, renovation sites, and new construction sites	Steel, glass, concrete, wood, etc.
6	Municipal waste	Landscaping parks, street cleaning, beaches, and wastewater treatment plants	Landscape and tree trimmings, beaches, park's common wastes, street sweepings, and sludge
7	Agriculture waste	Crops, farms, dairies, vineyards, orchards, and feedlots	Agricultural wastes, spoiled food wastes, and chemical waste like pesticides

is sorted, processed, and converted into materials suitable for manufacturing, and energy production. This stage involves recycling, composting, and waste-to-energy technologies. The remaining waste that cannot be recovered or recycled is disposed of in a sanitary landfill or incinerated.

The "7R" principles of "Refuse, Reduce, Reuse, Repair, Repurpose, Recycle, and Recover" play a vital role in effective waste management (Badola & Chauhan, 2022). Refuse and Reduce focus on minimizing waste by avoiding unnecessary purchases and reducing overall consumption. Reuse and Repair involve extending the life of existing products, either by using them as they are or by replacing specific components. Repurposing and recycling emphasize maximizing the use of materials in products, while recovering, which involves extracting embedded energy from waste, is considered the least favorable and least effective waste management method. For example, using waste as fuel for combustion can generate heat and electricity (Awino & Apitz, 2024).

The waste management industry can shift the narrative that waste solely produces GHGs by utilizing waste as an energy source. During recycling, valuable materials like paper, plastics, and metals are extracted and repurposed into new products. Recycling offers significantly greater climate benefits than other waste treatment methods. Additionally, composting is becoming increasingly popular as a means to convert organic waste into nutrient-rich compost for agricultural use. Various organizations and initiatives, such as Integrated Waste Management and Cleaner Production (IWMCP), emphasize the potential of enhanced waste management practices to address climate change. Furthermore, the Clean Development Mechanism (CDM) projects established by the Kyoto Protocol promote GHG reduction initiatives in the waste sector (UNEP, 2008).

Waste management procedures vary significantly between developing and developed countries, as well as between rural and urban areas, and across commercial and residential sectors, even within the same nation. In India, effective SWM has become a critical issue (Joshi & Ahmed, 2016). In response, the Indian government has initiated programs to enhance SWM, including the Swachh Bharat Mission and the Smart Cities Mission (Mokale, 2019). These initiatives promote the establishment of waste-to-energy plants, composting facilities, and campaigns for source segregation.

Effective waste management is essential for both developed and developing countries. It helps prevent pollution and mitigates its effects on ecosystems (John & Das, 2024; Prabhakar et al., 2019; Pradhan et al., 2023). It can help prevent the spread of diseases linked to unmanaged waste and safeguard public health. It encourages recycling and the reuse of materials, helping to conserve valuable resources. It minimizes the need for energy-intensive extraction and manufacturing processes. It ensures adherence to environmental regulations and laws, helping to avoid potential fines and penalties. It helps keep the environment clean and appealing, thereby improving the overall quality of life. It can help lower GHG emissions, contributing to the mitigation of climate change. Adopting sustainable waste management practices can generate job opportunities and create revenue through recycling and waste-to-energy initiatives. Embracing sustainable waste management practices advances the goal

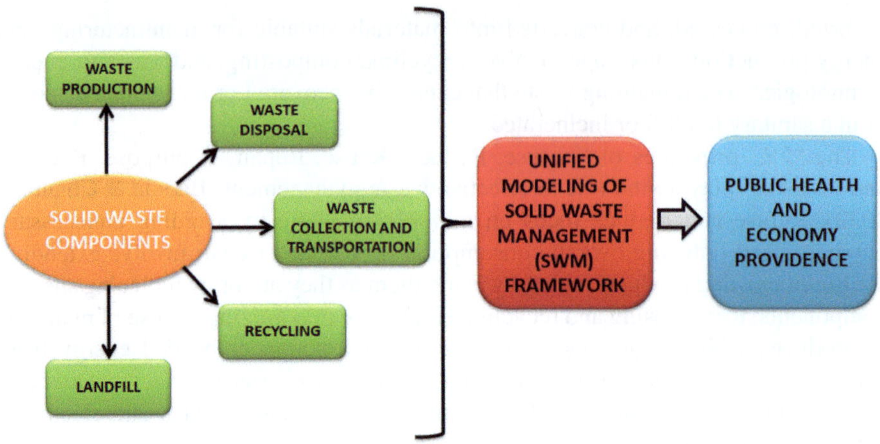

Fig. 2 Process of solid waste management (SWM)

of achieving long-term social, economic, and environmental sustainability. Waste management initiatives can engage the community, promoting a sense of ownership and responsibility for waste reduction and proper disposal. Improvements in waste management technologies can result in more efficient and sustainable methods for waste treatment and disposal (Das et al., 2025; Kanungo et al., 2024; Sahu et al., 2024). Figure 2 shows the process of solid waste management (SWM).

5 Challenges in Solid Waste Management (SWM)

Waste management is challenging globally because each type of waste presents its own unique difficulties during treatment and disposal. The waste management sector needs substantial improvements to address the increasing volume of waste being generated. A primary challenge is the insufficient infrastructure, which leads to practices such as open dumping and waste burning. These methods contribute to pollution and pose health risks. In developing countries, the management of MSW is particularly problematic due to rapid urbanization and population growth.

India faces significant challenges in MSW management, which are multifaceted (Kumar et al., 2017). These challenges include inadequate collection and transportation systems, a lack of waste segregation at the source, insufficient processing capacity, and reliance on unsanitary landfills for disposal. In response, the Indian government has launched various initiatives to promote source segregation and encourage recycling and composting techniques (Mani & Singh, 2016). However, implementing these measures on a large scale remains a major challenge. By addressing these issues, India can reduce the environmental and health impacts associated with improper waste disposal. While notable progress has been made in recent years, systemic changes are still needed in the Indian waste management sector, including improvements in waste segregation, enhancement of waste processing facilities, and formalization of the informal sector.

In India, public involvement and government action are essential for establishing a sustainable waste management system. Countries can collaborate by sharing knowledge, best practices, and resources to develop innovative solutions and enhance waste management practices globally in order to effectively tackle these challenges. This collaboration encompasses sharing research and technology, partnering on waste reduction and recycling initiatives, and enforcing international agreements and regulations to ensure proper waste disposal and reduce environmental impacts.

6 Impact of Solid Waste on the Environment and Human Health

Inadequate waste management and ineffective disposal methods significantly harm both human health and the environment (Sharma et al., 2018). The waste management infrastructure currently in place in developing countries is often insufficient, leading to pollution and health risks. The entire SWM process can be associated with health issues both directly and indirectly (Das et al., 2024; Dey et al., 2024; Kanungo et al., 2023). This connection occurs directly through the handling of solid waste and indirectly through the contamination of water, soil, and air. Leachate from landfills, produced by the decomposition of waste, can negatively impact surface water bodies and aquifers. Those living near waste disposal sites and waste pickers who sort and reclaim recyclable materials from open dumps are particularly affected by inadequate waste management and disposal practices.

The disposal of solid waste in open areas can lead to various diseases, including dengue fever, diarrhea, gastrointestinal issues, and malaria. Animals like dogs, calves, and pigs often scavenge through the waste and can transmit diseases to humans. Additionally, the burning of solid waste releases hazardous gases into the atmosphere. These gases can contribute to cancer and skin allergies. Furthermore, toxic chemicals from plastics, batteries, and industrial waste are released into the environment, posing significant health risks to humans (Bhardwaj & Jindal, 2019, 2020, 2022).

The use of certain novel chemicals and materials in advanced technologies poses health risks to workers and is not easily biodegradable. Solid waste releases acids, heavy metals, and oils, which can adversely affect soil fertility (Bhardwaj et al., 2023b). Additionally, solid waste significantly contributes to global warming primarily through the release of GHGs, such as carbon dioxide (CO_2) and methane (CH_4) (Bhardwaj, 2023). CH_4 is produced when biodegradable waste decomposes anaerobically in landfills, while carbon dioxide is released during the combustion of fossil fuels (Andres et al., 2012).

7 Technological Innovations in Solid Waste Management

With the increasing demand for sustainable and efficient solid waste management (SWM), technological advancements are playing a critical role in transforming how waste is collected, processed, and disposed of. These innovations aim to minimize environmental impacts while optimizing resource recovery and energy efficiency.

Smart Waste Management Systems

Smart technologies, powered by the Internet of Things (IoT) and data analytics, are revolutionizing waste management in urban areas. Some key innovations include:

- **Smart Bins**: IoT-enabled waste bins can monitor fill levels in real-time, alerting waste collectors when they need to be emptied. This optimizes collection schedules, reducing unnecessary trips and cutting down on fuel consumption and emissions.
- **Route Optimization Software**: Waste collection vehicles can use GPS-based software that analyzes traffic patterns and waste levels to plan the most efficient routes. This reduces operational costs, improves collection efficiency, and lowers greenhouse gas emissions.
- **Waste Tracking and Data Analytics**: Municipalities can collect and analyze data on waste generation patterns, recycling rates, and collection performance. This helps in policy formulation, identifying waste hotspots, and improving waste management services.

Automation and Robotics

Robotics and automation are being increasingly utilized in sorting, recycling, and treatment facilities to improve efficiency and accuracy:

- **Automated Sorting Machines**: Advanced mechanical sorting systems use sensors, artificial intelligence (AI), and robotics to automatically separate recyclables from the waste stream. This not only speeds up the recycling process but also improves material recovery rates.
- **Robotic Waste Collection**: Some cities are experimenting with robotic waste collection vehicles that can autonomously navigate and collect waste. While still in development, these systems could significantly reduce labor costs and improve collection efficiency.

Waste-to-Energy (WtE) Technologies

Waste-to-Energy (WtE) technologies convert non-recyclable waste into usable forms of energy, such as electricity, heat, or fuel. This reduces the amount of waste sent to landfills while generating renewable energy. Some popular WtE methods include:

- **Incineration**: Modern incinerators equipped with energy recovery systems can burn waste at high temperatures, converting it into energy while capturing harmful emissions. These plants are designed with pollution control systems to minimize air pollution.
- **Anaerobic Digestion**: This process involves the breakdown of organic waste (e.g., food waste, agricultural waste) by microorganisms in the absence of oxygen, producing biogas. The biogas can be used for electricity generation or as a renewable fuel source.
- **Pyrolysis and Gasification**: These advanced thermal processes convert organic waste into syngas, which can be used to generate energy or as a chemical feedstock. Unlike incineration, these methods operate at lower oxygen levels, resulting in fewer emissions.

8 The Future of Solid Waste Management

As global waste generation continues to rise, the future of solid waste management will require new innovations, policies, and cultural shifts. Some emerging trends include:

- **Circular Economy**: Moving toward a circular economy model, where products are designed for reuse, repair, and recycling, is critical for reducing waste generation.
- **Decentralized Waste Management Systems**: Smaller, community-based waste management systems can offer more localized and sustainable solutions, reducing the burden on large-scale facilities.
- **Carbon–Neutral Waste Solutions**: The integration of waste management into climate action plans, such as through carbon–neutral waste treatment facilities and the promotion of low-carbon technologies, will be essential in the fight against climate change.

9 Conclusion and Recommendations

Waste management aims to minimize the negative impacts of waste on human health, the environment, and natural resources. However, it has become a pressing global issue, particularly affecting many developing countries. Numerous researchers are addressing this challenge and working toward establishing sustainable waste management systems. Sustainable management of solid waste can facilitate a circular economy, where waste is viewed as a valuable resource that can be reused, recycled, or converted into energy. In recent years, there has been an increasing recognition of the importance of sustainable waste management practices. These practices focus on reducing waste generation, maximizing resource recovery, and minimizing the environmental impact of waste disposal. By collaborating, countries can pave the way for a more sustainable and environmentally friendly approach to waste management.

Population growth, along with the close link between economic growth and consumption, has led to unprecedented levels of waste production. Waste generation can be reduced by implementing strategies such as composting food waste, minimizing packaging, and encouraging the use of reusable products. SWM is essential for preserving urban environmental quality and protecting public health. To achieve sustainable SWM, there is an urgent need for infrastructure development, institutional enhancement, public education, and financial investment. Effective SWM is crucial for preventing environmental pollution, mitigating public health risks, and conserving natural resources. Safe disposal methods, such as landfilling and incineration, are necessary for waste that cannot be recycled or composted. Both approaches require careful planning and monitoring to reduce negative environmental effects.

We offer several recommendations that can assist authorities in managing solid waste effectively as follows:

- Effective SWM is crucial for environmental protection and can facilitate composting and energy production.
- Solid waste should be transported using a specially designed, covered system to prevent unpleasant odors and littering.
- Proper segregation and storage of waste are vital to avoid spoilage, littering, attracting pests, and reducing odors.
- Most waste should be sorted at the source.
- Markets for recycled products should be actively promoted.
- Waste found under bridges, in culverts, along roadways, and in drainage systems must be removed.
- Authorities should implement strict legislation and policies for waste management.
- Educational and awareness programs on solid waste disposal should be organized by authorities to inform the public about the importance of responsible waste management.
- During treatment, solid waste should be disposed of in sanitary landfills after extracting energy.
- Enhanced waste management techniques should be developed to achieve significant reductions in GHG emissions.
- Ongoing efforts and investments are essential to effectively tackle waste management issues and establish a robust waste management infrastructure.
- International cooperation and collaboration are necessary to address transboundary waste movements and prevent illegal dumping.

References

Abdel-Shafy, H. I., & Mansour, M. S. (2018). Solid waste issue: Sources, composition, disposal, recycling, and valorization. *Egyptian Journal of Petroleum, 27*(4), 1275–1290.

Akram, S. V., Singh, R., Gehlot, A., Rashid, M., AlGhamdi, A. S., Alshamrani, S. S., & Prashar, D. (2021). Role of wireless aided technologies in the solid waste management: A comprehensive review. *Sustainability, 13*(23), 13104.

Alam, S., Bhardwaj, L. K., Mallick, R., & Rai, S. (2023). Estimation of heavy metals and fluoride ion in vegetables grown nearby the stretch of River Yamuna, Delhi (NCR), India. *Indian Journal of Environmental Protection, 43*(1), 64–73.

Andres, R. J., Boden, T. A., Bréon, F. M., Ciais, P., Davis, S., Erickson, D., et al. (2012). A synthesis of carbon dioxide emissions from fossil-fuel combustion. *Biogeosciences, 9*(5), 1845–1871.

Arafat, H. A., Jijakli, K., & Ahsan, A. (2015). Environmental performance and energy recovery potential of five processes for municipal solid waste treatment. *Journal of Cleaner Production, 105*, 233–240.

Arena, U., & Di Gregorio, F. (2014). A waste management planning based on substance flow analysis. *Resources, Conservation and Recycling, 85*, 54–66.

Awino, F. B., & Apitz, S. E. (2024). Solid waste management in the context of the waste hierarchy and circular economy frameworks: An international critical review. *Integrated Environmental Assessment and Management, 20*(1), 9–35.

Badola, N., & Chauhan, J. S. (2022). Waste management: Challenges and opportunities. *Bioremediation of Environmental Pollutants: Emerging Trends and Strategies*, 1–23.

Bala, K., Anshul, Verma, S., Dogra, A., Das, A. P., Kumar, R., & Nandini. (2024). Sustainable management for urban plastic waste generation. *Sustainable Management of Urban Plastic Waste Through Circular Economic Approaches*, 211–236. https://doi.org/10.1201/9781003465447-13

Bhardwaj, L. K. (2022). Evaluation of bis (2-ethylhexyl) phthalate (DEHP) in the PET bottled mineral water of different brands and impact of heat by GC–MS/MS. *Chemistry Africa, 5*(4), 929–942.

Bhardwaj, L. K. (2023). *A comprehensive review on the climate change and its impact on health*. https://doi.org/10.20944/preprints202305.0159.v1

Bhardwaj, L. K., & Jindal, T. (2019). Contamination of lakes in Broknes peninsula, East Antarctica through the pesticides and PAHs. *Asian-Journal of Chemistry, 31*(7), 1574–1580.

Bhardwaj, L. K., & Jindal, T. (2020). Persistent organic pollutants in lakes of Grovnes Peninsula at Larsemann Hill area, East Antarctica. *Earth Systems and Environment, 4*, 349–358.

Bhardwaj, L., & Jindal, T. (2022). Polar ecotoxicology: Sources and toxic effects of pollutants. *New Frontiers in Environmental Toxicology*, 9–14.

Bhardwaj, L. K., Kumar, D., & Kumar, A. (2023b). Phytoremediation potential of Ocimum Sanctum: A sustainable approach for remediation of heavy metals. https://doi.org/10.20944/preprints202308.0593.v1

Bhardwaj, L. K., Sharma, S., & Jindal, T. (2023a). Estimation of physico-chemical and heavy metals in the lakes of Grovnes & Broknes Peninsula, Larsemann Hill, East Antarctica. *Chemistry Africa*, 1–18.

Central Pollution Control Board. (2013). Status report on municipal solid waste management. http://www.cpcb.nic.in/divisionsofheadoffice/pcp/MSW_Report.pdf. http://pratham.org/images/paper_on_ragpickers.pdf

Chandrappa, R., & Das, D. B. (2024). Legacy waste. *Solid waste management: Principles and practice* (pp. 639–656). Springer International Publishing.

Das, A., Dey, S., Bala, K., Gnanasekaran, R., & Das, A. P. (2024). Microbial bioremediation technology for sustainable treatment and management of synthetic microfiber waste. *Sustainable Microbial Technology for Synthetic and Cellulosic Microfiber Bioremediation, 221–238*. https://doi.org/10.1007/978-3-031-62600-5_12

Das, A., Dey, S., Gacem, A., Yadav, K. K., Nath, H., Ram, D. K., & Das, A. P. (2025). Revolutionizing solid hydrocarbon sourcing: Plastic's journey from waste to upcycled treasure. *Environmental Science and Engineering, Part, F218*, 331–351. https://doi.org/10.1007/978-3-031-81615-4_14

Das, A., & Ghosh, S. (2022). Role of microorganisms in extenuation of mining and industrial wastes. *Geomicrobiology Journal, 39*, 173–175. https://doi.org/10.1080/01490451.2022.2038953

Das, A., Ray, S. K., Mohanty, M., Mohanty, J., Dey, S., & Das, A. P. (2024). Ecotoxicity of microplastic wastes and their sustainable management: A review. *Environmental Chemistry and Ecotoxicology, 6*, 144–152. https://doi.org/10.1016/J.ENCECO.2024.05.003

Das, A. P., & Ghosh, S. (2018). Bioleaching of manganese from mining waste materials. *Materials Today: Proceedings, 5*(1), 2381–2390. https://doi.org/10.1016/J.MATPR.2017.11.459

David, A., Thangavel, Y. D., & Sankriti, R. (2019). *Recover, recycle and reuse: An efficient way to reduce the waste.*

Deng, D., Zhang, L., Dong, M., Samuel, R. E., Ofori-Boadu, A., & Lamssali, M. (2020). Radioactive waste: A review. *Water Environment Research, 92*(10), 1818–1825.

Dey, S., Das, A., Mallick, K., Sahu, A., & Das, A. P. (2023). Environmental petroleum waste: Pollution. *Toxicity, Sustainable Remediation, 159–175.* https://doi.org/10.1007/978-3-031-48220-5_7

Dey, S., Dash, A., & Das, A. P. (2024). Bioremediation as a sustainable and circular solution for mining waste management and bioenergy generation. *Sustainable Management of Mining Waste and Tailings: A Circular Economy Approach, 107–119.* https://doi.org/10.1201/9781003442455-6/BIOREMEDIATION-SUSTAINABLE-CIRCULAR-SOLUTION-MINING-WASTE-MANAGEMENT-BIOENERGY-GENERATION-SUDESHNA-DEY-AKANSHYA-DASH-ALOK-PRASAD-DAS

Gupta, M. J., Meenakshi, M., & Bakshi, R. (2022). Swachh Bharat Abhiyan: A catalyst for sustainable development. *Journal of Positive School Psychology, 6*(9), 4211–4221.

Hossain, M. S., Santhanam, A., Norulaini, N. N., & Omar, A. M. (2011). Clinical solid waste management practices and its impact on human health and environment–A review. *Waste Management, 31*(4), 754–766.

Iqbal, S., Naz, T., & Naseem, M. (2021). Challenges and opportunities linked with waste management under global perspective: A mini review. *Journal of Quality Assurance in Agricultural Sciences, 1*(01), 9–13.

John, A., & Das, A. P. (2024). Government strategies for achieving circular economy for plastic waste management in Southeast Asia. *Sustainable Management of Urban Plastic Waste through Circular Economic Approaches, 72–82.* https://doi.org/10.1201/9781003465447-6/GOVERNMENT-STRATEGIES-ACHIEVING-CIRCULAR-ECONOMY-PLASTIC-WASTE-MANAGEMENT-SOUTHEAST-ASIA-AKBAR-JOHN-ALOK-PRASAD-DAS

Joshi, R., & Ahmed, S. (2016). Status and challenges of municipal solid waste management in India: A review. *Cogent Environmental Science, 2*(1), 1139434.

Kanungo, J., Sahoo, T., Bal, M., & Behera, I. D. (2023). Performance of bioremediation strategy in waste lubricating oil pollutants: A review. *Geomicrobiology Journal.* https://doi.org/10.1080/01490451.2023.2245395

Kanungo, J., Sahoo, T., Bal, M., & Behera, I. D. (2024). Performance of bioremediation strategy in waste lubricating oil pollutants: A review. *Geomicrobiology Journal, 41*(4), 360–373. https://doi.org/10.1080/01490451.2023.2245395

Katinas, V., Marčiukaitis, M., Perednis, E., & Dzenajavičienė, E. F. (2019). Analysis of biodegradable waste use for energy generation in Lithuania. *Renewable and Sustainable Energy Reviews, 101*, 559–567.

Kaza, S., Yao, L., Bhada-Tata, P., & Van Woerden, F. (2018). *What a waste 2.0: A global snapshot of solid waste management to 2050.* World Bank Publications.

Khatun, R., Singh, S., Dubey, N. K., & Prasad Das, A. (2023). A review on marine-based phytochemicals and their application in biomedical research. *Recent Frontiers of Phytochemicals: Applications in Food, Pharmacy, Cosmetics, and Biotechnology, 383–395.* https://doi.org/10.1016/B978-0-443-19143-5.00013-X

Khatun, R., Singh, S., Dubey, N. K., & Prasad Das, A. (2023b). A review on marine-based phytochemicals and their application in biomedical research. In *Recent frontiers of phytochemicals: Applications in food, pharmacy, cosmetics, and biotechnology* (pp. 383–395). Elsevier. https://doi.org/10.1016/B978-0-443-19143-5.00013-X

Krishna, R. S., Mishra, J., Meher, S., Das, S. K., Mustakim, S. M., & Singh, S. K. (2020). Industrial solid waste management through sustainable green technology: Case study insights from steel and mining industry in Keonjhar, India. *Materials Today: Proceedings, 33*, 5243–5249.

Kumar, M., Pandit, S., Patel, V., Khanna, N., Nag, M., Lahiri, D., Ray, R. R., Das, A., & Das, D. (2023). Maximization of energy recovery from starch processing wastewater by thermophilic dark fermentation coupled with microbial fuel cell technology. *Geomicrobiology Journal, 1–12*. https://doi.org/10.1080/01490451.2023.2209555

Kumar, S., Smith, S. R., Fowler, G., Velis, C., Kumar, S. J., Arya, S., et al. (2017). Challenges and opportunities associated with waste management in India. *Royal Society Open Science, 4*(3), Article 160764.

Lino, F. A., Ismail, K. A., & Castañeda-Ayarza, J. A. (2023). Municipal solid waste treatment in Brazil: A comprehensive review. *Energy Nexus*, 100232.

Mahajan, R. (2019). Solid waste management: A prerequisite for achieving sustainable development goals in India. *International Journal of Research and Analytical Reviews (IJRAR), 633*.

Mani, S., & Singh, S. (2016). Sustainable municipal solid waste management in India: A policy agenda. *Procedia Environmental Sciences, 35*, 150–157.

Mishra, S., & Das, A. P. (2021). Current treatment technologies for removal of microplastic and microfiber pollutants from wastewater. *Wastewater Treatment: Cutting-Edge Molecular Tools, Techniques and Applied Aspects, 237–251*. https://doi.org/10.1016/B978-0-12-821881-5.00011-8

Mishra, S., & Das, A. P. (2022). Treatment of the wastewater polluted with synthetic microfiber released from washing machine. In B. B. Das, H. Hettiarachchi, P. K. Sahu, & S. Nanda (Eds.), *Lecture notes in civil engineering* (Vol. 207, pp. 109–117). Springer Science and Business Media Deutschland GmbH. https://doi.org/10.1007/978-981-16-7509-6_9/COVER

Mishra, S., & Das, A. P. (2023). Microbial remediation of synthetic microfiber contaminated wastewater. In M. P. Shah (Ed.), *Microbial technologies in industrial wastewater treatment* (pp. 337–350). Springer, Singapore. https://doi.org/10.1007/978-981-99-2435-6_17

Mishra, S., Dash, D., & Das, A. P. (2024). Isolation and molecular characterization of polyester-degrading bacterial strains from municipal wastewater and their application in microplastic degradation in contaminated wastewater. *Geomicrobiology Journal*. https://doi.org/10.1080/01490451.2024.2376101

Mishra, S., Ghosh, S., van Hullebusch, E. D., Singh, S., & Das, A. P. (2023). A critical review on the recovery of base and critical elements from electronic waste-contaminated streams using microbial biotechnology. *Applied Biochemistry and Biotechnology, 2023*, 1–30. https://doi.org/10.1007/S12010-023-04440-X

Mishra, S., Rout, P. K., & Das, A. P. (2019). Solar photovoltaic panels as next generation waste: A review. *Biointerface Research in Applied Chemistry, 9*(6).

Mohanty, S., Ghosh, S., Bal, B., & Das, A. P. (2018). A review of biotechnology processes applied for manganese recovery from wastes. *Reviews in Environmental Science and Bio/Technology 2018 17:4, 17*(4), 791–811. https://doi.org/10.1007/S11157-018-9482-1

Mohanty, S., Mishra, S., & Mohanty, A. (2022). Municipality solid waste management. A case study of smart city Bhubaneswar, Odisha. *Journal of Environmental Management & Tourism, 13*(5), 1361–1373.

Mokale, P. (2019). Smart waste management under smart city mission—Its implementation and ground realities. *International Journal of Innovative Technology and Exploring Engineering, 8*(12), 3095–3103.

Nanda, S., & Berruti, F. (2021). Municipal solid waste management and landfilling technologies: A review. *Environmental Chemistry Letters, 19*, 1433–1456.

Ogwueleka, T. (2009). Municipal solid waste characteristics and management in Nigeria. *Journal of Environmental Health Science & Engineering, 6*(3), 173–180.

Ojha, V., Sharma, A., Ranjan, V. P., Rautela, R., Dhawral, A., & Kumar, S. (2024). Resource recovery from legacy waste dumpsites in India: A path towards sustainable waste management. *Chemosphere*, 143337.

Padmanabhan, K. K., & Barik, D. (2019). Health hazards of medical waste and its disposal. In *Energy from toxic organic waste for heat and power generation* (pp. 99–118). Woodhead Publishing.

Prabhakar, A., Mishra, S., & Das, A. P. (2019). Isolation and identification of lead (Pb) solubilizing bacteria from automobile waste and its potential for recovery of lead from end of life waste batteries. *Geomicrobiological Journal, 36*(10), 894–903. https://doi.org/10.1080/01490451.2019.1654044

Pradhan, G., Tripathy, B., Ram, D. K., Digal, A. K., & Das, A. P. (2023). Bauxite mining waste pollution and its sustainable management through bioremediation. *Geomicrobiology Journal, 1–10*. https://doi.org/10.1080/01490451.2023.2235353

Priyadarsini, S., & Das, A. P. (2024). Analytical and structural characterization of waste lithium-ion batteries for their effective recycling strategy. *Environmental Chemistry and Ecotoxicology*. https://doi.org/10.1016/j.enceco.2024.12.004

Priyadarsini, S., & Das, A. P. (2025). Lithium bioleaching: A review on microbial-assisted sustainable technology for lithium bio-circularity. *Journal of Water Process Engineering, 69*, Article 106744. https://doi.org/10.1016/j.jwpe.2024.106744

Qu, X. Y., Li, Z. S., Xie, X. Y., Sui, Y. M., Yang, L., & Chen, Y. (2009). Survey of composition and generation rate of household wastes in Beijing, China. *Waste Management, 29*(10), 2618–2624.

Rajpoot, S. (2021). Impact of nanotechnology on environment—A review. *International Journal for Modern Trends in Science and Technology, 7*, 159–164. https://doi.org/10.46501/IJMTST0710026

Rajpoot, S. (2025). Smart and innovative nanotechnology applications for water purification: Exploring technologies for a cleaner future. *Emerging trends and technologies in water management and conservation* (p. 34). https://doi.org/10.4018/979-8-3693-6920-3.ch003

Ray, S. K., Dey, S., Das, A. P., & John, A. (2024). Microplastic waste management through greener technologies: A review. *Sustainable Management of Urban Plastic Waste through Circular Economic Approaches, 237–261*. https://doi.org/10.1201/9781003465447-14

Sadasivuni, K. K., Rattan, S., Waseem, S., Brahme, S. K., Kondawar, S. B., Ghosh, S., Das, A. P., Chakraborty, P. K., Adhikari, J., Saha, P., & Mazumdar, P. (2019). Silver nanoparticles and its polymer nanocomposites—Synthesis, optimization, biomedical usage, and its various applications. *Lecture Notes in Bioengineering, 331–373*. https://doi.org/10.1007/978-3-030-04741-2_11

Sahu, A., Dey, S., Mallick, K., Priyadarsini, S., Behera, I. D., & Das, A. P. (2024). A review on advanced technology for sustainable management of synthetic microplastic waste. *Sustainable Management of Urban Plastic Waste through Circular Economic Approaches, 262–279*. https://doi.org/10.1201/9781003465447-15

Sharma, B., Vaish, B., Srivastava, V., Singh, S., Singh, P., & Singh, R. P. (2018). An insight to atmospheric pollution-improper waste management and climate change nexus. *Modern age environmental problems and their remediation*, 23–47.

Taurino, R., Bondioli, F., & Messori, M. (2023). Use of different kinds of waste in the construction of new polymer composites. *Materials Today Sustainability, 21*, Article 100298.

Tayeh, H. N. A., Azaizeh, H., & Gerchman, Y. (2020). Circular economy in olive oil production–olive mill solid waste to ethanol and heavy metal sorbent using microwave pretreatment. *Waste Management, 113*, 321–328.

Tripathy, B., Dey, S., Dubey, N. K., Lo, W.-C., & Das, A. P. (2024). *An omics-based approach for managing the risk of COVID-19-generated wastes*, 51–66. https://doi.org/10.1007/978-3-031-50840-0_4

UNEP Sustainable Building, & Construction Initiative. (2008). *The Kyoto protocol, the clean development mechanism, and the building and construction sector: A report for the UNEP sustainable buildings and construction initiative.* UNEP/Earthprint.

Wan, C., Shen, G. Q., & Choi, S. (2019). Waste management strategies for sustainable development. *Encyclopedia of sustainability in higher education* (pp. 2020–2028). Springer International Publishing.

Wu, F., Liu, X., Qu, G., & Ning, P. (2022). A critical review on extraction of valuable metals from solid waste. *Separation and Purification Technology*, 122043.

Industrial Symbiosis and Circular Economy in Zero-Waste Management Approaches

H. G. T. P. Samarawickrama, Champika Liyanage, and K. P. P. Udayagee

1 Introduction

Rapid industrialization in past few decades made a huge concern in resource depletion, environmental degradation, and waste accumulation in all over the globe (Priyadarsini & Das, 2024, 2025). As a key approach in facing and mitigating these issues, the concept of zero-waste management is introduced as a part and a partial component of the circular economy (CE), that can be collaboratively facilitated by Industrial Symbiosis (IS) (Ntasiou & Andreou, 2017). On the one hand, while the CE focuses on regenerative design, resource recovery, waste minimization, remanufacturing, and recycling, on the other hand, the IS encourage collaborations between industries to exchange by-products, energy, and other resources considered as waste by mimicking natural ecosystems, aiming zero-waste targets. Figure 1 shows the zero-waste management model (Oughton et al., 2022).

The IS model aims to varolize wastes and by-products; thereby promoting the circular flow of materials, products and resources within the system. In this sight, materials, utilities, and by-products generated from one company's operation can be utilized as inputs for other processes, either internally or across different industries. Materials, utilities, and by-products interchange is a fundamental practice in the execution of the IS paradigm and calls for close collaboration between the different

H. G. T. P. Samarawickrama
Faculty of Graduate Studies, University of Sri Jayewardenepura, Nugegoda, Sri Lanka

C. Liyanage
School of Engineering, Faculty of Science and Technology, University of Central Lancashire, Preston, UK

K. P. P. Udayagee (✉)
Department of Biosystems Technology, Faculty of Technology, University of Sri Jayewardenepura, Pitipana, Homagama, Sri Lanka
e-mail: udayagee@sjp.ac.lk

A. P. Das et al. (eds.), *Sustainable Municipal Waste Management with a Zero Waste Approach*, Sustainable Environmental Waste Management Strategies,
https://doi.org/10.1007/978-3-032-05950-5_16

Fig. 1 Zero-waste model. *Source* https://www.researchgate.net/publication/381827546_Zero_Waste_Management_System

companies that are part of the IS. Thus, as anticipated in zero-waste management targets; these two approaches promotes the use of waste generated during intermediate stages of production in several other manufacturing lines, contributing to conervation of virgin resources and reduced environmental impact. Further, adopting IS model lessens the cost of product ion, cost of waste handling and management, creates additional profit and green jobs. This notion is being used in numerous sectors, including the management of waste and its treatment, manufacturing, and urban development. The zero-waste management concept is identified as an effective tool of resource recovery, which enables the substitution of raw materials, energy, and water while significantly reducing greenhouse gas emissions (Dey et al., 2023; Mishra et al., 2023). Thus, it has been identified as the path to achieving 2030 sustainable development goals, as it enhances the resource efficiency, economic resilience, reduce landfilling, and promotes the closed-loop system for sustainable production and consumption (Neves et al., 2019c).

This chapter explores the principles, applications, and benefits of these synergistic strategies, highlighting case studies and best practices that contribute to sustainable waste management and economic resilience.

2 Industrial Symbiosis: A Collaborative Approach to Waste Reduction

Industrial Symbiosis (IS) involves the exchange of waste materials, energy, water, and by-products between industries. IS facilitates to use waste from one process serves as a resource for another. While it has become trendy to talk about the globalization of economic activities and the relevant rise in importance of the *space of flows*, such activities necessitate a specific place where they are performed (Castellet-Viciano et al., 2022). Within the field of industrial ecology, this *space of place* has been created IS. Chertow, has well-defined IS as "engaging traditionally separate industries in a collective approach to competitive advantage involving physical exchange of materials, energy, water, and by-products." The key principles of IS lie in resource efficiency through maximizing the utility of raw materials and minimizing waste, collaboration of different businesses working together to optimize resource use and reduce waste generation and environmental and economic benefits achieved by cost savings, reduced emissions, and enhanced sustainability (Neves et al., 2019c).

Mainly, there are three types of IS; by-product exchange, shared resource utilization and energy cascading. In by-product exchange, one industry's waste is used as a raw material by another. The IS systems go with shared resource utilization use utilities shared such as water, steam, or logistics while energy cascading encourage reusing excess heat or energy across different industrial processes. However, there are systems or eco-industrial parts which in cooperated all these three types together and driving toward environmental sustainability. The collective approaches of IS systems reduces the waste generation, carbon footprint, while conserving of natural resources. Further the cost savings through resource optimization, new revenue streams from by-products, job creation, strengthening industrial networks, regional economic growth, and enhanced corporate social responsibility are instrumental benefits of IS systems (Castellet-Viciano et al., 2022).

Examples of Industrial Symbiosis: Case Studies from Industries Successfully Implementing the Model

Industrial Symbiosis is not limited to the Kalundborg case, where multiple industries share energy, water, and materials, reducing waste and operational costs. There are many instances of synergistic networks everywhere the world has been stated in the literature, with a varied variety in terms of the numbers of participants, sorts of economic activities, and how they are systematized (Oughton et al., 2022). In Europe, many foci of IS are spread across diverse countries, and most of the cases described in the literature are in northern and northwestern Europe, with the UK reporting the maximum number of cases. This is led by several programs that the governments have launched to assist companies to discover associates to use their waste as raw material. Finland also has numerous cases of IS, mainly arising from the robust existence of the pulp and paper industry, which has driven the formation of synergy relations. In Asia, a numeral of IS initiatives has also been reported, with the maximum number of cases in China, mainly due to restrictions on carbon dioxide emissions and the many plans and policies that have been executed to stand-in CE practices. In Japan, there have been cases of IS and industrial and urban symbiosis across numerous cities, driven by the Japanese Eco-Town Program that inspires the use of industrial, municipal, and commercial waste in industrial applications, with the goal of increasing the economy and reducing waste disposal. In North America, and predominantly in the US, there have been cases of IS that date back to the late 1970s. The case most often mentioned in publications is located in Barceloneta, Puerto Rico, where the robust occurrence of pharmaceutical companies has encouraged the creation and development of IS between numerous companies. A variey of cases of IS have also been stated in Australia, Brazil, Morocco, and Algeria (Castellet-Viciano et al., 2022).

Benefits of Incorporating IS in Sustainable Waste Management

In recent years, life cycle assessment (LCA) has been extensively used to measure environmental benefits of the symbiosis type, symbiosis process, and symbiosis network examining the lessening of numerous ecological influences during the life cycle. Carbon footprint that can alter diverse capitals into carbon dioxide equivalents is used to calculate the carbon emission reduction potential of the symbiosis. Material flow analysis is used to quantity the saving possibility of the symbiote by likening the energy and water flows in diverse models, revealing the effect of economic activities on the environmental load. Therein, LCA overlooks the influence of ecosystem goods, services, and natural capital. The carbon footprint approach captures only one dimension of climate change impacts, providing a limited perspective. Meanwhile, material flow analysis (MFA) fails to account for differences in material quality.

Based on the transformed input flows, there is an energy index system for evaluating the economic benefits, environmental impact, and sustainability of a system. As a technique that can consider economic matters the same as the environmental subject, energy analysis has been widely used to compare relevant indexes of diverse time scales or different technical levels to comprehensively assess the evolution of the economy and the environment (Lazar, 2024; Neves et al., 2019a, 2019c).

When teasing out symbiosis networks, symbiosis relationships are gathered into three categories—materials, water, energy—based on exchange substances. However, the fact that a symbiosis relationship can exchange more than two categories of resources—materials, water, and energy at the same time is ignored. For example, the exchanged water often simultaneously replaces the functions of energy and by-products (Shenyoputro & Jones, 2023).

3 Circular Economy in Waste Management

The CE is reinforced by three main principles, commonly known as the three R's: Reduce, Reuse, and Recycle. These principles work as the guiding framework for transforming our old-style linear economy, where goods are made, used, and disposed of, into a circular one that is restorative and re-forming by design. By mixing these principles into production and consumption processes, it turns out to be significantly less waste, lower greenhouse gas emissions, and protects natural resources. Each of these principles plays an exclusive but harmonizing role in achieving the goal of a sustainable CE (General, 2023; Ogunmakinde, 2019).

A serious component of this approach is extended producer responsibility (EPR), which holds manufacturers accountable for the entire lifecycle of their products, including post-consumer waste. By inspiring the use of recyclable materials and focusing on material recovery, EPR policies aim to minimize waste and reduce carbon emissions. In waste management systems, integrating CE principles can convert municipal solid waste or industrial waste into valuable resources. For example, a case study in a city might show how regulatory frameworks can be executed to support a CE, aiming to enlighten waste management practices and reduce waste disposal. By leveraging these plans, cities can considerably decrease their environmental impact, aligning waste management with broader sustainability goals (Das et al., 2024; Mishra et al., 2019). This approach also aids the textiles industry, where innovative recycling techniques can turn unwanted fabrics into new products, thus closing the loop on waste. Generally, adopting a CE approach in waste management not only supports environmental goals but also pushes economic growth by creating new markets and reducing the reliance on virgin resources (Valavanidis, 2018).

Circular Economy Business Models

CE business models arise in all shapes and sizes, reliant on where they happen in the value chain. This diversity is what makes a CE dynamic and openly circular. There are some key CE business models, such as Product-as-a-Service (PaaS), which encourage companies to provide access to products as a service instead of selling products. As an example, some companies offer lighting-as-a-service, where customers pay for light usage rather than owning bulbs, which enhances the product longevity and maintenance, reduces the use of raw materials, and minimizes the e-waste generation. Circular supply chains are another CE business model that runs on renewable, recycled, or biodegradable materials in production. The use of recycled fabrics for clothing and, production of carpets from recycled materials are some examples of this. Product life extension through repair, refurbishment, and remanufacture is another CE model aiming to extend of product lifespan, which also saves virgin resources and minimizes waste generation. Sharing economy, such as shared accommodation systems, ridesharing to reduce vehicle ownership, is another CE model which again reduces material consumption and lowers costs for consumers. Resource recovery and recycling, with the concept of capturing waste materials and transforming them into new resources is another business model in CE systems. Converting organic waste into biogas, recycles difficult-to-recycle waste (e.g., cigarette butts, plastic packaging) reduce the quantity of landfill waste and promote resources to stay within the cycle (Gaybor & Chavez, 2019).

4 Integrating IS with CE for Zero Waste

Connected to the idea of IS are eco-industrial parks, which denotes the practical application of IS. Although this has numerous explanations, the central idea in all of them is the same, referring to the development of a business community with material exchange networks in which the by-products or waste of a company or industrial process are transformed into products that can be reused by other processes, reducing the number of compulsory raw materials and the waste generated. This, together with the shared usage of infrastructures, services, information, or other logistical aspects, leads to an extra sustainable consumption and production system. By integrating IS within the framework of a CE, industries can create closed-loop systems that optimize the use of resources, minimize waste generation, and contribute to the overall goal of zero waste. Thus, the integration of IS the CE creates exciting opportunities to move toward a zero-waste future and net-zero targets (Das & Ghosh, 2022; Mishra & Das, 2023; Mohanty et al., 2018). By collaborating between industries, designing products for longevity, and recovering resources at every stage of the lifecycle, businesses can contribute significantly to sustainability while reducing costs (Castellet-Viciano et al., 2022; Oughton et al., 2022; Sgambaro et al., 2024). Figure 2 illustrates the benefits of collaborative approach of IS and CE.

Industrial parks where multiple companies collaborate to exchange waste, materials, and energy represent real-life applications of CE principles driven by IS are good examples that shows synergies of this integration. CE's principles of sustainable design and closed-loop systems encourage industries to see waste not as a by-product. It opens the eye to waste as a valuable resource to be recovered and reused, thus making IS more effective. By designing products with their end-of-life in mind, CE helps reduce the amount of waste entering the symbiosis network while increasing the range of materials that can be exchanged among industries. Companies in the electronics sector collaborate to recycle rare metals from discarded devices is one another good examples for this, therefore can then be reused in new products, reducing the need for mining and minimizing e-waste.

Development of eco-industrial parks promote collaboration among businesses within a specific geographical area, and help to create symbiotic relationships and exchange resources efficiently. Further, governments can play a key role by introducing policies that incentivize CE practices and industrial collaboration, such as tax breaks or subsidies for waste-to-resource projects. Digital tools, like blockchain or

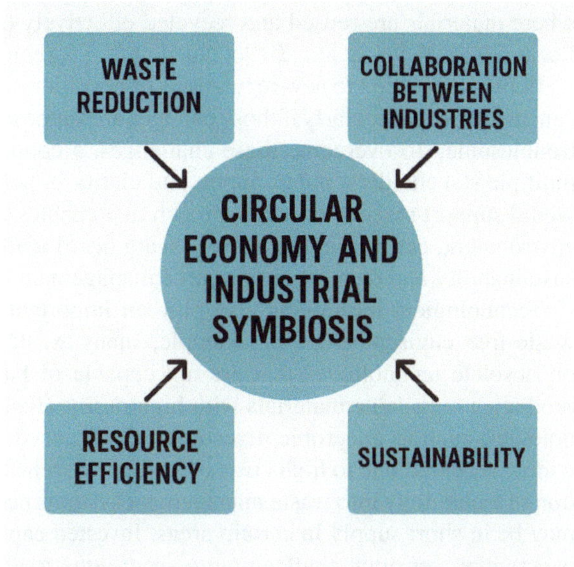

Fig. 2 Benefits of integration of IS and CE

material flow analysis, can help track materials, monitor resource use, and facilitate transactions between industries in a symbiotic network.

5 Barriers to Implementation and Overcoming Challenges

Regulatory and policy challenges denote important barriers to accomplishing a waste-free environment. Effective waste management necessitates a comprehensive framework of laws and regulations that stimulate sustainable practices and encourage waste reduction, recycling, and responsible disposal (Bala et al., 2024; Das & Ghosh, 2018; Mishra & Das, 2021; Ray et al., 2024). However, many governments fight with obsolete policies that do not speech the difficulties of modern waste management need. The absence of coordination across different authorities can lead to misunderstanding and discrepancy in waste regulations, making it problematic for businesses and individuals to comply. Furthermore, inadequate execution of present regulations often results in non-compliance and a culture of carelessness regarding waste management (Neves et al., 2019a, 2020; Song et al., 2015).

One of the major hurdles is the detach between environmental policy and economic incentives. Many regulatory frameworks absence the necessary mechanisms to support sustainable practices financially. For instance, there may be deficient incentives for businesses to capitalize in recycling technologies or innovative waste reduction strategies. Lacking financial backing or regulatory support, entities often highlight short-term savings over long-term sustainability. This gap not only hinders progress toward a waste-free environment but also hinders the creation of a CE where materials are reused and recycled effectively (Castellet-Viciano et al., 2022; Lazar, 2024; Neves et al., 2019c; Ntasiou & Andreou, 2017).

Public opposition to new regulations or deviations in waste management practices can also arise, particularly if those changes are supposed as problematic or financially troublesome. To overcome these challenges, a cooperative approach that includes multiple stakeholders, public input, and clarity in policy objectives is needed. This would support make a unified approach that enables the changeover to a waste-free environment, certifying that all voices are heard while maintaining an attention on sustainability and responsible resource management.

Technological limitations also play an important role in the journey toward a waste-free environment. For example, many waste processing facilities operate on obsolete technologies that are not capable of handling new waste streams or producing recyclable materials with high purity. Evolving waste management technologies, such as anaerobic digestion and advanced sorting systems, are often not widely accepted due to high costs or lack of awareness. Furthermore, the incorporation of technology into waste management systems needs a skilled workforce, which may be in short supply in certain areas. Invested capital in innovative technologies is essential, yet often challenging to secure due to challenging budgetary needs or economic uncertainty (Ntasiou & Andreou, 2017).

Barriers to realizing a waste-free environment also arise due to the interconnectedness of these infrastructure and technological challenges with broader issues such as regulatory frameworks and public engagement. For example, even if a public has access to the latest recycling technology, if there is no system in place to collect, sort, and transport materials efficiently, the benefits of that technology cannot be fully realized. Thus, addressing the infrastructure and technological limitations is not only a matter of improving facilities and introducing new tech; it needs a holistic approach that covers community engagement, investment in education, and supportive policy measures to create a favorable environment for innovation and sustainability. Reaching a waste-free society will need overcoming these barriers through collaborative efforts, strategic investment, and an unwavering commitment to change (Neves et al., 2019b, 2019c; Ogunmakinde, 2019).

Cultural attitudes and behavioral habits show a vital role in the drive toward a waste-free environment. These features are deeply ingrained in the daily routines and practices of individuals and communities, prompting how waste is perceived and managed. Many societies have established norms around consumption, convenience, and disposal that often prioritize immediate gratification over long-term sustainability. In certain cultures, for example, there might be a strong importance on consumerism, leading to rise in single-use products and packaging that contribute significantly to waste generation (Neves et al., 2019c).

Furthermore, cultural values can dictate the appropriateness of recycling or composting practices. In some cultures, there may be a lack of importance on environmental stewardship or personal responsibility headed for waste management, which can block efforts to stimulate waste reduction. Behavioral habits such as the trouble in breaking conventional patterns, reluctance to accept new practices, or simple suitability often outweigh the wish to involve in sustainable behaviors. For example, people may have a fondness to throw away items rather than invest time in sorting recyclables or composting organic waste, emphasizing a behavior that prioritizes ease over environmental consideration. Additionally, cultural stigma around assured waste management practices can also be a barrier. For example, in some communities, there may be negative insights related with second-hand goods or the act of repurposing materials, which can deter individuals from seeing the value in reducing waste through reuse. Addressing these cultural attitudes requires more than just education; it demands a comprehensive approach that includes engaging community leaders, utilizing local traditions, and creating incentives that align sustainable practices with cultural values. For a changeover to a waste-free environment to take place, it's necessary to promote a cultural shift that promotes sustainability as a desirable and normative behavior. This can be attained through community engagement initiatives, participatory programs, and campaigns that resonate with local values and lifestyles. Ultimately, by reshaping cultural attitudes and influencing behavioral habits, societies can move closer to achieving their waste reduction goals and foster a more sustainable relationship with their environment (Castellet-Viciano et al., 2022; Neves et al., 2019a).

6 Future Directions in Zero-Waste Management

Innovative Technologies

Around the world, businesses and waste management companies are already integrating smart automation and technology into their processes (Valavanidis, 2018). According to Romuno (2023), there are six of the most exciting emerging technologies for 2024.

Pello

Pello is a new technology that's been established to help businesses reduce their environmental impact and manage their waste collection more efficiently. Pello helps companies to accomplish these goals in a number of diverse ways. Firstly, the Pello system screens the fill level of your trash cans and offers real-time information on the dumpsters' contents and location. It also tells you if your container has been contaminated and directs pickup alerts when a collection is due.

Recycling Robots

Artificial Intelligence (AI) has been in the news a lot lately, and not always for the right reasons. However, one area where AI could make a real and appreciated difference is in the world of recycling. AI robots can be automated to quickly and precisely differentiate between different materials, and many recycling centers are already making use of the technology.

Introducing AI robotics into the process of waste management permits for more effective waste sorting in recycling centers and helps to divert as many recyclable materials away from landfills as possible. It also permits waste management companies to control longer hours or even stay open 24/7, dramatically increasing the quantity of waste processed (Jogarao et al., 2024; Kumar, 2025; Pal Singh, 2023).

Pneumatic Waste Pipes

As urban areas propagate and develop more densely populated, traditional waste disposal and collection systems come to be less practical. One of the best tech solutions to this difficulty is the pneumatic waste pipe. Pneumatic pipes can be fitted below public waste containers to transport waste straight to processing centers without the need for a trash pickup.

This system has two main benefits. The first is that it can dramatically decrease the number of garbage trucks on the roads. This can aid to cut unsafe emissions and lessen the number of vehicles clogging up our cities. Secondly, sending waste direct from dumpsters to waste management centers can help to avoid full containers overflowing. As well as being unsightly, overflowing dumpsters can cause an environmental health hazard and back the pollution of the local area.

Solar-Powered Trash Compactors

The more waste a trash truck can transport, the more it can collect on each round and the less trips it has to make. Solar-powered trash compactors wrapping trash as it collects inside a dumpster to increase capacity. This allows these smart containers to hold up to five times more than traditional trash bins.

As well as compressing waste, solar-powered trash compactors have built in waste level sensors. These sensors transmit data on the capacity of the bins, allowing users to schedule pickups and streamlining the collection process.

E-Waste Kiosks

Electronic waste, like old computers, phones, and TVs, contains a lot of harmful substances. If e-waste is inappropriately disposed of, these materials can escape out into the environment and pose a serious hazard to both human health and the local ecosystem. Despite of being sent to landfill, e-waste should be recycled or, even better, repaired, and reused. This will support to prevent harmful chemicals seeping into our soil and water and protect the local area.

One startup company has taken things into their own hands by initiation specialist e-waste recycling kiosks. EcoATM kiosks permit you to dispose of your unwanted electronics safely and easily. If your phone, tablet, or computer is still in good condition, they may even give you cash in exchange for the item, pleasing your efforts and helping to increase e-waste recycling rates in the process.

Recycling Apps

A new range of apps is now accessible to aid individuals and businesses direct the world of recycling. From donating your food waste to a well-meaning cause and giving away your clothes, to finding a nearby recycling center, these apps are a fanciful resource for those who want to retain their waste out of the landfill. Some apps also offer detailed information on precisely which materials can be recycled in your local area.

Because a lot of apps permit you to pass unwanted items on to other people, they're also a great method to develop a sustainable, CE.

Global Trends

The European Union produces more than 2.1 billion tons of waste every year. By updating its legislation on waste management, the EU desires to stimulate a shift to a more sustainable model—the circular economy.

CE plans can aid decrease global greenhouse gas (GHG) emissions by 40% by 2050, as estimated by the United Nations Development Program. Luckily, the changeover to circularity is already underway in many parts of the world. This global trend helping to drive this transition forward by switching to more sustainable raw materials, new legislation and regulation supporting circularity, tech-driven innovation for more circular solutions.

Table 1 Strategies and approaches for a zero waste vision

Strategies	Approaches
Integrated planning and programming for zero-waste city	Backing cities to develop their zero-waste strategy, roadmap, target, and indicators. Backing governance, policy, preparation of regulation and standards, design of waste sorting method, and recycling systems
Sustainable production and consumption	Stimulate corporate environmental and social governance (ESG) and sustainable consumption and production practices for circularity, escalation waste information transparency, identify circular business models
Financing instruments	Examine existing financing instruments to cover the cost of MSW management, aiming on ways to generate sustainable business models and public–private partnerships, covering investments in the key infrastructure and daily operational charges
Training, education, advocacy, and formalization	Backing communication for sustainable MSW through changed attitudes and behaviors of people and companies depending on local conditions and diverse cultural contexts

Opportunities for Expansion

UNDP purposes to present the zero-waste vision in 50 developing countries by 2025, improve sustainable waste management governance in cities, and finance the development of zero-waste policies and strategies in selected cities through gender responsive and all-encompassing approaches (Ogunmakinde & Maund, 2017). Table 1 showcases some highlights of strategies and approaches.

7 Conclusions

The successful transition to a waste-free environment can be achieved by the adoption of CE principles which emphasize sustainability through resource efficiency, reduction of waste, and the promotion of recycling and reuse. However, for CE to be effective, IS plays a critical role by encouraging collaboration between industries to create mutually beneficial relationships. This collaboration maximizes the value of resources, protection of virgin resources while minimizing environmental impact. However, transforming to this platform is demanding due to many reasons such as outdated regulatory frameworks, technological limitations, and cultural resistance. These highlight the need for a multifaceted approach. However, this is a consequential in achieving zero-waste targets and net-zero emissions which are key features in sustainable development goals. The approaches such as strengthening policy coordination, incentivizing technological innovations, and fostering public engagement are essential steps in overcoming these challenges. In particular, the integration of advanced waste management technologies, such as AI-powered robots and pneumatic waste pipes, can significantly improve efficiency and sustainability. Additionally, addressing cultural and behavioral attitudes through education and community engagement is critical for ensuring the long-term success of waste reduction strategies. By aligning CE and IS with technological advancements, regulatory support, and cultural transformation, a sustainable, waste-free society is within reach, offering tangible environmental and economic benefits for both industries and communities. Through continued collaboration and innovation, CE and IS can drive the change necessary for a circular, sustainable future.

References

Bala, K., Anshul, Verma, S., Dogra, A., Das, A. P., Kumar, R., & Nandini. (2024). Sustainable management for urban plastic waste generation. *Sustainable Management of Urban Plastic Waste Through Circular Economic Approaches*, 211–236. https://doi.org/10.1201/9781003465447-13

Castellet-Viciano, L., Hernández-Chover, V., Bellver-Domingo, Á., & Hernández-Sancho, F. (2022a). Industrial symbiosis: A mechanism to guarantee the implementation of circular economy practices. *Sustainability (Switzerland), 14*(23). https://doi.org/10.3390/su142315872

Das, A., & Ghosh, S. (2022). Role of microorganisms in extenuation of mining and industrial wastes. *Geomicrobiology Journal, 39*, 173–175. https://doi.org/10.1080/01490451.2022.2038953

Das, A., Ray, S. K., Mohanty, M., Mohanty, J., Dey, S., & Das, A. P. (2024). Ecotoxicity of microplastic wastes and their sustainable management: A review. *Environmental Chemistry and Ecotoxicology, 6*, 144–152. https://doi.org/10.1016/J.ENCECO.2024.05.003

Das, A. P., & Ghosh, S. (2018). Bioleaching of manganese from mining waste materials. *Materials Today: Proceedings, 5*(1), 2381–2390. https://doi.org/10.1016/J.MATPR.2017.11.459

Dey, S., Das, A., Mallick, K., Sahu, A., & Das, A. P. (2023). Environmental petroleum waste: Pollution. *Toxicity, Sustainable Remediation, 159–175.* https://doi.org/10.1007/978-3-031-48220-5_7

Gaybor, J., & Chavez, H. (2019). The Argentinean zero waste framework. *The circular economy and the Global South, November*, 95–112. https://doi.org/10.4324/9780429434006-6

General, D. (2023). *Circular economy: Definition, importance and benefits*, 1–5.

Jogarao, M., Lakshmanna, B. C., & Naidu, S. T. (2024). AI-enabled circular economy management for sustainable smart cities: Integrating artificial intelligence in resource optimization and waste reduction. *Smart cities and circular economy, September*, 83–96. https://doi.org/10.1108/978-1-83797-957-820241008

Kumar, A. (2025). *Articles from linear to circular: A comprehensive SWOT analysis of India's from linear to circular: A comprehensive SWOT analysis of India's circular economy potential, March.* https://doi.org/10.69968/ijisem.2025v4i1189-203

Lazar, N. (2024). The importance of Eu-funded projects for the sustainable development of local communities. *Agora International Journal of Economical Sciences, 18*(1), 111–123. https://doi.org/10.15837/aijes.v18i1.6715

Mishra, S., & Das, A. P. (2021). Current treatment technologies for removal of microplastic and microfiber pollutants from wastewater. *Wastewater Treatment: Cutting-Edge Molecular Tools,*

Techniques and Applied Aspects, 237–251. https://doi.org/10.1016/B978-0-12-821881-5.00011-8

Mishra, S., & Das, A. P. (2023). Microbial remediation of synthetic microfiber contaminated wastewater. In M. P. Shah (Ed.), *Microbial technologies in industrial wastewater treatment* (pp. 337–350). Springer, Singapore. https://doi.org/10.1007/978-981-99-2435-6_17

Mishra, S., Ghosh, S., van Hullebusch, E. D., Singh, S., & Das, A. P. (2023). A critical review on the recovery of base and critical elements from electronic waste-contaminated streams using microbial biotechnology. *Applied Biochemistry and Biotechnology, 2023*, 1–30. https://doi.org/10.1007/S12010-023-04440-X

Mishra, S., Rout, P. K., & Das, A. P. (2019). Solar photovoltaic panels as next generation waste: A review. *Biointerface Research in Applied Chemistry, 9*(6).

Mohanty, S., Ghosh, S., Bal, B., & Das, A. P. (2018). A review of biotechnology processes applied for manganese recovery from wastes. *Reviews in Environmental Science and Bio/Technology, 17*(4), 791–811. https://doi.org/10.1007/S11157-018-9482-1

Neves, A., Godina, R., Azevedo, S. G., & Matias, J. C. O. (2019a). Current status, emerging challenges, and future prospects of industrial symbiosis in Portugal. *Sustainability (Switzerland), 11*(19). https://doi.org/10.3390/su11195497

Neves, A., Godina, R., Azevedo, S. G., & Matias, J. C. O. (2019b). Environmental, economic, and social impact of industrial symbiosis: Methods and indicators review. *Springer Proceedings in Mathematics and Statistics, 281*(May), 157–165. https://doi.org/10.1007/978-3-030-14973-4_15

Neves, A., Godina, R., Azevedo, S. G., & Matias, J. C. O. (2020). A comprehensive review of industrial symbiosis. *Journal of Cleaner Production, 247*(November 2019). https://doi.org/10.1016/j.jclepro.2019.119113

Neves, A., Godina, R., Azevedo, S. G., Pimentel, C., & Matias, J. C. O. (2019c). The potential of industrial symbiosis: Case analysis and main drivers and barriers to its implementation. *Sustainability (Switzerland), 11*(24), 1–68. https://doi.org/10.3390/su11247095

Ntasiou, M., & Andreou, E. (2017). The standard of industrial symbiosis. Environmental criteria and methodology on the establishment and operation of industrial and business parks. *Procedia Environmental Sciences, 38*, 744–751. https://doi.org/10.1016/j.proenv.2017.03.157

Ogunmakinde, O. E. (2019). A review of circular economy development models in China, Germany and Japan. *Recycling, 4*(3). https://doi.org/10.3390/recycling4030027

Ogunmakinde, O. E., & Maund, K. (2017). Circular construction: Opportunities and threats. *Project Management Symposium, July.*

Oughton, C., Kurup, B., Anda, M., & Ho, G. (2022). Industrial symbiosis to circular economy: What does the literature reveal for a successful complex industrial area? *Circular Economy and Sustainability, 2*(4), 1317–1344. https://doi.org/10.1007/s43615-022-00153-1

Pal Singh, J. (2023). Artificial intelligence in circular economies: A pathway to sustainable resource management. *International Journal of Science and Research (IJSR), 12*(12), 1128–1131. https://doi.org/10.21275/sr231214040053

Priyadarsini, S., & Das, A. P. (2024). Analytical and structural characterization of waste lithium-ion batteries for their effective recycling strategy. *Environmental Chemistry and Ecotoxicology.* https://doi.org/10.1016/j.enceco.2024.12.004

Priyadarsini, S., & Das, A. P. (2025). Lithium bioleaching: A review on microbial-assisted sustainable technology for lithium bio-circularity. *Journal of Water Process Engineering, 69*, Article 106744. https://doi.org/10.1016/j.jwpe.2024.106744

Ray, S. K., Dey, S., Das, A. P., & John, A. (2024). Microplastic waste management through greener technologies: A review. *Sustainable Management of Urban Plastic Waste through Circular Economic Approaches, 237–261.* https://doi.org/10.1201/9781003465447-14

Romuno, J. (2023, May15). *6 Smart waste management technologies emerging in 2025: RTS.* Recycle Track Systems. https://www.rts.com/blog/smart-waste-management-technologies/

Sgambaro, L., Chiaroni, D., Lettieri, E., & Paolone, F. (2024). Exploring industrial symbiosis for circular economy: Investigating and comparing the anatomy and development strategies in Italy. *Management Decision*. https://doi.org/10.1108/MD-04-2023-0658

Shenyoputro, K., & Jones, T. E. (2023). Reflections on a two-decade journey toward zero waste: A case study of Kamikatsu town, Japan. *Frontiers in Environmental Science, 11*(April), 1–4. https://doi.org/10.3389/fenvs.2023.1171379

Song, Q., Li, J., & Zeng, X. (2015). Minimizing the increasing solid waste through zero waste strategy. *Journal of Cleaner Production, 104*(March), 199–210. https://doi.org/10.1016/j.jclepro.2014.08.027

Valavanidis, A. (2018). *Concept and practice of the circular economy concept and practice of the circular economy. Athanasios Valavanidis*, *July*, 1–30.

The Potential of Using Industrial Bio-based Sludge as a Feedstock for Sustainable Biochar Production

S. H. I. S. Gunasiri, Chamathi Sethara Ranasinghe, and K. P. P. Udayagee

1 Introduction

The rapid expansion of industrial activities has led to the generation of solid and semi-solid industrial waste worldwide (Ogunwumi et al., 2023). Such waste is a by-product of manufacturing process and varies depending on the type of industry, with classification typically based on physical state, composition, and environmental impact (Dey et al., 2023; Mishra et al., 2023). Bio-based sludge is a semi-solid or thick, mud-like residue that remains after treating industrial wastewater, that include biodegradable organic materials (Jiménez-Rosado et al., 2023). In contrast to toxic industrial sludge, bio-based sludge is more suitable for circular use as it holds a significant potential for upcycling and transforming into eco-friendly, valuable products (Das & Ghosh, 2022; Mohanty et al., 2018).

The industrial sector produces a vast amount of bio-based sludge, including agricultural residues, food processing by-products, and organic sludge from wastewater treatment plants (Bala et al., 2024; Das & Ghosh, 2018; Ray et al., 2024). According to the Food and Agriculture Organization (FAO), the global food industry generates over 1.6 billion tons of food waste annually, while the forestry sector produces 600 million m^3 of wood residues yearly (Francioso, 2024). However, a significant portion of this bio-based waste is not efficiently used, but practiced with conventional disposal methods such as landfilling, incineration, and wastewater discharge, leading to severe environmental issues (Ghosh & Das, 2017; Pradhan et al., 2023). Thus, sustainable management of bio-based sludge is a timely requirement that offers

S. H. I. S. Gunasiri
Faculty of Graduate Studies, University of Sri Jayewardenepura, Nugegoda, Sri Lanka

C. S. Ranasinghe · K. P. P. Udayagee (✉)
Faculty of Technology, University of Sri Jayewardenepura, Homagama, Sri Lanka
e-mail: udayagee@sjp.ac.lk

A. P. Das et al. (eds.), *Sustainable Municipal Waste Management with a Zero Waste Approach*, Sustainable Environmental Waste Management Strategies,
https://doi.org/10.1007/978-3-032-05950-5_17

a circular economy approach. This approach is critical to transforming these undervalued resources into valuable products like biofuels, biofertilizers, bioplastics, bio-based chemicals, and biochar etc. Thus, the efficient utilization of bio-based sludge can significantly reduce reliance on virgin materials and fossil-based energy sources, fostering a cleaner environment. These efforts lead the path to zero-emission targets by mitigating greenhouse gas emissions, lowering waste disposal costs, creating new revenue streams by selling bio-based products (Parikka, 2004). Moreover, the thermal conversion of industrial bio-based sludge into valuable products presents a promising approach to achieving multiple Sustainable Development Goals (SDGs), including SDG 6, clean water and sanitation; SDG 7, affordable and clean energy; SDG 9, industry, innovation, and infrastructure; SDG 12, responsible consumption and production; SDG 13, climate action; and SDG 15, life on land.

Biochar has emerged as a sustainable material that can be produced using organic waste. During its production process, it also yields bioenergy in the form of syngas and bio-oil. Biochar is a carbon-rich material produced by heating organic materials, such as agricultural residues, wood chips, manure, and industrial bio-based waste. The process of converting these materials into biochar is called pyrolysis or hydrothermal carbonization, depending on the method followed (Yaashikaa et al., 2020). Both processes are controlled thermal decomposition processes that prevent complete combustion and preserve carbon in a stable form. Biochar production methods primarily include pyrolysis, where biomass is heated to high temperatures in limited oxygen conditions. Hydrothermal carbonization is a process of converting heated biomass into hydrochar, which is a biochar-like material under pressurized water conditions. Biochar is highly beneficial because of its environmental and industrial applications. Applying of biochar in agricultural fields enhances soil fertility by improving water retention, nutrient availability, and microbial activity while reducing the need for chemical fertilizers (Wang et al., 2020). As a carbon sequestration strategy, biochar plays a huge role by locking carbon in a stable form for hundreds to thousands of years, lowering CO_2 emissions and climate change. Additionally, it contributes to waste management by converting bio-based waste into a valuable product, reducing landfill waste, methane emission, and pollution (Palansooriya et al., 2019). Beyond soil conditioning and carbon sequestration, biochar has many applications, such as water purification, air purification, livestock farming, construction, and electronic device material. It also has industrial applications, serving as a renewable energy source.

This chapter explores the potential of sludge-derived biochar as a sustainable solution for waste management, meeting zero-emission targets, and energy production. This chapter comprehensively analyzes industrial bio-based sludge characteristics, biochar production processes, its environmental and economic benefits, and associated challenges and risks, highlighting the need for further research and innovation to optimize its use.

2 Bio-based Sludge as Feedstock for Biochar Production

The pyrolysis process and the chemical nature of biomass feedstock are crucial factors for the final quality of biochar production, including its carbon content, surface area, porosity, and minerals. Therefore, selecting feedstock is important to obtain biochar with desired properties (Pradhan et al., 2024). Lignocellulosic feedstock, which includes cellulose, hemicellulose, and lignin, profoundly affects the biochar's final chemical composition and structural characteristics. Feedstocks rich in cellulose and hemicellulose and lignin produce biochar with high-carbon content due to the relatively high pyrolysis efficiency of these components with thermally stable and decomposes at higher temperatures (Kwon et al., 2019; Zhou et al., 2024). Feedstocks with high-volatile matter, such as food waste and manure, produce biochar with lower carbon content and higher gas yields, which can be captured for energy or syngas production (Amalina et al., 2022). However, plant-based feedstocks have lower volatile matter. During pyrolysis, feedstock with high-moisture content requires additional energy for drying. This reduces thermal efficiency and yield. Lower moisture content feedstock is optimal for achieving higher carbonization yields. According to recent research, fast pyrolysis processes with around 10% moisture content are desirable during the charcoal-making process (Peng et al., 2023).

Bio-based sludge is generated from biological processes in industries such as wastewater treatment, food processing, agriculture, or biotechnological operations (Romero et al., 2024). It is primarily composed of microbial biomass, plant residues, and other organic compounds that result from the breakdown of organic matter. Bio-based sludge typically has a high-water content, ranging from 50 to 80% depending on the type of sludge and its source (Škorjanc et al., 2024; Zhou et al., 2024). This high-moisture content can make handling, transporting, and processing challenging. In many cases, water removal through dewatering is required before the sludge can be further treated or converted into value-added products like biochar.

Food-Based Sludge

Food waste sludge refers to the organic waste material generated during the processing or treatment of food waste. A considerable portion of this waste occurs during food processing. The beverage industry produces about 26% of industrial food waste, followed by the dairy industries (21.3%), fruit and vegetable production (14.8%), manufacture of grain and starch products (12.9%), and meat production (8%) (Jaouhari et al., 2023). Food sludge waste consists of lipids, hemicellulose, cellulose, starch, lignin, and protein, similar to the other carbonaceous solid waste used in biochar production. Lignocellulosic food sludge includes by-products from beverage production, e.g., fruit and vegetable peels, stems, and pulp; cereal and grain processing, e.g., husks and straw, sugarcane bagasse, and molasses. These materials are high in lignin, a tough polymer that enhances biochar's stability and carbon sequestration potential. Feedstocks rich in lipids or proteins, like food waste, may produce biochar with different characteristics, such as higher nutrient content and increased bioavailability of minerals (Škorjanc et al., 2024). However, the presence of protein makes it prone to biodegradation and complicates long-term storage (Wang et al., 2022).

Paper and Pulp

During the process of paper and pulp manufacturing, about 40–50 kg of sludge is generated to produce 1 ton of paper at a paper mill. This sludge contains approximately 70% primary sludge and 30% secondary sludge, and includes 60–95% of organic matter and mineral filters (Novotný et al., 2023). In some cases, the paper mill sludge is considered toxic due to the presence of heavy metals, which may enter the process through pigments and dyes, filters and coatings and or as contaminants from recycled paper used in the paper processing industry. However, the pollutants load in paper mill sludge varies depending on the production proceses higher levels are typically assosiated with recycled paper production, while only minute levels are observed in virgin paper production. For sustainable waste management, this sludge can be converted into biochar. Traditional pyrolysis is less efficient for high-moisture feedstocks; instead, hydrothermal carbonization (HTC) offers a more effective alternative, enhancing process efficiency and producing hydrochar with superior sorption properties (Hämäläinen et al., 2022).

Wastewater Treatment Sludge from Organic-Based Industries

Wastewater treatment sludge can be divided into two categories: primary sludge and secondary sludge. Primary sludge, such as sedimentation and filtration, is produced during the primary wastewater treatment operation. Secondary sludge, such as activated sludge, is produced during secondary wastewater treatment (Devi & Saroha, 2017). Turning this waste into biochar is a feasible and sustainable strategy to reduce potential environmental risks and recover energy. Due to the high total solids content and carbon concentration in organic-based wastewater sludge, this feedstock is suitable for biochar production (Mian et al., 2022). According to past studies, applying wastewater sludge-derived biochar to soil has significantly enhanced soil electrical conductivity and increased phosphorus and nitrogen content. Additionally, research indicates that wastewater treatment plants processing predominantly domestic wastewater, with minimal industrial wastewater influence, produce the highest biochar yields (Lu et al., 2013).

3 Biochar Production Process from Industrial Bio-based Sludge

Thermochemical conversion techniques are the standard for biochar production using bio-based sludge. Pyrolysis, gasification, hydrothermal carbonization, and torrefaction are the main thermochemical conversion techniques for biochar production in low-oxygen environments (thermochemical-reduction processes). Figure 1 shows the process flow of biochar production from industrial bio-based sludge.

4 Pre-treatment Process

Dewatering and Drying

Industrial bio-based sludge often contains 70% water content. Pyrolysis of wet feedstock is inefficient and less feasible due to high energy consumption. To mitigate these challenges, a two-stage dewatering process is required. Initially, mechanical dewatering techniques, such as belt presses, centrifugation, or filter presses, significantly reduce the sludge's water content. This is followed by thermal drying methods, including solar drying (at 60 °C for 2 days), rotary dryers, or flash dryers, to further lower the moisture content to below 10%. Achieving this optimal moisture level (Carpenter et al., 2014; Jaouhari et al., 2023).

Blending with Other Biomass

Blending with other biomass is also called co-pyrolysis. Blending sludge with lignocellulosic biomass, such as wood chips, agricultural residues, or sawdust, enhances the structural and physicochemical properties of the resultant biochar. The lignocellulosic feedstock, rich in cellulose, hemicellulose, and lignin, facilitates the development of a well-structured carbon matrix, improving porosity and specific surface area. It was found that mixing wood chips could reduce the activation energy of sludge by pyrolysis and promote the charring of sludge. Another research mentioned mixing bio-based sludge with cotton straw enhances the cation exchange capacity of biochar (Hurley et al., 2017; Jaouhari et al., 2023).

Pelletizing

After the dewatering process and the blending with other feedstock, bio-based sludge could be formed as pellets using a mechanical pelletizing process (Jaouhari et al.,

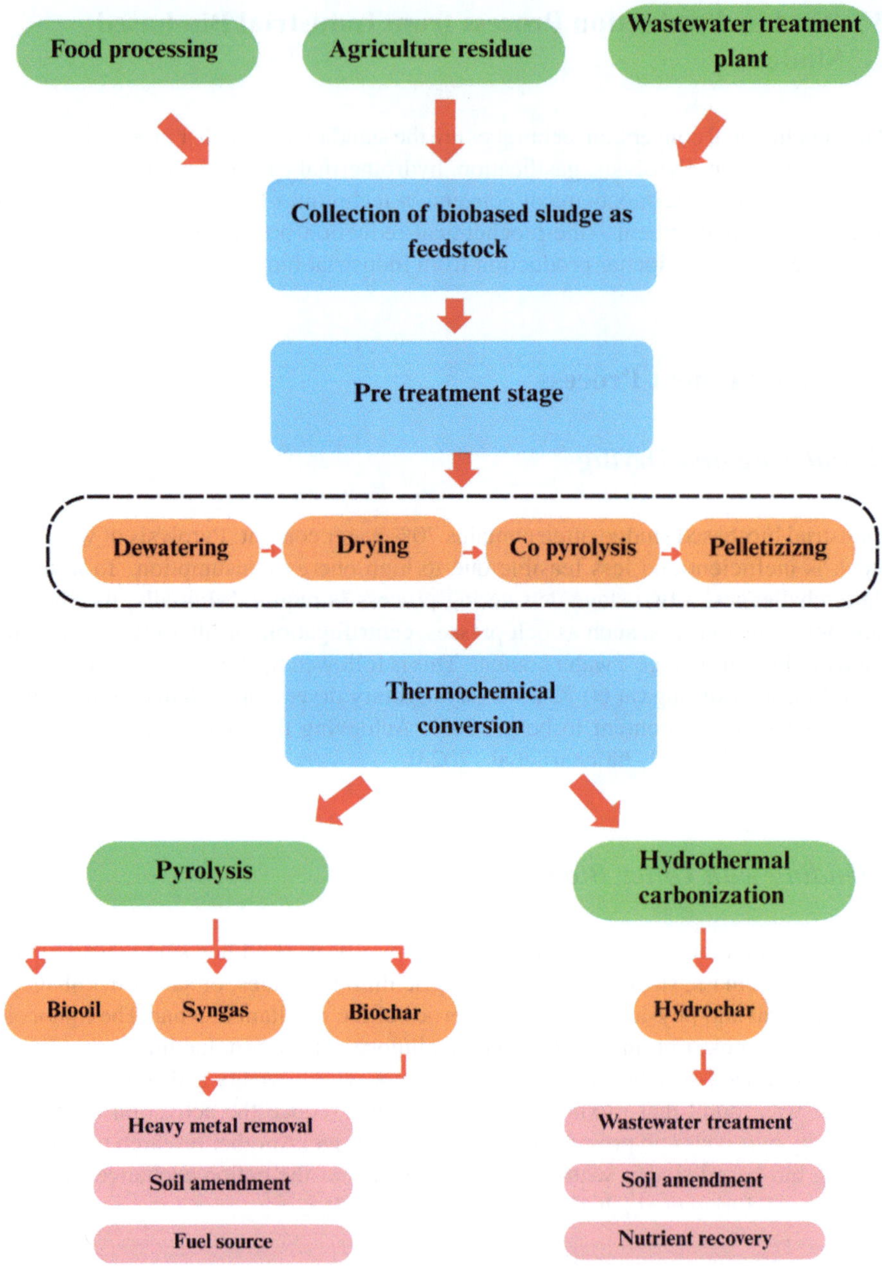

Fig. 1 Process flow of biochar production from industrial bio-based sludge

2023). Here, the prepared feedstock is fed into a ring die or flat die pelletizing machine, where high pressure and friction cause the material to compact into dense cylindrical pellets (Wang et al., 2021). The application of moderate heat during extrusion aids in pellet binding. Pelletizing enhances sludge bulk density. Meanwhile, it promotes the pyrolysis process.

5 Thermal Conversion Process of Bio-based Sludge

There are few different application types in converting bio-based sludge into biochar. Table 1 summarizes the comparison of thermochemical conversion techniques for biochar production.

Table 1 Comparison of thermochemical conversion techniques for biochar production

Conversion method	Temperature range (°C)	Residence time	Moisture tolerance	Main products	Advantages	Limitations
Slow pyrolysis	300–550	5 min–35 h	< 10%	Biochar, some bio-oil, syngas	High biochar yield, carbon sequestration	Longer processing time
Fast pyrolysis	650–1000	< 2 s	< 10%	High bio-oil, some syngas, low biochar	High energy production	Low biochar yield
Moderate pyrolysis	450–550	10–30 s	< 10%	Biochar, bio-oil, syngas	Balanced yield	Requires precise control
Hydrothermal carbonization (HTC)	180–250	1–12 h	High (up to 70%)	Hydrochar, liquid phase, gas phase	Can process wet sludge, lower energy demand	Lower carbon content than pyrochar
Gasification	800–1300	Seconds to minutes	< 30%	Syngas, biochar (low yield)	High energy conversion efficiency	Low biochar production
Torrefaction	200–300	10–60 min	< 30%	Biochar, volatile compounds	Mild process, improves fuel properties	Lower carbonization level

Pyrolysis

Pyrolysis is the thermochemical decomposition of organic matter in absent oxygen conditions at higher temperatures. It results in the formation of three types of energy sources: biochar, bio-oil, and syngas. The biochar yield typically ranges from 10 to 80%, depending on process conditions and feedstock types. Pyrolysis operates across a broad temperature range (100–1300 °C) and residence times varying from 0.05 s to 12 h, under low or atmospheric pressure. Low pyrolysis temperature and extended resident time may accelerate biochar production yield and quality. During the pyrolysis of feedstock, biochar is formed from a series of reactions, such as isomerization, dehydration, decarboxylation, depolymerization, and charring (Novotný et al., 2023). Pyrolysis is most effective when using relatively dry biomass, typically containing less than 30% moisture content, though an ideal level is around 10%. The biomass is also ground into different particle sizes depending on the type of pyrolysis being performed.

According to the temperature range, the classical pyrolysis process can be divided into three types: slow, fast, and moderate pyrolysis. Each type primarily produces different products. Slow pyrolysis operates at low heat (300–550 °C) and longer resident time (5–30 min or even 25–35 h) (Nzediegwu et al., 2021). Moderate pyrolysis conducted at 450 and 550 °C is more rapid than slow pyrolysis, takes 10–30 s to complete, and produces less charcoal than slow pyrolysis. Fast pyrolysis, which is performed at relatively high temperatures and in the shortest residence time, often avoids secondary reactions. The maximum temperature could be 650–1000 °C. Due to the unavailability of the secondary reaction, the produced biochar yield is significantly lower compared to other methods, while bio-oil production is high (Kumar et al., 2020). In slow pyrolysis, the large carbon bonds remain intact due to the low temperature, as there is insufficient energy to break them down. As a result, most of the sludge retains its original structure, leading to larger particle sizes, lower surface area, and fewer pores (5–50 mm) (Li et al., 2020; Pradhan et al., 2024).

Hydrothermal Carbonization

Hydrothermal carbonization is a thermochemical process for converting the organic compound to the structured carbon called hydrochar. HTC was conducted at a relatively low temperature varying from 180 to 250 °C (Škorjanc et al., 2024). HTC, also called wet pyrolysis, can convert wet feedstock like sludge into biochar without any energy-intensive and costly dewatering treatments. Industrial sludge, which usually contains higher moisture, can easily undergo this treatment. HTC includes several key chemical reactions. First, hydrolysis occurs, where water breaks down complex organic molecules like polysaccharides and proteins into simpler compounds such as sugars and acids. Then, water is removed from organic compounds during dehydration, increasing the biomass's carbon content, resulting in a more carbon-rich material. Decarboxylation follows, where carboxyl groups break down, releasing carbon dioxide and reducing the oxygen content, making the product more hydrophobic (Masoumi et al., 2021). Finally, the remaining molecules undergo polymerization and aromatization, forming stable aromatic carbon structures and producing hydrochar, a carbon-rich solid. The process pressure and temperature must be maintained optimally to obtain better results. According to the research findings, the maximum hydrochar yield was obtained at 250 °C and a pressure of 2 MPa. Optimal pressure helps maintain water in a liquid state at high temperatures, typically between 180 and 250 °C (Zhou et al., 2024).

However, due to distinct processes, some physical and functional differences exist between biochar and hydrochar. Hydrochar contains lower carbon content and pores than biochar due to incomplete carbonization. Hydrochar is better for applications requiring high reactivity, such as fertilizer, soil remediation, and energy production from wet biomass. Biochar is more stable and porous, making it ideal for carbon sequestration, water filtration, and long-term soil amendment.

6 Potentials of Using Industrial Bio-based Sludge as a Feedstock for Sustainable Biochar Production

The application of circular economic principles to achieve waste management strategies through innovative approaches has become a trend in this era. One such approach is utilizing industrial bio-based sludge to produce biochar. Those products grab the industrial attention due to their potential to address environmental and economic challenges. Figure 2 summarizes the potentials of using industrial bio-based sludge as feedstock for biochar production to address global challenges.

Industrial bio-based sludge is generally disposed of in landfills or incinerated, and these actions cause significant harm to the environment by producing greenhouse gases, soil and water contamination due to leachate formation, and odor problems (Salama et al., 2017). However, moving with bio-based biochar can be a quick and effective solution for this prevalent issue in many industries. For example, Finland's pulp and paper mill successfully implemented a biochar production system to process its sludge, reducing landfill dependency by 40%. This minimizes environmental pollution and aligns with stricter waste management regulations in many countries (Salcedo-Puerto et al., 2024).

Further, producing biochar helps sequester carbon in a stable form for centuries. When it comes to industrial waste, industries pyrolyze their biowaste into biochar, stabilizing the carbon in the waste. This prevents the carbon from being released into the atmosphere through decomposition, unlike other waste management strategies (Liu et al., 2023). By stabilizing carbon through these processes, greenhouse gas emissions from waste are significantly reduced. This helps lower the global warming

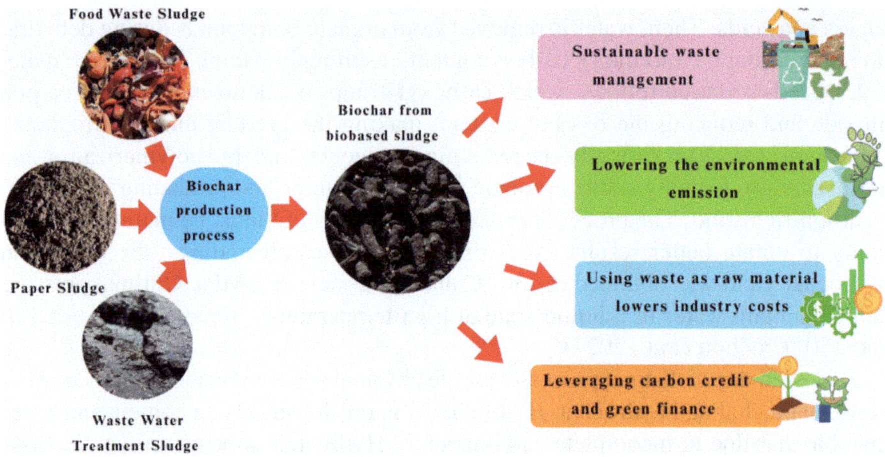

Fig. 2 Potentials of using industrial bio-based sludge as feedstock for biochar production to address global challenges

potential and contributes to global climate goals, including the Paris Agreement, while also assisting industries in meeting their sustainability targets (Margal et al., 2023).

Producing biochar from bio-based waste is a prime example of this circular economic approach (Júnior et al., 2022). Further, it is an environmentally responsible choice and a highly cost-effective solution for many industries. Since the raw material comes from waste, production costs remain low, turning a liability into a valuable resource. Large-scale adoptions like industrial symbiosis become even more feasible when industries utilize the excess heat generated during pyrolysis to power operations, further reducing energy costs. Thus, for reaching global sustainability goals and stricter regulations, investing in biochar technology offers industries a practical, scalable, and profitable way to manage their bio-based sludge. On the other hand, carbon credits, which function as rewards for reducing greenhouse gas emissions, are another approach to promote this. Biochar production becomes more applicable and economical when these credits are offered for extra income. Additionally, companies can get finance for environmentally sustainable initiatives through green financing alternatives such as government subsidies, green bonds, or impact investors. This helps pay for the setup of pyrolysis systems. Businesses may profit, save waste, and benefit the environment simultaneously by combining carbon credit with sustainable finance.

7 Challenges and Risks

The concept of converting industrial bio-based sludge into biochar or hydrochar must be handled appropriately to ensure its sustainability and safety. The existence of contaminants such as heavy metals, legal restrictions, technological limitations, and public opinion is among the primary main concerns. The existence of heavy metals in industrial waste, especially from pulp and paper sludge, which poses serious toxicity issues, is one of the main obstacles. The residues may remain in biochar, endangering crop safety, soil health, and human consumption. Other than heavy metals, the formulation of harmful by-products is another concern. Pyrolysis of bio-based sludge may generate hazardous pollutants such as polycyclic aromatic hydrocarbons (PAHs), dioxins, and furans, depending on feedstock composition and operating conditions. Thus, strict testing and treatment procedures are necessary to lessen this problem, but they can make scaling more challenging and raise manufacturing costs. For biochar to be accepted and used effectively in agricultural applications, its quality and safety must be guaranteed. The use of bio-based sludge for the synthesis of biochar is made more difficult by regulatory obstacles. Because of the possible health and environmental hazards, governments and environmental organizations frequently limit the use of waste-derived products in agriculture. It takes a lot of studies, adherence to safety laws, and cooperation with legislators to navigate these rules. The broad use of this technology could be impeded in the absence of transparent and encouraging regulatory frameworks.

The heterogenic nature of this bio-based sludge is another concern. The process is made more complicated when the feedstock time is different, leading to the selection of different time–temperature combinations from time to time. On the other hand, the high moisture content becomes a significant barrier, demanding several preliminary steps for dewatering and drying, which may also demand additional technical and financial resources.

Technological limitations also pose a significant challenge. While biochar production from bio-based sludge is technically feasible, scaling up the process to meet industrial and agricultural demands remains a hurdle. Current technologies may lack the efficiency or cost-effectiveness required for large-scale implementation. High investments are needed for the settings, such as pyrolysis or HTC facilities, yet the commercial viability of the biochar depends on several factors. Thus, profitability matters. Although producing biochar from industrial bio-based sludge has huge sustainability potential, solving issues with heavy metal pollution, regulations, technological limitations, and public opinion is essential. This strategy can help create a more sustainable and circular economy by addressing these problems via creativity, collaboration, and education.

8 Future Prospects and Recommendations

Converting industrial bio-based sludge into biochar or hydrochar holds significant potential for sustainable waste management and resource recovery. However, future advancements must focus on safety, efficiency, and market adoption to maximize their environmental and economic benefits. The improved pyrolysis and HTC facilities, in particular, need to be improved with the involvement of ongoing research aiming to optimize time–temperature combinations to facilitate the requirements needed for bio-based sludge pyrolysis. This can be further improved by incorporating hybrid pyrolysis and HTC systems to enhance carbon and energy efficiency. Predictive modeling developed with the inputs of AI tools further optimizes the application.

Further, biochar can be made safer through implementing pre-treatment and post-treatment methods to remove heavy metals and other impurities. Selective pyrolysis techniques also minimize the potential air pollution.

Furthermore, expanding the potential of biochar uses another approach to promote it. Expanding the applications beyond agriculture, such as environmental applications, energy storage and carbon merits, construction, and composite materials, and adding extra value to the biochar may enhance the investments in bio-based sludge-based biochar.

Most importantly, circular economy approaches can create closed-loop waste management systems, such as linking sludge-to-biochar processes with other industries (e.g., bioenergy, wastewater treatment). Further, the small-scale biochar production units at industrial sites following the concept of decentralized processing can minimize transportation costs and emissions.

Interdisciplinary research, pilot-scale demonstrations, and public–private partnerships are essential to implement this application collaboratively. Educating potential users on the benefits and safe use of sludge-derived biochar will increase acceptance. Further, developing a bio-based sludge-derived biochar quality certification system can help build consumer trust.

9 Conclusions

The transformation of industrial bio-based sludge into biochar presents a promising opportunity for sustainable waste management and resource recovery. This process mitigates environmental pollution and thereby supports global climate goals. Conversion of bio-based sludge into biochar is a critical strategy in carbon sequestration and reducing reliance on fossil-based resources. This approach aligns with industry-based circular economy initiations, which can be successfully done by integrating the industrial symbiosis concept. This process gives value to waste while minimizing disposal costs. However, the potential risk of having contaminants, regulatory limitations, and technological and financial barriers must be addressed to ensure the safe and efficient use of biochar generated through this transformation. Innovations in feedstock pre-treatment, stringent quality control measures, and policy-driven incentives are essential in meeting the true goals of the process, and it can be recommended to initiate collaborative works between researchers, policymakers, and industries. Green approaches such as carbon credits and green financing can further enhance biochar's feasibility on an industrial scale.

References

Amalina, F., Razak, A. S. A., Krishnan, S., Sulaiman, H., Zularisam, A. W., & Nasrullah, M. (2022). Biochar production techniques utilizing biomass waste-derived materials and environmental applications—A review. *Journal of Hazardous Materials Advances, 7*, Article 100134. https://doi.org/10.1016/j.hazadv.2022.100134

Bala, K., Anshul, Verma, S., Dogra, A., Das, A. P., Kumar, R., & Nandini. (2024). Sustainable management for urban plastic waste generation. *Sustainable management of urban plastic waste through circular economic approaches*, 211–236. https://doi.org/10.1201/9781003465447-13

Carpenter, D., Westover, T. L., Czernik, S., & Jablonski, W. (2014). Biomass feedstocks for renewable fuel production: A review of the impacts of feedstock and pretreatment on the yield and product distribution of fast pyrolysis bio-oils and vapors. *Green Chemistry, 16*(2), 384–406. https://doi.org/10.1039/C3GC41631C

Das, A. P., & Ghosh, S. (2018). Bioleaching of manganese from mining waste materials. *Materials Today: Proceedings, 5*(1), 2381–2390. https://doi.org/10.1016/J.MATPR.2017.11.459

Das, A., & Ghosh, S. (2022). Role of microorganisms in extenuation of mining and industrial wastes. *Geomicrobiology Journal, 39*, 173–175. https://doi.org/10.1080/01490451.2022.2038953

Devi, P., & Saroha, A. K. (2017). Utilization of sludge based adsorbents for the removal of various pollutants: A review. *Science of the Total Environment, 578*, 16–33. https://doi.org/10.1016/j.scitotenv.2016.10.220

Dey, S., Das, A., Mallick, K., Sahu, A., & Das, A. P. (2023). Environmental petroleum waste: Pollution. *Toxicity, Sustainable Remediation, 159–175.* https://doi.org/10.1007/978-3-031-48220-5_7

Francioso, O. (2024). Current and future perspectives for biomass waste management and utilization. *Scientific Reports, 14*(1), 9635. https://doi.org/10.1038/s41598-024-59623-1

Ghosh, S., & Das, A. P. (2017). Bioleaching of manganese from mining waste residues using Acinetobacter sp. *Geology, Ecology, and Landscapes, 1*(2), 77–83. https://doi.org/10.1080/24749508.2017.1332847

Hämäläinen, A., Kokko, M., Kinnunen, V., Hilli, T., & Rintala, J. (2022). Hydrothermal carbonization of pulp and paper industry wastewater treatment sludges—Characterization and potential use of hydrochars and filtrates. *Bioresource Technology, 355*, Article 127258. https://doi.org/10.1016/j.biortech.2022.127258

Hurley, S., Shrestha, P., & Cording, A. (2017). Nutrient leaching from compost: Implications for bioretention and other green stormwater infrastructure. *Journal of Sustainable Water in the Built Environment, 3*(3), 04017006. https://doi.org/10.1061/JSWBAY.0000821

Jaouhari, Y., Travaglia, F., Giovannelli, L., Picco, A., Oz, E., Oz, F., & Bordiga, M. (2023). From industrial food waste to bioactive ingredients: A review on the sustainable management and transformation of plant-derived food waste. *Foods, 12*(11), 2183. https://doi.org/10.3390/foods12112183

Jiménez-Rosado, M., Perez-Puyana, V., & Romero, A. (2023). The use of biowaste for the production of biodegradable superabsorbent materials. *Current Opinion in Food Science, 49*, Article 100975. https://doi.org/10.1016/j.cofs.2022.100975

Júnior, C. D. C., Pimenta, A. S., Souza, E. C. de, Pereira, A. K. S., & Júnior, A. F. D. (2022). Uso agrícola e florestal do biochar: Estado da arte e futuras pesquisas. *Research, Society and Development, 11*(2), Article 2. https://doi.org/10.33448/rsd-v11i2.25999

Kumar, M., Shukla, S. K., Upadhyay, S. N., & Mishra, P. K. (2020). Analysis of thermal degradation of banana (*Musa balbisiana*) trunk biomass waste using iso-conversional models. *Bioresource Technology, 310*, Article 123393. https://doi.org/10.1016/j.biortech.2020.123393

Kwon, G., Tsang, D. C. W., Oh, J.-I., Kwon, E. E., & Song, H. (2019). Pyrolysis of aquatic carbohydrates using CO_2 as reactive gas medium: A case study of chitin. *Energy, 177*(C), 136–143.

Li, Y., Xing, B., Ding, Y., Han, X., & Wang, S. (2020). A critical review of the production and advanced utilization of biochar via selective pyrolysis of lignocellulosic biomass. *Bioresource Technology, 312*, Article 123614. https://doi.org/10.1016/j.biortech.2020.123614

Liu, J., Zhang, W., Jin, H., Li, Z., Liu, G., Xing, F., & Tang, L. (2023). Exploring the carbon capture and sequestration performance of biochar-artificial aggregate using a new method. *Science of the Total Environment, 859*, Article 160423. https://doi.org/10.1016/j.scitotenv.2022.160423

Lu, H., Zhang, W., Wang, S., Zhuang, L., Yang, Y., & Qiu, R. (2013). Characterization of sewage sludge-derived biochars from different feedstocks and pyrolysis temperatures. *Journal of Analytical and Applied Pyrolysis, 102*, 137–143. https://doi.org/10.1016/j.jaap.2013.03.004

Margal, P. B., Gaikwad, A. S., & Titirmare, N. S. (2023). *Carbon sequestration.* Zenodo. https://doi.org/10.5281/ZENODO.8331178

Masoumi, S., Borugadda, V. B., Nanda, S., & Dalai, A. K. (2021). Hydrochar: A review on its production technologies and applications. *Catalysts, 11*(8), Article 8. https://doi.org/10.3390/catal11080939

Mian, M. M., Alam, N., Ahommed, M. S., He, Z., & Ni, Y. (2022). Emerging applications of sludge biochar-based catalysts for environmental remediation and energy storage: A review. *Journal of Cleaner Production, 360*, Article 132131. https://doi.org/10.1016/j.jclepro.2022.132131

Mishra, S., Ghosh, S., van Hullebusch, E. D., Singh, S., & Das, A. P. (2023). A critical review on the recovery of base and critical elements from electronic waste-contaminated streams using microbial biotechnology. *Applied Biochemistry and Biotechnology, 2023*, 1–30. https://doi.org/10.1007/S12010-023-04440-X

Mohanty, S., Ghosh, S., Bal, B., & Das, A. P. (2018). A review of biotechnology processes applied for manganese recovery from wastes. *Reviews in Environmental Science and Bio/Technology, 17*(4), 791–811. https://doi.org/10.1007/S11157-018-9482-1

Novotný, M., Marković, M., Raček, J., Šipka, M., Chorazy, T., Tošić, I., & Hlavínek, P. (2023). The use of biochar made from biomass and biosolids as a substrate for green infrastructure: A review. *Sustainable Chemistry and Pharmacy, 32*, Article 100999. https://doi.org/10.1016/j.scp.2023.100999

Nzediegwu, C., Arshad, M., Ulah, A., Naeth, M. A., & Chang, S. X. (2021). Fuel, thermal and surface properties of microwave-pyrolyzed biochars depend on feedstock type and pyrolysis temperature. *Bioresource Technology, 320*, Article 124282. https://doi.org/10.1016/j.biortech.2020.124282

Ogunwumi, O. T., Salami, L., Ogunwumi, O. T., & Salami, L. (2023). Perspective chapter: Industrial waste landfills. In *Solid Waste and Landfills Management—Recent Advances.* IntechOpen. https://doi.org/10.5772/intechopen.108787

Palansooriya, K. N., Ok, Y. S., Awad, Y. M., Lee, S. S., Sung, J.-K., Koutsospyros, A., & Moon, D. H. (2019). Impacts of biochar application on upland agriculture: A review. *Journal of Environmental Management, 234*, 52–64. https://doi.org/10.1016/j.jenvman.2018.12.085

Parikka, M. (2004). Global biomass fuel resources. *Biomass and Bioenergy, 27*(6), 613–620. https://doi.org/10.1016/j.biombioe.2003.07.005

Peng, X., Jiang, Y., Chen, Z., Osman, A. I., Farghali, M., Rooney, D. W., & Yap, P.-S. (2023). Recycling municipal, agricultural and industrial waste into energy, fertilizers, food and construction materials, and economic feasibility: A review. *Environmental Chemistry Letters, 21*(2), 765–801. https://doi.org/10.1007/s10311-022-01551-5

Pradhan, G., Tripathy, B., Ram, D. K., Digal, A. K., & Das, A. P. (2023). Bauxite mining waste pollution and its sustainable management through bioremediation. *Geomicrobiology Journal, 1–10*. https://doi.org/10.1080/01490451.2023.2235353

Pradhan, S., Parthasarathy, P., Mackey, H. R., Al-Ansari, T., & McKay, G. (2024). Food waste biochar: A sustainable solution for agriculture application and soil–water remediation. *Carbon Research, 3*(1), 41. https://doi.org/10.1007/s44246-024-00123-2

Ray, S. K., Dey, S., Das, A. P., & John, A. (2024). Microplastic waste management through greener technologies: A review. *Sustainable Management of Urban Plastic Waste through Circular Economic Approaches, 237–261*. https://doi.org/10.1201/9781003465447-14

Romero, L., Weng, S., Oulego, P., Collado, S., Marcet, I., & Díaz, M. (2024). Hydrolyzed sewage sludge as raw bio-based material for hermetic bag production. *Waste Management, 174*, 31–43. https://doi.org/10.1016/j.wasman.2023.11.022

Salama, Y., Chennaoui, M., Amraoui, M. E., & Mountadar, M. (2017). A review of compost produced from biological wastes: Sugarcane industry waste. *International Journal of Food Science and Biotechnology, 1*(1), Article 1. https://doi.org/10.11648/j.ijfsb.20160101.14

Salcedo-Puerto, O., Breton, M., Sermyagina, E., Gonçalves, L., Cardoso, M., Havukainen, J., Vakkilainen, E., & Mendoza-Martinez, C. (2024). Life cycle assessment of alternative pulp mill sludge treatment methods in Finland. *Waste Management (New York, N.Y.), 186*, 236–248. https://doi.org/10.1016/j.wasman.2024.06.025

Škorjanc, A., Goričanec, D., & Urbancl, D. (2024). Assessing energy potential and chemical composition of food waste thermodynamic conversion products: A literature review. *Energies, 17*(8), 1897. https://doi.org/10.3390/en17081897

Wang, D., Jiang, P., Zhang, H., & Yuan, W. (2020). Biochar production and applications in agro and forestry systems: A review. *Science of the Total Environment, 723*, Article 137775. https://doi.org/10.1016/j.scitotenv.2020.137775

Wang, X., Wei-Chung Chang, V., Li, Z., Song, Y., Li, C., & Wang, Y. (2022). Co-pyrolysis of sewage sludge and food waste digestate to synergistically improve biochar characteristics and heavy metals immobilization. *Waste Management, 141*, 231–239. https://doi.org/10.1016/j.wasman.2022.02.001

Wang, Z., Tian, Q., Guo, J., Wu, R., Zhu, H., & Zhang, H. (2021). Co-pyrolysis of sewage sludge/cotton stalks with K_2CO_3 for biochar production: Improved biochar porosity and reduced heavy metal leaching. *Waste Management, 135*, 199–207. https://doi.org/10.1016/j.wasman.2021.08.042

Yaashikaa, P. R., Kumar, P. S., Varjani, S., & Saravanan, A. (2020). A critical review on the biochar production techniques, characterization, stability and applications for circular bioeconomy. *Biotechnology Reports, 28*, Article e00570. https://doi.org/10.1016/j.btre.2020.e00570

Zhou, Y., Gao, J., Yang, X., Ni, H., Qi, J., Zhu, Z., Yang, Y., Fang, D., Zhou, L., & Li, J. (2024). Recent progress in sludge-derived biochar and its role in wastewater purification. *Sustainability, 16*(12), 5012. https://doi.org/10.3390/su16125012

Synergizing AI, Energy Recovery, and Sustainability in Solid Waste Management

Ankita Das, Satya Ranjan Dash, and Shikta Singh

1 Introduction

Imagine a world where every waste is just another resource—with solid waste generation expected to rise to 3.4 billion tons globally by 2050, this vision is no longer a dream, it is a necessity. Growing population, increased consumption, and rapid urbanization have resulted in massive waste production that damages the environment, raises costs, and causes social issues (Priyadarsini & Das, 2024, 2025). The old "take-make-dispose" waste management model is no longer environmentally friendly or sustainable, so we need to find a way to balance social, economic, and environmental demands (World Bank, 2018a).

Improper waste disposal currently accounts for 5% of all greenhouse gas emissions, mostly due to methane from landfills. Instead of being disposed of under control, over 90% of waste in developing nations is either burned or dumped in the open (World Bank, 2018b).

Open burning consequently emits toxic pollutants, contributing to air pollution. Similarly, open dumping is a major problem, with plastic making up 12% of the worldss solid waste. Every year, over 242 million tons of plastic end up in the ocean, polluting marine life. The damage to the environment will worsen as waste generation rises, endangering wildlife and accelerating climate change (Dey et al., 2023; Mishra et al., 2023).

A. Das · S. R. Dash
School of Computer Applications, KIIT University, Bhubaneswar, India
e-mail: dankita312@gmail.com

S. R. Dash
e-mail: sdashfca@kiit.ac.in

S. Singh (✉)
KIIT School of Management, KIIT University, Bhubaneswar, India
e-mail: shiktasingh@ksom.ac.in

A. P. Das et al. (eds.), *Sustainable Municipal Waste Management with a Zero Waste Approach*, Sustainable Environmental Waste Management Strategies,
https://doi.org/10.1007/978-3-032-05950-5_18

Additionally, solid waste management reinforces underlying social injustices and disparities, particularly in marginalized communities that are susceptible to various difficulties (Das et al., 2024; Mishra et al., 2019b). It is appalling that nearly 3 billion people across the world lack basic waste collection facilities. This, as a result, affects rural communities and the poorer urban settlements. Unsound waste management is also accountable for affecting public health and hampering sustainable development efforts by severely limiting economic opportunities among the affected population.

The economic cost of poor waste management is also daunting. In 2020, the direct costs of waste management worldwide were estimated to be USD 252 billion. However, the shadow costs, such as pollution-related health effects and climate change, pushed the estimate to USD 361 billion. If poor waste management practices continue, then these costs might almost double to USD 640 billion per annum by 2050 (UNEP Report, 2024).

To counter these issues, new technologies present revolutionary solutions. Artificial Intelligence (AI) can be utilized for developing smart waste management systems that maximize resource efficiency and minimize the extent of damage caused to nature. AI systems can change traditional waste management approaches through efficient collection routes, increasing the precision of waste sorting and forecasting the amount of waste likely to be generated with 90% accuracy. It can also help lower operational costs, reduce carbon footprints, and increase efficiency in overall waste management operations. The time is now; we must shift to a circular economy where waste is not only redefined as a resource but also guarantees a sustainable tomorrow for all.

Understanding Traditional Solid Waste Management

Collection, transportation, sorting, recycling, and disposal are some of the interconnected processes that make up standard solid waste management (Das & Ghosh, 2022; Mohanty et al., 2018). The traditional systems of manual labor and antiquated procedures result in hazardous environmental conditions and ineffective operations. Conventional waste disposal entails collecting waste from residences and commercial buildings and transporting it to landfills, recycling centers, or incinerators (Bala et al., 2024; Das & Ghosh, 2018; Tripathy et al., 2024). The primary causes of failure of these traditional methods are the limited recovery of reusable resources, contaminated recyclable materials, and insufficient waste segregation at the source.

Waste management is still reactive rather than proactive in many places, particularly in low- to middle-income nations. For example, due to inadequate infrastructure and a lack of knowledge about appropriate waste disposal, waste is still frequently burned or dumped in the open (Ghosh & Das, 2017; Ray et al., 2024; Sahu et al., 2024). These outdated methods eliminate resource recovery and further pollute the environment. Only about 13.5% of the total recyclable municipal solid waste gets recycled due to poor sorting and low investment in technology.

The lack of real-time data is another reason why traditional waste management systems don't work well. The majority of waste collection routes are static and ineffective, which results in excessive fuel consumption and higher operating expenses. Because of this, waste collection in crowded cities is frequently neglected, resulting in unsanitary conditions and a higher risk of disease transmission by pests. The absence of coordinated systems that can monitor waste generation patterns further limits the capacity to advance effective long-term solutions.

The Challenge of Rising Waste Volume in Traditional Solid Waste Management

The rapid increase in the amount of waste is one of the most challenging issues in traditional solid waste management systems. Global waste production is expected to increase from 2.1 billion tons per year in 2020 to an estimated 3.4 billion tons by 2050. Waste production is expected to triple in developing nations like South Asia and sub-Saharan Africa during the same time (Trends in Solid Waste, 2024).

In many regions, landfills are a widely used waste disposal method, but growing waste generation is causing them to run out of space rapidly (Prabhakar et al., 2019; Pradhan et al., 2023). In addition to consuming valuable land, this method of disposing of waste releases greenhouse gases or heat like methane (CH4), which can warm the atmosphere 84 times more quickly than carbon dioxide and impact climate change. Shockingly, landfills contribute to 15.1% of global methane emissions, significantly impacting climate change (Salem et al., 2023).

In urban areas with limited landfill space, incineration, a popular waste removal technique, is used to cut down waste production by up to 90%. However, this approach comes with its risks. Modern incinerators still produce residual ash, which is disposed in landfills even though they have sophisticated systems in place to control emissions like carbon dioxide (CO_2) and other pollutants.

The financial costs of conventional waste management are also high. The estimated global direct costs of municipal solid waste management in 2020 was USD 252 billion, nearly doubling in the years to follow (excluding hidden costs like health effects from pollution and climate change). These costs will place immense financial burdens on governments and municipalities.

Besides, the increasing volume of plastic waste production further adds up to these challenges. Plastic waste products represent roughly 12% of the world's solid waste (Dey et al., 2024; Mishra et al., 2020; Paul et al., 2024). By 2015 alone, more than 242 million tons of plastic were dumped in marine bodies, endangering marine ecosystems and overall biodiversity. The main causes of the plastic crisis are inadequate recycling infrastructure and inadequate waste segregation at the source, making prevention alone an inadequate solution (Mallick et al., 2024; Mishra et al., 2022; Mishra et al., 2019a).

Limitations of Traditional Solid Waste Management

Traditional solid waste management systems struggle with rising waste and environmental risks. The open dumping technique pollutes soil and water in many developing nations and draws disease-carrying rodents and mosquitoes. On the other hand, burning waste materials pollutes the air with harmful substances like furans and dioxins, endangering the health of the communities around it (Ferronato & Torretta, 2019).

Sanitary landfilling takes up a lot of land and requires careful leachate management to avoid harming the ecosystem, which are disadvantages of even the most sophisticated techniques. While technologies for landfill gas recovery recover methane for energy generation, their use is limited in many parts of the world where implementation requires high capital costs.

The incineration facilities suffer the same financial constraints because starting one and running it require a large capital outlay, thus rendering such technologies accessible only to wealthier nations or urban centers with high populations. Furthermore, during their operation, incinerators are blamed for relying on nonrenewable energy sources and for contributing to greenhouse gas emissions despite the technologies in place for the control of emissions.

A Call for Modernization

Even though the traditional solid waste management system has been a successful urban sanitation solution for many years, it is currently having trouble handling complex materials like e-waste, industrial by-products, and growing waste volumes. This system is inefficient and unreliable for sustainable waste disposal, necessitating a shift to advanced solutions.

To overcome these challenges and set cleaner prospects for tomorrow, it is necessary to develop sustainable practices. Cutting-edge recycling techniques like chemical recycling, pyrolysis, IoT-enabled smart bins, and AI-driven waste management are some innovative approaches that contribute to resource recovery and efficiency gains while lessening their negative effects on the environment.

The Role of Artificial Intelligence (AI) in Transforming Solid Waste Management

AI has become a game-changer in every sector, holding an open-ended promise in the arena of modern waste handling, collection, processing, and recycling. An AI waste management system is capable of processing large amounts of data, gaining

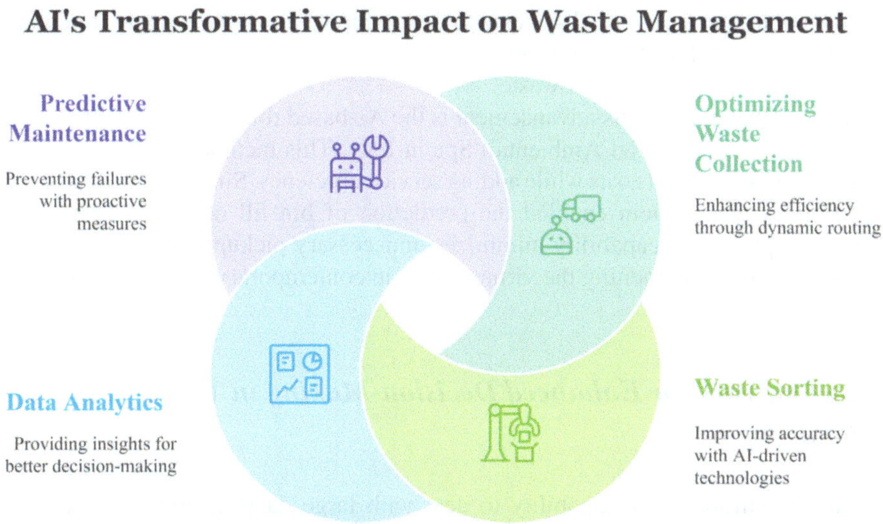

Fig. 1 This figure shows how AI transforms waste management by facilitating data-driven insights, advanced waste sorting for a sustainable future, predictive maintenance, and efficient collection

predictive insights, and automating processes that traditional systems cannot comprehend. It uses machine learning algorithms along with robotics and data analytics to optimize operations, eliminate inefficiencies, and improve sustainability.

As shown in Fig. 1, there are various ways of using AI in waste management. Among these, predictive maintenance stands out as the most promising application in solid waste management. This application enables the analysis of the data retrieved from equipment to prevent equipment failure and minimize operator maintenance expenses and downtime. Moreover, AI systems enhance operational efficiency by improving sorting accuracy and streamlining waste collection routes. These not only lead to cost reduction but also eliminate adverse environmental impacts, thus making waste management systems sustainable and efficient (Fang et al., 2023).

2 Optimizing Waste Collection and Waste Sorting

The traditional static routes of waste collection often lead to unnecessary trips and delays, making the process inefficient. AI is revolutionizing waste collection through dynamic route optimization and waste sorting. The dynamic route optimization of waste collection can be facilitated by the input of real-time information, such as traffic conditions, bin fill levels, weather forecasts, etc., into AI-powered algorithms (Olawade et al., 2024). Additionally, IoT-enabled smart bins fitted with fill-level sensors communicate with AI systems to determine when waste is ready for collection. This helps reduce fuel consumption, lower operational costs, reduce emissions, and provide timely removal of waste.

A prime example of this advancement is the AI-based route optimization method developed by Alia Servizi Ambientali Spa in Italy. This method resulted in a 30% decrease in fuel-related costs while adding service efficiency. Similarly, in Barcelona, an IoT-integrated system enabled the prediction of bin fill rates by a whopping 92%. This predictive capability minimizes unnecessary pickups, further enhancing operations and strengthening the virtues of AI in contemporary waste management.

Data Analytics for Enhanced Decision-Making in Waste Management

AI has an extraordinary capability to deal with large data sets that can generate insights regarding waste generation and disposal behaviors. The system also records the analysis of previous data with real-time input, which helps AI spot trends to guide decision-making in waste management organizations.

For example:

- **Identifying High-Waste Areas**: AI can pinpoint waste-producing regions, enabling targeted interventions such as public awareness campaigns or infrastructure upgrades.
- **Monitoring Recycling Program Success**: Artificial intelligence (AI) examines the composition of waste collected to find barriers to effective recycling, such as low participation rates or contamination, and suggests ways to overcome them.

Such information can help municipalities dynamically modify their strategies. Platforms like Pello provide real-time monitoring that instantaneously communicates via an easy-to-use dashboard update on the bin capacity. Such tools empower the operators to prevent overflow or contamination while optimizing resource allocation. Besides, AI has played a major role in facilitating circular economy goals by tracking the lifecycle of materials, which ensures maximum recovery of valuable resources with minimum dependency on landfills.

Predictive Maintenance and Operational Efficiency in Waste Management

AI-driven predictive maintenance has significantly increased the effectiveness of waste management by preventing unplanned equipment failures. Traditional waste collection vehicles and processing machinery experience wear and tear because of heavy use and exposure to corrosive substances. Reactive maintenance plans have also been observed to result in unexpected malfunctions and higher expenses. By integrating AI functionalities, waste management systems can now predict potential equipment problems ahead of time.

Artificial intelligence (AI) algorithms can monitor performance indicators such as vibration, patterns, or temperature changes to predict equipment issues before they lead to failures. This proactive approach allows for the preplanning of repairs, extending the life of equipment and cutting downtime by as much as 20%.

For example, self-diagnostic robotic sorting systems can identify equipment issues early on to guarantee seamless and continuous operation. These predictive analytics would help to combine dependability with lower operational costs related to emergencies.

3 Energy Recovery from Waste Through Artificial Intelligence (AI)

Artificial intelligence (AI) algorithms enable real-time waste composition analysis using sophisticated methods like Raman spectroscopy and laser-induced breakdown spectroscopy (LIBS). These methods quickly characterize the MSW feedstock in terms of its ash fusion temperature, moisture content, and calorific value. On top of that, AI systems can automatically modify combustion parameters, including temperature, oxygen levels, and feed rates, by integrating these data points with machine learning models. This allows the automated systems to optimize energy yield while minimizing harmful emissions such as CO_2, NO_x, and others.

Optimization of Waste-To-Energy Conversion

Waste-to-energy (WTE) technologies address both the generation of energy and the disposal of solid waste. With this technology, nonrecyclable waste can be converted into reusable resources such as electricity, heat, and biofuels. However, the diverse character of municipal solid waste (MSW) has historically made WTE procedures ineffective. Today, AI is enhancing these systems by adhering to the many steps shown below in Fig. 2. The AI-integrated WTE conversion process aims to increase energy recovery, streamline operations, and lessen the environmental effect of waste disposal.

Copenhagen's Amager Bakke (CopenHill) is an excellent example of how AI-powered incinerators improve waste-to-energy efficiency. With its AI-powered combustion system, energy extraction has increased, and emissions have decreased by more than 30% since 2023. This cutting-edge technology not only reduces its negative effects on the environment but also produces enough energy each year to heat 15,0000 homes, demonstrating a sustainable waste management future.

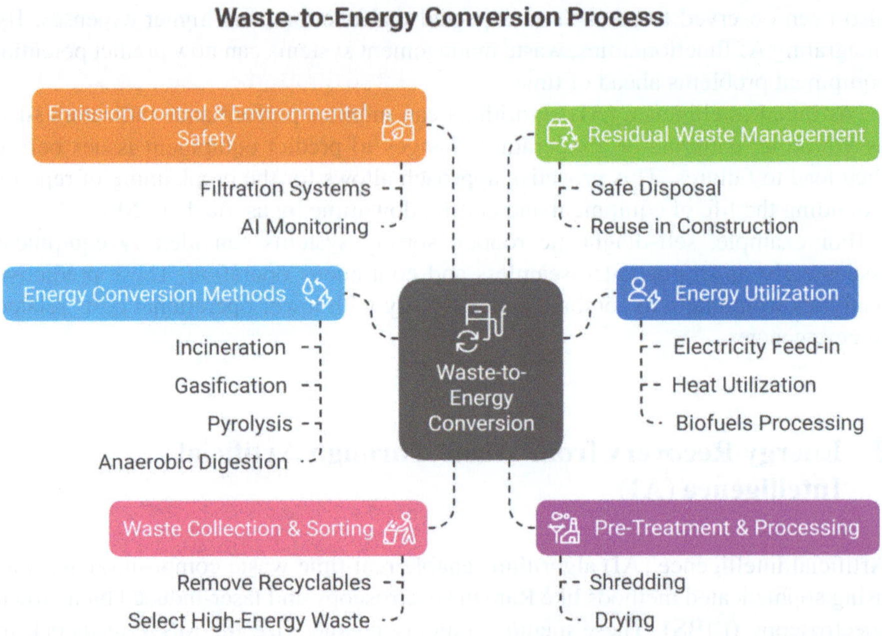

Fig. 2 This diagram illustrates the overview of the waste-to-energy conversion process. The processes of turning waste into energy, such as sorting treatment, energy production, and environmental safety precautions

Aside from incineration, AI significantly increases the effectiveness of pertinent decision-making systems, making it possible to determine which WTE technologies are most appropriate for various kinds of waste. For example, AI systems can use the input waste data to decide whether pyrolysis, gasification, or anaerobic digestion will extract the most energy at the lowest environmental cost. This optimization supports the sustainability objectives while increasing the plant's efficiency.

AI-Driven Anaerobic Digestion Processes

Anaerobic digestion is a process in which microorganisms convert organic matter, including food wastes and agricultural residues, into a renewable energy source called biogas, a mixture of methane (CH_4) and carbon dioxide (CO_2) (Cruz et al., 2022). While anaerobic digestion has provided renewable energy generation for a long time, it is one of the processes whose efficiency is quite strongly influenced by the maintenance of optimum parameters in the digester. In this case, AI comes in to assist in monitoring and optimizing the parameters such as temperature inside the digester, pH levels, retention time, and activity of microbes. Figure 3 illustrates

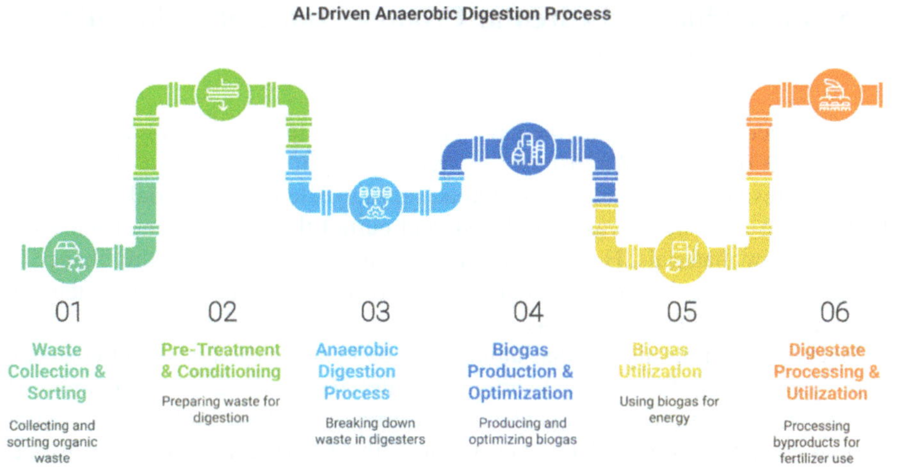

Fig. 3 This diagram illustrates the entire process of turning organic waste into biogas through AI-driven anaerobic digestion, including waste collection, digestate processing, and energy use

the entire process of turning organic waste into biogas through AI-driven anaerobic digestion, including waste collection, digestate processing, and energy use.

AI-based sensor systems interfaced with IoT devices provide real-time data for the anaerobic digesters. The data are processed through machine learning algorithms for pattern recognition and prediction of organic waste input variations (AI-Powered Waste Management, 2023). Changes in the quality or composition of input feedstock could promote imbalances in microbial ecosystems within the digester, hence compromising biogas yield. By predicting fluctuations in the input feedstock, operators are now able to modify feedstock ratios or include other materials to stabilize microbial activity.

The Swedish AI-driven anaerobic digestion plants have performed quite successfully, producing almost 25% more biogas with the kind of predictive intelligence described herein. In addition to boosting methane yields, the system predicts maintenance requirements for critical equipment—such as pumps and agitators—thus limiting downtime. The extra biogas is ultimately converted to electricity and renewable natural gas (RNG) for transportation or heating purposes.

The advantages of AI-enabled anaerobic digestion processes toward circular economy objectives are further amplified by converting organic waste into valuable by-products like digestate, which serves as an organic fertilizer supplement. Thus, this dual effect of energy recovery and resource regeneration represents the potent impact of AI on sustainable waste management.

4 Sustainability and Circular Economy Integration in Solid Waste Management

A circular economy can be defined as a new approach to transcending the age-long linear model of "take-make-dispose" by transforming it into a system of recycling, reuse, and resource efficiency. This approach aims to extend the material's longevity, diminish the amount of waste produced, and curtail the environmental impact of extracting raw materials. Here, AI is a powerful tool that accelerates this change by optimizing resource use, improving recycling processes, and promoting sustainable practices across industries (Fig. 4).

Machine learning algorithms, along with computer vision, can automate the identification, classification, and sorting of materials in mixed waste streams with 72.8–99.95% accuracy (Alsabt et al., 2024) Currently, such technologies are setting grounds for application in MRFs where robots with AI can recognize material distinction based on their type, color, and even brand. This approach helps eliminate contamination and ensures high-quality raw materials that can replace traditional ones.

Additionally, AI-based platforms are also making progress toward increased transparency and accountability in waste management. These systems can be used across industries to make better resource allocation and recycling strategy decisions through real-time analytical results. AI waste monitoring systems track materials from production to disposal, ensuring resources stay in a closed loop. This reduces the reliance on raw materials and minimizes the environmental impact.

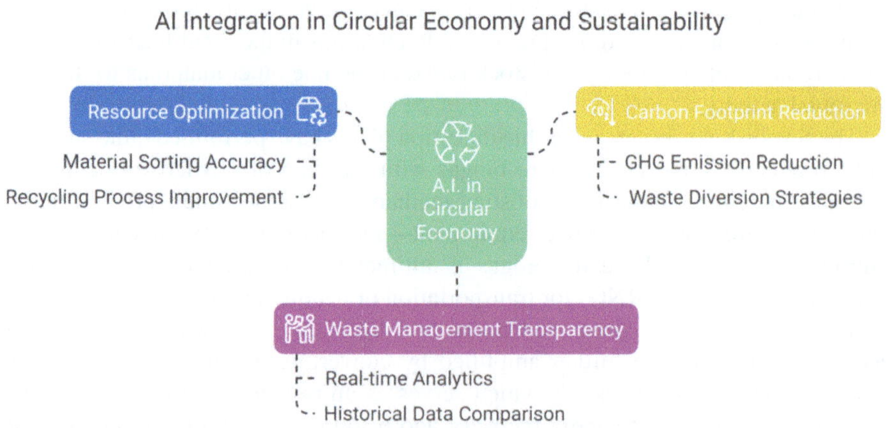

Fig. 4 This diagram illustrates how AI improves resource optimization, recycling accuracy, emission reduction, and waste management transparency

Reducing Carbon Footprint Through AI Applications

The role of AI in sustainability is not limited solely to the recovery of materials; it also plays a significant role in minimizing GHG emissions produced during conventional waste management processes. With the introduction of AI into waste management systems, one can greatly diminish landfill dependency through smarter diversion strategies.

Forecasting models based on AI can analyze historical data and real-time input on population growth, consumption patterns, etc., to provide insights into waste generation trends. These predictive mechanisms enable policymakers and waste management companies to work on preventive measures that can limit excess waste in landfills. Cities such as Seoul use AI-based forecasting systems to optimize waste collection schedules and redirect organic waste for composting or anaerobic digestion facilities instead of landfilling.

Besides waste diversion, AI also applies in supply chains to promote sustainability by reducing waste production right at the manufacturing stage. It leverages machine learning algorithms to analyze production processes and identify inefficiencies or risks of overproduction. This prevents surplus material from entering disposal systems, saving energy and resources.

AI systems play a pivotal role in recycling waste products by improving material recovery and reducing contamination levels of recyclables. This consequently helps prevent mixed recyclables from getting rejected due to contamination. For example, the vision AI technology implemented by AMCS detects contamination in real time, resulting in cleaner recycling streams and reducing energy-intensive reprocessing. Figure 5 illustrates the way AI improves waste sorting, maximizes energy recovery, forecasts waste generation, and increases public participation to lower carbon emissions.

5 Conclusion

AI in solid waste management is transforming waste from a problem into an opportunity for value creation and sustainability. By integrating AI with energy recovery approaches such as pyrolysis and anaerobic digestion, waste management can become more efficient while reducing environmental impact and boosting economic habits. These systems optimize collection schedules in real time, cutting fuel use and emissions. Meanwhile, the machine learning robotic sorters achieve nearly 100% accuracy in separating recyclables to reduce contamination and maximize resource recovery. Together, these innovations showcase a transformed shift toward more efficient and environmentally friendly solid waste management.

Beyond operational efficiency, AI waste systems track material lifecycles through predictive analytics to promote product reusability and lower the dependency on raw resources. The analytics also empower policymakers to take timely action

Fig. 5 This diagram illustrates the way AI improves waste sorting, maximizes energy recovery, forecasts waste generation, and increases public participation to lower carbon emissions

to minimize waste. Additionally, the integration of AI in blockchain technology ensures transparency in recycling, improving resource efficiency and public trust in sustainable waste management practices.

Despite the many benefits of AI in solid waste management, there are still barriers to its widespread adoption. High costs, technical complexities, lack of expertise, and data privacy concerns are some of the challenges that slow its widespread use (Adewuyi et al., 2024). Nevertheless, as technology advances and organizations invest in training and infrastructure, these barriers are expected to diminish over-the-time.

Integrating artificial intelligence (AI) in solid waste management is more than just technological progress—it's a significant step toward achieving sustainability at large. By integrating AI with energy recovery and circular economy principles, societies can produce valuable resources while minimizing environmental harm. As innovation and collaboration grow, AI will play a key role in building a cleaner and greener future for generations to come.

References

Adewuyi, A. Y., Adebayo, K. B., Adebayo, D., Kalinzi, J. M., Ugiagbe, U. O., Ogunruku, O. O., Oladipo, O. F., & Adeniyi, S. A. (2024). Application of big data analytics to forecast future waste trends and inform sustainable planning. *World Journal of Advanced Research and Reviews, 23*(1), 2469–2479.

Alsabt, R., Alkhaldi, W., Adenle, Y. A., & Alshuwaikhat, H. M. (2024). Optimizing waste management strategies through artificial intelligence and machine learning—An economic and environmental impact study. *Cleaner Waste Systems, 8*, Article 100158.

Bala, K., Anshul, Verma, S., Dogra, A., Das, A. P., Kumar, R., & Nandini. (2024). Sustainable management for urban plastic waste generation. *Sustainable Management of Urban Plastic Waste Through Circular Economic Approaches*, 211–236. https://doi.org/10.1201/9781003465447-13

Cruz, I. A., Chuenchart, W., Long, F., Surendra, K. C., Andrade, L. R. S., Bilal, M., Liu, H., & Ferreira, L. F. R. (2022). Application of machine learning in anaerobic digestion: Perspectives and challenges. *Bioresource Technology, 345*, Article 126433.

Das, A. P., & Ghosh, S. (2018). Bioleaching of manganese from mining waste materials. *Materials Today: Proceedings, 5*(1), 2381–2390. https://doi.org/10.1016/J.MATPR.2017.11.459

Das, A. P., & Ghosh, S. (2022). Role of microorganisms in extenuation of mining and industrial wastes. *Geomicrobiology Journal, 39*, 173–175. https://doi.org/10.1080/01490451.2022.2038953

Das, A., Ray, S. K., Mohanty, M., Mohanty, J., Dey, S., & Das, A. P. (2024). Ecotoxicity of microplastic wastes and their sustainable management: A review. *Environmental Chemistry and Ecotoxicology, 6*, 144–152. https://doi.org/10.1016/J.ENCECO.2024.05.003

Dey, S., Bissoyi, A., Koteswara Reddy, G., & Das, A. P. (2024). Synthetic microfibers: An urban emerging pollutant demanding for sustainable environmental solutions. *Sustainable Microbial Technology for Synthetic and Cellulosic Microfiber Bioremediation*, 107–121. https://doi.org/10.1007/978-3-031-62600-5_7

Dey, S., Das, A., Mallick, K., Sahu, A., & Das, A. P. (2023). *Environmental Petroleum Waste: Pollution, Toxicity, Sustainable Remediation*. 159–175. https://doi.org/10.1007/978-3-031-48220-5_7

Fang, B., Yu, J., Chen, Z., Osman, A. I., Farghali, M., Ihara, I., Hamza, E. H., Al-Fatesh, A. S., & Yap, P. S. (2023). Artificial intelligence for waste management in smart cities: A review. *Environmental Chemistry Letters, 21*(4), 1959–1989. https://doi.org/10.1007/s10311-023-01604-3

Ferronato, N., & Torretta, V. (2019). Waste mismanagement in developing countries: A review of global issues. *International Journal of Environmental Research and Public Health, 16*(6), 1060. https://doi.org/10.3390/ijerph16061060

Ghosh, S., & Das, A. P. (2017). Bioleaching of manganese from mining waste residues using Acinetobacter sp. *Geology, Ecology, and Landscapes, 1*(2), 77–83. https://doi.org/10.1080/24749508.2017.1332847

International Institute for Sustainable Development. (n.d.). UNEP report calls for shift to zero waste to prevent runaway negative impacts. https://sdg.iisd.org/news/unep-report-calls-for-shift-to-zero-waste-to-prevent-runaway-negative-impacts

Mallick, K., Sahu, S., Sahu, A., Dey, S., & Das, A. P. (2024). A critical review of marine microfiber pollution routes, toxicity, and its sustainable remediation. *Environmental Science and Engineering, Part, F2333*, 189–211. https://doi.org/10.1007/978-3-031-51792-1_10

Mishra, S., Dash, D., & Das, A. P. (2022). Detection, characterization and possible biofragmentation of synthetic microfibers released from domestic laundering wastewater as an emerging source of marine pollution. *Marine Pollution Bulletin, 185*, Article 114254. https://doi.org/10.1016/J.MARPOLBUL.2022.114254

Mishra, S., Ghosh, S., van Hullebusch, E. D., Singh, S., & Das, A. P. (2023). A critical review on the recovery of base and critical elements from electronic waste-contaminated streams using microbial biotechnology. *Applied Biochemistry and Biotechnology, 2023*, 1–30. https://doi.org/10.1007/S12010-023-04440-X

Mishra, S., Rath, C. charan, & Das, A. P. (2019a). Marine microfiber pollution: A review on present status and future challenges. *Marine Pollution Bulletin, 140*, 188–197. https://doi.org/10.1016/J.MARPOLBUL.2019.01.039

Mishra, S., Rout, P. K., & Das, A. P. (2019b). Solar photovoltaic panels as next generation waste: A review. *Biointerface Research in Applied Chemistry, 9*(6).

Mishra, S., Singh, R. P., Rath, C. C., & Das, A. P. (2020). Synthetic microfibers: Source, transport and their remediation. *Journal of Water Process Engineering, 38*, Article 101612. https://doi.org/10.1016/J.JWPE.2020.101612

Mohanty, S., Ghosh, S., Bal, B., & Das, A. P. (2018). A review of biotechnology processes applied for manganese recovery from wastes. *Reviews in Environmental Science and Bio/Technology, 17*(4), 791–811. https://doi.org/10.1007/S11157-018-9482-1

North Carolina State University College of Natural Resources. (2023, November). AI-powered waste management system to revolutionize recycling. https://cnr.ncsu.edu/news/2023/11/ai-waste-management

Olawade, D. B., Fapohunda, O., Wada, O. Z., Usman, S. O., Ige, A. O., Ajisafe, O., & Oladapo, B. I. (2024). Smart waste management: A paradigm shift enabled by artificial intelligence. Waste Management Bulletin.

Paul, A., Priyadarsini, S., Karakuş, S., Behera, J. K., Gnanasekaran, R., & Das, A. P. (2024). Fate and accumulation of synthetic microfibers in the environment. *Sustainable Microbial Technology for Synthetic and Cellulosic Microfiber Bioremediation*, 61–76. https://doi.org/10.1007/978-3-031-62600-5_4

Prabhakar, A., Mishra, S., & Das, A. P. (2019). Isolation and identification of lead (Pb) solubilizing bacteria from automobile waste and its potential for recovery of lead from end of life waste batteries. *Geomicrobiological Journal, 36*(10), 894–903. https://doi.org/10.1080/01490451.2019.1654044

Pradhan, G., Tripathy, B., Ram, D. K., Digal, A. K., & Das, A. P. (2023). Bauxite mining waste pollution and its sustainable management through bioremediation. *Geomicrobiology Journal*, 1–10. https://doi.org/10.1080/01490451.2023.2235353

Priyadarsini, S., & Das, A. P. (2024). Analytical and structural characterization of waste lithium-ion batteries for their effective recycling strategy. *Environmental Chemistry and Ecotoxicology.* https://doi.org/10.1016/j.enceco.2024.12.004

Priyadarsini, S., & Das, A. P. (2025). Lithium bioleaching: A review on microbial-assisted sustainable technology for lithium bio-circularity. *Journal of Water Process Engineering, 69*, Article 106744. https://doi.org/10.1016/j.jwpe.2024.106744

Ray, S. K., Dey, S., Das, A. P., & John, A. (2024). Microplastic waste management through greener technologies : A review. *Sustainable Management of Urban Plastic Waste Through Circular Economic Approaches*, 237–261. https://doi.org/10.1201/9781003465447-14

Sahu, A., Dey, S., Mallick, K., Priyadarsini, S., Behera, I. D., & Das, A. P. (2024). A review on advanced technology for sustainable management of synthetic microplastic waste. *Sustainable Management of Urban Plastic Waste Through Circular Economic Approaches*, 262–279. https://doi.org/10.1201/9781003465447-15

Salem, K. S., Clayson, K., Salas, M., Haque, N., Rao, R., Agate, S., Lucia, L., & Pal, L. (2023). A critical review of existing and emerging technologies and systems to optimize solid waste management for feedstocks and energy conversion. *Matter, 6*(10), 3348–3377.

Tripathy, B., Dey, S., Dubey, N. K., Lo, W.-C., & Das, A. P. (2024). *An Omics-Based Approach for Managing the Risk of COVID-19-Generated Wastes*. 51–66. https://doi.org/10.1007/978-3-031-50840

World Bank. (2018, September 20). Global waste to grow by 70 percent by 2050 unless urgent action is taken: World Bank report. https://www.worldbank.org/en/news/press-release/2018/09/20/global-waste-to-grow-by-70-percent-by-2050-unless-urgent-action-is-taken-world-bank-report

World Bank. (2018, September 20). What a waste: An updated look into the future of solid waste management. https://www.worldbank.org/en/news/immersive-story/2018/09/20/what-a-waste-an-updated-look-into-the-future-of-solid-waste-management

World Bank. (n.d.). Trends in solid waste management. What a Waste. https://datatopics.worldbank.org/what-a-waste/trends_in_solid_waste_management.html

Insights into the Advances in Microbial Technology for the Remediation of Aromatic Hydrocarbons of Petroleum Origin

Pritam Gyanaranjan Rout, Alok Prasad Das, and Sunil Kumar Behera

1 Introduction

Petroleum is a naturally occurring fossil fuel. Typically, petroleum appears as a yellowish-black liquid mainly composed of hydrocarbons. Crude petroleum undergoes refining and extraction processes to yield several products, including petrol, diesel, kerosene, jet fuel, and lubricants (Behera et al., 2020; Lahiri et al., 2021). Additionally, petroleum is a raw material in manufacturing plastics, textiles, refrigerants, synthetic rubber, solvents, fertilizers, pesticides, pharmaceuticals, and more. Petroleum is primarily used to refine fuel oil and gasoline as primary energy resources (Dey et al., 2023; Mohanta et al., 2024). Petroleum mainly contains hydrocarbon in alkenes, cycloalkenes, aromatic hydrocarbons, and other complex hydrocarbons. These hydrocarbons serve as the building blocks for the manufacture of various petroleum products. The primary elements present in petroleum are carbon (83–87%), hydrogen (10–14%), nitrogen (0.1–2%), oxygen (0.05–1.5%), sulfur (0.05–6.0%), and metals (<0.1%). Crude oil mainly consists of over 30% poly-aromatic hydrocarbons (PAHs) and other various hydrocarbons such as naphthenes, paraffin, and aromatic compounds, along with organic nitrogen compounds, sulfur compounds, and oxygen-containing hydrocarbons (phenols).

Petroleum contamination occurs when crude oil or its refined products (gasoline, diesel, lubricants, etc.) enter the environment, leading to soil, water, and air pollution. Some common causes include accidental release during petroleum transportation through tankers, pipeline leaks, ruptures in petroleum pipelines, poorly maintained fuel storage tanks corrode over time, releasing petroleum into the soil

P. G. Rout · S. K. Behera (✉)
Department of Biotechnology, Berhampur University, Berhampur, Odisha, India
e-mail: skbehera2020@gmail.com

A. P. Das
Department of Life Sciences, Rama Devi Women's University, Bhubaneswar, Odisha, India

A. P. Das et al. (eds.), *Sustainable Municipal Waste Management with a Zero Waste Approach*, Sustainable Environmental Waste Management Strategies,
https://doi.org/10.1007/978-3-032-05950-5_19

and groundwater (Behera et al., 2021, 2022a, 2022b). Petroleum and petroleum byproducts released from refineries, petrochemical plants, and manufacturing industries may release petroleum residues into the environment through improper waste disposal (Behera et al., 2021). Further, discarding used engine oil and industrial lubricants into drains, soil, or water bodies can lead to petroleum contamination. Natural calamities like earthquakes and floods can damage storage facilities and pipelines, leading to petroleum leaks. The presence of sulfur in petroleum makes it even more hazardous due to toxic gas emissions during combustion (Das et al., 2025; Kanungo et al., 2023). Various remediation methods have been developed to mitigate the environmental pollution caused by petroleum contamination. Several physical and chemical remediation methods for petroleum contamination have been developed. However, these approaches are always associated with significant drawbacks such as environmental harm, high costs, and secondary pollution. In this context, remediation processes catalyzed by biological agents such as microbial bioremediation, phytoremediation, and rhizoremediation, which use microorganisms, plants, or a combination of both to prevent petroleum contamination, have been developed. These biological methods are environmentally friendly, cost-effective, and efficient, without producing harmful secondary products, even at low pollutant concentrations. Hence, this chapter focuses on the advancements in microbial technology in the remediation of aromatic hydrocarbons of petroleum origin.

2 Petroleum Hydrocarbons

Petroleum hydrocarbons (PHCs) are categorized into two major groups: aliphatic and aromatic. Aliphatic mainly contain single-bonded or double-bonded carbon atoms arranged in straight chains and cyclic structures (Mohanta et al., 2024; Reddy Koteswara et al., 2025). Major aliphatic groups contain alkanes, alkenes, and cycloalkanes. On the other hand, aromatics bear one or more benzene rings in their chemical structures, contributing to their strong chemical integrity. One benzene ring containing aromatics is known as monoaromatics, while polycyclic aromatic hydrocarbons (PAHs) constitute two or more bonded benzene rings within their chemical structures, respectively. A brief list of petroleum hydrocarbons, their nomenclature, and their structures are described in the Table 1.

3 Remediation of Petroleum Hydrocarbons

The remediation of PHCs involves a series of steps to transform them into less harmful and non-hazardous components that can be assimilated into the biogeochemical cycles. Mostly, the chemical transformation processes such as oxidation–reduction, dehalogenation, emulsification, activated carbon treatment, dispersion, and photocatalytic process are used for the remediation of petroleum hydrocarbon contaminants.

Table 1 The list of hydrocarbon compounds found in the petroleum

IUPAC name	Molecular formula	Chemical structure
Benzene	C_6H_6	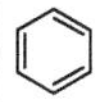
Ethyl benzene	C_8H_{10}	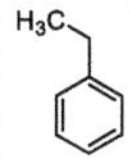
Methylbenzene	C_7H_8	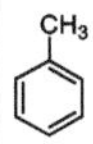
1,3-dimethylbenzene	C_8H_{10}	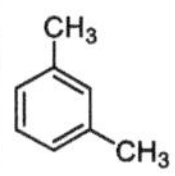
1,2-dimethylbenzene	C_8H_{10}	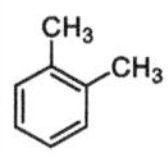
1,4-dimethylbenzene	C_8H_{10}	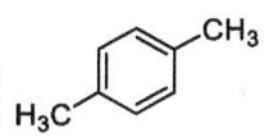
1,2-dihydroacenaphthylene	$C_{12}H_{10}$	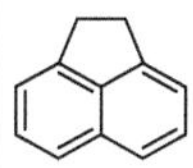
Acenaphthylene	$C_{12}H_8$	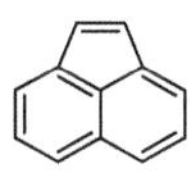
Anthracene	$C_{14}H_{10}$	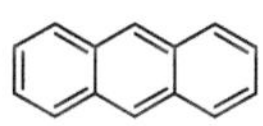
Benz[a]anthracene	$C_{18}H_{12}$	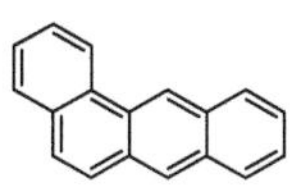
Benzo[a]tetraphene	$C_{20}H_{12}$	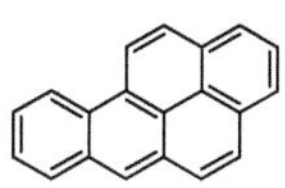
Benzo[b]fluoranthene	$C_{20}H_{12}$	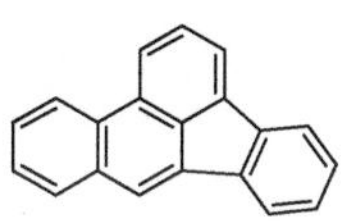

(continued)

Table 1 (continued)

IUPAC name	Molecular formula	Chemical structure
Benzo[g,h,i]perylene	$C_{22}H_{12}$	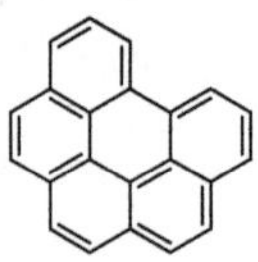
Benzo[k]fluoranthene	$C_{20}H_{12}$	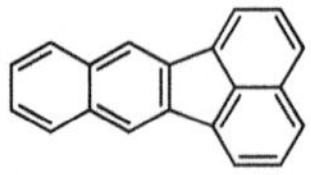
Chrysene	$C_{18}H_{12}$	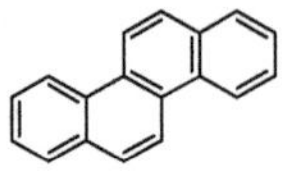
Fluoranthene	$C_{16}H_{10}$	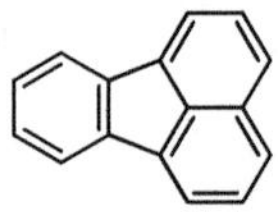
9H-fluorene	$C_{13}H_{10}$	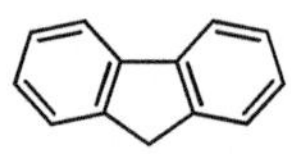
Indeno[1,2,3-c,d]pyrene	$C_{22}H_{12}$	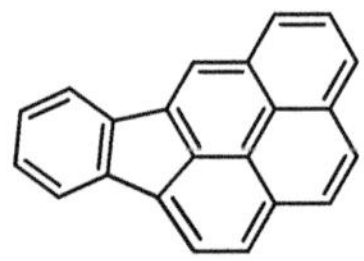
Naphthalene	$C_{10}H_8$	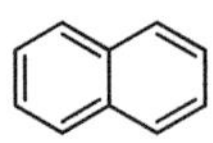
Phenanthrene	$C_{14}H_{10}$	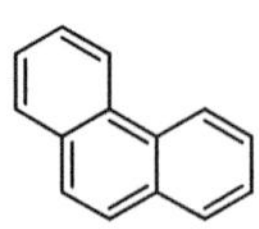
Pyren	$C_{16}H_{10}$	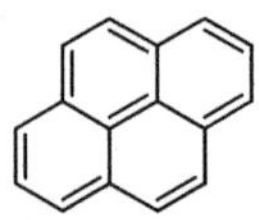

The chemical transformation involves hydrolysis and oxidation–reduction reactions, which convert them into simpler molecules (Maletic et al., 2013). However, the chemical transformation of petroleum hydrocarbons is directly and indirectly associated with toxic emissions. In addition, these processes are termed as more complex and energy-intensive, involving high operational costs. Therefore, petrochemical remediation methods operated under ambient conditions catalyzed by microorganisms are evolving as alternatives to the chemical remediation processes.

Bioremediation of Petroleum Hydrocarbons

The bioremediation approach has been introduced as a green approach for alleviating petroleum hydrocarbon contaminants in the environment (Das et al., 2024a, 2015; Paul et al., 2023; Pradhan et al., 2023). In general, the bioremediation process involves biodegradation and biotransformation of the hydrocarbons of petroleum origin. The biochemical reactions can occur under aerobic and anaerobic conditions (Das et al., 2024a; Kanungo et al., 2024). The bioremediation approaches have been studied using various types of microorganisms. Microorganisms are potentially responsible for the remediation process by consuming petroleum hydrocarbons as the source of the electrons required in their metabolic activities (Bhattacharjee et al., 2021; Das et al., 2023; Dey et al., 2024; Nag et al., 2023). The microorganisms can remediate or degrade petroleum hydrocarbons and various organic contaminants to simpler and non-toxic substances without any adverse effect on the contaminated environments (Lim et al., 2016). Potential microbes like *Pseudomonas, Bacillus,* and *Rhodococcus* have been used to degrade petroleum hydrocarbons (Das & Chandran, 2011). The microbes perform oxidation as their metabolic activities to remove and neutralize the contaminants into non-toxic or simpler compounds specifically CO_2 and H_2O (Behera, 2024; Das & Chandran, 2011). The microbial metabolism very easily degrades the aliphatic hydrocarbon compounds. However, the long chain and the branched or cyclic chain hydrocarbon are more resistant to most bioremediation approaches (Al-Hawash et al., 2018). Metabolizing petroleum hydrocarbon compounds reveals that the microorganisms consume the carbon compounds as their sole energy source. Some of the most commonly isolated petroleum hydrocarbon-degrading bacteria belong to the genus *Pseudomonas* which efficiently degrade petroleum hydrocarbon into simpler compounds (Wang et al., 2011; Wongsa et al., 2004). In addition, fungi species such as *Penicillium, Fusarium,* and *Rhizopus* species have been isolated and utilized in bioremediation of petroleum hydrocarbon-contaminated soil and sediments. There are several types of biodegradation approaches for petroleum hydrocarbons developed, which include bio-attenuation, biostimulation, rhizoremediation, bioaugmentation, bioventing, biosparging, bioslurry, biopiling, composting, bioreactors, windrows, biotransformation, mycoremediation, etc. (Chukwunonso Ossai et al., 2022).

Bio-Attenuation

This includes the utilization of a variety of naturally occurring processes such as physical and biochemical processes (reduction, neutralization, and transformation of the mass, volume, concentration, and toxicity) for the transformation of petroleum contaminants by microorganisms. Bio-attenuation can occur through dispersion, sorption, dissolution, volatilization, biological transformation, stabilization, biodegradation, etc. (Das & Chandran, 2011) to manage the contamination level up to a certain limit. The bio-attenuation process is observed to be applicable for contaminated sites where the petroleum contaminants are dominated by a lower concentration, and where other remediation methods can't be reached easily (Das & Chandran, 2011; Sharma, 2021).

Biostimulation

The biostimulation process enhances the bioremediation of petroleum hydrocarbons by improving the natural biodegradation process by promoting the growth of microorganisms that can degrade petroleum hydrocarbons selectively. In general, biostimulation is a bioremediation technique that involves adding essential nutrients, oxygen, biosurfactants, biopolymers, or other components to the contaminated environment to stimulate and support the microbial growth and the enzymatic activities of Indigenous microorganisms, which can be able to degrade the hydrocarbons extensively (Wu et al., 2017). Biostimulation can be achieved by adding the limiting factors, micronutrients, and electron acceptors like carbon, oxygen, nitrogen, phosphorous, and potassium, which are generally available in low levels in the contaminated soil, and these factors are responsible for constraining the microbial growth. This remediation mode is known as the most efficient bioremediation method compared with other *in-situ* remediation applied for petroleum hydrocarbon-contaminated soil (Simpanen et al., 2016).

Rhizoremediation

The rhizoremediation process is a cooperative manner of petroleum hydrocarbons present in the rhizospheres of the plant roots. This type of phytoremediation uses plant roots to extensively degrade organic pollutants. The remediation process is carried out by the actions of the plant and the microorganisms present in the roots together. The plants in the contaminated soil absorb the petroleum hydrocarbon compounds and release organic compounds, creating a favorable environment for microbial growth. Therefore, the microorganisms colonize the rhizospheres and break down the petroleum hydrocarbons using enzymes and biochemical pathways (Viesser et al., 2020). The potential microorganisms, such as *Actinobacteria* and *Proteobacteria, Alphaproteobacteria, Betaproteobacteria, Gammaproteobacteria*, and *Acidobacteria* were reported to degrade aliphatic and polycyclic aromatic hydrocarbons present in the contaminated soil (Pagé et al., 2015; Yergeau et al., 2012, 2014).

Bioaugmentation

Bioaugmentation is a reliable method applied to the bioremediation of petroleum hydrocarbons. This method includes introducing selective microbial agents to the sites contaminated with petroleum hydrocarbons. These microorganisms have high adaptability and specific catabolic activities against the hydrocarbons associated with the soil environment. The general purpose of bioaugmentation is to supplement the existing microbial community with the optimum environmental conditions so that microbial activities can be accelerated toward the extensive remediation of hydrocarbon contaminants extensively (El Fantroussi & Agathos, 2005). It is important to note that the microorganisms have variable modes of hydrocarbon degradation rates. Depending upon this fact, the selection and inoculation of the Indigenous microbial strains into the contaminated soil samples are very essential for a successful remediation. The most studied bacterial strains used in the bioaugmentation process include *Acinetobacter sp., Arthrobacter, Pseudomonas, Flavobacterium, Sphingomonas,* and *Achromobacter, Mycobacterium, Bacillus, and Rhodococcus* (El Fantroussi & Agathos, 2005; Gentry et al., 2004; Throne-Holst et al., 2007). Furthermore, the fungal strains such as *Aspergillus, Penicillium, and Verticillium, Cephalosporium, Penicillium, Graphium,* etc. have been utilized for the remediation of soil contaminated with petroleum hydrocarbons to a large extent (Chaillan et al., 2004; Singh, 2006; Mt & Zaida, 2018).

Bioventing

Bioventing is an in-situ biological process for the remediation of petroleum hydrocarbon contaminants by introducing oxygen through direct air injection into the contaminated soil. The introduction of oxygen creates the aerobic condition which in turn enhances the growth and biological activities of the indigenous microbial population; thus, degrading petroleum hydrocarbons very effectively. Bioventing process is applicable for the remediation of both the water as well as the soil samples contaminated with petroleum hydrocarbons.

Biosparging

Biosparging is a bioremediation approach for petroleum hydrocarbon contents, which involves injection of oxygen and other nutrients into the saturated zone under pressure to increase groundwater oxygen level. As a result, this could stimulate the biological activities of the Indigenous microorganisms for the extensive degradation of petroleum hydrocarbons (Chukwunonso Ossai et al., 2022; Kao et al., 2008). This bioremediation approach is helpful for the reduction of contaminants associated with the soil and the groundwater as well. However, the process's effectiveness depends on soil permeability and pollutant degradability by the active participation of potential microbes (Philp & Atlas, 2014; Sharma, 2021). Biosparging methods generally work by introducing air into the groundwater table through wells to increase the dissolved oxygen content within the aqueous phase to facilitate microbial growth.

Bioslurry

The bioslurry method involves the treatment of contaminated soil using bioreactors such as batch, fed-batch, continuous, and multistage bioreactors. The bioreactors are equipped with optimal nutrient resources to meet the microbial nutrient demand; thus, the microbial metabolic activities are enhanced along with the metabolism of petroleum hydrocarbons. The reactors are designed to keep the process controls in an optimal condition so that the microorganisms in the reactors can survive in the environment efficiently by achieving maximum removal efficiency. Bioreactors are operated at certain advantages such as little space requirement, rapid reaction kinetics, better control over the reaction kinetics, and removal efficiency. However, this approach has certain limitations such as a longer treatment period, soil excavation followed by transport to the treatment facility, required control of volatile emissions, etc.

Biopiling

Biopiling is an effective ex-situ bioremediation method for soil contaminated with petroleum hydrocarbons. This method is ideal for the aerobic removal of organic matter. The process of piling excavated contaminated soil follows biostimulation and aeration to enhance the microbial biochemical activities to degrade contaminants effectively. This method has been observed to be suitable for the treatment of large volumes of contaminated soil and sediments (present in extreme environments) in a confined space (Whelan et al., 2015). This remediation contains major components such as the addition of air (oxygen), moisture (water), nutrients, and bulking agents (organic materials), a leachate collection system, and a treatment bed.

Biotransformation

Biotransformation represents the biotechnological process involving microbial-mediated structural modifications in the chemical components to form highly polar molecules (Singh, 2017). This process involves chemically modifying the organic compounds by microbial enzymes to reduce their toxic effects. Biotransformation can occur in various forms such as oxidation, reduction, condensation, isomerization, hydrolysis, denitrification, sulfidogenesis, methanogenesis, etc.

Several groups of microorganisms consist of potential enzyme systems responsible for the easy transformation of petroleum hydrocarbons into simple biodegradable components. The potential microorganisms with high affinity for degrading hydrocarbons of petroleum origin include *Achromobacter, Acinetobacter, Alkanindiges, Alteromonas, Arthrobacter, Burkholderia, Dietzia, Enterobacter, Kocuria, Marinobacter, Mycobacterium, Pandoraea, Pseudomonas, Staphylococcus, Streptococcus, Streptobacillus, Rhodococcus,* etc. (Das & Chandran, 2011).

Biocomposting

This is a microbial-assisted aerobic biochemical degradation of organic waste materials and their conversion into simple and stabilized organic materials that can be effectively used as soil conditioners for the hydrocarbon remediation process (Cai et al., 2021). Composting involves adding nutrients, tilling, watering, and adding potential microbial consortia as forms of organic waste to improve bioremediation (Sarika Saxena, 2015). Moreover, composting requires thermophilic conditions of 50–60 °C to degrade the petroleum hydrocarbon compounds properly. Composting is a promising method of degrading and converting contaminants into harmless substances with substantial potential for the remediation of petroleum hydrocarbon-contaminated soil samples.

Mycoremediation

The method of myco-remediation involves the utilization of potential fungal microorganisms to degrade petroleum hydrocarbon contaminants to less toxic or non-toxic simple components, thereby eliminating environmental contaminants effectively. The fungi secrete extracellular enzymes like peroxidizes, which can selectively break down lignin and cellulose; thereby, they can degrade many environmental recalcitrant pollutants (Lee et al., 2014). The most prominent fungal agents, such as *Polyporus* sp. and *Phanaerochaete chrysosporiumare*, are known to be very important for the degradation of a large group of toxic pollutants. A study reported that *Penicillium sp.* was able to degrade polyethylene effectively (Yamada-Onodera et al., 2002). Several reports have demonstrated the potential degradation of petroleum hydrocarbons by the rot fungal strains due to their high synthesis of oxidative enzymes, extracellular enzymes, chelators, and organic acids. In a study, it was demonstrated that keratinolytic fungi *Trichophyton ajelloi* was able to remove hexadecane and pristine from crude oil (Yamada-Onodera et al., 2002). Furthermore, a fungal isolate *Pleurotospulmonarius* was used to treat a soil sample contaminated with a petroleum hydrocarbon mixture of petrol, diesel, spent engine oil, and spent diesel engine oil (at a ratio of 1:1:1:1) at various concentrations of 2.5%, 5%, 10%, and 20%. The authors have demonstrated that the soil with a petroleum hydrocarbon mixture at 10% was able to remove 68.34% of total petroleum hydrocarbons (Njoku et al., 2017).

Windrows

This method is applied for the bioremediation of soil contaminated with petroleum hydrocarbons. This method involves periodic rotation of piled polluted soil to encourage bioremediation by enhancing the biodegradation activities of indigenous microorganisms present in polluted soil. The increase in biodegradation activities of the microorganisms that exist in the contaminated soil can speed up the rate of degradation which is accomplished by biotransformation, assimilation, and mineralization (Azubuike et al., 2016; Jiang et al., 2016; Whelan et al., 2015). This type of bioremediation can achieve more hydrocarbon removal rate as compared to biopiling. However, the soil-contaminated volatile compounds are not removed effectively. This process can produce greenhouse gases like methane due to the creation of anaerobic zones within the piled heap. In a study, the authors conducted the degradation of petroleum-contaminated soil using a windrow soil system for about ten months. They have demonstrated that 60% of the total petroleum hydrocarbons were removed in the first eight months (Al-Daher et al., 1998).

Phycoremediation

Bioremediation of petroleum hydrocarbons also involves phycoremediation using algal species (microalgae and macroalgae). The algal species such as *Chlamydomonas*, *Chlorella*, *Botryococcus*, and *Phormidium* have been potentially used to sequester, remove, break down, biotransformation, and metabolize pollutants associated with water bodies (Ossai et al., 2020). Algal organisms efficiently metabolize and degrade toxic organic pollutants like hydrocarbons, biphenyls, pesticides, and phenolics (Abed et al., 2008). Algae are very adaptable to the environment in many aspects; therefore, they are categorized into autotrophic, mixotrophic, or heterotrophic groups. Mixotrophic algal microbes are potential agents in bioremediation and carbon sequestration processes (Subashchandrabose et al., 2013). Their environmental adaptability makes them ideal for growth under autotrophic, mixotrophic, or heterotrophic conditions. Algal species have the potential for mineral uptake by physical adsorption onto the mineral ions, followed by selective separation via chemisorption (Fomina & Gadd, 2014). The use of algae in the remediation approach reveals that the microorganisms can oxidize petroleum hydrocarbon constituents into less noxious compounds (Hussein et al., 2022).

Petroleum hydrocarbon pollutants have been observed to seriously threaten human health by polluting the soil, sediments, surface water, and groundwater. The impacted environments such as soils, surface water, sediments and groundwater, which are usually contaminated with high and low molecular weight petroleum hydrocarbon compounds, volatile organic compounds (VOCs), semi-volatile organic compounds (SVOCs), polycyclic aromatic hydrocarbons (PAHs), persistent organic pollutants (POPs), polychlorinated biphenyls (PCBs), organochlorinated pesticides (OCPs), hydrophobic organic compounds (HOCs), non-aqueous phase liquids (NAPL), xenobiotic organics and heavy metals need to be cleaned and free from these recalcitrant components. Several microbial studies have been conducted to improve the biological process towards the reduction of petroleum hydrocarbons. Most importantly, remediation approaches have been focused on aromatic hydrocarbons of petroleum origin.

Biomolecular Mechanisms Associated with Petroleum Hydrocarbon Metabolism

Microbial Degradation of Aliphatic Hydrocarbons

The potential microorganisms that have been associated with the metabolism of petroleum hydrocarbons have shown the oxidation of both aliphatic and aromatic components (Nzila, 2018). A range of microbial communities is responsible for carrying out the biological transformation of hydrocarbon contaminants present in petroleum and other mineral-contaminated environments (Das & Ghosh, 2022, 2023; Das et al., 2024b; Priyadarsini & Das, 2025). The microorganisms not only possess hydrocarbon metabolism but also help in the selective extraction of valuable metals for the beneficiation of hydrometallurgy (Behera et al., 2011, 2012, 2018, 2022a, 2022b). It has been observed that the microorganisms isolated from hydrocarbon and effluent-enriched environments show better degradation ability. As an example, *Acidithiobacillus ferrooxidans* is a gram-negative, chemolithotrophic bacteria possessing the degradation of both aliphatic and aromatic hydrocarbon moieties present in the minerals (Priyadarsini & Das, 2024; Rout et al., 2022, 2024). Most of the microorganisms explored so far have been observed to oxidize the hydrocarbon contents for their metabolism. Potential microorganisms use soluble or integral-membrane-bound enzymes such as non-heme iron monooxygenases. These enzymes, specifically alkane hydroxylases, hydroxylate the hydrocarbon substrates (Rojo, 2009). Generally, the biodegradation of alkanes begins with the microbial oxidation of the terminal methyl groups by forming a primary alcohol. The oxidation products, along with the alcohol, are further microbially oxidized by alcohol and aldehyde dehydrogenases, forming the corresponding aldehyde molecule. The resulting molecule is further transformed into fatty acid catalyzed by microbial oxidation (Kotani et al., 2007). The fatty acid molecule is further coupled with CoA and transitioned into the β-oxidation pathway as acetyl-CoA (Fig. 1). In the case of microbial oxidation, the sub-terminal chains of long-chain alkanes generate secondary alcohols. These secondary alcohols are biotransformed into the ketone, which is again converted to an ester molecule catalyzed by the microbial monooxygenase enzyme. Following microbial metabolic activity, the ester molecule is finally hydrolyzed to an alcohol and a fatty acid by an enzyme, esterase (Fig. 1).

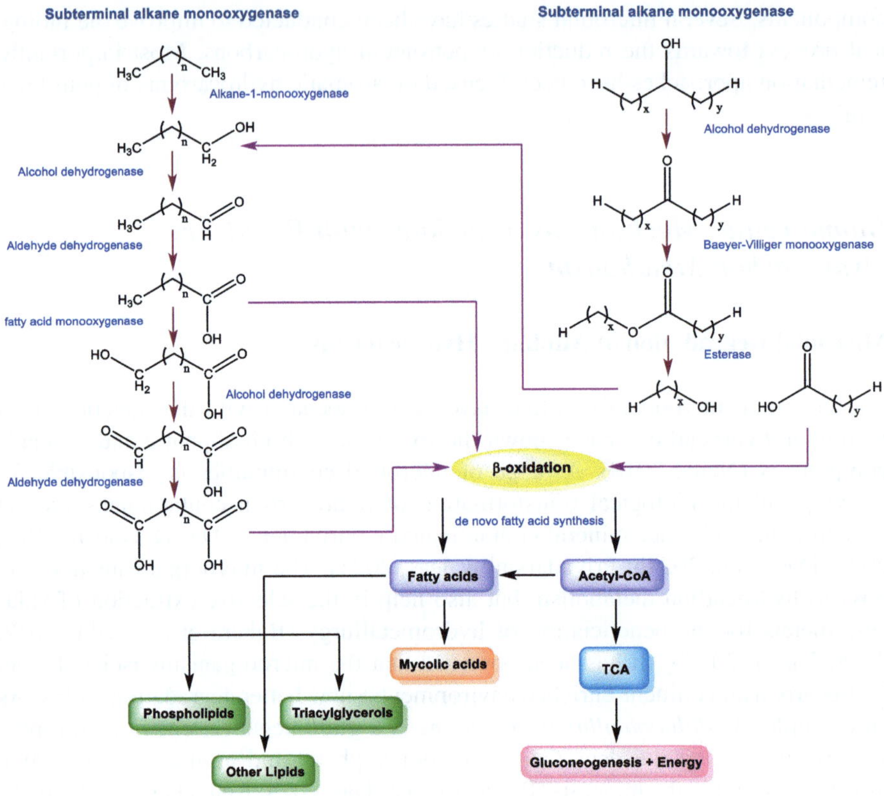

Fig. 1 Microbial n-alkane degradation pathways (terminal and subterminal oxidation)

Microbial Degradation of Aromatic Hydrocarbons

Petroleum hydrocarbon-degrading microbial communities belong to the genera *Pseudomonas*, *Proteus*, *Rhodococcus*, *Sphingobium*, *Bacillus*, *Arthrobacter*, *Mycobacterium*, and *Sphingomonas* spp. are ubiquitous in the ecosystem (Deborah Gnana Selvam & Thatheyus, 2018; Rout et al., 2024; Sylvester et al., 2025). These microbes have the potential to metabolize and transform aromatic hydrocarbons present in petroleum and other petrochemicals; the catabolism process is associated with an oxidation step of one of the aromatic rings along with the generation of PAH metabolites and CO_2 (Rout et al., 2022, 2024; Rout & Behera, 2024). In general, the microbial metabolism of aromatic hydrocarbon is strictly catalyzed by aromatic hydrocarbon ring hydroxylating enzymes, which form *cis*-dihydrodiols. These dihydrodiols undergo an oxidation reaction catalyzed by a microbial dehydrogenase enzyme system to produce PAH dihydroxy derivatives, which further undergo ring cleavage by the dioxygenases. These aromatic ring-cleavage reactions of the dihydroxy derivative molecules are favored by incorporating a pair of oxygen atoms (O_2).

Further, dihydroxylated intermediate aromatics are then followed by either an intradiol manner (*ortho*) cleavage pathway, or through the intradiol ring (*meta*) cleavage pathway, with the formation of catechols, which are finally metabolized to carbon dioxide and water through the TCA (tricarboxylic acid cycle) cycle (Muratova et al., 2014) (Fig. 2). Several bacterial communities oxidize PAHs via the cytochrome P450 monooxygenase, producing trans-dihydrodiols (Bamforth & Singleton, 2005).

The microbial metabolic processes for PAH degradation are well documented. Additionally, reports demonstrate that fungi are capable of metabolizing PAH contaminants. Two different metabolic pathways are associated with the fungal metabolism of PAH. Non-ligninolytic fungi metabolize PAHs to arene oxides (the

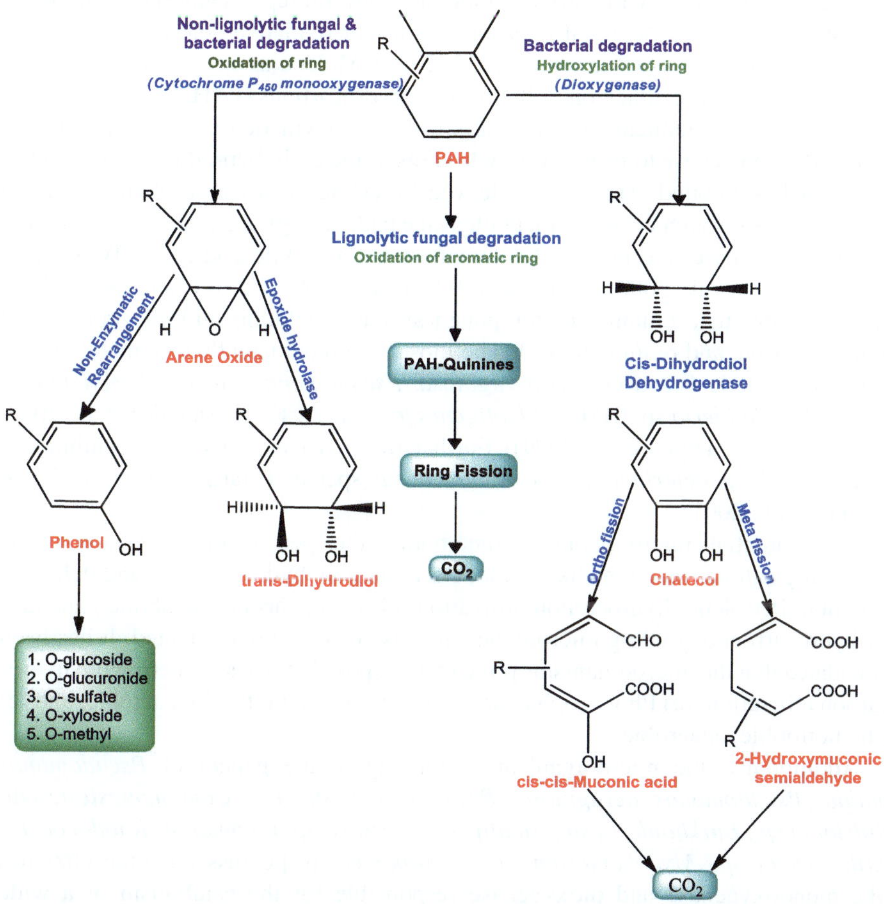

Fig. 2 Microbial breakdown pathways for Aromatic hydrocarbons. Redrawn from (Haritash & Kaushik, 2009)

primary products of PAH metabolism) by using the cytochrome P450 monooxygenase pathway via the integration of an oxygen atom into the cyclic rings of the substrates (Fig. 2).

On the other hand, white-rot fungi (ligninolytic fungi) can metabolize PAHs by using extracellular ligninolytic enzymes like manganese peroxidase, laccases, and lignin peroxidase. Some fungi, such as *Pleurotus ostreatus* and *Phanerochaete chrysosporium*, are reported to produce non-ligninolytic and ligninolytic enzymes. The aerobic transformation of aromatic hydrocarbons of petroleum assisted by bacterial and fungal communities is accomplished through three principal pathways (Fig. 2).

Several microorganisms possess monooxygenase enzymes that catalyze potential oxidation of aromatic hydrocarbon compounds. Several reports have been published describing the use of potential microorganisms for oxidative degradation of poly-aromatic hydrocarbons (Rout et al., 2022, 2024). Molecular oxygen plays a crucial role in the enzymatic attack on poly-aromatic hydrocarbons (PAHs) by the microorganisms. The biochemical pathway starts with the enzymatic oxidation of arenes of PAHs by dioxygenase to form *cis*-dihydrodiols as the early byproducts. These dehydroxylated compounds are further cleaved by estradiol or estradiol ring-cleaving dioxygenases through ortho- or meta-cleavage pathways giving rise to the formation of catechols. These catechols further transform tricarboxylic acid cycle (TCA) intermediates (Cao et al., 2009; Cerniglia, 1992; Husain, 2008). Dioxygenases are abundant in bacteria that contain multicomponent systems, including multiple proteins and non-heme iron, and require NADH. The most prominent naphthalene metabolizing microbes include *Pseudomonas aeruginosa, Pseudomonas stutzeri, Pseudomonas putida, Burkholderia* sp. RP007, *Alcaligenes faecalis* AFK2 (Fernández et al., 2012; Qi et al., 2015; Vázquez et al., 2009)). Further, the microbes metabolizing anthracene belong to *Mycobacterium* sp. and *Rhodococcus* species (Martínková et al., 2009; Moody et al., 2001).

The potential microorganisms metabolize the petroleum hydrocarbons by catalyzing cellular reactions like oxidation, reduction, hydroxylation, and dehydrogenation. Petroleum hydrocarbon utilization following various metabolic pathways by hydrocarbon-degrading microorganisms has been well documented. It has been elucidated that the microorganisms possess three possible ways for petroleum hydrocarbon utilization: (a) Phototrophic, anoxygenic; (b) Chemotrophic, aerobic; and (c) Chemotrophic, anaerobic.

Additionally, the microorganisms belonging to the genera of *Pseudomonas putida, Pseudomonas aeruginosa, Pseudomonas stutzeri, Comamonastesteroni, Ralstonia sp., Burkholderia sp., Alcaligenes faecalis, Sphingomonas, Rhodococcus, Arthrobacter sp., Mycobacterium sp., Terrabacter* sp. possess potential enzymes like monooxygenase and dioxygenase responsible for the catabolism of a wide range of the substrates such as biphenyl, (substituted) naphthalene(s), fluorene, (substituted) phenanthrene(s), pyrene, (chlorinated) diphenylether(s), (chlorinated) furan(s), (chlorinated) dibenzo-p-dioxin(s), carbazole, estradiol, polyethylene glycols, chlorinated phenols. The presence of monooxygenase enzyme is responsible for the incorporation of oxygen atoms into the hydrocarbon rings by which the poly-aromatic hydrocarbons are converted into arene oxides (the primary products of PAH metabolism). In contrast, the fungal species (ligninolytic fungi) mineralize PAHs using a soluble extracellular ligninolytic enzyme such as manganese peroxidase, laccases, and lignin peroxidase (Kumar & Chandra, 2020). Several microorganisms discovered from different petroleum-contaminated sites such as *Acinetobacter lwoffii, Aeromonas hydrophila, Arthrobacter sp., Alcaligenes faecalis, Bacillus licheniformis, Bacillus marcescens, Corynebacterium sp., Flavobacterium sp., Klebsiella pneumonia, Klebsiella aerogenes, Moraxella sp., Micrococcus sp., Mycobacterium sp., Proteus vulgaris, Pseudomonas alcaligenes, P. putida, P. putrefaciens, P. aeruginosa and Vibrio sp.* have shown to dispose the petroleum pollutants (Adebusoye et al., 2007; Geetha et al., 2013; Ghazali et al., 2004; Hamzah & Tavakoli, n.d.; Jesubunmi et al., 2022; Jobson et al., 1974; Mittal & Singh, 2009; Olubunmi, 2014; Raza et al., 2013; Sathishkumar et al., 2008; Thenmozhi et al., 2013; Varjani, 2017). Cook and Westlake discovered that *Achromobacter sp., Acinetobacter sp., Alcaligenes sp., Flavobacterium sp., Pseudomonas sp.,* and *Xanthomonas sp.* are capable of metabolizing the crude oil (Fedorak & Westlake, 1981). In another study, thermophilic bacteria (*Thermomicrobium* sp.) were observed to be capable of hydrocarbon utilization. Rosenberg, (2006), reported on the utilization of petroleum hydrocarbons by the microorganisms belonging to *Alcaligenes sp., Arthrobacter sp., Brevibacterium sp., Pseudomonas sp., Spirillum sp.,* and *Xanthomonas sp.* from asphaltic flow in Alaska. Interestingly, bacteria and yeast populations in aquatic environments have been reported as hydrocarbon-utilizing microbes that contain genera of *Pseudomonas, Achromobacter, Arthrobacter, Micrococcus, Nocardia, Vibrio, Acinetobacter, Brevibacterium, Corynebacterium, Flavobacterium, Candida, Rhodotorula, and Sporobolomyces* (Bartha & Atlas, 1977). Austin et al., (1977), demonstrated petroleum-degrading bacteria belonging to *Actinomycetes, Enterobacteriaceae, coryneform, Klebsiella aerogenes, Micrococcus spp., Nocardia spp., Pseudomonas spp.,* and *Sphaerotilusnatans.*

4 Conclusion

The persistent contamination of the environment by petroleum hydrocarbons, especially aromatic compounds, poses significant ecological and health hazards. Conventional physicochemical remediation methods, while effective to some extent, often involve high operational costs and can lead to secondary environmental pollution. This chapter highlights the potential of microbial technologies as sustainable and eco-friendly alternatives for the bioremediation of aromatic hydrocarbons. Various microbial strategies such as bio-attenuation, biostimulation, bioaugmentation, rhizoremediation, and mycoremediation have been explored and demonstrated significant potential in degrading complex hydrocarbon pollutants into harmless end products like carbon dioxide and water. The metabolic versatility of diverse microbial genera—bacteria, fungi, and algae—under aerobic and anaerobic conditions underlines their value in mitigating petroleum pollution. Continued research into microbial pathways, enzymatic systems, and optimal environmental conditions will enhance the efficacy and scalability of these bioremediation approaches for field applications.

References

Adebusoye, S. A., Ilori, M. O., Amund, O. O., Teniola, O. D., & Olatope, S. O. (2007). Microbial degradation of petroleum hydrocarbons in a polluted tropical stream. *World Journal of Microbiology and Biotechnology, 23*(8), 1149–1159. https://doi.org/10.1007/s11274-007-9345-3

Al-Daher, R., Al-Awadhi, N., & El-Nawawy, A. (1998). Bioremediation of damaged desert environment using the windrow soil pile system in Kuwait. *Environment International, 24*(1–2), 175–180. https://doi.org/10.1016/S0160-4120(97)00133-5

Al-Hawash, A. B., Dragh, M. A., Li, S., Alhujaily, A., Abbood, H. A., Zhang, X., & Ma, F. (2018). Principles of microbial degradation of petroleum hydrocarbons in the environment. *Egyptian Journal of Aquatic Research, 44*(2), 71–76. https://doi.org/10.1016/j.ejar.2018.06.001

Austin, B., Calomiris, J. J., Walker, J. D., & Colwell, R. R. (1977). Numerical taxonomy and ecology of petroleum-degrading bacteria. *Applied and Environmental Microbiology, 34*(1), 60–68. https://doi.org/10.1128/aem.34.1.60-68.1977

Azubuike, C. C., Chikere, C. B., & Okpokwasili, G. C. (2016). Bioremediation techniques–classification based on site of application: Principles, advantages, limitations and prospects. *World Journal of Microbiology and Biotechnology, 32*(11), 180. https://doi.org/10.1007/s11274-016-2137-x

Bamforth, S. M., & Singleton, I. (2005). Bioremediation of polycyclic aromatic hydrocarbons: Current knowledge and future directions. *Journal of Chemical Technology & Biotechnology, 80*(7), 723–736. https://doi.org/10.1002/jctb.1276

Bartha, R., & Atlas, R. M. (1977). The microbiology of aquatic oil spills. In *Advances in applied microbiology* (vol. 22, pp. 225–266). Elsevier. https://doi.org/10.1016/S0065-2164(08)70164-3

Behera, I. D., Basak, G., Kumar, R. R., Sen, R., & Meikap, B. C. (2020). Treatment of petroleum refinery sludge by petroleum degrading bacterium Stenotrophomonas pavanii IRB19 as an efficient novel technology. *Journal of Environmental Science and Health, Part A, 56*(2), 226–239. https://doi.org/10.1080/10934529.2020.1866924

Behera, I. D., Nayak, M., Biswas, S., Meikap, B. C., & Sen, R. (2021). Enhanced biodegradation of total petroleum hydrocarbons by implementing a novel two-step bioaugmentation strategy using indigenous bacterial consortium. *Journal of Environmental Management, 292*, Article 112746. https://doi.org/10.1016/J.JENVMAN.2021.112746

Behera, I. D., Nayak, M., Mishra, A., Meikap, B. C., & Sen, R. (2022a). Strategic implementation of integrated bioaugmentation and biostimulation for efficient mitigation of petroleum hydrocarbon pollutants from terrestrial and aquatic environment. *Marine Pollution Bulletin, 177*, Article 113492. https://doi.org/10.1016/J.MARPOLBUL.2022.113492

Behera, S. K. (2024). Microbial remediation of microplastics and integration into the circular economy. In A. P. Das, I. D. Behera, & D. Bhanja, (Eds.), *Sustainable microbial technology for synthetic and cellulosic microfiber bioremediation*. Springer, Cham. https://doi.org/10.1007/978-3-031-62600-5_11

Behera, S. K., Manjaiah, M., Sekar, S., Panda, S. K., Mavumengwana, V., & Mulaba-Bafubiandi, A. F. (2018). Optimization of microbial leaching of base metals from a South African sulfidic nickel ore concentrate by *Acidithiobacillus ferrooxidans*. *Geomicrobiology Journal, 35*(6), 447–459. https://doi.org/10.1080/01490451.2017.1326543

Behera, S. K., Panda, P. P., Singh, S., Pradhan, N., Sukla, L. B., & Mishra, B. K. (2011). Study on reaction mechanism of bioleaching of nickel and cobalt from lateritic chromite overburdens.

International Biodeterioration & Biodegradation, 65(7), 1035–1042. https://doi.org/10.1016/j.ibiod.2011.08.004

Behera, S. K., Panda, S. K., & Mulaba-Bafubiandi, A. (2022b). Valorization of copper smelter slag through the recovery of metal values by a synergistic bioprocess system of bio-flotation and bioleaching. *Environmental Quality Management, 32*(1), 233–241. https://doi.org/10.1002/tqem.21885

Behera, S. K., Panda, S. K., Pradhan, N., Sukla, L. B., & Mishra, B. K. (2012). Extraction of nickel by microbial reduction of lateritic chromite overburden of Sukinda, India. *Bioresource Technology, 125*, 17–22. https://doi.org/10.1016/j.biortech.2012.08.076

Bhattacharjee, J., Mishra, S., & Das, A. P. (2021). Recent advances in sensor-based detection of toxic dyes for bioremediation application: A review. *Applied Biochemistry and Biotechnology, 2021*, 1–20. https://doi.org/10.1007/S12010-021-03767-7

Cai, Y., Wang, R., Rao, P., Wu, B., Yan, L., Hu, L., Park, S., Ryu, M., & Zhou, X. (2021). Bioremediation of petroleum hydrocarbons using Acinetobacter sp. SCYY-5 isolated from contaminated oil sludge: Strategy and effectiveness study. *International Journal of Environmental Research and Public Health, 18*(2), 819. https://doi.org/10.3390/ijerph18020819

Cao, B., Nagarajan, K., & Loh, K.-C. (2009). Biodegradation of aromatic compounds: Current status and opportunities for biomolecular approaches. *Applied Microbiology and Biotechnology, 85*(2), 207–228. https://doi.org/10.1007/s00253-009-2192-4

Cerniglia, C. E. (1992). Biodegradation of polycyclic aromatic hydrocarbons. *Biodegradation, 3*(2–3), 351–368. https://doi.org/10.1007/BF00129093

Chaillan, F., Le Flèche, A., Bury, E., Phantavong, Y., Grimont, P., Saliot, A., & Oudot, J. (2004). Identification and biodegradation potential of tropical aerobic hydrocarbon-degrading microorganisms. *Research in Microbiology, 155*(7), 587–595. https://doi.org/10.1016/j.resmic.2004.04.006

Chukwunonso Ossai, I., Shahul Hamid, F., & Hassan, A. (2022). Biological treatments for petroleum hydrocarbon pollutions: The eco-friendly technologies. In R. Banu Jeyakumar, K. Sankarapandian, & Y. Kannah Ravi (Eds.), *Hazardous Waste management*. IntechOpen. https://doi.org/10.5772/intechopen.102053

Das, A., Dey, S., Bala, K., Gnanasekaran, R., & Das, A. P. (2024a). Microbial bioremediation technology for sustainable treatment and management of synthetic microfiber waste. *Sustainable Microbial Technology for Synthetic and Cellulosic Microfiber Bioremediation*, 221–238. https://doi.org/10.1007/978-3-031-62600-5_12

Das, A., Dey, S., Gacem, A., Yadav, K. K., Nath, H., Ram, D. K., & Das, A. P. (2025). Revolutionizing solid hydrocarbon sourcing: plastic's journey from waste to upcycled treasure. *Environmental Science and Engineering, Part, F218*, 331–351. https://doi.org/10.1007/978-3-031-81615-4_14

Das, A., & Ghosh, S. (2022). Role of microorganisms in extenuation of mining and industrial wastes. *Geomicrobiology Journal, 39*, 173–175. https://doi.org/10.1080/01490451.2022.2038953

Das, A. P., & Ghosh, S. (2023). Metagenomic insights into unculturable microbial diversity. *Manganese Mining Microorganisms*, 63–76. https://doi.org/10.1016/B978-0-12-822147-1.00002-9

Das, A. P., Ghosh, S., Mohanty, S., & Sukla, L. B. (2015). Advances in manganese pollution and its bioremediation. In L. B. Sukla, N. Pradhan, S. Panda, & B. K. Mishra (Eds.), *Environmental microbe biotechnology* (1st ed., pp. 313–328). Springer, Cham. https://doi.org/10.1007/978-3-319-19018-1_16

Das, A. P., Mohanty, A. K., van Hullebusch, E. D., Figueiredo, G., Das, A. P., Mohanty, A. K., van Hullebusch, E. D., & Figueiredo, G. (2023). Bioremediation as an emerging technology for the removal of synthetic microplastic pollutants from marine ecosystem. *MarPB, 194*, Article 115297. https://doi.org/10.1016/J.MARPOLBUL.2023.115297

Das, A. P., Nayak, B., & Behera, I. D. (2024b). Applications of microorganisms in advanced geomicrobiological and biotechnological exploration. *Geomicrobiology Journal, 41*(4), 309–311. https://doi.org/10.1080/01490451.2024.2327059

Das, N., & Chandran, P. (2011). Microbial degradation of petroleum hydrocarbon contaminants: An overview. *Biotechnology Research International, 2011*, 1–13. https://doi.org/10.4061/2011/941810

Deborah Gnana Selvam, A., & Thatheyus, A. J. (2018). Microbial degradation of petroleum hydrocarbons: An overview. In V. Kumar, M. Kumar, & R. Prasad (Eds.), *Microbial action on hydrocarbons* (pp. 485–503). Springer. https://doi.org/10.1007/978-981-13-1840-5_19

Dey, S., Dash, A., & Das, A. P. (2024). Bioremediation as a sustainable and circular solution for mining waste management and bioenergy generation. *Sustainable Management of Mining Waste and Tailings: A Circular Economy Approach*, 107–119. https://doi.org/10.1201/9781003442455-6

Dey, S., Das, A., Mallick, K., Sahu, A., & Das, A. P. (2023). *Environmental Petroleum Waste: Pollution, Toxicity, Sustainable Remediation.* 159–175. https://doi.org/10.1007/978-3-031-48220-5_7

El Fantroussi, S., & Agathos, S. N. (2005). Is bioaugmentation a feasible strategy for pollutant removal and site remediation? *Current Opinion in Microbiology, 8*(3), 268–275. https://doi.org/10.1016/j.mib.2005.04.011

Fedorak, P. M., & Westlake, D. W. S. (1981). Microbial degradation of aromatics and saturates in Prudhoe Bay crude oil as determined by glass capillary gas chromatography. *Canadian Journal of Microbiology, 27*(4), 432–443. https://doi.org/10.1139/m81-066

Fernández, M., Niqui-Arroyo, J. L., Conde, S., Ramos, J. L., & Duque, E. (2012). Enhanced tolerance to naphthalene and enhanced rhizoremediation performance for pseudomonas putida KT2440 via the NAH7 catabolic plasmid. *Applied and Environmental Microbiology, 78*(15), 5104–5110. https://doi.org/10.1128/AEM.00619-12

Fomina, M., & Gadd, G. M. (2014). Biosorption: Current perspectives on concept, definition and application. *Bioresource Technology, 160*, 3–14. https://doi.org/10.1016/j.biortech.2013.12.102

Geetha, S. J., Joshi, S. J., & Kathrotiya, S. (2013). Isolation and characterization of hydrocarbon degrading bacterial isolate from oil contaminated sites. *APCBEE Procedia, 5*, 237–241. https://doi.org/10.1016/j.apcbee.2013.05.041

Gentry, T., Rensing, C., & Pepper, I. (2004). New approaches for bioaugmentation as a remediation technology. *Critical Reviews in Environmental Science and Technology, 34*(5), 447–494. https://doi.org/10.1080/10643380490452362

Ghazali, F. M., Rahman, R. N. Z. A., Salleh, A. B., & Basri, M. (2004). Biodegradation of hydrocarbons in soil by microbial consortium. *International Biodeterioration & Biodegradation, 54*(1), 61–67. https://doi.org/10.1016/j.ibiod.2004.02.002

Hamzah, A., & Tavakoli, A. (n.d.). *Detection of toluene degradation in bacteria isolated from oil contaminated soils.*

Haritash, A. K., & Kaushik, C. P. (2009). Biodegradation aspects of Polycyclic Aromatic Hydrocarbons (PAHs): A review. *Journal of Hazardous Materials, 169*(1–3), 1–15. https://doi.org/10.1016/j.jhazmat.2009.03.137

Husain, S. (2008). Literature overview: Microbial metabolism of high molecular weight polycyclic aromatic hydrocarbons. *Remediation Journal, 18*(2), 131–161. https://doi.org/10.1002/rem.20165

Hussein, Z., Hamido, N., Hegazy, A. K., El-Dessouky, M. A., Mohamed, N., & Safwat, G. (2022). Phytoremediation of crude petroleum oil pollution: A Review. *Egyptian Journal of Botany.* https://doi.org/10.21608/ejbo.2022.136551.1980

Jesubunmi, C. O., Phil, E. K., & Chigbu, C. C. (2022). Isolation and optimization of hydrocarbon—Degrading bacteria. *Asian Journal of Biotechnology and Bioresource Technology*, 46–54. https://doi.org/10.9734/ajb2t/2022/v8i130119

Jiang, Y., Brassington, K. J., Prpich, G., Paton, G. I., Semple, K. T., Pollard, S. J. T., & Coulon, F. (2016). Insights into the biodegradation of weathered hydrocarbons in contaminated soils by bioaugmentation and nutrient stimulation. *Chemosphere, 161*, 300–307. https://doi.org/10.1016/j.chemosphere.2016.07.032

Jobson, A., McLaughlin, M., Cook, F. D., & Westlake, D. W. S. (1974). Effect of amendments on the microbial utilization of oil applied to soil. *Applied Microbiology, 27*(1), 166–171. https://doi.org/10.1128/am.27.1.166-171.1974

Kanungo, J., Sahoo, T., Bal, M., & Behera, I. D. (2024). Performance of bioremediation strategy in waste lubricating oil pollutants: A review. *Geomicrobiology Journal, 41*(4), 360–373. https://doi.org/10.1080/01490451.2023.2245395

Kanungo, J., Sahoo, T., Swain, L. P., & Behera, I. D. (2023). *Toxicity of Persistent Hydrocarbon Pollutants, Sources and Sustainable Remediation Process*. 39–65. https://doi.org/10.1007/978-3-031-48220-5_2

Kao, C. M., Chen, C. Y., Chen, S. C., Chien, H. Y., & Chen, Y. L. (2008). Application of in situ biosparging to remediate a petroleum-hydrocarbon spill site: Field and microbial evaluation. *Chemosphere, 70*(8), 1492–1499. https://doi.org/10.1016/j.chemosphere.2007.08.029

Kotani, T., Yurimoto, H., Kato, N., & Sakai, Y. (2007). Novel acetone metabolism in a propane-utilizing bacterium, *Gordonia* sp. Strain TY-5. *Journal of Bacteriology, 189*(3), 886–893. https://doi.org/10.1128/JB.01054-06

Kumar, A., & Chandra, R. (2020). Ligninolytic enzymes and its mechanisms for degradation of lignocellulosic waste in environment. *Heliyon, 6*(2), Article e03170. https://doi.org/10.1016/j.heliyon.2020.e03170

Lahiri, D., Nag, M., Dey, A., Sarkar, T., Joshi, S., Pandit, S., Das, A. P., Pati, S., Pattanaik, S., Tilak, V. K., & Ray, R. R. (2021). biofilm mediated degradation of petroleum products. *Geomicrobiology Journal*, 1–10. https://doi.org/10.1080/01490451.2021.1968979

Lee, H., Jang, Y., Choi, Y.-S., Kim, M.-J., Lee, J., Lee, H., Hong, J.-H., Lee, Y. M., Kim, G.-H., & Kim, J.-J. (2014). Biotechnological procedures to select white rot fungi for the degradation of PAHs. *Journal of Microbiological Methods, 97*, 56–62. https://doi.org/10.1016/j.mimet.2013.12.007

Lim, M. W., Lau, E. V., & Poh, P. E. (2016). A comprehensive guide of remediation technologies for oil contaminated soil—Present works and future directions. *Marine Pollution Bulletin, 109*(1), 14–45. https://doi.org/10.1016/j.marpolbul.2016.04.023

Maletic, S., Dalmacija, B., & Roncevic, S. (2013). Petroleum hydrocarbon biodegradability in soil—Implications for bioremediation. In V. Kutcherov (Ed.), *Hydrocarbon*. InTech. https://doi.org/10.5772/50108

Martínková, L., Uhnáková, B., Pátek, M., Nešvera, J., & Křen, V. (2009). Biodegradation potential of the genus Rhodococcus. *Environment International, 35*(1), 162–177. https://doi.org/10.1016/j.envint.2008.07.018

Mittal, A., & Singh, P. (2009). A feasibility study for assessment of in-situ bioremediation potential of a crude oil degrading pseudomonas consortium. *Journal of Scientific Research, 2*(1), 127–137. https://doi.org/10.3329/jsr.v2i1.2601

Mohanta, S., Pradhan, B., & Behera, I. D. (2024). Impact and remediation of petroleum hydrocarbon pollutants on agricultural land: A review. *Geomicrobiology Journal, 41*(4), 345–359. https://doi.org/10.1080/01490451.2023.2243925

Moody, J. D., Freeman, J. P., Doerge, D. R., & Cerniglia, C. E. (2001). Degradation of phenanthrene and anthracene by cell suspensions of *Mycobacterium* sp. Strain PYR-1. *Applied and Environmental Microbiology, 67*(4), 1476–1483. https://doi.org/10.1128/AEM.67.4.1476-1483.2001

Mt, P., & Zaida Z, N. (2018). Effectiveness of single and microbial consortium of locally isolated beneficial microorganisms (LIBeM) in bioaugmentation of oil sludge contaminated soil at different concentration levels: A laboratory scale. *Journal of Bioremediation & Biodegradation, 09*(02). https://doi.org/10.4172/2155-6199.1000430

Muratova, A., Pozdnyakova, N., Makarov, O., Baboshin, M., Baskunov, B., Myasoedova, N., Golovleva, L., & Turkovskaya, O. (2014). Degradation of phenanthrene by the rhizobacterium Ensifer meliloti. *Biodegradation, 25*(6), 787–795. https://doi.org/10.1007/s10532-014-9699-9

Nag, M., Lahiri, D., Ghosh, S., Sarkar, T., Pati, S., Das, A. P., Ram, D. K., Bhattacharya, D., & Ray, R. R. (2023). Application of microorganisms in biotransformation and bioremediation of

environmental contaminant: A review. *Geomicrobiology Journal*. https://doi.org/10.1080/01490451.2023.2261443
Njoku, K., Yussuf, A., Akinola, M., Adesuyi, A., Jolaoso, A., & Adedokun, A. (2017). Mycoremediation of petroleum hydrocarbon polluted soil by *Pleurotus pulmonarius*. *Ethiopian Journal of Environmental Studies and Management, 9*(1), 865. https://doi.org/10.4314/ejesm.v9i1.6s
Nzila, A. (2018). Current status of the degradation of aliphatic and aromatic petroleum hydrocarbons by thermophilic microbes and future perspectives. *International Journal of Environmental Research and Public Health, 15*(12), 2782. https://doi.org/10.3390/ijerph15122782
Olubunmi, C. (2014). Isolation of oil—Degrading microorganisms in spent engine oil—Contaminated soil. *Journal of Biology*.
Ossai, I. C., Ahmed, A., Hassan, A., & Hamid, F. S. (2020). Remediation of soil and water contaminated with petroleum hydrocarbon: A review. *Environmental Technology & Innovation, 17*, Article 100526. https://doi.org/10.1016/j.eti.2019.100526
Pagé, A. P., Yergeau, É., & Greer, C. W. (2015). Salix purpurea stimulates the expression of specific bacterial xenobiotic degradation genes in a soil contaminated with hydrocarbons. *PLoS ONE, 10*(7), Article e0132062. https://doi.org/10.1371/journal.pone.0132062
Paul, A., Dey, S., Ram, D. K., & Das, A. P. (2023). Hexavalent chromium pollution and its sustainable management through bioremediation. *Geomicrobiology Journal*, 1–11. https://doi.org/10.1080/01490451.2023.2218377
Philp, J. C., & Atlas, R. M. (2014). Bioremediation of contaminated soils and aquifers. In R. M. Atlas & J. Philip (Eds.), *Bioremediation* (pp. 139–236). ASM Press. https://doi.org/10.1128/9781555817596.ch5
Pradhan, G., Tripathy, B., Ram, D. K., Digal, A. K., & Das, A. P. (2023). Bauxite mining waste pollution and its sustainable management through bioremediation. *Geomicrobiology Journal*, 1–10. https://doi.org/10.1080/01490451.2023.2235353
Priyadarsini, S., & Das, A. P. (2024). Analytical and structural characterization of waste lithium-ion batteries for their effective recycling strategy. *Environmental Chemistry and Ecotoxicology*. https://doi.org/10.1016/j.enceco.2024.12.004
Priyadarsini, S., & Das, A. P. (2025). Lithium bioleaching: A review on microbial-assisted sustainable technology for lithium bio-circularity. *Journal of Water Process Engineering, 69*, Article 106744. https://doi.org/10.1016/j.jwpe.2024.106744
Qi, J., Wang, B., Li, J., Ning, H., Wang, Y., Kong, W., & Shen, L. (2015). Genetic determinants involved in the biodegradation of naphthalene and phenanthrene in Pseudomonas aeruginosa PAO1. *Environmental Science and Pollution Research, 22*(9), 6743–6755. https://doi.org/10.1007/s11356-014-3833-4
Raza, M., Zakaria, M. P., Hashim, N. R., Yim, U. H., Kannan, N., & Ha, S. Y. (2013). Composition and source identification of polycyclic aromatic hydrocarbons in mangrove sediments of Peninsular Malaysia: Indication of anthropogenic input. *Environmental Earth Sciences, 70*(6), 2425–2436. https://doi.org/10.1007/s12665-013-2279-1
Reddy Koteswara, G., Bhargavi, S., Yamuna, D., Anjani Devi, M., Jeevana Lakshmi Kumar, G., Manas, T., Adarsh, S., Bhavya Sai, V., Nath, H., Alreshidi, M. A., Togun, H., Ram, D. K., & Das, A. P. (2025). Recent advancements in the remediation of marine hydrocarbon and plastic pollutants. *Environmental Science and Engineering, Part F218*, 127–142. https://doi.org/10.1007/978-3-031-81615-4_7
Rojo, F. (2009). Degradation of alkanes by bacteria. *Environmental Microbiology, 11*(10), 2477–2490. https://doi.org/10.1111/j.1462-2920.2009.01948.x
Rosenberg, E. (2006). Hydrocarbon-Oxidizing bacteria. In M. Dworkin, S. Falkow, E. Rosenberg, K.-H. Schleifer, & E. Stackebrandt (Eds.), *The Prokaryotes* (pp. 564–577). Springer. https://doi.org/10.1007/0-387-30742-7_18
Rout, P. G., & Behera, S. K. (2024). Evaluation of organosulfur compound degradation property of a noble bacterial strain of Proteus mirabilis. *Environmental Chemistry and Ecotoxicology, 6*, 223–228. https://doi.org/10.1016/j.enceco.2024.06.004

Rout, P. G., Mohanty, A. K., Pradhan, N., Biswal, S. K., & Behera, S. K. (2022). Study on the reaction mechanism of oxidative microbial desulfurization of organic sulfur-rich coal. *Geomicrobiology Journal, 39*(3–5), 210–218. https://doi.org/10.1080/01490451.2021.1967523

Rout, P. G., Singh, A., & Behera, S. K. (2024). Insights into the kinetics of the microbial degradation of dibenzothiophene (DBT) by *Acidithiobacillus ferrooxidans*. *Geomicrobiology Journal, 41*(4), 451–458. https://doi.org/10.1080/01490451.2023.2245378

Sarika Saxena, V. P. (2015). Treatment of oil sludge contamination by composting. *Journal of Bioremediation & Biodegradation, 06*(03). https://doi.org/10.4172/2155-6199.1000284

Sathishkumar, M., Binupriya, A. R., Baik, S., & Yun, S. (2008). Biodegradation of crude oil by individual bacterial strains and a mixed bacterial consortium isolated from hydrocarbon contaminated areas. *CLEAN—Soil, Air, Water, 36*(1), 92–96. https://doi.org/10.1002/clen.200700042

Sharma, I. (2021). Bioremediation techniques for polluted environment: concept, advantages, limitations, and prospects. In M. Alfonso Murillo-Tovar, H. Saldarriaga-Noreña, & A. Saeid (Eds.), *Trace Metals in the Environment—New Approaches and Recent Advances*. IntechOpen. https://doi.org/10.5772/intechopen.90453

Simpanen, S., Dahl, M., Gerlach, M., Mikkonen, A., Malk, V., Mikola, J., & Romantschuk, M. (2016). Biostimulation proved to be the most efficient method in the comparison of in situ soil remediation treatments after a simulated oil spill accident. *Environmental Science and Pollution Research, 23*(24), 25024–25038. https://doi.org/10.1007/s11356-016-7606-0

Singh, H. (2006). *Mycoremediation: Fungal bioremediation* (1st ed.). Wiley. https://doi.org/10.1002/0470050594

Singh, R. (2017). Microbial biotransformation: A process for chemical alterations. *Journal of Bacteriology & Mycology: Open Access, 4*(2). https://doi.org/10.15406/jbmoa.2017.04.00085

Subashchandrabose, S. R., Ramakrishnan, B., Megharaj, M., Venkateswarlu, K., & Naidu, R. (2013). Mixotrophic cyanobacteria and microalgae as distinctive biological agents for organic pollutant degradation. *Environment International, 51*, 59–72. https://doi.org/10.1016/j.envint.2012.10.007

Sylvester, S. M. S., Muthusamy, S., & Alagarsamy, P. (2025). Biodegradation and bioremediation efficacy of mangrove-associated Bacteria. In S. Muthusamy, R. Manikkam, & G. Venugopal (Eds.), *Mangrove microbiome* (pp. 171–186). Springer Nature. https://doi.org/10.1007/978-981-96-2602-1_8

Thenmozhi, R., Arumugam, K., Nagasathya, A., Thajuddin, N., & Paneerselvam, A. (2013). *Studies on Mycoremediation of used engine oil contaminated soil samples.*

Throne-Holst, M., Wentzel, A., Ellingsen, T. E., Kotlar, H.-K., & Zotchev, S. B. (2007). Identification of novel genes involved in long-chain *n*-alkane degradation by *Acinetobacter* sp. Strain DSM 17874. *Applied and Environmental Microbiology, 73*(10), 3327–3332. https://doi.org/10.1128/AEM.00064-07

Varjani, S. J. (2017). Microbial degradation of petroleum hydrocarbons. *Bioresource Technology, 223*, 277–286. https://doi.org/10.1016/j.biortech.2016.10.037

Vázquez, S., Nogales, B., Ruberto, L., Hernández, E., Christie-Oleza, J., Lo Balbo, A., Bosch, R., Lalucat, J., & Mac Cormack, W. (2009). Bacterial community dynamics during bioremediation of diesel oil-contaminated antarctic soil. *Microbial Ecology, 57*(4), 598–610. https://doi.org/10.1007/s00248-008-9420-9

Viesser, J. A., Sugai-Guerios, M. H., Malucelli, L. C., Pincerati, M. R., Karp, S. G., & Maranho, L. T. (2020). Petroleum-tolerant rhizospheric bacteria: isolation, characterization and bioremediation potential. *Scientific Reports, 10*(1), 2060. https://doi.org/10.1038/s41598-020-59029-9

Wang, Q., Zhang, S., Li, Y., & Klassen, W. (2011). Potential approaches to improving biodegradation of hydrocarbons for bioremediation of crude oil pollution. *Journal of Environmental Protection, 02*(01), 47–55. https://doi.org/10.4236/jep.2011.21005

Whelan, M. J., Coulon, F., Hince, G., Rayner, J., McWatters, R., Spedding, T., & Snape, I. (2015). Fate and transport of petroleum hydrocarbons in engineered biopiles in polar regions. *Chemosphere, 131*, 232–240. https://doi.org/10.1016/j.chemosphere.2014.10.088

Wongsa, P., Tanaka, M., Ueno, A., Hasanuzzaman, M., Yumoto, I., & Okuyama, H. (2004). Isolation and characterization of novel strains of pseudomonas aeruginosa and serratia marcescens possessing high efficiency to degrade gasoline, kerosene, diesel oil, and lubricating oil. *Current Microbiology, 49*(6), 415–422. https://doi.org/10.1007/s00284-004-4347-y

Wu, M., Li, W., Dick, W. A., Ye, X., Chen, K., Kost, D., & Chen, L. (2017). Bioremediation of hydrocarbon degradation in a petroleum-contaminated soil and microbial population and activity determination. *Chemosphere, 169*, 124–130. https://doi.org/10.1016/j.chemosphere.2016.11.059

Yamada-Onodera, K., Mukumoto, H., Katsuyama, Y., & Tani, Y. (2002). Degradation of long-chain alkanes by a polyethylene-degrading fungus, Penicillium simplicissimum YK. *Enzyme and Microbial Technology, 30*(6), 828–831. https://doi.org/10.1016/S0141-0229(02)00065-0

Yergeau, E., Sanschagrin, S., Beaumier, D., & Greer, C. W. (2012). Metagenomic analysis of the bioremediation of diesel-contaminated canadian high arctic soils. *PLoS ONE, 7*(1), Article e30058. https://doi.org/10.1371/journal.pone.0030058

Yergeau, E., Sanschagrin, S., Maynard, C., St-Arnaud, M., & Greer, C. W. (2014). Microbial expression profiles in the rhizosphere of willows depend on soil contamination. *The ISME Journal, 8*(2), 344–358. https://doi.org/10.1038/ismej.2013.163

Petroleum Hydrocarbon Pollutants from Waste Engine Oil in Municipal Waste: Concern and Remediation Strategies

Rajalaxmi Badra, Mitrasen Majhi, Binod Bihari Panda, Ipsita Dipamitra Behera, Alok Prasad Das, and Manisha Bal

1 Introduction

Waste engine oil is highly polluting material that must be handled carefully before it is dumped on the ground or into water ways such as sewers. This could lead to contamination of soil and ground water such as contaminated materials being recycled to reduce engine oil costs (Demirbas et al., 2015). Waste engine oil contains a number of poisonous compounds that are extremely harmful to both human health and the environment. The main cause of these components is breakdown of engine oils. Different heavy metals exist. Lead is a serious pollutant that can harm the nervous system and is frequently present in high concentrations. Different levels of chromium have been found and it is connected to skin irritation and lung problems. There are also other different metals that are known to cause cancer, such as arsenic and cadmium (Dhiman, 2024). Engine oil waste contains harmful organic compounds. Hydrocarbon polycyclic aromatic hydrocarbons found in waste engine oil are hazardous

R. Badra · M. Majhi · B. B. Panda · I. D. Behera (✉)
Department of Chemical Engineering, Indira Gandhi Institute of Technology, Sarang, Dhenkanal, Odisha, India
e-mail: ipsitadbehera@gmail.com

M. Majhi
State Pollution Control Board, Bhubaneswar, Odisha, India

Biju Patnaik University of Technology, Rourkela, Odisha, India

A. P. Das
Department of Life Science, Rama Devi Women's University, Bhubaneswar, Odisha, India

M. Bal
Department of Chemical Engineering, National Institute of Technology, Durgapur, West Bengal, India

A. P. Das et al. (eds.), *Sustainable Municipal Waste Management with a Zero Waste Approach*, Sustainable Environmental Waste Management Strategies,
https://doi.org/10.1007/978-3-032-05950-5_20

and could have long-term effects on health (Mishra et al., 2021). When oil breaks down, volatile organic compounds like benzene and mercaptans are released into the atmosphere. Toxic compounds like anti-wear agents and antioxidants are frequently included in engine lubricants, which can make the waste more dangerous (Sánchez-Alvarracín et al., 2021). Waste motor oil enters the environment primarily through improper disposal, leakage and runoff, all of which have detrimental impacts on the ecosystem and public health. Polycyclic aromatic hydrocarbon and heavy metals are among the harmful compounds found in this oil that pollute the air, water and soil and also endanger both people and animals (Behera et al., 2020). Through the food chain, the contamination eventually affects human health by upsetting ecosystems, influencing microbial communities and perhaps causing bioaccumulation in aquatic ecosystems (Das & Mishra, 2008; Dey et al., 2024; Kanungo et al., 2023a, 2023b; Mallick et al., 2023). On land, waste engine oil is frequently disposed of incorrectly which contaminates the soil. Toxic compounds are introduced, disrupting the microbial populations and soil structure. Toxic bioaccumulation in aquatic life and disturbance of aquatic ecosystems are caused by runoff and direct discharge into water bodies. Incomplete combustion and volatilization release toxic gases and particle matter which aggravate respiratory conditions and contribute to air pollution (Mishra et al., 2021). Oxidative stress, immunotoxicity, genotoxicity and cancer can result from exposure to used motor oil. Because heavy metal bioaccumulate, it can also lead to organ problems and skin discomfort (Priyadarsani & Das, 2024, 2025). When animals are exposed to spent crankcase oil, their body weight decreases, their blood parameters change and their enzyme levels rise, all of which are signs of liver and kidney damage. Reduced microbial development and higher mortality among earthworms and other soil microorganisms result in changed soil functionality (Paul et al., 2024). Both human and animals are at serious risk for serious health problems when exposed to spent crankcase oil, which is a waste engine oil. Such exposure can cause a wide range of symptoms that impact different physiological systems and result in both acute and long-term health problems (Das et al., 2023; Dey et al., 2023b; Kanungo et al., 2024a, 2024b; Reddy Koteswara et al., 2025). Direct contact with contaminated surfaces frequently results in skin discomfort and rashes. Breathing problems and neurological issues might result from inhaling volatile chemicals. Prolonged exposure can lead to immunological malfunction, stress and possibly even cancer (Bekibele et al., 2022). Animals exposed to spent crankcase oil exhibit substantial decreases in haemoglobin levels and red blood cell counts. Vital organs including the liver and kidneys have lesions found by histopathological investigations. Physiological problems result from animals exposed to contaminated settings having increased levels of dangerous metals including cadmium and lead (Kuppusamy et al., 2020). Research on Swiss albino mice has demonstrated that exposure to water-soluble components of waste engine oil causes the liver to significantly accumulate metals like iron, zinc, copper and nickel in addition to raising levels of renal biomarkers like creatinine and urea. Similarly male Sprague Dawley rats exposed to used engine oil experienced a decline in sperm count and a rise in abnormalities suggesting negative effects on reproduction (Akintunde et al., 2015). Research conducted after the deepwater horizon oil spill found that individuals exposed to oil suffered from multiple organ dysfunction with symptoms such as immune system impairment and oxidative stress. Diesel engine emissions have been associated with respiratory problems and potential carcinogenic risks in human although evidence of genotoxic effects remains inconsistent (Babalola & Oni, 2018). The toxicity levels of fresh and waste engine oil differ notably due to variations in their chemical composition and environmental impact. Fresh engine oil typically has fewer contaminants and retains its original properties, whereas waste engine oil accumulates hazardous substances over time making it more toxic. Fresh engine oil improves engine performance and protect its components. With a higher flashpoint and lower viscosity, it demonstrates greater stability and reduced toxicity. Over time, waste engine oil gathers pollutants such as heavy metals, oxidized byproducts and degraded additives, all of which contribute to its increased toxicity. Research indicates that exposure to waste oil can result in oxidative stress and other toxicological effects in both humans and animals (Sulayman, 2019).

Polycyclic aromatic hydrocarbons are heavy metals like cadmium, benzene, arsenic, lead, toluene and other neurotoxic compounds which are dangerous materials found in waste motor oil (Behera et al., 2020; Lahiri et al., 2021). Prolonged exposure to these substances can have detrimental effects on the nervous system. These compounds can enter the body by the mouth, nose or skin and progressively accumulate toxicity. An elevated risk of neurological disorders including Parkinson's and Alzheimer's disease that has been associated with long exposure to heavy metals and PAHs found in old motor oil (Wami-Amadi 2025). Neuronal damage and oxidative stress can result from the buildup of lead, cadmium and manganese in the brain. Prolonged exposure can cause nerve injury which can cause pain, tingling, numbness and paralysis in the extremities. Waste engine oil contains certain heavy metals that can disrupt the transmission of nerve signals. Chronic exposure often results in memory loss, concentration problems and diminished problem-solving skills. Exposure to benzene and toluene has been linked to behavioural abnormalities, learning disabilities and a lower IQ. Because it affects neurotransmitters like dopamine and serotonin, prolonged exposure can lead to mood changes, irritation, melancholy and anxiety. Certain compounds in waste engine oil have the potential to alter brain chemistry and hormone balance. Impaired coordination, tremors, muscle weakness and issues with balance and fine motor abilities can all result from damage to the central nervous system. Vascular damage brought on by heavy metals and PAHs raises the risk of stroke, brain ischemia and persistent inflammation of the brain's tissues (Das et al., 2014). Lead, arsenic and cadmium are among the hazardous heavy metals found in waste engine oil, along with benzene, polycyclic aromatic hydrocarbons and other hydrocarbons that can contaminate land, water and air. New lubricants, fuels or industrial oils can be made from used oil once it has been cleaned and processed. Refining reduces pollution and conserves resources by eliminating harmful pollutants. Avoid disposing of waste engine oil in water sources, drains or soil. Authorized recycling facilities or hazardous waste disposal locations to dispose of spent oil. Waste oil collection facilities should be set up at community centres, vehicle repair shops and petrol stations by government and business (Mong'ina 2018). Hydrocarbons are broken down into non-toxic chemicals by oil degrading bacteria and fungus such as Pseudomonas, Mycobacterium and Aspergillus, utilized in soil remediation and oil spill cleanup operations. Hydrocarbons are absorbed and broken down by some plants such as vetiver grass, willows and sunflowers, beneficial for the restoration of agriculture and contaminated wetlands. Natural breakdown is accelerated by the addition of oil degrading enzymes such as laccases and peroxidases (Chunyan et al., 2023). Garages and businesses use sealed storage tanks, oil sumps and spill containment trays. To clean up spills use environmentally friendly absorbents like sawdust, peatmoss and coconut husk. Regular inspections and waste oil management procedures should be implemented by industries. Make use of bio-based oils that break down more quickly than petroleum-based ones, such as castor, soybean or rapeseed oil. Lessens environmental toxicity over time. By extending engine oil's lifespan and advanced oil filtering systems help cut down on waste. Minimizes the need for disposal practices. Promoted corporate social responsibility initiatives for environmentally good behaviour (Das et al., 2014).

Edible oil is in greater demand as of rising living standards and population growth since it supplies essential energy and minerals for everyday requirements of everyone. Triacyl glycerides, which are made up of several fatty acids, make up the majority of edible oils. Phytosterols, phospholipids, tocopherols, free fatty acids and other antioxidants are among the molecules or groups of substances present in animal and plant life. Because it contains so many hazardous substances, eating so called waste cooking oil or its derivatives can be harmful to our health (Wee et al., 2016). Since some of the Taiwanese company's lard products were derived from lard for animal feed or collected waste oils, they might have been contaminated in this food catastrophe. However unlawful disposal of waste cooking oil may result in environmental degradation including vapour explosion, water pollution and stink. Importantly, WCO provides renewable resources for fuel oil synthesis and alternative feedstocks for petroleum-based chemicals (Karmee 2017). Triglycerides are formed when carbon, hydrogen and oxygen combine to generate oils like in other organic compounds. These are ester molecule compound for one glycerol and three fatty acids. Based on the fatty acid structure, cooking oil can be classified as either saturated or unsaturated. The majority of animal fats contain saturated fatty acids which have longer shelf life than unsaturated ones because they are more stable and less likely to turn rancid (Gunstone 2013). WCO includes a lot of free fatty acids, it smells terrible and corrodes concrete and metal components. More significantly, residential waste and related commercial, industrial and institutional wastes are categorized as municipal wastes. It may result in significant environmental issues. However, WCO is a neglected technology. Bioresource in urban settings with potential for material and energy related value addition (Panadare, 2015).

2 Sources of WEO

WEO is a major component of municipal waste generated from industrial machinery vehicles and other mechanical systems that rely on lubricating oils. Over time, engine oil degrades due to thermal breakdown, oxidation and contamination with heavy metals, fuel residues and other toxic substances (Kanungo et al., 2023a, 2023b). Once it is no longer suitable for lubrication it becomes waste and if not properly managed, poses serious environmental and health risks. Improper disposal of waste engine oil such as disposing of WEO in soil, water and landfills that cause extensive contamination. It contaminates soil, reducing water supplies. When burned under uncontrolled conditions, waste engine oil releases harmful emissions that contribute to air pollution and respiratory diseases. Due to its hazardous nature, waste engine oil is classified as a significant environmental pollutant, requiring effective waste management strategies. Managing waste engine oil as municipal waste requires sustainable approaches including recycling, re-refining, bioremediation and strict regulatory policies. Proper disposal and treatment can minimize its harmful impact while promoting resources recovery and environmental protection (Nutongkaew et al., 2014). The demand for production of petroleum and consumption is rising as a result of its growing significance in the advancement of society that large amount of petroleum waste is dumped into the environment without being properly treated (Behera et al., 2022). Petroleum contamination is a global issue that endangers both environment and human health. Major source of petroleum hydrocarbon pollution is the buildup of oily sludge at the bottom of crude oil storage tanks. Along with other materials like sand, clay, dirt and heavy metals, this oil tank bottom sludge has large quantities of petroleum hydrocarbons. Oil tank sludge must be appropriately treated and disposed of in an environmentally friendly manner due to its dangerous nature (Liu et al., 2018). Environmental groups around the world have adopted a variety of technologies to handle various kinds of oily sludge, including petroleum hydrocarbon residue reduction (Dey et al., 2023a; Mohanta et al., 2024). Compost, food waste and sludge can be providing microbes, nutrients and organic substances are easily biodegradable despite being toxic. The aeration and water penetration of the composting mixture can be enhanced by adding bulking agents such as green trash which will enhance the removal of petroleum hydrocarbons. There are two steps in the bioremediation of petroleum pollution. Rapid decomposition is the initial stage where readily biodegradable components (Behera et al., 2021, 2022). As these are depleted the process slows down in the second stage due to the remaining compounds being more resistant to degradation (Nutongkaew et al., 2014). Oil is vital globally, contributing significantly to economic expansion and escalating international rivalry. Both residential and commercial applications including manufacturing, transportation and heating depend on its energy. Revenue from crude oil exports significantly boostsGDP and supports industrial expansions, particularly in the petrochemical sector. To harness these benefits, various processes such as exploration, extraction, refining and transportation are undertaken. These activities contribute to environmental pollution, raising global concerns. Crude oil is a complex mixture of hydrocarbons that include aromatic, cyclic, branching and linear molecules (Fig. 1). Nonhydrocarbon components like nitrogen, sulphur and oxygen are present in certain hydrocarbons such as asphaltenes and resins (Tang & Angela, 2019). This technique is known as hydraulic fracturing or fracking that is used to release crude oil and natural gas from dense underground rock formations. Using this technique, high-pressure injections of a water-based fluid are used to fracture the reservoir rock. Larger fractures that radiate outward from the wellbore are created when the injected fluid widens the fissures in the rock. The flow and extraction of oil and gas are improved by these fractures because they increase the reservoir's surface area and hydraulic conductivity. In several phases of the hydraulic fracturing process, gelling polymers whether linear or crosslinked in structure are essential. Adding polymers to drilling muds to serve as lubricants and promote smoother operations is one of the first uses of polymers in well drilling. Polymers are used in gravel packing to stabilize sand and prevent the buildup of formation sand which could damage the well (Mouser et al., 2016). Polymers and gelling agents used to frack fluids minimize fluid loss by forming a filter cake that retains the fluid inside the fractures. Because the filter cakes are useless, the fracking fluids can seep into the nearby rock. Unwanted internal filter cakes may form within the fractures as a result of the fluid's polymers being absorbed. Fracking fluids are treated with oxidizers and acids such as ammonium persulfates and hydrogen chloride that are activated with catalysts such as iron, acetoacetate or amine that break down any residual polymers in the reservoir. These oxidative breakers produce free radicals that can damage filter cakes by rupturing carbon bonds in the polymers. Because oxidizers are not very selective in their breakdown of polymers, high concentrations work best. Their use can also result in the formation of unwanted by products which may cause corrosion reduce conductivity, pose environmental risks and create hazardous conditions for workers (Bunger et al., 2013). Waste oil presents a valuable and economical resource for biodiesel production, sourced from various materials like used cooking oil, animal fats and byproducts from vegetable oil refining. The crankcase and lubricating wastes produced during routine maintenance of motorized vehicles and machinery which are the source of waste oil and petroleum products. One major source is the cooking oil that is discarded from homes and businesses (Musa et al., 2023).

Although used cooking oil and animal fats share a similar basic chemical structure, difference in their fatty acid makeup especially after frying. A solution to waste disposal problems is to turn used cooking oil into biodiesel. Animal fats which are often solid or thick at room temperature include higher levels of sulphur, nitrogen and saturated fatty acids. These components can negatively affect biodiesel's performance in cold climates and increase sulphur dioxide emissions (Panadare 2015). Waste oils can combine with fresh fuel oil and be used in boilers to produce heat or electricity or they can be utilized straight away without processing as road oils to reduce dust. It is crucial to carefully evaluate the emissions of heavy metals and other environmental concerns prior to burning or recycling waste oil in any other manner (Musa et al., 2023). Polychlorinated biphenyls, lubricants, aliphatic and aromatic hydrocarbons, and the heavy metal pollutants are dangerous and carcinogenic compounds included in used oil (Das et al., 2025; Reddy Koteswara

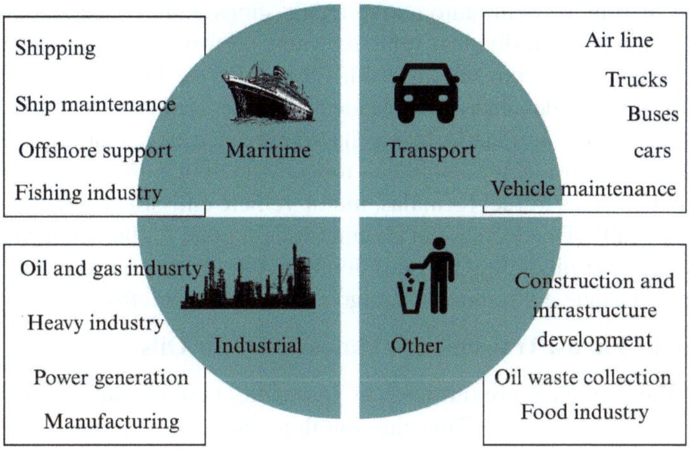

Fig. 1 Sources of waste oil

et al., 2025). waste engine oil contains highest concentration of heavy metals which are created engine parts deteriorate and include aluminium, chromium, tin, lead, manganese copper zinc, silicon, etc. (Xiaojie et al., 2022; Prabhakar et al., 2019; Raj & Das, 2023). Waste oil comes in the form of cooking oil, animal fat, yellow or brown grease and sludge oil or soap stock from the refining of vegetable oil. Two biggest sources of cooking oil are industrial and house hold waste. Spent cooking oil and animal fat have similar amounts of acylglycerol to vegetable oil (Panadare 2015). Oil waste contributes the most toxic and dangerous compounds and is one of the wastes produced in the greatest quantities worldwide. Waste oil from domestic and industrial sources has detrimental effect on the environment because it is poorly soluble and develops a film that hinders oxygen from diffusing into the water and damages the aquatic biota, waste cooking oil pollutes the water when it is disposed of in the open area. Waste oil has been considered as a possible feedstock for chemical or biological processes that produce biodiesel. Carbon sources for microbial PHA synthesis are abundant in oil waste. In terms of cost and PHA generation in microbial fermentation, lipid-rich oil was found to be more advantageous than a traditional substrate (Musa et al., 2023).

Automative and Mechanical Sources

A. Use Motor Oil

Used motor oil is a major source of waste, originates from cars, trucks, motorcycles and other vehicles during oil changes. It is produced by industrial machinery, heavy equipment and various types of engines. A significant portion comes from routine oil changes in vehicles with automotive repair shops and service stations serving as key collection point. Individual vehicle owners performing their own oil changes contribute to the generation of waste oil. Pb and Br are the conventional marker elements for automobiles that run on leaded gasoline. However, as leaded gasoline is phased out, lead is gradually losing its significance as a marking element for motor vehicles. Due to factors including modern catalytic converters, alterations to motor oil and changes in fuel formulas, motor vehicle emission characteristics have varied dramatically. The decrease in Pb emissions is good for human health, it could complicate receptor modelling and cast doubt on the source composition profiles of present motor vehicle emissions (Kanungo et al., 2024a, 2024b).

B. **Hydraulic Fluids, Transmission Fluids and Gear Oils**

Hydraulic fluids which are utilized in hydraulic systems in automobiles and machinery become waste oil. Transmission fluids need to be replaced on a regular basis just like motor oil which produces waste oil. Of all lubricants, hydraulic oils are the most significant. Greater viscosity index hydraulic oils are for mobile applications, greater temperature and lower ambient temperatures in particular hydraulic systems (Bock 2017). Petroleum-based lubricants have become commonplace in the majority of industrial and automotive applications during the past century. The complexity of lubricants' compositions has gradually increased over the past few decades as their performance and operating needs have become more demanding (Mannekote et al., 2018).

C. **Industrial Sources**

Industries utilize a variety of oils for processing, cooling and lubricating. These oils can all end up as waste. Waste oil is produced when heavy machinery and equipment used in construction, manufacturing and other industries need to be lubricated. Waste hydraulic oil is produced by industrial establishments that use hydraulic systems. Waste engine oil is a consequence of maintaining large engines in ships, airplanes and generators. As a highly polluting substance, used lubricant oil needs to be handled carefully. When old motor oil is spilled into the ground or into streams of water including sewers. It can impact the ecosystems (Mekonnen & Ababa, 2014).

D. **Metal Working Fluids**

Metal working fluids are utilized in metal production and machining and they eventually get polluted. Lawnmowers and other tiny engines run on oil. Oil from power steering and transmission. Because of the possible environmental risks, it is imperative that waste engine oil be handled and disposed properly. Metal working lubricants are essential for controlling the parameters during the process, providing cooling, lubrication and corrosion prevention to boost productivity and efficiency. There should be some metal-to-metal contact during metalworking operations. The quality of the lubricating layer that forms is crucial. Cutting fluids depend heavily on viscosity, viscosity temperature and viscosity pressure characteristics that they are associated with the thickness of the film that is created (Osama et al., 2017).

E. **Food Industry and Household Sources of Oil**

Cooking oil waste is produced by frying and meal preparation in homes, restaurants and food processing facilities. Municipal oil waste is a serious environmental challenge, especially when it comes to spent lubricating and leftover cooking oils. Effective management strategies are required to lessen the risks of improper disposal, which can cause significant environmental contamination. Products left over after animals are processed for food. From the animal fats, biodiesel can be made. Hand-me-down cooking oils are fats and oils that are used for cooking or frying at eateries, fast food restaurants and residences owned by consumers. Under the legislation, they are categorized as municipal wastes such as home garbage and comparable commercial, industrial and institution wastes by the European garbage catalogue, including fractions that are collected separately. Similar to grease and oil mixtures from oil or water separation that contain edible oil and fats, used cooking oil derived from waste treatment facilities is similarly regarded as non-hazardous (Anekwe et al., 2022). WCOs are bio-based oils that have been used in homes, restaurants, fast food chains and the food production sector for cooking and other processing applications. Cooking oil will undergo chemicals reactions such as hydrolysis, heat deterioration, oxidation and polymerization during the procedures resulting in both physical and chemical components and an unpleasant odour. More significantly because it might result in major environmental issues, this disposal has been categorized as a municipal trash. Such as home waste and comparable commercial, industrial and institutional waste. WCOs are an underutilized bioresource in urban settings that have the potential to be valuable in the material and energy sectors (Mekonnen & Ababa, 2014).

Composition of Waste Engine Oil

Waste engine oil contains a complex blend of many contaminants (Table 1). These pollutants come from a variety of sources, such as the oil's natural degradation, the engine's working conditions and component wear and tear. Among them include polycyclic aromatic hydrocarbons, heavy metals and others (Patel & Shadangi, 2020).

A. **Heavy Metals**

The environment contains dangerous and enduring heavy metals (Das et al., 2011; Das & Ghosh, 2023; Ghosh et al., 2016; Mohanty et al., 2016). They are among the most alarming constituents in spent motor oil. Typical heavy metals in waste engine oil include arsenic, cadmium, lead, etc. (Panadare 2015). The wear of engine bearings and other parts are made of lead contamination. Leaded gasoline is still present in engine oil due to legacy contamination and specific additives, despite the fact that its use has drastically declined (Tang, 2024). Engine oil can include cadmium due to the deterioration of metal coating and specific additives. At low quantities, this

Table 1 Composition of waste engine oil (Patel & Shadangi, 2020)

S.l. No.	Composition	Waste engine oil (%)
1	Naphthalene	2.6
2	Acenaphthylene	0.7
3	2- bromonaphthalene	2.9
4	Acenaphthene	2.6
5	Fluorene	4.2
6	Phenanthrene	4.3
7	Anthracene	2.9
8	Fluoranthene	3.4
9	Pyrene	3.9
10	Benzo anthracene	4.2
11	Chrysene	11.6
12	Benzo fluoranthene	4.1
13	Benzopyrene	12.9
14	Indenopyrene	13.5
15	Dibenzo anthracene	3.4
16	Benzo perylene	6.3
17	Iron	325.4
18	Manganese	1.2
19	Zinc	5.4
20	Copper	3.3
21	Chromium	2.3
22	Cadmium	2.4
23	Lead	1.2
24	Nickel	1.1
25	Vanadium	0.8

metal is extremely poisonous and presents significant risks to the environment and human health (Patel & Shadangi, 2020). Although arsenic is less frequent, some metal components and impurities in fuels and additives can cause arsenic contamination, because of its high toxicity and carcinogenic nature, its presence in waste oil is very concerning (Mekonnen & Ababa, 2014).

B. **Additive and Degradation Products**

As the engine runs, the additive alters chemically and to used oil's overall contamination profile. Notable categories include engine part wear and tear, oxidation products and nitration and sulfation products. Prolonged exposure to high temperatures and oxygen produce oxidation products which include varnish, sludge and acidic chemicals. Corrosive acids and other degradation products can be produced by reactions with the nitrogen and sulphur components in the fuel known as nitration and sulfation products. The accumulation of wear debris particles in the oil is caused by mechanical wear on metal surfaces such as copper, aluminium and iron. Several important engine system interactions and activities are the source of the pollutants found in waste engine oil. Metallic particles are produced during this wear process and end up suspended in the engine oil. These particles primarily originate from bearings, valve trains, pistons and rings which are made by lead, copper and tin (Chunyan et al., 2023).

3 Effects of Waste Engine Oil

Petroleum hydrocarbon pollution includes things like gasoline, diesel, crude oil, creosote and other petroleum distillates. Oil spills, underground storage tank breaches, the construction of petrochemical processing facilities and transportation and improper disposal of waste containing petrochemicals are some of the ways that TPHs can contaminate soil. TPH buildup has a detrimental effect on the environment and the health of living things, resulting in water pollution, soil erosion, plant growth suppression and delayed seed germination. Food can get contaminated with hydrocarbons when TPH from soil and water enters the food chain. TPHs, especially lighter ones, have the ability to volatilize and reach the atmosphere, raising the possibility of inhalation exposure for humans. Exposure to TPH has been linked to a number of health problems such as lung irritation, liver and kidney damage and impacts on development and reproduction (Tang, 2024). Waste engine oil-related contaminates can be hazardous to the environment if they are burned or carelessly disposed of in rivers or land. Environmentally acceptable disposal is necessary for dirty clays and acid sludge that can be produced even during refining process. Negligent motor oil disposal seriously contaminates the soil, affecting microbiological activity and earthworm populations. Lower microbial growth and higher earthworm mortality rates have been associated with higher oil concentrations. This can compromise soil health and agricultural output. Every year, Nigeria produces more than 87 million litres of used engine oil waste. Gravity causes the oil to seep into the soil when it spills and it spreads laterally by capillary action and changes in the nature of the soil until it reaches an impermeable layer like bedrock, waterproof clay or an aquifer (Nutongkaew et al., 2014). The compounds may have detrimental impacts on biological systems if they are absorbed by live cells and metabolized. Pollutants from petroleum hydrocarbons are exposed to people, plants and animals through ground water. The chemical ingredients in wasted engine oil influence exposure to the oil affects living things. The waste engine oil pollutants buildup in the kidney and liver. These are highly harmful to living systems. Cadmium is the most dangerous elements in the environment because of its extraordinary long half-life. Exposure to dangerous heavy metals and petroleum hydrocarbons is another risk factor for renal function impairment. Renal failure results in a decreased capacity to remove portentous catabolites, which raises urea, serum creatine and urea are known markers of renal function due to the marked drop up in glomerular filtration rate (Chunyan et al., 2023).

Water Contamination by Waste Engine Oil

Waste engine oils are dangerous oils that come from heavy duty industrial machines and the transportation industry. Significant pollution of the air, soil, sediments, surface water and ground water are caused by improper handling, dumping and other misuses. Long term or acute and long term or chronic toxicities can develop from occupational exposure through repetitive, sustained contact which might offer direct or indirect health concerns. In addition to causing phytotoxicity by affecting plant development, physiology and metabolism, soil pollution also produces genotoxicity by altering the physicochemical properties and biocenosis of the soil.

Aquatic habitats and biodiversity are impacted by surface water pollution (Chunyan et al., 2023). Waste engine oil spills are a noticeable issue that are not only caused by auto mechanics who remove oil from cars but also by generator sets which contribute to some of the WEO that is released into the environment. Particularly Nigeria's public power supply is epileptic and has led to rise in use of generators and the need for engine lubricants. There are other uses that release a lot of waste motor oil released into the environment (Obinia & Afiukwaa, 2013). Water bodies get contaminated via runoff and direct discharge, which can upset aquatic ecosystems, lead to bioaccumulation in aquatic life and pose a health risk to humans. Runoff and leaching are two ways that spent motor oil can get into water bodies. When it rains, oil-contaminated soil can add to surface runoff which can carry oil residues into lakes, rivers and streams. Leaching also has the ability to move oil pollutants into ground water through the soil profile. Improper disposal of dumping spent motor oil into landfills, drains or straight into the ground can contaminate water supplies in the area. Large amounts of used oil can enter aquatic bodies directly through spills and leaks that occur during oil changes, transportation or storage. Used motor oil has negative impact on aquatic life and water quality when it is present in aquatic habitats (Nutongkaew et al., 2014). Decreased dissolved oxygen levels result from oil coatings on water surfaces that hinder oxygen exchange. This may result in hypoxic conditions that are harmful to aquatic life. Aquatic organisms are more vulnerable to the negative effects of PAHs and heavy metals. The acute and long-term toxicity of contaminants can affect fish, invertebrate and rates of reproduction, growth and survival of other aquatic species. Pollutants from used motor oil can buildup in aquatic life and spread up the food chain, endangering the health of human and other predatory animals (Obinia & Afiukwaa, 2013).

Air Contamination

Incineration pollutes the air by releasing toxic chemicals, particle matter that causes lung illness and greenhouse gases that promote global warming (Fig. 2). Total petroleum hydrocarbons which include polycyclic aromatic hydrocarbons, benzene, toluene, ethylbenzene, xylene, congeners of polychlorinated biphenyls, organometallic compounds and toxic chemicals additives are the cause of waste engine oil's toxicity. The noxiousness of waste engine oil is caused by total petroleum hydrocarbons which include polycyclic aromatic hydrocarbons, benzene, toluene, xylene, congeners of polychlorinated biphenyls, organometallic compounds and toxic chemical additives (Bekibele et al., 2022). Numerous volatile organic compounds found in used engine oil have the ability to volatize into the atmosphere. Organic substances are readily evaporated into the atmosphere due to their high vapour pressure at ambient temperature. These substances may come from fuel residues, combustion leftovers or deteriorated additives among other sources found in old oil. VOCs can contribute to air pollution by evaporating from water surfaces. Soil has been contaminated by oil or carelessly stored spent oil. These substances can be discharged into the atmosphere even at room temperature due to their volatile nature. VOCs are released together with other dangerous pollutants including particulate matter and PAHs when spent engine oil is not completely burned, whether it is being burned as fuel or as disposal strategy (Dhiman 2024). VOCs are released into the atmosphere after combustion and have the potential to generate particulate matter and ground level ozone. Used engine oil releases volatile organic compounds and other pollutants that impact air quality and lead to more significant atmospheric changes including smog formation. Main ingredients of smug, ground level ozone is created when VOCs combine with nitrogen oxides. In the presence of sunlight. Smog can worsen respiratory disorders, make asthma worse and make it harder to see. The main components of smug, ground level ozone are nitrogen oxides and volatile organic compounds. While sunshine is present, smog can extract respiratory conditions, exacerbate asthma and impair vision. Ground level ozone is strong respiratory irritant and a serious public health concern, especially in cities with heavy traffic and industrial activity (Bekibele et al., 2022).

Toxic pollutants like benzene and PAHs, which are carcinogenic substances that can be inhaled by humans and animals. These contaminants can be found in old motor oil emissions. Chronic exposure to these harmful air pollutants can cause serious health problems. Some volatile organic compounds and polycyclic aromatic hydrocarbons emitted from spent motor oil are greenhouse gases or precursors to secondary pollutants that affect climatic patterns and cause global warming (Sulayman 2019).

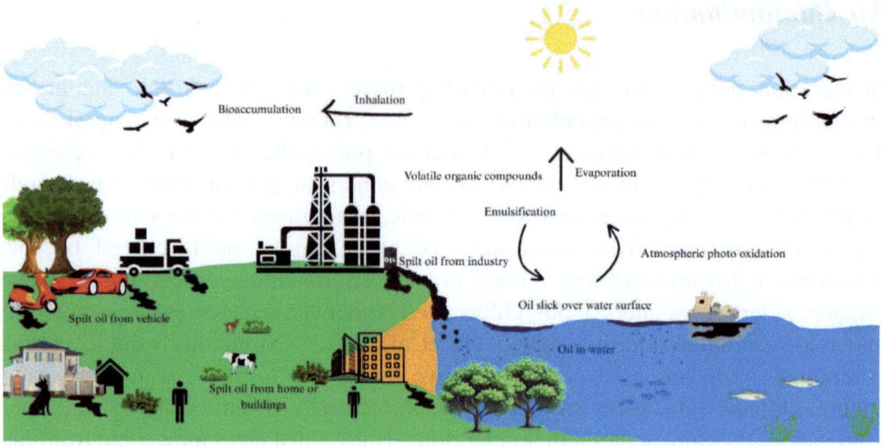

Fig. 2 Environment polluted by waste oil

Soil Contamination

Over 75% of Nigeria's total used oil generating capacity in 2004 came from used crankcase engine oil, and the remaining 25% coming from industry-based used oil. A survey of the majority of Nigerian cities revealed that there were little to no organized disposal practices for waste engine oil. The prevalent methods of disposal included careless spraying on land, pouring down sewers and storing in plastic containers like kegs, jerry cans and drums. The country lacks an organized facility for recycling waste oil and the amount of waste oil produced annually, along with the careless and intentional methods of disposal, necessities a regular assessment of the waste oil's effects on the environment and human health. Engine oil waste has little organized disposal methods according to a survey conducted in most Nigerian cities. Careless spraying on land, pouring down sewers and storing in plastic containers like drums and jerry cans were the most common disposal techniques. The volume of waste oil produced annually, the lack of saturated infrastructure for recycling and the deliberate and negligent disposal techniques necessitate a frequent evaluation of waste oil's impact on the environment and public health. These include preserving wood from termites using waste engine oil combined with bitumen for road construction and using it as a weed killer and dust suppressant by spraying into the soil surface (Obinia & Afiukwaa, 2013).

Oil pollution can be caused by toxic substances that disrupt microbial populations, soil structure and nutrient cycles. Used engine oil contains a complex mix of contaminants to soil and alters its physical and chemical characteristics with changing its structure and permeability. Used engine oil contaminate or can decrease soil porosity and permeability which can impact water infiltration and retention. Erosion and surface runoff may rise. Used motor oil's hydrocarbon have the capability to attribute to soil particles and decrease the availability of vital nutrients like phosphate and nitrogen. Plant development and soil fertility may suffer (Karmee 2017). Depending on the particular contaminants and interaction with the soil's constituents, used engine oil can change the pH of the soil making it more acidic and alkaline. The modifications may interfere with microbes and plant's ability to absorb nutrients. Plant life and soil microbes are especially susceptible to the harmful impacts of pollutants from used motor oil. The variety and activity of soil microorganisms can be inhibited by heavy metals and PAHs (Fig. 2). Because used engine oil contains harmful components, it can cause physiological stress, slow plant growth and lower germination rates. Heavy metals can bioaccumulate of plant absorption, endangering herbivores and higher trophic levels. Long-term contamination can reduce soil biodiversity which will impact the soil ecosystems resilience and functionality (Ebenezer & Abiodun, 2015).

Health Effect

Human exposure to waste engine oil can occur through the submission of waste engine oil in hair creams to protect the scalp from the burning sensation of hair relaxers, as well as through the application of WEO to open wounds or cuts on the skin with the simple belief that it aids in the healing process. This is in addition to the inhalation and dermal contact of fumes from waste engine oil, particularly by automobile repairs who handle this substance frequently. All of these actions and behaviours are the result of improper treatment and disposal of spent motor oil, which has led to some ecological damage and environmental deterioration in the Nigeria ecosystem (Obinia & Afiukwaa, 2013).

4 Remediation of Waste Engine Oil

Many locations presently use the technically feasible method for turning waste oil into fuel. Viable process of refining waste oil to produce fuel. In the past, the widespread practice of recycling has been hindered by environmental rules, product specification and challenges in eliminating contaminants such as dirt, metals, oxidation products and water. Used engine oil pollutes the environment, damaging both human well-being and ecosystems. Increased soil deterioration and the associated loss of crucial resources necessary for life are the results of unsustainable ecosystem management and use. Marinating ecosystems for continuity and long-term health is essential and needs immediate care and especially considering the growing anthropogenic pressure on the soil brought on by people extensive material consumption. Waste motor oil treatment is a significant environmental issue and several strategies are investigated to lessen its negative impacts. Particular bacterial and fungal strains microbial bioremediation have become a viable method for breaking down the hydrocarbons in wasted engine oil (Fig. 3). Bio-stimulation techniques which increase microbial

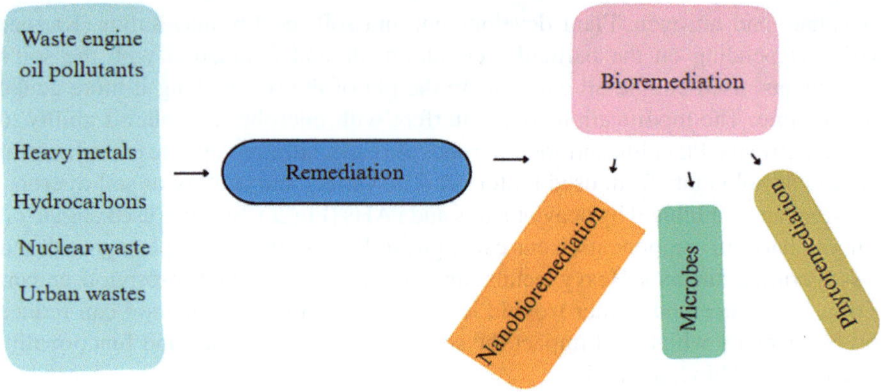

Fig. 3 Remediation techniques of waste engine oil pollutants

activity by adding organic modifications like pig dung and poultry manure are used in conjunction with this strategy (Behera et al., 2022).

Microbial Bioremediation

Numerous studies have examined the potential of microorganisms in decomposing petroleum compounds in soil (Behera et al., 2021). The bioremediation of WEO has advanced in recent years due to continuous attempts to improve biodegradation efficiency, optimize microbial degradation routes and create new bioremediation methods. The solubilization and increased bioavailability of hydrophobic hydrocarbons in used motor oil have been made possible by the development of bio-based surfactants and microorganisms that produce them, which has improved microbial access and breakdown rates. Microbial strains like *Bacillus, Acinetobacter* and *Pseudomonas* (Table 2) have demonstrated a great deal of promise in breaking down spent engine oil with degradation rates as high as 86.7% over 60 days (Salisu & Ibrahim, 2024). When immobilized, the Pseudomonas oleovorans strain showed an astounding 98–99% breakdown efficiency in just 7 days. Microbial bioremediation has a number of benefits over traditional techniques such as being economical, environmentally friendly and capable of fully mineralized organic contaminants. With benefits including non-invasiveness, ease of maintenance, large area application and little environmental disturbance, this method uses the metabolic powers of microorganisms to collapse and purify the hydrocarbon contaminants found in waste engine oil. The bioremediation of WEO has significantly in recent years due to continuous attempts to improve biodegradation efficiency, optimize microbial degradation routes and create new bioremediation methods (Ossai et al., 2023). The identifications and isolation of native hydrocarbon debasing microorganisms, the improvement of environmental conditions for microbial growth and activity and the investigation of genetic and metabolic engineering techniques to improve biodegradation capabilities are some of many facets of these developments. Number of studies have shown that different microbial species have the capacity to break down hydrocarbons. Vermicast was found to be helpful in improving soil quality and promoting plant growth when used to remediate spent engine oil-contaminated soil. Amaranthus viridis plant performance discovered that the engine oil had negative impact on both validating the necessity of efficient remediation techniques (Ekpo et al., 2023).

The ability of Sorghum bicolour, Helianthus annus and Telifaria occidentalis to phyto remediate different textures of spent engine oil polluted-soils that these plants may considerably lower spent engine oil contamination in a range of soil types. The phyto extraction and survival capacities of Nauclea diderrichii under SEO polluted environment, demonstrated the plant's capacity to both survive and efficiently extract heavy metals. The influence of ginger on growth indices and lead bioaccumulation in tomato fruits cultivated in soils contaminated with lead and discovering that ginger

Table 2 Microorganism and plants are used in remediation

S.l. No.	Pollutant treated	Plants used/ microorganisms	Method of phytoremediation	Reference
1	Crude oil	Moringa oleifera	Phytoremediation, hydrocarbon degrading microorganisms	Ekpo et al. (2023)
2	Wate engine oil, heavy metals	Rhizophora racemosa	Soil amendment with charcoal	Ossai et al. (2023)
3	Waste engine oil	Erusa sativa	Waste engine oil caused significant anatomical changes in leaf analysis, indicating stress	Salisu and Ibrahim (2024)
4	Waste engine oil, industrial effluent	Air potato	Waste engine oil and industrial effluent reduced soil quality and plant growth	Salisu and Ibrahim (2024)
5	Waste engine oil	Maize	Phytoremediation using maize leaves enhanced waste engine oil degradation in soil	Salisu and Ibrahim (2024)
6	Waste engine oil, heavy metals	Nuclea diderrichii phyto extraction	Plant survived and extracted amounts of heavy metals	Bolanle-Ojo and Sridhar (2021)
7	Crude oil spill	Sphingomnoas microorganism	Microbial consortia involving biosurfactant producing microbes highly efficient in crude oil removal	Ramdass and Rampersad (2023)
8	Waste lubricants	Ochrobactrum thiophenivorans	Positive degradation of waste lubricating oil	Bhattacharya et al. (2015)
9	Waste petroleum hydrocarbon pollutants	Aspergillus candida	Capable of removing pollutants	Obayori et al. (2010)

may lessen the negative consequence of these pollutants (Bolanle-Ojo & Sridhar, 2021). The effect of charcoal on Rhizophora racemosa growth in soil contaminated with SEO is that finding the addition of charcoal greatly enhanced plant growth. The phytoremediation potential of Vigna unguiculata and Pennisetum glaucum demonstrates that both plants have the capacity to lower waste engine oil levels in contaminated soil. The anatomical alterations brought on by spent engine oil pollution on Erusa sativa leaves, observing notable signs of stress and damage. The application of hydrocarbon degrading microbes and leaf extracts from Moringa oleifera for green remediation in soils affected by crude oil showing improved remediation and metal accumulation. Water hyacinth biomass works very well as an environmentally benign sorbent for cleaning up petroleum oil spills (Bhattacharya et al., 2015). The impact of spent engine oil-induced toxicity on Pennisetum glaucum germination, development and yield that demonstrating notable adverse effects. Seaweed extract and rice husk manure were found to greatly improve remediation when used to increase the phytoremediation potential of Guinea grass (Bolanle-Ojo & Sridhar, 2021).

Bio-Stimulation Techniques

Organic amendments have been demonstrated that adding pig dung and poultry manure to soil can improve the biodegradation process and increase remediation efficiency by up to 50.62% Crude oil is a very complex mixture of hydrocarbons that includes solid asphaltene leftovers, lubricating oil, kerosene, gasoline and volatile components of gasoline. It has been demonstrated that using microbes to break down petroleum hydrocarbons in the environment is a cost-effective, adaptable and environmentally benign treatment. By increasing the variety of bacteria that use hydrocarbons, addition of organic waste can accelerate the breakdown rates (Agarry et al., 2010).

Physical Chemical Methods

Clay purification efficiencies of 72.6% and 70.12% were attained by using raw and acid-treated clay to purify waste engine oil respectively. This technique enhances the quality of oil while successfully lowering harmful heavy metals (Ramdass & Rampersad, 2023).

5 Conclusion

When motor oil is improperly liable of as municipal waste, it can cause major environmental issues such contaminated soil, poisoned water and declining air quality. Hydrocarbons of pollutants that can persist in ecosystems and pose a threat to biodiversity and human health. In order to reduce these threats, waste management techniques frequently fall short, calling for stronger laws and better recycling plans. Techniques for remediation including chemical treatment, improved filtration and bioremediation are viable ways to lessen environmental harm. Nevertheless, prevention by appropriate disposal, public education and the use of sustainable substitutes such as waste oil reuse and refining continues to be the most successful strategy. By combining these strategies, we can lessen the negative effects of WEO and SEO on the environment and promote a cleaner, more sustainable world.

References

Agarry, S. E., Owabor, C. N., & Yusuf, R. O. (2010). Bioremediation of soil artificially contaminated with petroleum hydrocarbon oil mixtures: Evaluation of the use of animal manure and chemical fertilizer. *Bioremediation Journal, 14*(4), 189–195. https://doi.org/10.1080/10889868.2010.514965

Akintunde, W. O., Olugbenga, O. A., & Olufemi, O. O. (2015). Some adverse effects of used engine oil (common waste pollutant) on reproduction of male sprague dawley rats. *Open Access Macedonian Journal of Medical Sciences, 3*(1), 46. https://doi.org/10.3889/oamjms.2015.035

Anekwe, I. M. S., Armah, E. K., & Tetteh, E. K. (2022). Bioenergy production: Emerging technologies. *Biomass, Biorefineries and Bioeconomy, 225*. https://doi.org/10.5772/intechopen.102692

Babalola, S. O., & Oni, A. A. (2018). Hepatotoxic and renal effects of the water-soluble fractions of spent engine oil in Swiss albino mice. *International Journal of Biological and Chemical Sciences, 12*(2), 650–658. https://doi.org/10.4314/ijbcs.v12i2.3

Behera, I. D., Basak, G., Kumar, R. R., Sen, R., & Meikap, B. C. (2020). Treatment of petroleum refinery sludge by petroleum degrading bacterium Stenotrophomonas pavanii IRB19 as an efficient novel technology. *Journal of Environmental Science and Health, Part A Toxic/Hazardous Substances and Environment Engineering, 56*(2), 226–239. https://doi.org/10.1080/10934529.2020.1866924

Behera, I. D., Nayak, M., Biswas, S., Meikap, B. C., & Sen, R. (2021). Enhanced biodegradation of total petroleum hydrocarbons by implementing a novel two-step bioaugmentation strategy using indigenous bacterial consortium. *Journal of Environmental Management, 292*, Article 112746. https://doi.org/10.1016/J.JENVMAN.2021.112746

Behera, I. D., Nayak, M., Mishra, A., Meikap, B. C., & Sen, R. (2022). Strategic implementation of integrated bioaugmentation and biostimulation for efficient mitigation of petroleum hydrocarbon pollutants from terrestrial and aquatic environment. *Marine Pollution Bulletin, 177*, Article 113492. https://doi.org/10.1016/J.MARPOLBUL.2022.113492

Bekibele, G. E., Anacletus, F. C., & Patrick-Iwuanyanwu, K. C. (2022). The toxic effects of spent crankcase oil exposures: Systematic review and meta-analysis. *Toxicological Research*, 1–23. https://doi.org/10.1007/s43188-021-00093-2

Bhattacharya, M., Biswas, D., Sana, S., & Datta, S. (2015). Biodegradation of waste lubricants by a newly isolated Ochrobactrum sp. C1. *3 Biotech, 5*, 807–817. https://doi.org/10.1007/s13205-015-0282-9

Bock, W. (2017). Hydraulic oils. *Lubricants and Lubrication*, 345–420. https://doi.org/10.1002/9783527645565.ch11

Bolanle-Ojo, O. I., & Sridhar, M. K. C. (2021). Survival and phyto-extraction of heavy metals by Nauclea diderrichii (De Wild. & T. Durand) Merr. plantlets in spent engine oil medium. *Journal of Agriculture, Forestry and the Social Sciences, 19*(1), 1–7. https://doi.org/10.4314/joafss.v19i1.1

Bunger, A. P., McLennan, J., & Jeffrey, R. (2013). *Effective and Sustainable Hydraulic Fracturing*. https://doi.org/10.5772/45724

Chunyan, X., Qaria, M. A., Qi, X., & Daochen, Z. (2023). The role of microorganisms in petroleum degradation: Current development and prospects. *Science of the Total Environment, 865*, Article 161112. https://doi.org/10.1016/j.scitotenv.2022.161112

Das, A., Dey, S., Gacem, A., Yadav, K. K., Nath, H., Ram, D. K., & Das, A. P. (2025). Revolutionizing solid hydrocarbon sourcing: Plastic's journey from waste to upcycled treasure. *Environmental Science and Engineering, Part, F218*, 331–351. https://doi.org/10.1007/978-3-031-81615-4_14

Das, A. P., & Ghosh, S. (2023). Microorganisms in manganese biomining processes. *Manganese Mining Microorganisms*, 55–61. https://doi.org/10.1016/B978-0-12-822147-1.00013-3

Das, A. P., Mallavarapu, M., & Ghosh, S. (2023). Ecotoxicity of mining pollutants on the environment and their remediation. *Environmental Chemistry and Ecotoxicology, 5*, 165–167. https://doi.org/10.1016/J.ENCECO.2023.08.002

Das, A. P., & Mishra, S. (2008). Hexavalent Chromium (VI): Health hazards and environmental pollutant. *Journal of Environmental Research and Development, 2*(3), 386–392.

Das, A. P., Sukla, L. B., Pradhan, N., & Nayak, S. (2011). Manganese biomining: A review. *Bioresource Technology, 102*(16), 7381–7387. https://doi.org/10.1016/J.BIORTECH.2011.05.018

Das, S., Raj, R., Mangwani, N., Dash, H., & Chakraborty, J. (2014). Heavy metals and hydrocarbons: Adverse effects and mechanism of toxicity. *Microbial Biodegradation Bioremediation*. https://doi.org/10.1016/b978-0-12-800021-2.00002-9

Demirbas, A., Baluabaid, M. A., Kabli, M., & Ahmad, W. (2015). Diesel fuel from waste lubricating oil by pyrolitic distillation. *Petroleum Science and Technology, 33*(2), 129–138. https://doi.org/10.1080/10916466.2014.955921

Dey, S., Bissoyi, A., Koteswara Reddy, G., & Das, A. P. (2024). Synthetic microfibers: An urban emerging pollutant demanding for sustainable environmental solutions. *Sustainable Microbial Technology for Synthetic and Cellulosic Microfiber Bioremediation*, 107–121. https://doi.org/10.1007/978-3-031-62600-5_7

Dey, S., Das, A., Mallick, K., Sahu, A., & Das, A. P. (2023a). *Environmental Petroleum Waste: Pollution, Toxicity, Sustainable Remediation*. 159–175. https://doi.org/10.1007/978-3-031-48220-5_7

Dey, S., Tripathy, B., Kumar, M. S., & Das, A. P. (2023b). Ecotoxicological consequences of manganese mining pollutants and their biological remediation. *Environmental Chemistry and Ecotoxicology*. https://doi.org/10.1016/J.ENCECO.2023.01.001

Dhiman, V. (2024). Heavy metal toxicity and their effects on environment. In *Heavy metal toxicity: Human health impact and mitigation strategies* (pp. 1–23). Springer Nature Switzerland. https://doi.org/10.1007/978-3-031-56642-4_1

Ebenezer, A. B., & Abiodun, O. O. (2015). Effects of soil amendment on the toxicity of a two-month old waste engine oil-polluted soil on Solanum esculentum Linn. *Journal of Environment and Earth Science, 5*(15), 1–6.

Ekpo, P. B., Ekpo, I. P., Ifon, H. T., Ubi, G. M., Ikpeme, E. V., Ekaluo, U. B.,. & Ekanem, A. P. (2023). Green remediation and enhanced metal accumulation using moringa oleifera leaf extracts and hydrocarbon-degrading microorganisms in crude oil-impacted soils of the niger delta, nigeria. *Global Journal of Geological Sciences, 21*(2), 215–227. https://doi.org/10.4314/gjgs.v21i2.5

Ghosh, S., Mohanty, S., Nayak, S., Sukla, L. B., & Das, A. P. (2016). Molecular identification of indigenous manganese solubilising bacterial biodiversity from manganese mining deposits. *Journal of Basic Microbiology, 56*(3), 254–262. https://doi.org/10.1002/JOBM.201500477

Gunstone, F. D. (2013). Composition and properties of edible oils. *Edible Oil Processing*, 1–39. https://doi.org/10.1002/9781118535202.ch1

Kanungo, J., Sahoo, S., Bal, M., & Behera, I. D. (2023a). Performance of bioremediation strategy in waste lubricating oil pollutants: A review. *Geomicrobiology Journal*. https://doi.org/10.1080/01490451.2023.2245395

Kanungo, J., Sahoo, T., Bal, M., & Behera, I. D. (2024a). Performance of bioremediation strategy in waste lubricating oil pollutants: A review. *Geomicrobiology Journal, 41*(4), 360–373. https://doi.org/10.1080/01490451.2023.2245395

Kanungo, J., Sahoo, T., Swain, L. P., & Behera, I. D. (2023b). *Toxicity of Persistent Hydrocarbon Pollutants, Sources and Sustainable Remediation Process*. 39–65. https://doi.org/10.1007/978-3-031-48220-5_2

Kanungo, J., Sahoo, T., Swain, L. P., & Behera, I. D. (2024b). Toxicity of persistent hydrocarbon pollutants, sources and sustainable remediation process. In *Impact of petroleum waste on environmental pollution and its sustainable management through circular economy* (pp. 39–65). Springer Nature Switzerland. https://doi.org/10.1007/978-3-031-48220-5_2

Karmee, S. K. (2017). Fuel not food—Towards sustainable utilization of gutter oil. *Biofuels, 8*(3), 339–346. https://doi.org/10.1080/17597269.2016.1231952

Kuppusamy, S., Maddela, N. R., Megharaj, M., Venkateswarlu, K. (2020). Impact of total petroleum hydrocarbons on human health. *Total Petroleum Hydrocarbons: Environmental Fate, Toxicity, and Remediation*, 139–165. https://doi.org/10.1007/978-3-030-24035-6_6

Lahiri, D., Nag, M., Dey, A., Sarkar, T., Joshi, S., Pandit, S., Das, A. P., Pati, S., Pattanaik, S., Tilak, V. K., & Ray, R. R. (2021). Biofilm mediated degradation of petroleum products. *Geomicrobiology Journal*, 1–10. https://doi.org/10.1080/01490451.2021.1968979

Liu, C., Zhang, Y., Sun, S., Huang, L., Yu, L., Liu, X., Lai, R., Luo, Y., & Zhang, Z. (2018). Oil recovery from tank bottom sludge using rhamnolipids. *Journal of Petroleum Science and Engineering, 170*, 14-20. https://doi.org/10.1016/j.petrol.2018.06.031

Mallick, K., Sahu, A., Dubey, N. K., & Das, A. P. (2023). Harvesting marine plastic pollutants-derived renewable energy: A comprehensive review on applied energy and sustainable approach. *Journal of Environmental Management, 348*, Article 119371. https://doi.org/10.1016/J.JENVMAN.2023.119371

Mannekote, J. K., Kailas, S. V., Venkatesh, K., & Kathyayini, N. (2018). Environmentally friendly functional fluids from renewable and sustainable sources—A review. *Renewable and Sustainable Energy Reviews, 81*, 1787–1801. https://doi.org/10.1016/j.rser.2017.05.274

Mekonnen, H. A., & Ababa, A. (2014). Recycling of used lubricating oil using acid-clay treatment process. *Addis Ababa*. https://doi.org/10.35940/ijeat.d8420.049420

Mishra, A., Siddiqi, H., Kumari, U., Behera, I. D., Mukherjee, S., & Meikap, B. C. (2021). Pyrolysis of waste lubricating oil/waste motor oil to generate high-grade fuel oil: A comprehensive review. *Renewable and Sustainable Energy Reviews, 150*, Article 111446. https://doi.org/10.1016/j.rser.2021.111446

Mohanta, S., Pradhan, B., & Behera, I. D. (2024). Impact and remediation of petroleum hydrocarbon pollutants on agricultural land: A review. *Geomicrobiology Journal, 41*(4), 345–359. https://doi.org/10.1080/01490451.2023.2243925

Mohanty, S., Ghosh, S., Nayak, S., & Das, A. P. (2016). Isolation, identification and screening of manganese solubilizing fungi from low-grade manganese ore deposits. *Geomicrobiological Journal, 34*(4), 309–316. https://doi.org/10.1080/01490451.2016.1189016

Mong'ina, M. K. (2018). *Waste oil management by petroleum business operators: An analysis of the legal and institutional framework* (Doctoral dissertation, University of Nairobi). https://doi.org/10.21203/rs.3.rs-475257/v1

Mouser, P. J., Borton, M., Darrah, T. H., Hartsock, A., & Wrighton, K. C. (2016). Hydraulic fracturing offers view of microbial life in the deep terrestrial subsurface. *FEMS microbiology ecology, 92*(11), fiw166. https://doi.org/10.1093/femsec/fiw166

Musa, B., Muhammad, B. A., & Shehu, P. (2023). Impacts and management of vehicular used lubricating oil: A threat to environmental sustainability in Kaduna Metropolis, Nigeria. In *Conference: 62nd Annual National Conference of The Association of Nigerian Geographers, Hosted by Department of Geography At: University of Nigeria, Nsukka.* https://doi.org/10.35912/jshe.v1i3.535

Nutongkaew, T., Duangsuwan, W., Prasertsan, S., & Prasertsan, P. (2014). Physicochemical and biochemical changes during composting of different mixing ratios of biogas sludge with palm oil mill wastes and biogas effluent. *Journal of Material Cycles and Waste Management, 16*, 131–140. https://doi.org/10.1007/s10163-013-0165-2

Obayori, O. S., Salam, L. B., & Ogunwumi, O. S. (2014). Biodegradation of fresh and used engine oils by Pseudomonas aeruginosa LP5. https://doi.org/10.4172/2155-6199.1000213

Obinia, U., & Afiukwaa, J. N. (2013). Environmental and health impact of waste engine oil disposal in Nigeria: A review. *Education and Science Journal of Policy Review and Curriculum Development, 3*(3), 54–61. https://doi.org/10.1111/pce.12315

Osama, M., Singh, A., Walvekar, R., Khalid, M., Gupta, T. C. S. M., & Yin, W. W. (2017). Recent developments and performance review of metal working fluids. *Tribology International, 114*, 389–401. https://doi.org/10.1016/j.triboint.2017.04.050

Ossai, D., Akonye, L. A., & Okonwu, K. (2023). Effect of charcoal and spent engine oil on some growth indices of Rhizophora racemosa g. Mayer grown in garden soil. *Scientia Africana, 23*(1), 219–226. https://doi.org/10.4314/sa.v23i1.19

Panadare, D. C. (2015). Applications of waste cooking oil other than biodiesel: A review. *Iranian Journal of Chemical Engineering (IJChE), 12*(3), 55–76. https://doi.org/10.1016/j.molcatb.2017.05.003

Patel, N., & Shadangi, K. P. (2020). Characterization of waste engine oil (WEO) pyrolytic oil and diesel blended oil: Fuel properties and compositional analysis. *Materials Today: Proceedings, 33*, 4933–4936. https://doi.org/10.1016/j.matpr.2020.02.679

Paul, A., Dey, S., Ram, D. K., & Das, A. P. (2024). Hexavalent chromium pollution and its sustainable management through bioremediation. *Geomicrobiology Journal, 41*(4), 324–334. https://doi.org/10.1080/01490451.2023.2218377

Prabhakar, A., Mishra, S., & Das, A. P. (2019). Isolation and identification of lead (Pb) solubilizing bacteria from automobile waste and its potential for recovery of lead from end of life waste batteries. *Geomicrobiological Journal, 36*(10), 894–903. https://doi.org/10.1080/01490451.2019.1654044

Priyadarsini, S., & Das, A. P. (2024). Analytical and structural characterization of waste lithium-ion batteries for their effective recycling strategy. *Environmental Chemistry and Ecotoxicology.* https://doi.org/10.1016/j.enceco.2024.12.004

Priyadarsini, S., & Das, A. P. (2025). Lithium bioleaching: A review on microbial-assisted sustainable technology for lithium bio-circularity. *Journal of Water Process Engineering, 69*, Article 106744. https://doi.org/10.1016/j.jwpe.2024.106744

Raj, K., & Das, A. P. (2023). Lead pollution: Impact on environment and human health and approach for a sustainable solution. *Environmental Chemistry and Ecotoxicology, 5*, 79–85. https://doi.org/10.1016/J.ENCECO.2023.02.001

Ramdass, A. C., & Rampersad, S. N. (2023). Naturally-occurring microbial consortia for the potential bioremediation of hydrocarbon-polluted sites in Trinidad. *Bioremediation Journal, 27*(4), 443–452. https://doi.org/10.1080/10889868.2022.2043235

Reddy Koteswara, G., Bhargavi, S., Yamuna, D., Anjani Devi, M., Jeevana Lakshmi Kumar, G., Manas, T., Adarsh, S., Bhavya Sai, V., Nath, H., Alreshidi, M. A., Togun, H., Ram, D. K., & Das, A. P. (2025). Recent advancements in the remediation of marine hydrocarbon and plastic pollutants. *Environmental Science and Engineering, Part F218*, 127–142. https://doi.org/10.1007/978-3-031-81615-4_7

Salam, L. B. (2016). Metabolism of waste engine oil by Pseudomonas species. *3 Biotech, 6*(1), 98. https://doi.org/10.1007/s13205-016-0419-5

Salisu, B. U., & Ibrahim, F. (2024). Microbial bioremediation of spent engine oil: Current advances, challenges, and future directions. *UMYU Scientifica, 3*(4), 260–274. https://doi.org/10.56919/usci.2434.020

Sánchez-Alvarracín, C., Criollo-Bravo, J., Albuja-Arias, D., García-Ávila, F., & Pelaez-Samaniego, M. R. (2021). Characterization of used lubricant oil in a Latin-American medium-size city and analysis of options for its regeneration. *Recycling, 6*(1), 10. https://doi.org/10.3390/recycling6010010

Santos, J. C. O. (2018). Recovery of used lubricating oils—A brief review. *Progress in Petrochemical Science, 1*(4), 000516. https://doi.org/10.31031/pps.2018.01.000516

Sulayman, H. O. (2019). *Bioremediation of spent engine oil in some selected contaminated soils* (Master's thesis, Kwara State University (Nigeria)). https://doi.org/10.21467/ajgr.8.1.91-104

Tang, K. H. D. (2024). Phytoremediation of petroleum hydrocarbons: An update of its recent progress. *Tropical Environment, Biology, and Technology, 2*(2), 106–123. https://doi.org/10.1088/1757-899x/495/1/012054

Tang, K. H. D., & Angela, J. (2019, June). Phytoremediation of crude oil-contaminated soil with local plant species. In *IOP Conference series: materials science and engineering* (vol. 495, p. 012054). IOP Publishing. https://doi.org/10.56919/usci.2434.020

Wami-Amadi, C. F. (2025). The impact of air borne toxins from gas flaring on cardiopulmonary and other systemic functions. *Scholars International Journal of Anatomy and Physiology, 8*(1), 12–28. https://doi.org/10.36348/sijap.2025.v08i01.003

Wee, H. M., Budiman, S. D., Su, L. C., Chang, M., & Chen, R. (2016). Responsible supply chain management—An analysis of Taiwanese gutter oil scandal using the theory of constraint. *International Journal of Logistics Research and Applications, 19*(5), 380–394. https://doi.org/10.1080/13675567.2015.1090964

Xiaojie, Z., Mukherjee, K., Manna, S., Das, M. K., Kim, J. K., & Sinha, T. K. (2022). Efficient management of oil waste: Chemical and physicochemical approaches. In *Advances in Oil-Water Separation* (pp. 439–467). Elsevier. https://doi.org/10.1016/b978-0-323-89978-9.00027-6

MIX
Papier aus verantwortungsvollen Quellen
Paper from responsible sources
FSC® C105338

If you have any concerns about our products,
you can contact us on
ProductSafety@springernature.com

In case Publisher is established outside the EU,
the EU authorized representative is:
Springer Nature Customer Service Center GmbH
Europaplatz 3, 69115 Heidelberg, Germany

Printed by Libri Plureos GmbH
in Hamburg, Germany